Mathematical Sciences Research Institute
Publications

66

Lectures on
Mathematical Relativity

Mathematical Sciences Research Institute Publications

This series is based on work undertaken at the Simons Laufer Mathematical Sciences Institute (SLMath), formerly the Mathematical Sciences Research Institute (MSRI), in Berkeley, California. It publishes surveys and workshop proceedings of long-lasting value, as well as lecture notes and monographs by visitors to the Institute. The volumes below are published by Cambridge University Press; earlier ones may be available from Springer-Verlag.

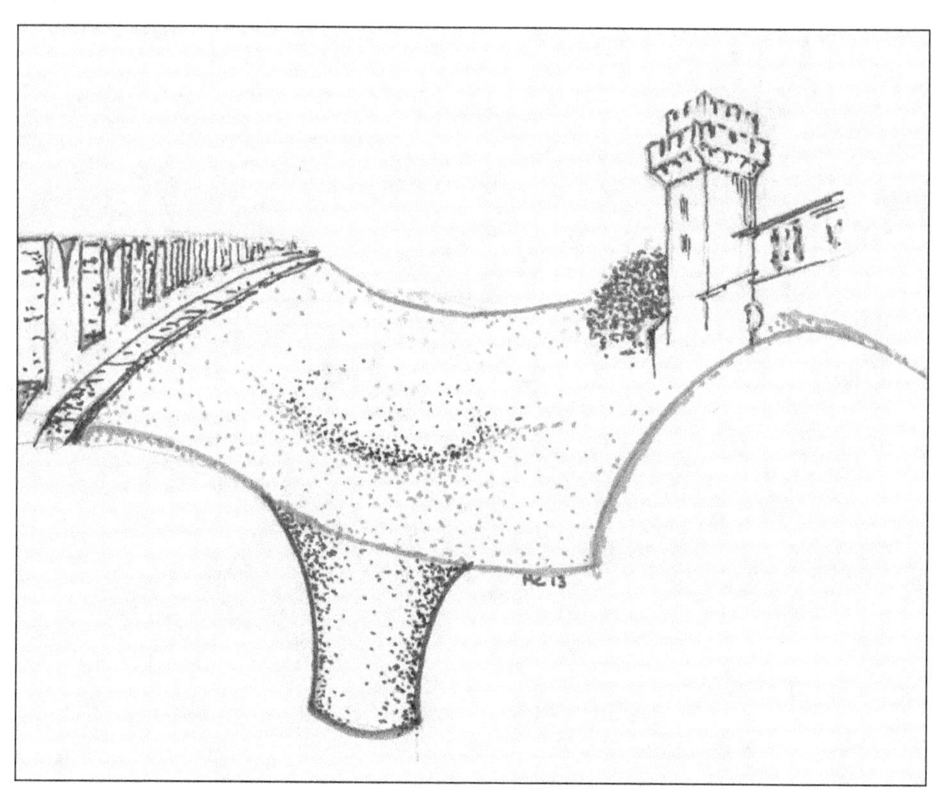

Schwarzschild meets Cortona: with thanks to Mauro Carfora

Lectures on Mathematical Relativity

Justin Corvino

Lafayette College

Pengzi Miao

University of Miami

with additional chapters by

Lan-Hsuan Huang

University of Connecticut, Storrs

Brian Allen

Lehman College, CUNY

Fernando Schwartz

University of Tennessee, Knoxville

Justin Corvino
Lafayette College
Easton, PA 18042
United States
corvinoj@lafayette.edu

Pengzi Miao
University of Miami
Coral Gables, FL 33146
United States
pengzim@math.miami.edu

Silvio Levy (*series editor*)
levy@msp.org

The Simons Laufer Mathematical Sciences Institute wishes to acknowledge support by the National Science Foundation and the *Pacific Journal of Mathematics* for the publication of this series.

CAMBRIDGE
UNIVERSITY PRESS

Shaftesbury Road, Cambridge CB2 8EA, United Kingdom
One Liberty Plaza, 20th Floor, New York, NY 10006, USA
477 Williamstown Road, Port Melbourne, VIC 3207, Australia
314-321, 3rd Floor, Plot 3, Splendor Forum, Jasola District Centre, New Delhi - 110025, India
103 Penang Road, #05-06/07, Visioncrest Commercial, Singapore 238467

Cambridge University Press is part of Cambridge University Press & Assessment, a department of the University of Cambridge. We share the University's mission to contribute to society through the pursuit of education, learning and research at the highest international levels of excellence.

www.cambridge.org
Information on this title: www.cambridge.org/9781107079939
DOI: 10.1017/9781139942300

When citing this work, please include a reference to the DOI 10.1017/9781139942300

First published 2025

A catalogue record for this publication is available from the British Library

A Cataloging-in-Publication data record for this book is available from the Library of Congress

ISBN 978-1-107-07993-9 Hardback
ISBN 978-1-107-43925-2 Paperback

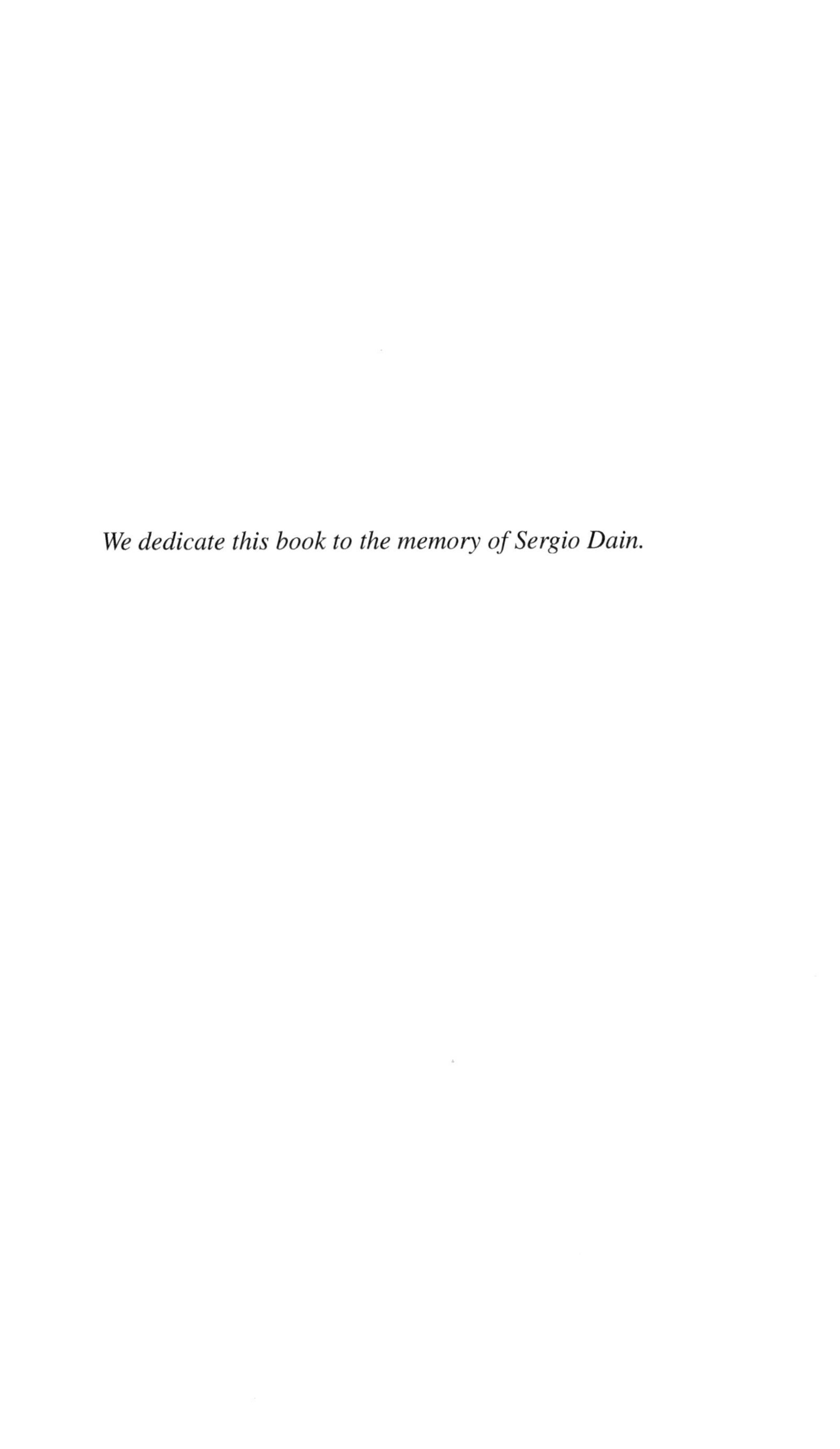

We dedicate this book to the memory of Sergio Dain.

Preface

This volume arose from the Summer Graduate Workshop in Mathematical General Relativity at the Mathematical Sciences Research Institute (MSRI, now renamed the Simons Laufer Mathematical Sciences Institute) in Berkeley, CA in 2012, and the subsequent summer school in Cortona, Italy, in 2013. The editors of the volume served as scientific organizers for the summer schools. The contributions to the volume grew out of lectures given at one or both of the schools.

We have endeavored to enhance the presentation of the material covered in the two-week summer schools to make it suitable for reading in book form, while at the same time remaining faithful to the spirit of those schools.

The advertised prerequisites for the schools, and hence this volume, included a standard first-year graduate analysis course, with elements of real and functional analysis as might be found in *Real analysis* by H. Royden [193] and *Real and complex analysis* by W. Rudin [194]. We also assumed introductory graduate courses in differential and Riemannian geometry, at the level of the following texts: *An introduction to smooth manifolds* and *Riemannian manifolds*, by J. M. Lee [141; 140]; *Riemannian geometry* by M. P. do Carmo [41]; *Riemannian geometry* by P. Peterson [182]; and of particular relevance to the summer schools, *Semi-Riemannian geometry* by B. O'Neill [174]. A graduate course in partial differential equations (PDE), at the level of *Partial differential equations* by L. C. Evans [86] and the first half of *Elliptic partial differential equations of second order* by D. Gilbarg and N. Trudinger [107], was not a requirement for the schools, and although some of the lectures needed to draw on some PDE results, students without such background could profit from the bulk of the material discussed at the schools. For this volume, however, certain PDE details in some of the presentations have been fleshed out, so those sections would be better approached with this background in hand. We have endeavored to bridge the gap by including a section introducing and motivating some of the PDE tools.

The students came to the schools with a wide range of backgrounds in mathematics, from those who had nearly completed their doctoral dissertations to those who came without the prerequisite geometry background. Through exercises and tutorial sessions, students were able to build enough intuition and computational skills to understand much of the material presented. While we decided a primer section on elliptic PDE was essential for the flow of this book, we resisted the temptation to add further sections on background geometry. That said, we do recall or develop some foundational material where needed, and some startup notations and conventions are reviewed starting on p. xix. We include exercises that were assigned before and during the schools, both to give readers a feel for the tutorials and to help focus those who are learning the topics for the first time or reviewing on the fly. We added many exercises as well, some collected at the end of chapters, some interspersed in the text. Of particular note, many exercises in Chapters 1 and 2 serve to review and extend background in geometry.

Strictly speaking, no physics background is required. We assume, as we did at the schools, a nodding acquaintance with pre-relativity physics, enough so that students can approach the development of the theory of special and general relativity with context from which to appreciate the rudiments of spacetime structure and the line of thought from Galileo to Newton to Einstein, and to motivate why the Einstein equations and the initial value constraints were to receive so much of their attention. In part for this reason, the first chapter contains some very basic material that would be included in an undergraduate course in special relativity, but we found it to be a fun way to start each school, engendering some interesting discussion amongst participants without needing much in the way of background. The first two chapters on special and general relativity may seem somewhat chatty, including some discussion of physics without always being mathematically efficient or fastidious, but we hope it helps to frame the mathematical theory. We could have cut the physics discussion short by formulating the mathematical postulates from the start with a small amount of motivation, but we decided, given the audience, to put some more time into developing these ideas from their genesis in physics. Even giving ourselves some leeway, the presentation is not too leisurely, and the lecture schedule at the schools called for covering the physics background reasonably efficiently at the beginning of the first week.

The mathematical and physical foundations of relativity have been an active topic of discussion and research for over a hundred years, and we have not tried to approach the scope of the debate (for instance, we chose not to discuss Mach's principle in depth), nor have we tried to use too fine a brush in painting the logical and philosophical distinctions, nor strained to give a serious historical account of the development of the theory. Interested readers can follow up with references such as [82; 83; 85; 161; 169; 170; 171], and with a wealth of material available online.

Even while starting off in an elementary fashion, and keeping in mind the range of student backgrounds represented, we were able to cover a reasonable amount of ground at each school. During the MSRI workshop, Pengzi Miao covered sufficient elements of causal theory to present the proof of the Penrose singularity theorem. Justin Corvino developed enough background in scalar curvature and asymptotically flat solutions of the constraint equations to be able to present a proof of the Riemannian positive mass theorem in three dimensions, while Lan-Hsuan Huang and Fernando Schwartz were able to build on this to discuss advanced aspects of the geometry of initial data sets, with Lan-Hsuan discussing constant mean curvature surfaces and the notion of center of mass, and with Fernando outlining multiple approaches to the Riemannian Penrose inequality. This volume reflects essentially the material covered during the MSRI workshop.

At Cortona, in lieu of Pengzi's lectures, Mauro Carfora (Università di Pavia) presented an engaging and marvelously illustrated development connecting the constraint equations (elliptic PDE governing initial values for the Einstein evolution) and the Ricci flow,[1] while Michael Eichmair (ETH Zürich, now at the University of Vienna) developed connections between the positive mass theorem and the geometry of initial data sets (including isoperimetry of large spheres),[2] which dovetailed beautifully with the lectures of Huang and Schwartz.

With all this background to present, the organizers decided to focus the topics lectures on the Einstein constraint equations which govern the initial data for the Einstein evolution, at the expense of not including advanced and/or current topics on the evolution problem. While this is a reasonable basis for criticism (of the schools and hence this volume), the field has developed to a point where there is room for multiple programs on each of these topics, and the relations between them; articles such as [64] and volumes such as [11; 51] indicate the considerable breadth and depth of the field.

The years just after the workshops witnessed a flurry of activity in general relativity. The centennial year of 2015 marked the hundredth anniversary of Einstein's formulation of a geometric theory of gravity governed by the Einstein equation, and was capped off with the excitement over the detection by LIGO of gravitational waves generated from black hole mergers — the discovery of which led to the 2017 Nobel Prize in physics. Roger Penrose shared the 2020

---

[1] A full treatment of the topic in Mauro's lectures can be found in the recent monograph [40].
[2] For this material see [32; 33; 79; 80; 81].

Nobel Prize in physics for his work on singularity formation and black holes, some of which we discuss. We hope the field will continue to develop in a robust manner, and that this work will be of some value in introducing graduate students to the field, and showing them some aspects of more advanced topics. Along these lines, we enthusiastically point the reader to the graduate text *Geometric Relativity* by Dan A. Lee (Queens College, CUNY), which has appeared recently [142], and would surely have been a recommended text for the schools.

The first two chapters of this volume present the basic background, from Minkowski spacetime and special relativity, to Einstein's equation and general relativity. Chapters 3 and 4 treat causality and the Penrose singularity theorem. Chapter 5 on the Einstein constraint equations rounds out the basic background from general relativity. Starting from Chapter 6 the text takes a sharp turn in the direction of geometric analysis. Chapter 6 includes some background motivation on elliptic PDE, with some applications to the constraint equations and scalar curvature; of note, there is an excursus on the first and second variations of area, which will appear throughout the rest of the text. Chapters 7–9 are written as topical chapters and are largely independent of each other, though one might find utility in referring to Chapter 7 for some properties of asymptotically flat spaces. That said, on a first pass, some readers might find themselves giving some of the more technical discussions in Chapter 7 a light read.

We would like to thank the graduate students for their hard work and enthusiasm at the summer schools, and in particular Alan Parry and Xin Zhou, as well as Peter McGrath and Andrea Santi, for their work as graduate assistants at the MSRI and Cortona schools, respectively. During one tutorial session, Alan introduced us to his research area, by presenting work of his thesis advisor Hubert Bray (Duke University), which modifies the Einstein–Hilbert action of general relativity with a goal to model dark matter; while we do not treat this topic in the text, we refer the interested reader to [27]; see also [30]. It has been inspiring to the scientific organizers to see so many of the students producing a staggering amount of interesting theses and papers in the years since the summer schools were held, and many have moved on to postdocs and faculty positions. In particular, Brian Allen, currently in the Department of Mathematics at Lehman College, CUNY, attended the MSRI summer school as a graduate student, and is a coauthor on Chapter 9 in this volume.

There are many people to thank for helping this project along. Giorgio Patrizio (Università di Firenze) first broached the idea of a volume after the Cortona summer school. We thank Heléne Barcelo (MSRI) for her enthusiastic support throughout the process. We also thank all the great staff at MSRI, and in particular Chris Marshall, for their support before, during and after the school, and likewise at Cortona, in particular Silvana Boscherini and Cinzia Benedetti. Funding for the schools was provided in part by National Science Foundation, the Clay Foundation, and INdAM (Istituto Nazionale di Alta Matematica), and we thank them for their generous support. Likewise we thank our respective home institutions, Lafayette College and the University of Miami. The editors shaped the book in part during their invited mini-course at the 2013 Taiwan International Conference on Geometry, at the National Taiwan University, and we would like to extend our thanks to Yng-Ing Lee for that opportunity. JC thanks Lehigh University, and especially Huai-Dong Cao, for inviting him to teach a graduate course in mathematical relativity in 2011, an experience that helped frame the approach to some of the material. JC would also like to acknowledge invitations from the Park City Math Institute, the Erwin Schrödinger Institute in Vienna, as well as from the Ravello Summer School, where he delivered mini-courses in the summers of 2013, 2014 and 2015, respectively, at which some of the presentation was honed. Of particular note is the support of Tommaso Ruggeri (Università di Bologna) for both the Cortona and Ravello summer schools. We thank Greg Galloway for reading Chapters 3 and 4 and offering some helpful feedback. JC thanks former student Kevin Manogue (Lafayette College) for feedback on Chapters 1 and 2, David Maxwell (University of Alaska, Fairbanks) for discussions on the conformal method, Farhan Abedin (Lafayette College) for reading parts of several chapters, and whose critical feedback led to a reorganization of Chapters 5–7, and finally John D. Norton (University of Pittsburgh) for several enlightening email exchanges on the foundations of general relativity. In addition to lecturing in Cortona, Mauro Carfora read several chapters in detail and offered critical advice from a physics perspective; in addition, his beautiful sketch of the palace at which the school was held adorns this volume. A huge thank you goes out to the editor Silvio Levy not only for his advice and encouragement, but for his calm patience while this project took longer than anticipated.

This book is dedicated to our friend and colleague Sergio Dain, who passed away in February 2016 at the age of 46. Sergio was an inspiration — through his work and his talks, he shared his deep insights into mathematical relativity and inspired you to be a better mathematician, while through his friendly and generous personality, interacting with him inspired you to be a better person. We lack the words to express how much he is missed.

Notation and conventions

We will often indicate conventions when they appear in the text (sometimes repeatedly), but we will mention a few here, just to get started.

While we generally use the term *smooth* to mean C^∞ (partly for definiteness), we note that often it will be obvious that a certain C^k-smoothness level is *sufficiently smooth* for the context under consideration. Subset notation $A \subset B$ also allows for $A = B$. Vectors will be denoted in various ways; standard basis vectors in coordinates x^i will be often written as partial derivative operators $\partial/\partial x^i$, so that a vector V can be written as a linear combination $V = V^i \, \partial/\partial x^i$. Here we have used the *Einstein summation convention* of summing over repeated upper and lower indices. While this convention will be in force unless otherwise noted, we will repeat it on occasion for the sake of clarity.

The term *manifold* will generally refer to a smooth manifold without boundary. A *closed manifold* will refer to a compact manifold (again, without boundary). While we assume the standard topological conditions that manifolds are Hausdorff and second countable, we are ambivalent about whether to restrict to connected manifolds: many results will not require connectedness, and for certain results that do, it is rather obvious that a statement as written would only hold on each component separately. We will try to point out where connectedness is assumed, but we trust the reader can discern if we have missed such an instance. A submanifold of codimension one is a *hypersurface*, which will generally be taken to be smoothly embedded, though we will try to point out when we allow it to be immersed, or weaken the regularity assumption (as in Chapter 3).

We will work with *semi-Riemannian* (also called *pseudo-Riemannian*) metrics on M, mostly Lorentzian or Riemannian; our signature for Lorentzian metrics is $(-, +, +, \ldots, +)$. When the spacetime is the focus, it may be given as a Lorentzian manifold (M, g), whereas at some point, the focus in the book will shift primarily to Riemannian manifolds, often construed as Riemannian hypersurfaces in a spacetime, so that the Riemannian manifold might then be given as (M, g), and the corresponding spacetime (if referenced) by $(\mathcal{S}, \bar{g})$, for example. Pay close attention to this, and also to the dimension of the spacetime.

This will be made clear in each situation, but just keep it in mind when cross-referencing formulae across chapters and sections.

When dealing with tensors, we sometimes just need the value of the tensor at a point, and sometimes we are referring to a *tensor field*; this will not always be explicitly stated, but should be clear in context. If a formula refers to derivatives of the tensor field, we will assume, unless stated otherwise, that the tensor field is smooth, or, at least smooth enough to do the indicated computations. For example, "consider a one-form θ" might really mean "consider a smooth one-form field θ". At various points we will consider fields that have less regularity (e.g., Sobolev spaces of tensor fields), and that will be made clear when needed; in particular, we will be more deliberate about emphasizing the regularity when it comes to the fore starting with the PDE discussion in Chapter 6.

Recall that a connection on the tangent bundle TM (an *affine connection*) assigns to vector fields X and Y a vector field $\nabla_X Y$, which is $C^\infty(M)$-linear (and hence tensorial) in X and $\mathbb{R}$-linear in Y, and satisfies the product rule $\nabla_X(fY) = (\nabla_X f)Y + f\nabla_X Y$ for $f \in C^\infty(M)$, where $\nabla_X f = X[f]$ is the directional derivative of f; the value of $(\nabla_X Y)|_p$ depends only on $X|_p$ and the values of Y along a curve tangent to $X|_p$. One can extend the connection to tensor fields T, defining $\nabla_X T$ by applying a product rule; e.g., if T is a one-form, $\nabla_X(T(Y)) = (\nabla_X T)(Y) + T(\nabla_X Y)$. In general, $\nabla_X T$ is a tensor of the same rank as T, and it follows easily from the definition that $\nabla_X T$ is tensorial in X. Hence we can construe ∇T as a tensor with rank higher by one: if T is an (r, s)-tensor, producing a scalar from a tuple of r one-forms and s vectors, then ∇T is an $(r, s+1)$-tensor. On a semi-Riemannian (M, g), there is a unique connection, called the *Levi-Civita connection* and denoted by ∇ (among other notations you might see in the text), which is torsion-free ($\nabla_X Y - \nabla_Y X = [X, Y]$) and satisfies $\nabla g = 0$; this will generally be the connection employed unless stated otherwise.

A metric g will often be written in bracket notation: $g(X, Y) = \langle X, Y \rangle$. In coordinates, g is given by a symmetric matrix of components g_{ij}, so that locally $g = g_{ij}\, dx^i \otimes dx^j = g_{ij}\, dx^i dx^j$, where for one-forms θ and η we define $\theta\eta = \frac{1}{2}(\theta\otimes\eta + \eta\otimes\theta)$ (whereas the *wedge product* is given by $\theta\wedge\eta = \theta\otimes\eta - \eta\otimes\theta$). Thus the Euclidean metric $g_{\mathbb{E}^n}$ on $\mathbb{R}^n$, for which the component functions x^i are *Cartesian* coordinates, is then expressed as $g_{\mathbb{E}^n} = \delta_{ij}\,dx^i dx^j$, for example. The nondegeneracy of g corresponds in components to the invertibility of the matrix (g_{ij}), and we write $(g^{ij}) = (g_{ij})^{-1}$, i.e., $g^{ij}g_{jk} = \delta^i_k$. There is a natural volume measure dv_g associated to g, which in local coordinates takes the form $dv_g = \sqrt{|\det(g_{ij})|}\, dx$, where dx is the Euclidean (Lebesgue) volume measure in coordinates; dv_g corresponds to a volume form ω_g in case M is orientable. We sometimes let $\det g = \det(g_{ij})$ for abbreviation, and we often let $d\sigma$ or $d\sigma_g$ be the volume measure induced on a semi-Riemannian submanifold.

Since at each point on M the metric g is nondegenerate, it can be used to change the tensor type, e.g., a vector X is associated to a dual form $X^\flat$ by $g(X, Y) = X^\flat(Y)$, and likewise a one-form α can be associated to its vector dual $\alpha^\sharp$ by $g(\alpha^\sharp, Y) = \alpha(Y)$. It is easy to check in a basis v_j for T_pM with dual basis θ^i for T_p^*M (so $\theta^i(v_j) = \delta^i{}_j$) that if $X = X^j v_j$ then $X^\flat = X_i \theta^i$ with $X_i = g_{ij} X^j$, where $g_{ij} = g(v_i, v_j)$; similarly, if $\alpha = \alpha_i \theta^i$, then $\alpha^\sharp = \alpha^j v_j$ with $\alpha^j = g^{ij}\alpha_i$. This kind of operation, known as *raising* and *lowering* of indices from the way the notation is arranged, can be performed on more general tensors T, with the positions of the indices generally indicating tensor type in lieu of the musical $\sharp$ and $\flat$ notation.

We remark on the consistency of the raising/lowering notation: if T is a $(0, 2)$-tensor with components T_{ij}, then $T^{ij} = g^{ik} g^{j\ell} T_{k\ell}$ give the components of the tensor obtained by type-changing using g, so that if $T = g$, then in fact we see $T^{ij} = g^{ij}$ (the components of the inverse matrix). Furthermore, we can extend g as a bilinear form on more general tensors, defining $\langle S, T \rangle$ to be an appropriate metric contraction of $S \otimes T$; e.g., if S and T are $(1, 2)$-tensors, then in a local basis $\langle S, T \rangle = g_{i\ell} g^{js} g^{km} S^i_{jk} T^\ell_{sm}$. We may write this in various ways, depending on context: $\langle S, T \rangle = \langle S, T \rangle_g = S \cdot_g T = S \cdot T$, and we let $|T|_g^2 = \langle T, T \rangle_g$ (this is in fact nonnegative when g is Riemannian). Note that if h is a $(0, 2)$-tensor, then $\langle g, h \rangle_g = g^{ij} h_{ij} = \mathrm{tr}_g h$, and similarly if h is a $(2, 0)$-tensor.

The *Riemann curvature tensor* will be defined via the vector field

$$R(X, Y, Z) = \nabla_X \nabla_Y Z - \nabla_Y \nabla_X Z - \nabla_{[X,Y]} Z = R(X, Y)Z,$$

with index conventions $R^\ell_{ijk} \frac{\partial}{\partial x^\ell} = R\big(\frac{\partial}{\partial x^i}, \frac{\partial}{\partial x^j}, \frac{\partial}{\partial x^k}\big)$, in which R is a $(1, 3)$-tensor, while the components of the corresponding $(0, 4)$-tensor are given by $R_{ijk\ell} = g_{\ell m} R^m_{ijk} = \langle R\big(\frac{\partial}{\partial x^i}, \frac{\partial}{\partial x^j}, \frac{\partial}{\partial x^k}\big), \frac{\partial}{\partial x^\ell}\rangle$. Different books use different conventions, so be alert! The curvature tensor enjoys a number of symmetries. Clearly, $R(X, Y, Z) = -R(Y, X, Z)$; slightly less obvious is *symmetry-by-pairs* $\langle R(V, W, Y), Z \rangle = \langle R(Y, Z, V), W \rangle$. Thus we have the component identities: $R_{k\ell ij} = R_{ijk\ell} = -R_{jik\ell} = R_{ji\ell k}$. For a nondegenerate two-plane $\Pi \subset T_pM$, the following expression is independent of basis $\{V, W\}$ for Π, and defines the *sectional curvature* $K(\Pi)$:

$$K(\Pi) = \frac{\langle R(V, W, W), V \rangle}{\langle V, V \rangle \langle W, W \rangle - \langle V, W \rangle^2}. \tag{0.0.1}$$

For given X and Y, $R(\,\cdot\,, X, Y)$ is a linear transformation, whose trace is defined to be $\mathrm{Ric}(X, Y)$, the *Ricci curvature*. The Ricci tensor Ric (alternatively, $\mathrm{Ric}(g)$

or Ric_g) is a *symmetric* $(0, 2)$-tensor (via the preceding curvature component identities), and it is generally the same tensor across texts (though a notable exception is [221], where the sign differs from ours), which means the way it is defined from the Riemann tensor may differ to account for sign. In our convention,

$$\mathrm{Ric}(X, Y) = dx^\ell\big(R\big(\tfrac{\partial}{\partial x^\ell}, X, Y\big)\big) = g^{k\ell}\big\langle R\big(\tfrac{\partial}{\partial x^\ell}, X, Y\big), \tfrac{\partial}{\partial x^k}\big\rangle,$$
$$R_{ij} = \mathrm{Ric}\big(\tfrac{\partial}{\partial x^i}, \tfrac{\partial}{\partial x^j}\big) = R^\ell_{\ell ij} = g^{k\ell} R_{\ell ijk}.$$

The *scalar curvature* is the metric trace of the Ricci tensor, and is given in components by $R(g) = g^{ij} R_{ij}$.

A comma is used to denote a partial derivative, whereas a semicolon is used to denote components of the covariant derivative of a tensor. For example, with $T_{ijk} = g_{km} T^m_{ij}$, we have $(\nabla T)_{ijk\ell} = T_{ijk;\ell} = (g_{km} T^m_{ij})_{;\ell} = g_{km} T^m_{ij;\ell}$, since $\nabla g = 0$. While the covariant derivative of a function f is naturally a one-form df, i.e., $\nabla f(X) = \nabla_X f = X[f] = df(X)$, sometimes ∇f is instead taken to be the vector $(df)^\sharp = \mathrm{grad}_g f$ dual to df, i.e., the *gradient* of f with respect to the metric g, so that $df(X) = g(X, \mathrm{grad}_g f)$; the meaning should be clear in context.

The *Christoffel symbols* Γ^k_{ij} for a coordinate frame are defined by

$$\nabla_{\frac{\partial}{\partial x^i}} \frac{\partial}{\partial x^j} = \Gamma^k_{ij} \frac{\partial}{\partial x^k},$$

and can be computed in terms of the metric as $\Gamma^k_{ij} = \tfrac{1}{2} g^{km}(g_{mj,i} + g_{im,j} - g_{ij,m})$.

If u is a smooth function on M, the *Hessian* of u is defined by $\mathrm{Hess}_g u = \nabla(du)$. It is a $(0, 2)$-tensor, with $(\mathrm{Hess}_g u)_{ij} = u_{;ij}$ in components, and moreover it is symmetric (Exercise 1-9). The *Laplacian* is the trace of the Hessian:

$$\Delta_g u = \mathrm{tr}_g(\mathrm{Hess}_g u) = g^{ij} u_{;ij}.$$

In some texts, the term *Laplacian* is reserved for the case (M, g) is Riemannian, and may be defined as the *negative* of our definition. When (M, g) is Lorentzian, the trace of the Hessian is often called (again, up to a sign) the *wave operator* $\Box_g$.

Geodesic normal coordinates at a point $p \in M$ can be useful in computations. In such a coordinate system, $g_{ij}(p) = \pm\delta_{ij}$ and $\nabla_{\frac{\partial}{\partial x^i}} \frac{\partial}{\partial x^j}\big|_p = \Gamma^k_{ij}\big|_p \frac{\partial}{\partial x^k}\big|_p = 0$, the latter condition being equivalent to the vanishing $g_{ij,k}(p) = 0$ of all the partial derivatives of the components of g at p. Thus, for example, if T is a $(1, 2)$-tensor field, then $T^k_{ij;\ell} = T^k_{ij,\ell} + \Gamma^k_{\ell m} T^m_{ij} - \Gamma^m_{\ell i} T^k_{mj} - \Gamma^m_{\ell j} T^k_{im}$, which greatly simplifies at a point p in normal coordinates. When we use an expression like "at a point in normal coordinates", we generally imply evaluating at the point p around which the normal coordinates chart is centered.

We will sometimes use "big O" notation: $f = O(h)$ means that $|f| \leq C|h|$ for some $C > 0$, where the quantities may be tensors, with corresponding norms. Generally one must pay attention to the dependence of C. If f and h are functions of x, then C might be uniformly chosen for x in a compact subset, or possibly f is a function of a tensor h, and so the C might depend on the set of tensors under consideration. Sometimes this notation also implies some bounds on derivatives of f as well, which will have to be specified in context.

Various function spaces will play a role in some of the analysis herein. We will in Chapter 6 recall basic definitions of Sobolev and Hölder spaces, and we encourage the reader to review their basic properties from references such as [2; 86; 107; 144]. We let Ω be an open subset of $\mathbb{R}^n$, sometimes called a *domain in* $\mathbb{R}^n$. For k a nonnegative integer, we let $C^k(\Omega)$ be the set of all functions u on Ω such that u and all its partials up through order k are continuous, and we let $C^\infty(\Omega) = \bigcap_{k=0}^\infty C^k(\Omega)$.

CHAPTER 1

Special relativity and Minkowski spacetime

This chapter is a brief introduction to Minkowskian geometry, highlighting the physical motivation for this model of spacetime. This mathematical model resolves a number of perplexing issues that had arisen in experimental and theoretical physics by the dawn of the twentieth century, and it led to surprising predictions for physics, which have been confirmed in the laboratory. While we review several well-known features of spacetime, we certainly cannot do the topic justice (from the point of view of mathematics or physics) in this brief introduction. Rather, we will focus on what we need for our purposes, and refer the reader to the bibliography, on which we have drawn heavily for both inspiration and details.

1.1. Lorentz transformations

We begin by discussing some of the physical underpinnings of the Minkowski spacetime model of physics, following in spirit the ideas of Einstein. We make the mathematical assumption that events in space and time form a four-dimensional continuum (manifold) S, which for now we take to be $\mathbb{R}^4$ (with the standard topology). We examine the role of a distinguished class of physical *observers*, corresponding to *inertial frames of reference*, and giving rise to a family of coordinate charts for Minkowski spacetime. The transformations between such charts form a group of diffeomorphisms of $\mathbb{R}^4$, and from the Kleinian perspective, the invariants of this group yield *geometric* quantities of interest. Einstein echoed the *principle of relativity* of Galileo and Newton in asserting the equivalence of inertial observers from the point of view of physics, extending it from the realm of mechanics to include electromagnetism. There should be no preferred inertial frame that could be distinguished by experiment, and thus the laws of mechanics and electromagnetism should take the same form in any inertial frame (*special covariance* of the laws of physics). This being so, quantities of interest in physics would be invariants of the associated transformation group. In this respect, then, the departure from the *Galilean invariance* of Newtonian mechanics is that Einstein identified a different group of transformations between inertial

1

frames — in essence, a different geometric structure, moving from Euclidean to Minkowskian geometry — which allowed him to incorporate the laws of electromagnetic phenomena, such as the speed of light in vacuum, in a satisfactory manner. The physics and the associated Minkowskian geometry distinguish a special class of frames, the inertial frames to which the principle of relativity is applied, and as such the resulting theory is known as *special relativity*.

1.1.1. *Galilean transformations.* Inertial frames of reference (and corresponding inertial observers) are those for which Newton's first law, *the law of inertia*, holds: objects will move with constant velocity unless acted upon by a force. As this can interpreted to be the *definition* of an inertial frame, Newton's first law can be read as asserting the *existence* of an inertial frame of reference. Newton's law of inertia incorporates Descartes' first two laws of motion, which had in turn modified Galileo's law of inertia, and while all of these appear to echo an idea from Aristotle's *Physics* (IV, 8), the law of inertia in fact overturns Aristotelian law, which maintained that objects which are not acted upon by a force should naturally come to *rest*.

The principle of relativity of Galileo and Newton asserts that mechanics should look the same to all inertial observers: all inertial systems are equivalent for the formulation of the laws of mechanics. From Newton's law of inertia, it is clear that an observer in uniform motion with respect to an inertial observer is also an inertial observer. As reference frames yield coordinate charts for spacetime, we are interested in those coordinate changes that correspond to comparisons of measurements made by two inertial observers.

If we assume that the universe is endowed with some *Newtonian time function* that measures absolute *time intervals* between events, then making a simple time translation to coordinate the time for two observers, we can arrange that their time functions agree. A *rigid frame of reference* then corresponds to a family of curves foliating spacetime, (the world lines of) *observers* at rest with respect to the frame, so that the points on the curves at any fixed time form a three-dimensional Euclidean space, with distances between observers independent in time. If we pick an observer and choose Cartesian axes at a given time, we can build a coordinate chart for spacetime adapted to the frame in a natural way, with the chosen observer at the origin of spatial coordinates at each time. The coordinate transformation relating two such charts obtained by two rigid frames $\mathcal{O}$ and $\widetilde{\mathcal{O}}$ in uniform motion with respect to one another is a *Galilean transformation*, with one frame being inertial if and only if the other one is.

We can write the transformation that relates the coordinates $(t, \boldsymbol{x})$ of an *event* (a point in spacetime) in one frame $\mathcal{O}$ to the coordinates $(\tilde{t}, \tilde{\boldsymbol{x}})$ in a frame $\widetilde{\mathcal{O}}$

moving at constant velocity v with respect to $\mathcal{O}$, arranging that the spacetime origins agree, and the Cartesian axes coincide at time $t = 0$:

$$\tilde{t} = t, \qquad \tilde{x} = x - vt.$$

If we also arrange the relative velocity to lie along the x-axis, then the transformation becomes, with $v = v\,\partial/\partial x$,

$$\tilde{t} = t, \qquad \tilde{x} = x - vt, \qquad \tilde{y} = y, \qquad \tilde{z} = z. \tag{1.1.1}$$

Moreover, if $\widehat{\mathcal{O}}$ is an observer moving with constant velocity w with respect to $\widetilde{\mathcal{O}}$, then it is elementary to obtain the *Galilean law of addition of velocities*:

$$x = \tilde{x} + vt = \hat{x} + wt + vt = \hat{x} + (v + w)t. \tag{1.1.2}$$

We see that relative velocities satisfy a very simple addition rule.

Consider a curve $\gamma : I \subset \mathbb{R} \to S$ parametrized by Newtonian time. If we let $x(t)$ be the spatial components of $\gamma(t)$ in $\mathcal{O}$, and let $\tilde{x}(t)$ be likewise in $\widetilde{\mathcal{O}}$, then $\tilde{x}(t) = x(t) - vt$. Therefore, if a prime denotes a time derivative, $\tilde{x}'(t) = x'(t) - v$, and $\tilde{x}''(t) = x''(t)$. Hence the acceleration $a(t) = x''(t)$ of the path γ is the same as measured in either frame.

Suppose γ is the path of an object of mass m, an observer-independent quantity that we take to be independent of t, so that the *momentum* $p(t) = mx'(t)$ in $\mathcal{O}$ differs from $\tilde{p}(t) = m\tilde{x}'(t)$ in $\widetilde{\mathcal{O}}$ by a constant. Newton's second law of motion states that the net *force* F on the object due to physical interactions equals the time rate of change of its momentum, as measured in an inertial frame, which becomes the familiar $F = ma$; to obtain an analogous equation in a non-inertial frame, a *fictitious force* must be added to balance the frame acceleration. It seems reasonable that the net interaction forces should be observer-independent, as would be the case when the force between objects is a function of their relative separation and relative velocity. Therefore Newton's second law of motion holds in all inertial frames if it holds in one inertial frame.

Einstein's foundational 1905 paper "On the electrodynamics of moving bodies" [82] emphasizes how the incompatibility of electromagnetism and the Galilean transformations led to the reformulation of mechanics. There are of course very fundamental issues in interpreting electromagnetism. Nineteenth-century experiments revealed that a magnetic field is generated by charges in motion. The Lorentz force law $F = q(E + \frac{v}{c} \times B)$ (written in *Gaussian* or cgs units, with c the speed of light in vacuum) determines the force on a charge q moving at velocity v in an electromagnetic field. For example, consider a charge q which moves across the field lines of a stationary magnet. The moving charge experiences a magnetic force from the Lorentz force law. On the other hand, if we switch to a frame *moving with the charge*, then the charge q to which we apply the Lorentz force law is stationary and the magnet is moving. As such, the charge experiences a force from an electric field induced (Faraday's law) by the *changing* magnetic field moving past q. In the end, of course, the physical predictions are the same in each case, it is only the *interpretation* that differs. Einstein looked for a fundamental explanation of this in terms of relativity, that the laws of electromagnetism should have the same form in all inertial frames.

A consequence of Maxwell's equations for electromagnetism is that light travels according to a wave equation, the speed of which can be determined. Of course, this raises the question: the speed relative to what? And what would be the medium capable of transmitting electromagnetic disturbances at such a great speed, while seeming transparent to the motion of the earth through it? Attempts such as the Michelson–Morley experiment in the late nineteenth century failed to find the medium, a *preferred* reference frame (which was called the *ether frame*), with respect to which light in vacuum travels at speed c, roughly 3×10^8 meters per second. Under the Galilean transformations, inertial observers in relative motion with respect to the ether frame would have different measurements of the value of the speed of light. That this was not observed in experiments caused quite a quandary. Classical results on stellar aberration along with the Fizeau experiment supplied evidence against an *ether drag theory* that the ether moves along with massive bodies. Other experiments ruled out theories that were consistent with the null result of Michelson–Morley, for example the *Lorentz contraction hypothesis*, which on its own cannot account for the result of the Kennedy–Thorndike experiment; see [96; 137; 188], for instance, for more details on these experiments. The ether theory embraced the notion that the principle of relativity did not apply to electromagnetism, in the sense that the ether frame is a preferred frame of reference for the theory. That there were problems with this led Einstein to postulate that relativity does apply to electromagnetism. Thus, although Newtonian dynamics works well with the Galilean transformations, for relativity to apply to electromagnetism, the Galilean transformations required modification.

For some foreshadowing, consider frames of reference $\mathcal{O}$ and $\tilde{\mathcal{O}}$ with constant relative velocity, with respective coordinates related as in (1.1.1). Suppose a function $\psi : \mathbb{R}^4 \to \mathbb{R}$ satisfies the wave equation (with wave speed c) in $\mathcal{O}$, in the sense that

$$\frac{1}{c^2}\frac{\partial^2 \psi}{\partial t^2} = \frac{\partial^2 \psi}{\partial x^2} + \frac{\partial^2 \psi}{\partial y^2} + \frac{\partial^2 \psi}{\partial z^2}.$$

Then since

$$\frac{\partial}{\partial t} = \frac{\partial \tilde{t}}{\partial t}\frac{\partial}{\partial \tilde{t}} + \frac{\partial \tilde{x}}{\partial t}\frac{\partial}{\partial \tilde{x}} = \frac{\partial}{\partial \tilde{t}} - v\frac{\partial}{\partial \tilde{x}}$$

and

$$\frac{\partial}{\partial x} = \frac{\partial \tilde{t}}{\partial x}\frac{\partial}{\partial \tilde{t}} + \frac{\partial \tilde{x}}{\partial x}\frac{\partial}{\partial \tilde{x}} = \frac{\partial}{\partial \tilde{x}},$$

we see that $\Psi(\tilde{t}, \tilde{x}) := \psi(t, x)$ satisfies

$$\frac{1}{c^2}\left(\frac{\partial}{\partial \tilde{t}} - v\frac{\partial}{\partial \tilde{x}}\right)^2 (\Psi) = \frac{\partial^2 \Psi}{\partial \tilde{x}^2} + \frac{\partial^2 \Psi}{\partial \tilde{y}^2} + \frac{\partial^2 \Psi}{\partial \tilde{z}^2}.$$

This can be rewritten as (assuming ψ is C^2)

$$\frac{1}{c^2}\frac{\partial^2 \Psi}{\partial \tilde{t}^2} - 2\frac{v}{c^2}\frac{\partial^2 \Psi}{\partial \tilde{t}\,\partial \tilde{x}} = \left(1 - \frac{v^2}{c^2}\right)\frac{\partial^2 \Psi}{\partial \tilde{x}^2} + \frac{\partial^2 \Psi}{\partial \tilde{y}^2} + \frac{\partial^2 \Psi}{\partial \tilde{z}^2}.$$

We see that the wave equation is not invariant under Galilean transformations. This is perfectly reasonable for mechanical waves, but the homogeneous wave equation governing light propagation in vacuum is a consequence of Maxwell's equations, and we have seen that experiments indicate that light travels at the same speed in vacuum for all inertial observers.

However, it is not too hard to play around with the transformation so as to coax the preceding equation into the standard wave equation form. Namely, let

$$\tilde{t} = \frac{1}{\sqrt{1 - (v/c)^2}}\left(t - \frac{v}{c^2}x\right), \tag{1.1.3}$$

$$\tilde{x} = \frac{1}{\sqrt{1 - (v/c)^2}}(x - vt) \tag{1.1.4}$$

replace the Galilean coordinate change. Then, as we did above, we obtain

$$\frac{\partial}{\partial t} = \frac{\partial \tilde{t}}{\partial t}\frac{\partial}{\partial \tilde{t}} + \frac{\partial \tilde{x}}{\partial t}\frac{\partial}{\partial \tilde{x}} = \frac{1}{\sqrt{1 - (v/c)^2}}\left(\frac{\partial}{\partial \tilde{t}} - v\frac{\partial}{\partial \tilde{x}}\right),$$

$$\frac{\partial}{\partial x} = \frac{\partial \tilde{t}}{\partial x}\frac{\partial}{\partial \tilde{t}} + \frac{\partial \tilde{x}}{\partial x}\frac{\partial}{\partial \tilde{x}} = \frac{1}{\sqrt{1 - (v/c)^2}}\left(\frac{\partial}{\partial \tilde{x}} - \frac{v}{c^2}\frac{\partial}{\partial \tilde{t}}\right).$$

Thus we find

$$\frac{1}{c^2}\frac{\partial^2}{\partial t^2} = \frac{1}{c^2}\cdot\frac{1}{(1 - (v/c)^2)}\left(\frac{\partial^2}{\partial \tilde{t}^2} - 2v\frac{\partial^2}{\partial \tilde{t}\,\partial \tilde{x}} + v^2\frac{\partial^2}{\partial \tilde{x}^2}\right),$$

$$\frac{\partial^2}{\partial x^2} = \frac{1}{(1 - (v/c)^2)}\left(\frac{\partial^2}{\partial \tilde{x}^2} - 2\frac{v}{c^2}\frac{\partial^2}{\partial \tilde{t}\,\partial \tilde{x}} + \frac{1}{c^2}\left(\frac{v}{c}\right)^2\frac{\partial^2}{\partial \tilde{t}^2}\right)$$

as operators on C^2 functions, and therefore,

$$\frac{1}{c^2}\frac{\partial^2}{\partial t^2} - \frac{\partial^2}{\partial x^2} = \frac{1}{c^2}\frac{\partial^2}{\partial \tilde{t}^2} - \frac{\partial^2}{\partial \tilde{x}^2}.$$

We have found a coordinate change for which the wave equation is preserved, but at what cost? Especially disturbing is that $\tilde{t}$ depends not only on t, but on x and v as well! In fact, we will derive these equations from applying a few simple fundamental principles, as Einstein did. Namely, we combine the principle of relativity (the Galilean/Newtonian principle for mechanics, extended by Einstein to encompass electromagnetism), that physical laws should have the same form in all inertial frames, along with Einstein's postulate that the speed of light in a vacuum is a physical law, and thus must be the same for all inertial observers. This immediately mitigates the need for an ether, and gives rise to striking predictions that are consistent with experimental results. For example, *Lorentz contraction* can account for the result of Michelson–Morley, and together with *time dilation* can explain the result of Kennedy–Thorndike; we will derive both of these predictions in Section 1.2.3. (For a derivation of the Lorentz transformation based directly on experimental results, see [192].)

1.1.2. *Deriving the Lorentz transformations.* The above coordinate change (1.1.3)–(1.1.4) was motivated by preserving the wave equation for electromagnetic propagation. We will now derive this Lorentz transformation of coordinates between inertial observers from a more fundamental standpoint, as well as indicate the transformation rule for electromagnetic fields, from which one may confirm that Maxwell's equations are invariant under change of inertial frames. As such, Maxwell's equations for electromagnetism are seen to be physical laws which are in accordance with the principle of relativity. As we will see, however, the laws of mechanics require some modification from those of Newton.

We assume there exists an inertial frame of reference, in which the law of inertia holds. We assume in particular that using such a frame, we construct a coordinate chart for spacetime $\mathcal{S}$, a diffeomorphism between $\mathcal{S}$ and $\mathbb{R}^4$, assigning time and space coordinates $(t, \boldsymbol{x})$ to points in spacetime, so that the Euclidean distance between two points in any $\{t_0\} \times \mathbb{R}^3$ is construed as measuring the spatial distance between the corresponding events, while $\Delta t = t_1 - t_0$ is the time difference between events with respective coordinates in $\{t_0\} \times \mathbb{R}^3$ and $\{t_1\} \times \mathbb{R}^3$. For each fixed $\boldsymbol{x}_0 \in \mathbb{R}^3$, we obtain an observer at rest in the frame, given by events corresponding to points $(t, \boldsymbol{x}_0)$. In fact, for each such observer, we can construct a coordinate chart (in fact many, with freedom in choosing a Cartesian frame). For any inertial frame, and for any of these observers, the law of inertia holds, and so we refer to these observers as *inertial observers*, and we refer to any such set of coordinates built from an inertial frame as *inertial coordinates*. (Note that we have chosen to restrict inertial coordinates to be Cartesian on constant time slices $\{t_0\} \times \mathbb{R}^3$, whereas one might naturally consider inertial spherical or cylindrical coordinates on such slices.) Given a coordinate chart, we will often blur the distinction between points (or displacements) in S and their coordinates (or their coordinate displacement vectors) in $\mathbb{R}^4$. Given two inertial observers at rest in this frame, there are inertial coordinates for each for which the change of coordinates is a simple spacetime translation. We build in the assumption that at all points in an inertial frame, clocks at rest in the frame run at the same rate, and that any two inertial observers agree on the future direction of time.

Inherent in this framework, deriving from the equivalence of inertial observers, is the *homogeneity* and *isotropy* of this spacetime: no point is intrinsically different from any other, and there are no preferred spacetime directions. (See [174, p. 257–260] for a careful treatment of these notions and relations between them.) Indeed, with respect to any inertial observer, the spaces $\{t\} \times \mathbb{R}^3$ have no preferred direction, consistent with the isotropy of light propagation; moreover, the direction of the spacetime path of any given inertial observer should not be preferred over that of any other, whether at rest or in uniform motion relative to each other. To compare two inertial frames, then, we might as well arrange that the origins in each inertial coordinate system (and we reiterate that each inertial coordinate system is built on Cartesian coordinates for its constant-time spatial slices) correspond to the same point in S. Uniform motion with respect to an inertial observer is given by a straight-line spacetime path (with nonconstant time) in the corresponding coordinate system, and should therefore be a straight line in the other coordinate system (by the law of inertia and the principle of relativity); actually, as we will see, there are many such spacetime lines that cannot represent paths of physical particles. If nonetheless we propose that all lines map to lines under the coordinate change, and if we assume the coordinate transformation is continuous, we can conclude with a little work that the transformation must be linear (recall that the origins match up, and see Exercise 1-4). Another way to argue for linearity is based on the homogeneity and isotropy of spacetime: there should be no preferred directions or points. A nonlinear change of coordinates would not preserve displacements along some line parallel to an axis (space or time), which would violate both of these properties. One could also postulate linearity for simplicity's sake, or in the sense of a Taylor approximation, and see what happens; as we are about to see, linear maps exist which do the job!

There are various subtleties in physically setting up an inertial coordinate system, especially as regards synchronization of clocks (which at rest all run at the same rate), but we do not go into details here. Clearly, however, if two inertial observers are *not* in relative motion, then after synchronization of clocks, and up to an application of a spatial orthogonal transformation, the coordinates should simply differ by a spatial translation. On the other hand, when there is nontrivial relative motion, we will soon see that two inertial observers will generally *not* agree whether two events are simultaneous.

That notwithstanding, we note that if in one inertial frame $\mathcal{O}$, two events E_1 and E_2 are simultaneous and with spatial displacement orthogonal to the direction of motion of an inertial frame $\widetilde{\mathcal{O}}$, then those events are also simultaneous in $\widetilde{\mathcal{O}}$. Indeed, one can imagine a clock in $\widetilde{\mathcal{O}}$ which will pass halfway between the two events in space; from this clock, send light signals (at the same time) which will arrive at the events E_1 and E_2 and be reflected back. Such signals will arrive back at the clock (moving *orthogonally* to the separation of the two events in $\mathcal{O}$) at the *same event*, and so $\widetilde{\mathcal{O}}$ will also measure E_1 and E_2 as being simultaneous. These events should also have the same distance measurement in each frame. Indeed, suppose the two points differ by a spatial displacement of unit size in one frame of reference $\mathcal{O}$. If an inertial observer from $\widetilde{\mathcal{O}}$ moving orthogonally to this displacement were to measure the spatial displacement of the points to be, say, less than 1, then by reversing the roles (with no preferred direction, the results should agree), we would have to conclude that $\mathcal{O}$ should measure the displacement to be less than that measured by $\widetilde{\mathcal{O}}$, and that cannot be the case! Note that embedded in this is a notion of how to measure spatial displacement between events that are *simultaneous*, as is the case with E_1 and E_2 in both frames.

The paths of light rays emanating from the origin comprise the *lightcone* or *nullcone*, which can be represented in coordinates as $\Sigma := \{(t, \boldsymbol{x}) : (ct)^2 = |\boldsymbol{x}|^2\}$. Lightcones can emanate from any point, and in fact you might instead consider the displacement $(\Delta t, \Delta \boldsymbol{x})$ between two points in spacetime along a light ray, which must satisfy $\Delta s^2 := -c^2(\Delta t)^2 + |\Delta \boldsymbol{x}|^2 = 0$. Applying the Einstein postulate that the speed of light in vacuum is the same for all inertial observers, we require that the linear transformation between inertial coordinates should *preserve lightcones*. If we set $\Delta x^0 = c\Delta t$ and $\Delta \boldsymbol{x} = (\Delta x^1, \Delta x^2, \Delta x^3)$, and let $(\Lambda^{\mu}_{\ \nu})$ be the matrix for the linear transformation between inertial coordinates, with the rescaled time coordinate $x^0 = ct$, we have, using the *Einstein summation convention* of summing over repeated upper and lower indices, $\Delta x^{\mu} = \Lambda^{\mu}_{\ \nu}\Delta \tilde{x}^{\nu}$. Then with

$\eta_{00} = -1$, $\eta_{0i} = 0$ and $\eta_{ij} = \delta_{ij}$ for i, $j \in \{1, 2, 3\}$, we have

$$\Delta s^2 = \Delta x^\mu \eta_{\mu\nu} \Delta x^\nu = \Delta \tilde{x}^\lambda \Lambda^\mu_{\ \lambda} \eta_{\mu\nu} \Lambda^\nu_{\ \sigma} \Delta \tilde{x}^\sigma = \Delta \tilde{x}^\lambda \tilde{\eta}_{\lambda\sigma} \Delta \tilde{x}^\sigma,$$

where we let $\tilde{\eta}_{\lambda\sigma} = \Lambda^\mu_{\ \lambda} \eta_{\mu\nu} \Lambda^\nu_{\ \sigma}$; note immediately that $\tilde{\eta}_{\lambda\sigma} = \tilde{\eta}_{\sigma\lambda}$. We know that Δs^2 vanishes whenever $\Delta \tilde{s}^2 = 0$; for example, if $\Delta \tilde{x}^0 = 1$ and $\Delta \tilde{x}^1 = \pm 1$, with $\Delta \tilde{x}^2 = 0 = \Delta \tilde{x}^3$, we get $0 = \tilde{\eta}_{00} + \tilde{\eta}_{11} \pm 2\tilde{\eta}_{01}$. Thus $\tilde{\eta}_{01} = 0$; likewise $\tilde{\eta}_{02} = 0 = \tilde{\eta}_{03}$ and moreover $\tilde{\eta}_{00} = -\tilde{\eta}_{11} = -\tilde{\eta}_{22} = -\tilde{\eta}_{33}$. In general for $\Delta \tilde{x}^0 = |\Delta \tilde{\boldsymbol{x}}|$, we have

$$0 = \Delta s^2 = \tilde{\eta}_{00}(\Delta \tilde{x}^0)^2 + \tilde{\eta}_{ij} \Delta \tilde{x}^i \Delta \tilde{x}^j = \eta_{00}((\Delta \tilde{x}^0)^2 - |\Delta \tilde{\boldsymbol{x}}|^2) + \sum_{i \neq j} \tilde{\eta}_{ij} \Delta \tilde{x}^i \Delta \tilde{x}^j,$$

so that $0 = \sum_{i \neq j} \tilde{\eta}_{ij} \Delta \tilde{x}^i \Delta \tilde{x}^j$. By choosing $\Delta \tilde{x}^1 = 1 = \Delta \tilde{x}^2$ and $\Delta \tilde{x}^3 = 0$, and $\Delta \tilde{x}^0 = |\Delta \tilde{\boldsymbol{x}}|$, we infer that $\tilde{\eta}_{12} = 0$, and likewise that $\tilde{\eta}_{ij} = 0$ for $i \neq j$. We thus conclude in general that $\Delta \tilde{s}^2 = -\tilde{\eta}_{00} \Delta s^2$. In other words, that lightcones must be preserved implies that the spacetime intervals Δs^2 in two inertial coordinate systems must agree up to a multiplicative constant. By the principle of relativity this multiplicative constant must equal ± 1 (as it must be the same in going from frame $\mathcal{O}$ to $\tilde{\mathcal{O}}$ or going in reverse: only the relative motion should matter). On the other hand, we know that certain non-vanishing spacetime intervals are preserved from one observer to another (such as simultaneous events orthogonal to the direction of motion, as discussed above), so that we must have $-\tilde{\eta}_{00} = 1$. We see that $\Delta s^2 = \Delta \tilde{s}^2$, and $\tilde{\eta}_{\mu\nu} = \eta_{\mu\nu}$: the coordinate change is a linear transformation that preserves $\eta = -(dx^0)^2 + \sum_{i=1}^3 (dx^i)^2 = -c^2 dt^2 + dx^2 + dy^2 + dz^2$. Such linear maps are *Lorentz transformations*, and we want to get our hands on them by writing some down explicitly.

We can set up axes with the direction of relative motion along the x-axis in each coordinate system, oriented so that a photon path along the x-axis (fixed y and z) given by $x = ct$ corresponds to $\tilde{x} = c\tilde{t}$. Based on the assumption that there is no preferred direction orthogonal to the direction of motion, and recalling the above argument regarding such orthogonal spatial displacements, it follows that we can arrange $\tilde{y} - y$ and $\tilde{z} - z$. Similar symmetry considerations show that the coordinates $\tilde{t}$ and $\tilde{x}$ of an event should be independent of y and z, and thus depend on t and x only. Arguing from the principle of relativity, if $\tilde{\mathcal{O}}$ is moving with velocity $\boldsymbol{v} = v \, \partial/\partial x$ with respect to $\mathcal{O}$, then $\mathcal{O}$ is moving with velocity $-\boldsymbol{v}$ with respect to $\tilde{\mathcal{O}}$ — which can also be readily seen by letting $x = 0$ and using (1.1.5) below. As the transformation between coordinates should depend only on the relative velocity, if T_v maps the coordinates of $\tilde{\mathcal{O}}$ to those of $\mathcal{O}$, then we should have $T_v^{-1} = T_{-v}$ (compare Einstein's derivation in [82]).

We have reduced the problem to a two-dimensional one, with relative motion in the x-direction, and with the change of coordinates giving a map between the

$(\tilde{t}, \tilde{x})$ and (t, x) coordinates of events in spacetime. Moreover, a vector along the lightcone in the $(\tilde{t}, \tilde{x})$-plane should map to the lightcone in the (t, x)-plane. We let the reduced two-dimensional mapping with $\boldsymbol{v} = v\,\partial/\partial x$ be T_v. Using standard bases we represent the linear transformation by a matrix $[T_v] = \begin{bmatrix} a_{11} & a_{12} \\ a_{21} & a_{22} \end{bmatrix}$, so that

$$t = a_{11}\tilde{t} + a_{12}\tilde{x},$$
$$x = a_{21}\tilde{t} + a_{22}\tilde{x}.$$

From here, we could use that this reduced map must preserve η (a corollary of the fact shown above that the full mapping preserves η) to get relations between the coefficients; see [188], for instance, and compare [84, Appendix 1]. Instead we will follow [38], proceeding to highlight the preservation of lightcones, and in the process showing directly that η is preserved in spacetime dimension two.

Now $\tilde{x} = 0$ is the path of the observer moving with respect to $\mathcal{O}$ with velocity v, and is given by $t = a_{11}\tilde{t}$, $x = a_{21}\tilde{t}$, so that $v = a_{21}/a_{11}$. Now we apply the invariance of the lightcone: the path $x = \pm ct$ should map to $\tilde{x} = \pm c\tilde{t}$ (with respective signs corresponding by orientation), which implies that the vectors $\begin{bmatrix} 1 \\ c \end{bmatrix}$ and $\begin{bmatrix} 1 \\ -c \end{bmatrix}$ are eigenvectors of T_v; by orientation again, the eigenvalues should be positive. This implies there are respective eigenvalues $\lambda_\pm > 0$ such that

$$a_{11} \pm ca_{12} = \lambda_\pm,$$
$$a_{21} \pm ca_{22} = \pm c\lambda_\pm.$$

Thus $ca_{11} + c^2 a_{12} = a_{21} + ca_{22}$, and $ca_{11} - c^2 a_{12} = -a_{21} + ca_{22}$. Adding and subtracting these two equations yields $a_{11} = a_{22}$ and $a_{21} = c^2 a_{12}$, from which we deduce $v/c^2 = a_{12}/a_{11}$. We then see the matrix for T_v has the following form, with $\alpha(v) := a_{11}$:

$$[T_v] = \begin{bmatrix} a_{11} & a_{12} \\ c^2 a_{12} & a_{11} \end{bmatrix} = \alpha(v)\begin{bmatrix} 1 & v/c^2 \\ v & 1 \end{bmatrix}. \tag{1.1.5}$$

Now $0 < \lambda_+\lambda_- = \det[T_v] = (\alpha(v))^2(1 - (v/c)^2)$, so $|v| < c$: the relative speed of two inertial observers is less than that of light. We also see $\alpha(v) > 0$, because $2\alpha(v) = \operatorname{tr}[T_v] = \lambda_+ + \lambda_- > 0$.

We now apply a symmetry argument similar to one we used earlier. Start with adapted coordinates (t, x) and $(\tilde{t}, \tilde{x})$ as above for two inertial frames of reference $\mathcal{O}$ and $\widetilde{\mathcal{O}}$, respectively. We also consider adapted coordinates $(t', x') = (t, -x)$ and $(\hat{t}, \hat{x}) = (\tilde{t}, -\tilde{x})$ related to the original coordinates by the parity operator $P(\tau, \xi) = (\tau, -\xi)$. The axes in these coordinates are consistently aligned, and frame $\widetilde{\mathcal{O}}$ moves with velocity $\boldsymbol{v} = v\,\partial/\partial x = -v\,\partial/\partial x'$ with respect to $\mathcal{O}$. So the transformation from $(\hat{t}, \hat{x})$ to (t', x'), which like T_v represents the identity map on spacetime, should be given by T_{-v} (consistent with isotropy and the principle of relativity). Thus if $[T_{-v}]$ is the matrix representing this coordinate transformation in the respective standard coordinate bases, then with $[P] = \begin{bmatrix} 1 & 0 \\ 0 & -1 \end{bmatrix}$, we see $[P][T_{-v}][P] = [T_v]$. From this we conclude $\det[T_{-v}] = \det[T_v]$, but from $[T_v]^{-1} = [T_{-v}]$, we see $\det[T_v] = \pm 1$, and thus $\det[T_v] = 1$ since we know the determinant is positive.

Together with the result of the last paragraph, we get

$$\alpha(v) = \frac{1}{\sqrt{1 - (v/c)^2}},$$

and we note then that $\alpha(v) = \alpha(-v)$. One could have argued $\alpha(v) = \alpha(-v)$ from the principle of relativity, from which the value of $\alpha(v)$ then follows using $[T_v]^{-1} = [T_{-v}]$ and (1.1.5). In any case, we have

$$[T_v] = \frac{1}{\sqrt{1-(v/c)^2}} \begin{bmatrix} 1 & v/c^2 \\ v & 1 \end{bmatrix}, \qquad [T_{-v}] = \frac{1}{\sqrt{1-(v/c)^2}} \begin{bmatrix} 1 & -v/c^2 \\ -v & 1 \end{bmatrix}.$$

Hence we have arrived at the Lorentz transformation

$$t = \frac{1}{\sqrt{1-(v/c)^2}} \left(\tilde{t} + \frac{v}{c^2}\tilde{x} \right), \qquad x = \frac{1}{\sqrt{1-(v/c)^2}} (v\tilde{t} + \tilde{x}),$$

$$\tilde{t} = \frac{1}{\sqrt{1-(v/c)^2}} \left(t - \frac{v}{c^2} x \right), \qquad \tilde{x} = \frac{1}{\sqrt{1-(v/c)^2}} (-vt + x). \tag{1.1.6}$$

Note that if we change the first variable to $x^0 := ct$, so that the coordinates have the same units, we have

$$x^0 = \frac{1}{\sqrt{1-(v/c)^2}} \left(\tilde{x}^0 + \frac{v}{c}\tilde{x} \right), \qquad x = \frac{1}{\sqrt{1-(v/c)^2}} \left(\frac{v}{c}\tilde{x}^0 + \tilde{x} \right),$$

$$\tilde{x}^0 = \frac{1}{\sqrt{1-(v/c)^2}} \left(x^0 - \frac{v}{c} x \right), \qquad \tilde{x} = \frac{1}{\sqrt{1-(v/c)^2}} \left(-\frac{v}{c}x^0 + x \right).$$

Let $\beta = v/c$, so that the matrix for T_v relative to the bases $\{ \frac{1}{c}\frac{\partial}{\partial t}, \frac{\partial}{\partial x} \}$ and $\{ \frac{1}{c}\frac{\partial}{\partial \tilde{t}}, \frac{\partial}{\partial \tilde{x}} \}$ is then just

$$[T_v] = \frac{1}{\sqrt{1-\beta^2}} \begin{bmatrix} 1 & \beta \\ \beta & 1 \end{bmatrix}, \qquad \text{and likewise} \quad [T_{-v}] = \frac{1}{\sqrt{1-\beta^2}} \begin{bmatrix} 1 & -\beta \\ -\beta & 1 \end{bmatrix}.$$

For $-1 < \beta = v/c < 1$, there is a unique θ such that

$$\sinh\theta = \frac{\beta}{\sqrt{1-\beta^2}} = \frac{v/c}{\sqrt{1-(v/c)^2}}.$$

Then $\cosh\theta = 1/\sqrt{1-\beta^2} = 1/\sqrt{1-(v/c)^2}$ (since $\cosh^2\theta - \sinh^2\theta = 1$ and $\cosh\theta > 0$). Thus we can write from the above

$$[T_{-v}] = \begin{bmatrix} \cosh\theta & -\sinh\theta \\ -\sinh\theta & \cosh\theta \end{bmatrix},$$

so that if θ corresponds to v, $-\theta$ corresponds to $-v$. You might observe the appearance of the hyperbolic functions is not surprising, given that the level sets of the function $-c^2 t^2 + x^2$ are *invariant hyperbolae*: a Lorentz transformation preserves the value of this function, i.e., $-c^2\tilde{t}^2 + \tilde{x}^2 = -c^2 t^2 + x^2$ when $(\tilde{t}, \tilde{x})$ and (t, x) are related by a Lorentz transformation.

We can use this to find the velocity addition rule in special relativity: if θ_1 and θ_2 correspond to v_1 and v_2, then we note that if we consider frame $\tilde{\mathcal{O}}$ moving along the x-axis of $\mathcal{O}$ at velocity v_1, and $\widehat{\mathcal{O}}$ moving along the $\tilde{x}$ axis of $\tilde{\mathcal{O}}$ at velocity v_2, then the transformation T which takes coordinates measured in $\widehat{\mathcal{O}}$ to those in $\mathcal{O}$ satisfies the following:

$$[T] = [T_{v_1}][T_{v_2}] = \begin{bmatrix} \cosh\theta_1 & \sinh\theta_1 \\ \sinh\theta_1 & \cosh\theta_1 \end{bmatrix} \begin{bmatrix} \cosh\theta_2 & \sinh\theta_2 \\ \sinh\theta_2 & \cosh\theta_2 \end{bmatrix}$$

$$= \begin{bmatrix} \cosh(\theta_1 + \theta_2) & \sinh(\theta_1 + \theta_2) \\ \sinh(\theta_1 + \theta_2) & \cosh(\theta_1 + \theta_2) \end{bmatrix} = [T_v],$$

where v corresponds to $\theta = \theta_1 + \theta_2$. Thus the transformation between frames is precisely that which has $\widehat{\mathcal{O}}$ moving along the x-axis relative to $\mathcal{O}$ at velocity v. We see that the set of Lorentz transformations corresponding to inertial observers in relative motion along a common axis has an elementary group structure. Moreover, we also have from the above,

$$\frac{v}{c} = \tanh\theta = \tanh(\theta_1 + \theta_2) = \frac{\tanh\theta_1 + \tanh\theta_2}{1 + \tanh\theta_1 \tanh\theta_2} = \frac{\dfrac{v_1}{c} + \dfrac{v_2}{c}}{1 + \dfrac{v_1 v_2}{c^2}}.$$

Thus the Galilean rule (1.1.2) for the addition of velocities, $v = v_1 + v_2$, is replaced by

$$v = \frac{v_1 + v_2}{1 + \dfrac{v_1 v_2}{c^2}}.$$

Note that for $|v_1|, |v_2| < c$, we have $(1 \pm v_1/c)(1 \pm v_2/c) > 0$, from which it follows that $|v| < c$. As expected, for $|v| \ll c$, the velocity addition rule, and in fact the Lorentz transformations, approximate their Galilean counterparts.

1.1.3. *Electromagnetism.* As noted above, electromagnetism played a central role in the development of special relativity. We saw that the wave equation, which governs the propagation of the electric and magnetic fields in vacuum, is invariant under Lorentz transformations. Moreover, it is a fundamental fact of physics that the electromagnetic fields behave in such a way that the laws which govern electromagnetism, Maxwell's equations, are invariant under Lorentz transformations (sometimes referred to as *special covariance* of the equations). Thus the laws of electromagnetism take the same form in all inertial frames, in accordance with the principle of relativity.

While we will not develop the foundations of electromagnetic theory in a relativistic framework, we should at least indicate how the electromagnetic fields as measured in two inertial frames compare, i.e., how they transform under a Lorentz transformation. As we saw earlier, we expect the electric and magnetic fields to somehow transform together, since charges in motion induce and are affected by a magnetic field, but motion is relative to the frame of reference. In fact, the fields do transform together as an anti-symmetric two-tensor, called the *Faraday tensor*. In an inertial frame $\mathcal{O}$ (and using $x^0 = ct$), where the electric field has components E^i and the magnetic field B^i, the Faraday tensor has a component matrix

$$[F^{\mu\nu}] = \begin{bmatrix} 0 & E^1 & E^2 & E^3 \\ -E^1 & 0 & B^3 & -B^2 \\ -E^2 & -B^3 & 0 & B^1 \\ -E^3 & B^2 & -B^1 & 0 \end{bmatrix}. \tag{1.1.7}$$

We can write F as a two-form $F^\flat$ (with components down) as (where we have $E_i = E^i$, $B_i = B^i$ in inertial coordinates)

$$F^\flat = (E_1\,dx^1 + E_2\,dx^2 + E_3\,dx^3) \wedge dx^0$$
$$+ (B_1\,dx^2 \wedge dx^3 + B_2\,dx^3 \wedge dx^1 + B_3\,dx^1 \wedge dx^2).$$

How do the fields transform exactly? For example, let us compute the magnetic field component $\tilde{B}^3$ in another inertial frame $\widetilde{\mathcal{O}}$ moving along the x-axis at velocity v relative to $\mathcal{O}$. The associated Lorentz transformation has matrix elements $\Lambda^\mu_{\ \nu}$ given by

$$[\Lambda^\mu_{\ \nu}] = \begin{bmatrix} \dfrac{1}{\sqrt{1-(v/c)^2}} & \dfrac{-v/c}{\sqrt{1-(v/c)^2}} & 0 & 0 \\ \dfrac{-v/c}{\sqrt{1-(v/c)^2}} & \dfrac{1}{\sqrt{1-(v/c)^2}} & 0 & 0 \\ 0 & 0 & 1 & 0 \\ 0 & 0 & 0 & 1 \end{bmatrix}.$$

Hence, using the Einstein summation convention, we have

$$\tilde{B}^3 = \tilde{F}^{12} = \Lambda^1_{\ \mu} F^{\mu\nu} \Lambda^2_{\ \nu} = \Lambda^1_{\ 0} F^{02} \Lambda^2_{\ 2} + \Lambda^1_{\ 1} F^{12} \Lambda^2_{\ 2}$$

$$= -\frac{v/c}{\sqrt{1-(v/c)^2}} E^2 + \frac{1}{\sqrt{1-(v/c)^2}} B^3$$

$$= \frac{1}{\sqrt{1-(v/c)^2}} \left(B^3 - \frac{v}{c} E^2 \right).$$

Similarly we find

$$\tilde{E}^1 = \tilde{F}^{01} = \Lambda^0_{\ \mu} F^{\mu\nu} \Lambda^1_{\ \nu} = \Lambda^0_{\ 0} F^{01} \Lambda^1_{\ 1} + \Lambda^0_{\ 1} F^{10} \Lambda^1_{\ 0} = E^1.$$

One can check the other transformation rules:

$$\tilde{B}^1 = B^1,$$

$$\tilde{B}^2 = \frac{1}{\sqrt{1-(v/c)^2}} \left(B^2 + \frac{v}{c} E^3 \right),$$

$$\tilde{E}^2 = \frac{1}{\sqrt{1-(v/c)^2}} \left(E^2 - \frac{v}{c} B^3 \right),$$

$$\tilde{E}^3 = \frac{1}{\sqrt{1-(v/c)^2}} \left(E^3 + \frac{v}{c} B^2 \right).$$

In a region with electromagnetic fields but free of charges and hence current (for simplicity), Maxwell's equations are as follows (in inertial coordinates, using spatial divergence and curl):

$$\text{div } \boldsymbol{E} = 0, \qquad -\frac{1}{c}\frac{\partial \boldsymbol{E}}{\partial t} = \text{curl } \boldsymbol{B}, \tag{1.1.8}$$

$$\text{div } \boldsymbol{B} = 0, \qquad -\frac{1}{c}\frac{\partial \boldsymbol{B}}{\partial t} = -\text{curl } \boldsymbol{E}. \tag{1.1.9}$$

Maxwell's equations in any inertial frame will take the same familiar form, consistent with the principle of relativity; indeed, one can easily check, using the above transformation rules for the fields, that the form of Maxwell's equations in inertial coordinates is preserved by Lorentz transformations.

Having said that, it is instructive to write Maxwell's equations directly in terms of F. Equations (1.1.8) are equivalent to $F^{\mu\nu}_{\ \ ,\nu} = 0$, with Gauss's law $\text{div } \boldsymbol{E} = 0$ for $\mu = 0$, and with $\frac{1}{c}\frac{\partial \boldsymbol{E}}{\partial t} = \text{curl } \boldsymbol{B}$ corresponding to the remaining three components — and we recall that $\frac{\partial}{\partial x^0} = \frac{1}{c}\frac{\partial}{\partial t}$. Of course, $F^{\mu\nu}_{\ \ ,\nu}$ gives the components of $\text{div}_\eta F$ in inertial coordinates, so that (1.1.8) can be expressed as $\text{div}_\eta F = 0$. Similarly, expressing $dF^\flat = 0$ in inertial coordinates is equivalent to (1.1.9).

Since, as stated above, the electromagnetic field transforms according to the two-tensor F under change of inertial frame, Maxwell's equations (being expressible in terms of $\operatorname{div}_\eta F$ and $dF^\flat$) have the same familiar form in any inertial frame. By tensoriality, we can readily express F and $F^\flat$ in *any* coordinate system in terms of the explicit expressions of F and $F^\flat$ in an inertial coordinate system. This is a purely *mathematical* fact; that the Lorentz force law (p. 32) with this F should hold in any coordinate system, including one somehow adapted to a non-inertial frame, must derive from some principle of *physics*; see Section 1.2.4. Assuming that, writing Maxwell's equations in tensorial form makes manifest their *general covariance* with respect to all spacetime coordinate systems. All frames of reference are in principle equivalent from the point of view of formulating the physical law (consistent with the principle of *general* relativity). While the component form of Maxwell's equations in inertial coordinates in Minkowski spacetime seems particularly convenient, we shall see that in Einstein's theory of gravitation, (locally) inertial frames are not determined *a priori*, though in such frames the laws of physics should reduce to their special relativistic form (in accordance with a version of Einstein's *equivalence principle*). But this is getting slightly ahead of the story at this point!

1.2. Kinematics in Minkowski spacetime

In the remainder of the chapter, we will use Roman indices such as i, j, k to label spatial components, whereas Greek indices will continue to be used for space and time components. The Einstein convention will remain in force, summing over the relevant index range, as we will remind the reader sporadically throughout.

1.2.1. *The Minkowski metric.* Consider the matrix $\Lambda = \begin{bmatrix} \cosh\theta & \sinh\theta \\ \sinh\theta & \cosh\theta \end{bmatrix}$. Since

$$\Lambda^T \begin{bmatrix} -1 & 0 \\ 0 & 1 \end{bmatrix} \Lambda = \begin{bmatrix} -1 & 0 \\ 0 & 1 \end{bmatrix},$$

we again see that each element of the group of Lorentz transformations in one space dimension preserves the bilinear form $-(dx^0)^2 + dx^2 = -c^2 dt^2 + dx^2$. Returning to three spatial dimensions, the relevant bilinear form is

$$\eta = -(dx^0)^2 + dx^2 + dy^2 + dz^2 = -c^2 dt^2 + dx^2 + dy^2 + dz^2,$$

which as we have seen is preserved by the (four-dimensional) transformations between two sets of inertial coordinates (corresponding to rigid reference frames for two inertial observers with the same origin of spacetime coordinates). The set of linear maps that preserve the quadratic form η — the set of Lorentz transformations — forms a group, the *Lorentz group*; the *proper* Lorentz group is the component of the identity map, chosen to ensure observers use a consistent spatiotemporal orientation (and in particular keep the same arrow of time). The Lorentz group is a subgroup of the Poincaré group of maps which preserve η, which also includes translations.

We let $\langle v, w \rangle := \eta(v, w)$. The manifold $\mathbb{R}^4$ together with the semi-Riemannian metric η is four-dimensional Minkowski spacetime $\mathbb{M}^4$, or $\mathbb{R}_1^4$, indicating the signature of the metric, and similarly for $\mathbb{M}^n = \mathbb{R}_1^n$, for $n \geq 2$.

1.2.2. *Causal nature of vectors.* (See Figure 1.) Vectors $w = (c\Delta t, \Delta x)$ with $\langle w, w \rangle < 0$ are called *timelike*, since $|\Delta x| < c|\Delta t|$. Then $|v| = |\Delta x|/|\Delta t| < c$. Such vectors may thus represent the tangents to spacetime paths of material particles. We note in (1.2.1) below how such vectors represent the spacetime displacements between pairs of events that in some inertial frame share the same spatial location. Similarly, *null vectors* $w \neq 0$ satisfy $\langle w, w \rangle = 0$; these are tangent to the lightcone and represent paths of light rays. Finally, vectors with $\langle w, w \rangle > 0$ are called *spacelike*, and represent displacements between pairs of events which are simultaneous (i.e., share the same time coordinate) in some inertial frame; see (1.2.2) below. The zero vector is also defined to be spacelike.

Suppose that $w = (c\Delta t, \Delta x, 0, 0)$ is timelike, and let $v = \Delta x / \Delta t$, so that $|v| < c$. The Lorentz transformation T_{-v} satisfies

$$[T_{-v}] \begin{bmatrix} c\Delta t \\ \Delta x \end{bmatrix} = \frac{1}{\sqrt{1 - (v/c)^2}} \begin{bmatrix} 1 & -v/c \\ -v/c & 1 \end{bmatrix} \begin{bmatrix} c\,\Delta t \\ \Delta x \end{bmatrix} = \begin{bmatrix} * \\ 0 \end{bmatrix}. \qquad (1.2.1)$$

Similarly if $|\Delta x| > c|\Delta t|$, then let $v/c = c\Delta t / \Delta x$, so that $|v| < c$. Then

$$[T_{-v}] \begin{bmatrix} c\Delta t \\ \Delta x \end{bmatrix} = \frac{1}{\sqrt{1 - (v/c)^2}} \begin{bmatrix} 1 & -v/c \\ -v/c & 1 \end{bmatrix} \begin{bmatrix} c\Delta t \\ \Delta x \end{bmatrix} = \begin{bmatrix} 0 \\ * \end{bmatrix}. \qquad (1.2.2)$$

Thus in either case, there is a Lorentz transformation which maps the timelike or spacelike vector to align with a timelike or spacelike axis for an appropriate observer.

1.2.2.1. *Twin paradox.* It turns out that the familiar triangle inequality for vectors in Euclidean geometry is *reversed* for timelike vectors in Minkowski spacetime.

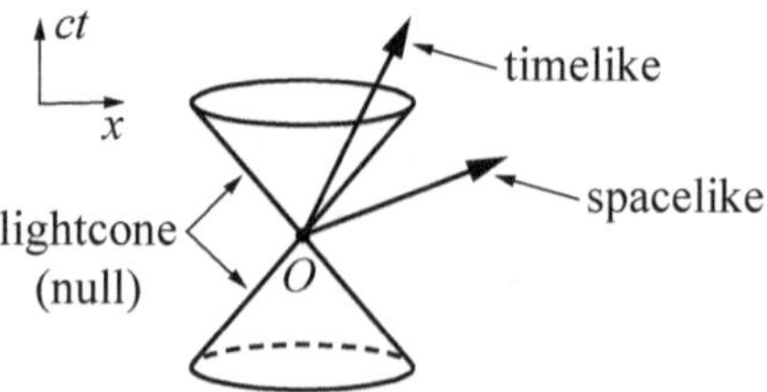

Figure 1. Causal nature of vectors.

Given a smooth timelike curve $\gamma(\lambda)$ (meaning that $\gamma'(\lambda)$ is timelike for all λ), we define the *proper time* $\Delta\tau$ along a portion of γ as

$$\Delta\tau = c^{-1}\int_{\lambda_0}^{\lambda_1}\sqrt{-\langle\gamma'(\lambda),\gamma'(\lambda)\rangle}\,d\lambda,$$

and the proper time function as

$$\tau(\lambda) = c^{-1}\int_{\lambda_0}^{\lambda}\sqrt{-\langle\gamma'(\underline{\lambda}),\gamma'(\underline{\lambda})\rangle}\,d\underline{\lambda}.$$

If we reparametrize γ by proper time τ by inverting to get $\lambda = \lambda(\tau)$, so that $\tilde{\gamma}(\tau) := \gamma(\lambda(\tau))$, we have

$$\tilde{\gamma}'(\tau) = \gamma'(\lambda(\tau))\frac{d\lambda}{d\tau} = \frac{c}{\sqrt{-\langle\gamma'(\lambda(\tau)),\gamma'(\lambda(\tau))\rangle}}\gamma'(\lambda(\tau)).$$

Thus we see the tangent vector has constant length $\sqrt{-\langle\tilde{\gamma}'(\tau),\tilde{\gamma}'(\tau)\rangle} = c$. This means that the parameter $\tau - \tau(\lambda_0)$ is indeed the proper time elapsed along γ from λ_0 to $\lambda(\tau)$.

We now consider the reversed triangle inequality. If the displacement vector $\overrightarrow{OB}$ from O to B is timelike, let

$$|\overrightarrow{OB}| = \sqrt{-\langle\overrightarrow{OB},\overrightarrow{OB}\rangle},$$

which is, up to a factor of c, the proper time along a straight-line path from O to B (the elapsed time measured by the inertial observer passing through O and B).

Definition 1-1. A vector is *causal* if it is timelike or null. A path is *causal* if its tangent vector at each point is causal. A causal vector w is *future-pointing* if $\langle w, \partial/\partial t\rangle < 0$, and analogously for *past-pointing*.

Proposition 1-2. *If $\overrightarrow{OB}$ is future-pointing and timelike, and $\overrightarrow{OA}$ and $\overrightarrow{AB}$ are future-pointing and causal, then $|\overrightarrow{OB}| \geq |\overrightarrow{OA}| + |\overrightarrow{AB}|$, with equality only in case O, A and B are collinear.*

Proof. By applying a Lorentz transformation as in (1.2.1), we may arrange that $\overrightarrow{OB}$ has components $(t_B, 0, 0, 0)$, for $t_B > 0$, and $\overrightarrow{OA}$ has components $(t_A, x_A, 0, 0)$, with $|x_A| \leq ct_A$. Then $\overrightarrow{AB}$ has components $(t_B - t_A, -x_A, 0, 0)$, with $|x_A| \leq c(t_B - t_A)$. Therefore,

$$|\overrightarrow{OA}|^2 = (ct_A)^2 - x_A^2 \quad\text{and}\quad |\overrightarrow{AB}|^2 = c(t_B - t_A)^2 - x_A^2,$$

so that

$$|\overrightarrow{OA}| + |\overrightarrow{AB}| = \sqrt{(ct_A)^2 - x_A^2} + \sqrt{c(t_B - t_A)^2 - x_A^2}$$
$$\leq ct_A + c(t_B - t_A) = ct_B = |\overrightarrow{OB}|.$$

The only way equality holds is if $x_A = 0$. $\square$

The reversed triangle inequality has the following analogue for piecewise smooth paths.

Proposition 1-3. *Suppose O and B are two points in Minkowski spacetime so that the displacement vector $\overrightarrow{OB}$ is future-pointing and timelike. Then amongst all piecewise smooth future-pointing causal paths from O to B, the one of maximal proper time interval is the straight-line path from O to B.*

Proof. By applying a Lorentz transformation as in (1.2.1), we may choose an inertial frame so that O is the origin and B lies on the positive t-axis. If γ is a timelike curve from O to B parametrized by proper time τ, with coordinates $(t(\tau), x(\tau), y(\tau), z(\tau))$, then along γ,

$$\Delta \tau = c^{-1} \int_0^{\Delta \tau} \sqrt{-\langle \gamma'(\tau), \gamma'(\tau) \rangle} \, d\tau$$
$$= c^{-1} \int_0^{\Delta \tau} \sqrt{c^2 \left(\frac{dt}{d\tau}\right)^2 - \left(\frac{dx}{d\tau}\right)^2 - \left(\frac{dy}{d\tau}\right)^2 - \left(\frac{dz}{d\tau}\right)^2} \, d\tau$$
$$\leq \int_0^{\Delta \tau} \frac{dt}{d\tau} d\tau = \Delta t.$$

We used the fact that $dt/d\tau > 0$. Also, if the curve were piecewise smooth, we could break it up into finitely many intervals and apply the above analysis on each interval. The inequality is clearly strict unless x, y and z are constant (equal to 0), in which case the curve is indeed along the straight line path. $\square$

The title "twin paradox" of the subsection refers to the following interpretation of the reversed triangle inequality. Suppose two twins are together at O, at which time they are inertial observers moving relative to each other along their x-axes with velocity 80% of the speed of light. Suppose that from the point of view of one of the twins who remains an inertial observer at rest in an inertial frame $\mathcal{O}$, the other twin travels for five years to arrive at spacetime point A, quickly turns around and returns to join the other twin at spacetime point B after a total travel time of ten years as determined in $\mathcal{O}$. In other words, the proper time elapsed on a straight path from O to B is ten years. On the other hand, the proper time that elapses on the other twin's path from O to A to B is strictly less than ten years, by the previous corollary: the other twin is *younger*! How can this be if both are moving relative to each other: are the situations not symmetric? Well, no: the twin moving to point A "and back" must accelerate to turn around. This acceleration means that this twin does not remain at rest with respect to a single inertial frame the entire time. In other words, two rigid frames of reference adapted to these observers (and respective coordinate charts for Minkowski spacetime) do not correspond by a simple Lorentz transformation.

Before we move on, let us compute how much time passes for the younger twin. Suppose the twins mark each passing year by sending a light signal to each other. From the point of view of $\mathcal{O}$, a signal sent at year $t = k$ will be received by the other twin on the *outgoing* part of the journey at time t_k determined by $vt_k = c(t_k - k)$, which determines how long it will take the light to catch up to the moving twin. Now, $v/c = 0.8$, so $t_k = 5k$. Thus the first signal ($k = 1$) will be received just as the second twin turns around to begin the return trip. Only one signal from $\mathcal{O}$ will be received on the outgoing portion of the trip, and so the other nine signals will be received by the other twin on the return portion of the journey.

What about signals sent from the twin moving relative to $\mathcal{O}$? The first signal is sent after one unit of proper time has passed along the path from O to A, which is at time $t = 1/\sqrt{1 - (v/c)^2}$ as measured in $\mathcal{O}$. The distance between the twins at this instant, as measured in $\mathcal{O}$ is just $v/\sqrt{1 - (v/c)^2}$. Thus the time at which the signal will arrive at $x = 0$ is

$$\frac{1}{\sqrt{1 - (v/c)^2}} + \frac{1}{c}\frac{v}{\sqrt{1 - (v/c)^2}} = \frac{\sqrt{1 + v/c}}{\sqrt{1 - v/c}}.$$

Similarly the signal sent after ℓ years of proper time have elapsed on the outward journey from O to A will arrive at $x = 0$ at

$$t = \frac{\sqrt{1 + v/c}}{\sqrt{1 - v/c}}\,\ell. \tag{1.2.3}$$

The coefficient of ℓ in (1.2.3) gives the time interval between the reception of successive signals which were emitted one unit in time apart, and thus it is the ratio between the frequency of emission of signals, as measured by the emitter, and the frequency of reception of signals, as measured by the receiver. As such, it gives a measure of the *relativistic Doppler shift*. When $v/c = 0.8$, this Doppler factor equals 3. The numerology works out so that the third signal sent along the path from O to A occurs at $t = 3/\sqrt{1 - (0.8)^2} = 5$, so three signals are sent along the outward journey, and the third signal arrives at $x = 0$ at $t = 9$. The twin at A has sent three signals back to his twin, but has received only one signal. On the "homeward" journey, this twin will send three more signals (the last just as the twins are back together again), and will receive a total of nine signals from the twin with frame $\mathcal{O}$, the last upon arrival.

1.2.3. *Simultaneity.* The worldlines of inertial observers are special paths, namely timelike *geodesics*. Moreover, inertial observers correspond to certain coordinate charts on Minkowski spacetime $\mathbb{M}$. Mathematically it is not a big deal when a point in $\mathbb{M}$ has two different sets of coordinates in two different charts. However, interpreting the coordinates in the physical model yields some interesting results: the coordinates are not merely labels, but rather they are supposed to be the results of physical measurements.

The first observation is that *simultaneity is relative*: two different observers adapted to respective inertial frames $\mathcal{O}$ and $\widetilde{\mathcal{O}}$ moving relative to each other will not agree in general on whether two events occur at the same time. Imagine we synchronize the observers at a common origin O, and that they are moving along their respective x-axes. The Lorentz transformations tell us that the events which $\widetilde{\mathcal{O}}$ charts as occurring simultaneously at $\tilde{t} = 0$ correspond to $t = vx/c^2$ in $\mathcal{O}$. Different points in this set have different t-coordinates, and so $\mathcal{O}$ will not agree that they occur at the same time; cf. (1.2.2).

Though simultaneity is relative, both observers will calculate the same value of $-(ct)^2 + x^2 = -(x^0)^2 + x^2 = -(\tilde{x}^0)^2 + (\tilde{x})^2 = -(c\tilde{t})^2 + (\tilde{x})^2$. This observation can be used to obtain two interesting conclusions that can be verified experimentally: *time dilation* and *Lorentz contraction*.

Consider the event A which has coordinates $\tilde{t}_A = 1$, $\tilde{x}_A = 0$ (we suppress the other spatial dimensions, whose coordinates we take to be 0). The Lorentz transformation gives us the coordinates (t_A, x_A) for A in $\mathcal{O}$, and in particular

$$t_A = \frac{\tilde{t}_A + v\tilde{x}_A/c^2}{\sqrt{1 - (v/c)^2}} = \frac{1}{\sqrt{1 - (v/c)^2}} > 1.$$

$\mathcal{O}$ measures more time to have elapsed from O to A, and so concludes that the moving clock in $\widetilde{\mathcal{O}}$'s frame *runs slow*. Another way to see this from the invariant hyperbola is to note that since $-(c\tilde{t}_A)^2 + \tilde{x}_A^2 = -c^2$, A is on the hyperbola $-(ct)^2 + x^2 = -c^2$; but $x_A = vt_A \neq 0$, so that we must have $t_A > 1 = \tilde{t}_A$! You can see this with a simple picture: the invariant hyperbola through A hits the t-axis at the point $(t, x) = (1, 0) = (\tilde{t}_A, 0)$, with t-coordinate clearly lower than t_A. (This point is labeled B in Figure 2, left.) In general, if time $\Delta\tilde{t}$ is measured between events at a fixed $\tilde{x}$ value, the time between the events as measured in $\mathcal{O}$ will be $\Delta t = \Delta\tilde{t}/\sqrt{1 - (v/c)^2} > \Delta\tilde{t}$; cf. (1.2.1).

Similarly, moving objects contract along the direction of motion. To be precise, consider a rod along the $\tilde{x}$-axis, whose rest length measured in $\widetilde{\mathcal{O}}$ is L, and which is moving with velocity $v > 0$ along the x-axis. This means that the ends of the rod are measured simultaneously in $\widetilde{\mathcal{O}}$ at, say, O given by $(\tilde{t}, \tilde{x}) = (0, 0)$ and A

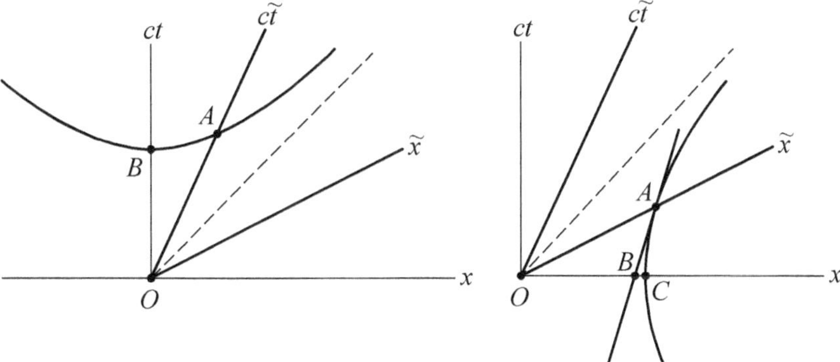

Figure 2. Left: time dilation. Right: Lorentz contraction.

given by $(\tilde{t}, \tilde{x}) = (0, L)$. The ends of the rod make paths in spacetime, one given by $\tilde{x} = 0$, the other by $\tilde{x} = L$. (See Figure 2, right.)

We need to find the coordinates of the point B where $\tilde{x} = L$ intersects $t = 0$, since then both O and B will be simultaneous with respect to $\mathcal{O}$. By the Lorentz transformation (1.1.6), the event B will have $\tilde{t} = -vL/c^2$, so that $x = \sqrt{1 - (v/c)^2}\, L < L$. In $\mathcal{O}$, the rod is measured to have length $\sqrt{1 - (v/c)^2}\, L$, since determining the length of the rod amounts to finding the spatial separation between the ends *at the same time*. It is this simultaneity that is relative. This length contraction is a necessary consequence of time dilation and the agreement by the observers on their relative speed, and can readily be seen geometrically in terms of the invariant hyperbola. Indeed, the hyperbola $-(c\tilde{t})^2 + \tilde{x}^2 = L^2$ through the point A lies to the right of the line $\tilde{x} = L$, touching only at the point of tangency at A. Therefore, if C is the event given by the intersection of the hyperbola and the line $t = 0$, the event B which occurs on $\tilde{x} = L$ is between the origin O and the point C along the line $t = 0$. Thus C must have coordinate $(t, x) = (0, L)$, since it lies on the hyperbola. Hence the point B must have x-coordinate *less than L*.

1.2.3.1. *Pole and barn paradox.* Consider a barn of rest length 10 and a pole of rest length 20. Suppose they are at rest in respective inertial frames moving along the x-axis with respect to each other with relative velocity v given by $v/c = \frac{\sqrt{3}}{2}$, so that $\sqrt{1 - (v/c)^2} = \frac{1}{2}$. From the point of view of the barn, the pole contracts along the direction of motion to half its rest length, so that it can fit entirely in the barn as it moves through. From the point of view of the pole, the barn is moving toward it, and thus it contracts to length 5; thus in the rest frame of the pole, it can never fit entirely inside the barn. Can both viewpoints of reality be correct?

The answer of course is *yes*! We analyze this first from the point of view of the barn frame $\mathcal{O}$, in which the one end of the barn has worldline $x = 0$, and the other end $x = 10$. The pole is moving in the positive x-direction, and at time $t = 0$ in $\mathcal{O}$, the ends of the pole are at $x = 0$ (front) and $x = -10$ (back). The worldlines for the front and back ends of the pole are respectively $x = vt/c = \frac{\sqrt{3}}{2}\, t$ and $x = -10 + \frac{\sqrt{3}}{2}\, t$. At time $t = \frac{20}{\sqrt{3}}$, the front end of the pole is at $x = 10$ (let this be spacetime point A), and the back end at $x = 0$ (at spacetime point B, say).

From the pole frame $\widetilde{\mathcal{O}}$, the barn never contains the pole. That the two observers disagree is not a paradox, since simultaneity is relative. The issue is simply that in $\widetilde{\mathcal{O}}$, the events A and B are not simultaneous as they are in $\mathcal{O}$. This is easy to compute by using the Lorentz transformation (1.1.6) on the points with coordinates $(t, x) = \left(\frac{20}{\sqrt{3}}, 10\right)$ (point A) and $(t, x) = \left(\frac{20}{\sqrt{3}}, 0\right)$ (point B).

1.2.4. Acceleration. If γ is a future-pointing timelike curve parametrized by proper time τ, then for each τ, there is a *momentarily comoving* inertial reference frame, or *local rest frame*, giving coordinates (t, x^k) for Minkowski spacetime for which $\gamma(\tau)$ corresponds to the origin and $\gamma'(\tau) = \partial/\partial t$ at $\gamma(\tau)$. Since $\langle \gamma'(\tau), \gamma'(\tau)\rangle$ is constant, the (covariant) acceleration $\gamma''(\tau) = D\gamma'(\tau)/d\tau$ is orthogonal to $\gamma'(\tau)$. At $\gamma(\tau)$ in the momentarily comoving inertial frame, then, $\gamma''(\tau)$ is purely spatial, and thus $\langle \gamma''(\tau), \gamma''(\tau)\rangle \geq 0$. In fact, in this frame *at* $\gamma(\tau)$, we have $dt/d\tau = 1$ and $dx^k/dt = 0$, so that

$$\frac{d^2 x^k}{d\tau^2} = \frac{d}{d\tau}\left(\frac{dt}{d\tau}\frac{dx^k}{dt}\right) = \frac{d^2 x^k}{dt^2}$$

holds here, from which we conclude

$$\gamma''(\tau) = \left.\frac{d^2 x^k}{dt^2}\right|_0 \left.\frac{\partial}{\partial x^k}\right|_{\gamma(\tau)}.$$

The spacetime norm of the covariant acceleration $\mathbb{A} = \gamma''(\tau)$ is thus the norm of the spatial acceleration $\boldsymbol{a} = (d^2 x^k/dt^2)\partial/\partial x^k$ as measured in the momentarily comoving inertial frame. (Recall that the Einstein convention applies here, even though $\partial/\partial x^k$ is written with a slash rather than in stacked notation.)

Consider an inertial frame $\mathcal{O}$ with respect to which the instantaneous velocity of $\gamma(\tau)$ is $\boldsymbol{v} = v^i \partial/\partial x^i$, with speed $v = |\boldsymbol{v}|$, and spatial acceleration $\boldsymbol{a} = (dv^i/dt)\partial/\partial x^i$. Then, since $dt/d\tau = 1/\sqrt{1 - (v/c)^2}$, we obtain

$$\mathbb{U} = \gamma'(\tau) = \frac{1}{\sqrt{1 - (v/c)^2}}\left(\frac{\partial}{\partial t} + \boldsymbol{v}\right), \tag{1.2.4}$$

$$\mathbb{A} = \frac{1}{\sqrt{1 - (v/c)^2}}\frac{d}{dt}\left(\frac{1}{\sqrt{1 - (v/c)^2}}\right)\left(\frac{\partial}{\partial t} + \boldsymbol{v}\right) + \frac{1}{1 - (v/c)^2}\boldsymbol{a}.$$

A simple exercise for the reader is to verify that $\mathbb{A} = 0$ if and only if $a = 0$.

It is interesting to consider how an accelerating observer can set up spacetime coordinates to make physical measurements, and then compare the coordinates to that of an inertial observer; for examples, see [38, Chapter 4] for a uniformly rotating frame, and a frame with constant acceleration. Facilitating this is the *clock hypothesis*, consistent with empirical evidence, which asserts that relative to an inertial frame, the rate of a clock depends on its instantaneous velocity, but not its acceleration. Consequently, spacetime measurements by an observer should be in accordance locally with those from the local rest frames along the worldline, and should yield the components of the Minkowski metric as appropriately transformed from those of an inertial frame, consistent with the extension of tensorial laws from inertial to non-inertial coordinates. In particular, the proper time along a worldline should agree with the time elapsed on a clock moving along the worldline.

1.2.4.1. *Constant acceleration: hyperbolic motion.* Suppose $\gamma(\tau)$ is a future-pointing timelike curve parametrized by proper time τ, and that for some inertial observer, $\gamma(\tau)$ has parametrization $(t(\tau), x(\tau), 0, 0)$. We consider the situation that $\langle \gamma''(\tau), \gamma''(\tau) \rangle = a^2$ is constant, for some $a > 0$. We can determine the coordinate functions using the two conditions

$$c^2 \left(\frac{dt}{d\tau} \right)^2 - \left(\frac{dx}{d\tau} \right)^2 = c^2, \quad -c^2 \left(\frac{d^2 t}{d\tau^2} \right)^2 + \left(\frac{d^2 x}{d\tau^2} \right)^2 = a^2.$$

Since $dt/d\tau > 0$ (γ is future-pointing), by the above we have $dt/d\tau \geq 1$, and there is a unique function $f(\tau)$ such that $dt/d\tau = \cosh f(\tau)$ and $dx/d\tau = c \sinh f(\tau)$, from which we that see $f(\tau)$ is smooth. Using the acceleration condition, we obtain

$$c^2 (f'(\tau))^2 = a^2.$$

Therefore $f(\tau) = \pm \frac{a}{c}(\tau + \tau_0)$ for some constant τ_0. Hence, by integration, there are constants t_0 and x_0 such that

$$ct(\tau) = \frac{c^2}{a} \sinh \left(\frac{a}{c}(\tau + \tau_0) \right) + ct_0, \quad x(\tau) = \pm \frac{c^2}{a} \cosh \left(\frac{a}{c}(\tau + \tau_0) \right) + x_0.$$

Thus the curve is a hyperbola in the tx-plane.

We finish this section with an interesting example; cf. [174, p. 181–183]. Built into special relativity is the feature that all inertial observers measure the same speed of light in vacuum. However, one might wonder if by accelerating, they could catch up with light. So, consider for $b > 0$ a curve $\gamma(\tau)$ with uniform acceleration $a = bc$, whose components in an inertial frame are given by $t(\tau) = \frac{1}{b}\sinh(b\tau)$, $x(\tau) = \frac{c}{b}\cosh(b\tau)$. In the tx-plane the path forms a hyperbola, and as one can infer from a sketch of the trajectory, any light signal sent from a point along γ will reach the inertial observer at $x = 0$ at some time $t > 0$, but if reflected, will *never* reach γ. Moreover, suppose that at $\gamma(0)$, a photon of light is emitted and moves in the positive x-direction. The accelerating observer γ is also moving in the positive x-direction for $\tau > 0$, so what speed does γ measure for the photon? Well, since $\gamma(\tau)$ is orthogonal to $\gamma'(\tau)$, a simple Lorentz transformation argument shows that the ray ρ_τ from the origin through $\gamma(\tau)$ is comprised of events which γ will deem (via radar) to be simultaneous with $\gamma(\tau)$. If we consider the path of the photon as parametrized by $(t(s), x(s)) = \left(s, \frac{c}{b} + cs\right)$, then it intersects ρ_τ at some point $A_\tau = h(\tau)\gamma(\tau)$. Thus $s = \frac{h(\tau)}{b}\sinh(b\tau)$, and

$$\frac{c}{b} + c\frac{h(\tau)}{b}\sinh(b\tau) = h(\tau)\frac{c}{b}\cosh(b\tau),$$

from which we see $h(\tau) = e^{b\tau}$. The displacement vector from $\gamma(\tau)$ to A_τ is spacelike, and its length is the distance from $\gamma(\tau)$ to the photon after τ units of time have elapsed as measured along γ. This displacement has length $|e^{b\tau} - 1| \cdot |\overrightarrow{O\gamma(\tau)}| = \frac{c}{b}(e^{b\tau} - 1)$. This is the distance from $\gamma(\tau)$ to the photon, the derivative of which with respect to τ is $ce^{b\tau}$. So, not only is the acceleration not helping the observer γ make any progress on catching the photon, but as measured along γ, the photon is accelerating *away* from γ. This is counterintuitive, but is a consequence of how γ makes measurements along its worldline, and furthermore, keep in mind that in any local rest frame, the photon has speed c.

1.3. Energy and momentum

We begin with a thought experiment due to Einstein. Imagine a box (and its contents) of total mass M and length L, at rest. Suddenly from inside the left side of the box some photons of total energy E are emitted in the direction toward the right side of the box. The formula for photon momentum p is $E = cp$. By conservation of momentum, then, the box should acquire a net momentum $-p = Mv$ (to the left). When the photons reach the right side of the box and stop, the motion ceases, with the box having moved to the left with a net displacement $\Delta x < 0$. (You might argue that by causality, what happens at one end cannot instantaneously effect the other end; this can be taken into account, still arriving at the conclusion below; see [96, p. 27–28] or [188, p. 138–143].)

Einstein argues that there is no reason why in this closed system the center of mass should have changed from the start to the end of the process. He suggests that the photons must have carried a mass m to the right side of the box to balance out the center, i.e., $m(L + \Delta x) + (M - m)\Delta x = 0$. The velocity of the box is determined by $(M - m)v = -p$ (conservation of momentum), and the elapsed time during the photon motion is $(L + \Delta x)/c = \Delta t = \Delta x/v$. Putting these together we obtain $m = p/c = E/c^2$, or

$$E = mc^2.$$

This is the famous formula relating mass m to energy.

1.3.1. *Energy-momentum four-vector.* Associated to any massive particle is a quantity called its *rest mass*. The rest mass is a measure of *inertia*, and while a discussion of the subtleties involved in the concepts of mass and inertia would be germane in the context of Einstein's theory of gravitation (especially *Mach's principle*), we will work with these notions operationally through Newton's second law $F = \frac{d}{dt}(mv)$ (so $F = ma$ if the mass of the object is constant). Thus one can imagine measuring the inertial mass by applying Newton's law in a frame of reference in which the massive particle is momentarily at rest. Of course, special relativity rewrites the kinematical and dynamical laws, but there is still an analogous concept of mass. The term *rest mass* suggests that in relativity, the quantity *mass* may itself be relative. However, the rest mass is by definition an invariant, since it is associated to the object itself, as measured in a frame adapted to the object. Our present goal is to tie together the concepts of mass, momentum and energy.

If $\gamma(\tau)$ is a timelike curve parametrized by proper time and representing the worldline of a particle of rest mass m_0, the *energy-momentum four-vector* is $\mathbb{P} = m_0 \gamma'(\tau) = m_0 \mathbb{U}$. At the point $\gamma(\tau)$ in a momentarily comoving inertial frame $\widetilde{O}$, we have $\mathbb{P} = m_0 \partial/\partial \tilde{t} = m_0 c \, \partial/\partial \tilde{x}^0$, where $\tilde{x}^0 = c\tilde{t}$. Note that $-\langle \mathbb{P}, \mathbb{P} \rangle = (m_0 c)^2 = (E_0/c)^2$, where $E_0 = m_0 c^2$ is the *rest energy*. If instead we now use an inertial frame O in which the particle is instantaneously moving with velocity v along the x-axis, we get from (1.2.4)

$$\mathbb{P} = \frac{m_0}{\sqrt{1-(v/c)^2}} \frac{\partial}{\partial t} + \frac{m_0 v}{\sqrt{1-(v/c)^2}} \frac{\partial}{\partial x} = \frac{m_0 c}{\sqrt{1-(v/c)^2}} \frac{\partial}{\partial x^0} + \frac{m_0 v}{\sqrt{1-(v/c)^2}} \frac{\partial}{\partial x}.$$

We identify

$$E := \frac{m_0}{\sqrt{1-(v/c)^2}} c^2 \quad \text{and} \quad p := \frac{m_0 v}{\sqrt{1-(v/c)^2}} \frac{\partial}{\partial x}$$

as, respectively, the particle's *energy* and *spatial momentum* as measured by the inertial frame O. Since the spatial velocity is $v = v \, \partial/\partial x$, from the point of view of the inertial frame $\mathcal{O}$ with respect to which the particle is moving, the mass might be construed to be

$$m(v) = \frac{m_0}{\sqrt{1 - (v/c)^2}}.$$

Note that $\lim_{v \nearrow c} m(v) = +\infty$, which indicates that inertia increases with speed; this is consistent with the fact that a force cannot accelerate any particle to or above the speed of light. Note also that by Taylor (binomial) expansion,

$$E = m(v)c^2 = m_0 c^2 \left(1 + \tfrac{1}{2}(v/c)^2 + O((v/c)^4)\right) = m_0 c^2 + \tfrac{1}{2}m_0 v^2 + O(v^4/c^2).$$

The first two terms are the rest energy and the kinetic energy. Furthermore if $\mathbb{U}_{\mathrm{obs}}/c$ is a timelike unit vector tangent to the path of an observer, the observed energy of the particle with momentum $\mathbb{P}$ is just $E_{\mathrm{obs}} = -\langle \mathbb{P}, \mathbb{U}_{\mathrm{obs}} \rangle$. Finally, we note that

$$\mathbb{P} = \frac{E_0}{c^2}\frac{\partial}{\partial \tilde{t}} = \frac{E}{c^2}\frac{\partial}{\partial t} + \boldsymbol{p} = \frac{E}{c}\frac{\partial}{\partial x^0} + \boldsymbol{p},$$

$$E_0^2 = -c^2 \langle \mathbb{P}, \mathbb{P} \rangle = E^2 - c^2|\boldsymbol{p}|^2 = (m(v)c^2)^2 - c^2|\boldsymbol{p}|^2.$$

In units where $c = 1$, we have $\mathbb{P} = E\,\partial/\partial x^0 + \boldsymbol{p}$ and $m_0^2 = E^2 - |\boldsymbol{p}|^2$.

Photons can carry momentum too, as we noted in the introductory thought experiment for this section. The energy and momentum of a photon are related by $E = cp$. The energy-momentum vector for a photon must be null, tangent to the spacetime trajectory of the photon, and the rest mass of a photon is zero. To define the energy-momentum four-vector, the trajectory of the photon is given as a geodesic γ, with affine parameter s, so that $\gamma''(s) = D\gamma'/ds = 0$. The energy-momentum four-vector is the null vector $\gamma'(s)$, which in a given inertial frame is written $\gamma'(s) = \frac{E}{c^2}\frac{\partial}{\partial t} + \boldsymbol{p} = \frac{E}{c^2}\frac{\partial}{\partial t} + p^i \frac{\partial}{\partial x^i}$. In an inertial frame $\tilde{\mathcal{O}}$ which has velocity v with respect to the original frame (along the aligned x-axes of the frames, say), the four-momentum of the photon will thus be

$$\gamma'(s) = \frac{\left(\frac{E}{c^2} - \frac{v}{c^2}\,p^1\right)}{\sqrt{1 - (v/c)^2}}\frac{\partial}{\partial \tilde{t}} + \frac{\left(-\frac{vE}{c^2} + p^1\right)}{\sqrt{1 - (v/c)^2}}\frac{\partial}{\partial \tilde{x}^1} + p^2 \frac{\partial}{\partial \tilde{x}^2} + p^3 \frac{\partial}{\partial \tilde{x}^3}.$$

Thus the measured photon energy in $\tilde{\mathcal{O}}$ will be

$$\widetilde{E} = \frac{E - vp^1}{\sqrt{1 - (v/c)^2}}. \tag{1.3.1}$$

1.3.2. The stress-energy tensor. One of the constituents of the Einstein equation in general relativity is the *stress-energy* or *energy-momentum* tensor, which encodes the energy and momentum fluxes associated to the physical matter or fields in the spacetime. This is a generalization to spacetime of the spatial stress tensor of classical mechanics. We introduce it here, just enough to motivate its place in the formulation of general relativity, for further reading see, e.g., [94; 161; 174; 189; 207]. In the next chapter, we will define the stress-energy tensor via the Lagrangian formulation of the Einstein equation.

We define the stress-energy tensor as a $(2, 0)$-tensor. Consider a one-form θ at a point P in Minkowski spacetime $\mathbb{M}^4$ (this can be readily generalized). We assume that θ is dual to a nonzero vector, which is either timelike or spacelike. In particular, it has a nonzero metric norm, which we normalize to be $\langle \theta, \theta \rangle = \pm 1$. For example, $\theta = dx^0 = c\, dt$, or $\theta = dx^i, i = 1, 2, 3$. In a region of spacetime, we imagine there is a collection of particles — maybe dust particles, or elements of a fluid or field — each possessing an energy-momentum four-vector. We consider $\theta \neq 0$ as a linear functional operating on the tangent space at P. Its nullspace W is three-dimensional, and by assumption on the causal nature of θ, it must be either spacelike or timelike (Lorentzian), and we consider it as an affine subspace at P. Let $B \subset W$ be a region inside W about P. Let $\Delta\mathbb{P}$ be the sum (or integral) of the four-vectors $\mathrm{sgn}(\theta(\widetilde{\mathbb{P}})) \cdot \widetilde{\mathbb{P}}$, summed over each of the energy-momentum four-vectors $\widetilde{\mathbb{P}}$ associated to the particles in the spacetime region B. The sign accounts for the direction of flow across the corresponding spatial boundary of B, or in case θ is timelike, the sign keeps track of the direction of time to the future or past. We remark that under standard energy conditions, the energy-momentum four-vector elements $\widetilde{\mathbb{P}}$ are future-pointing timelike, so that if θ is dual to a timelike vector, then $\theta(\widetilde{\mathbb{P}})$ does not change sign as $\widetilde{\mathbb{P}}$ varies over B.

Let ΔV be the volume of B with respect to the metric, and define the vector field

$$T(\theta, \cdot)|_P := \lim_{B \to \{P\}} \frac{c\Delta\mathbb{P}}{\Delta V}.$$

If ξ is another one-form at P, we define $T(\theta, \xi) = \xi(T(\theta, \cdot))$. Now we extend the definition of T by allowing scaling in θ; given θ which is timelike or spacelike, we define $\hat{\theta}$ by $\theta = b\hat{\theta}$, where $b = \sqrt{|\langle \theta, \theta \rangle|} > 0$. We then define $T(\theta, \zeta)$ as $bT(\hat{\theta}, \xi)$, where the latter quantity has been defined above.

Clearly, $T(\theta, \xi)$ is linear in ξ. Physical reasoning as in classical mechanics may be used to argue that T should also be linear in θ. What may be more surprising is that when ξ is also either timelike or spacelike, we can argue that $T(\theta, \xi) = T(\xi, \theta)$ (see [189] or [207, Chapter 4]). We then see that T can be extended to all forms by polarization, and yields a *symmetric* $(2, 0)$-tensor.

Let us interpret the components of this tensor in some inertial frame; this is often how the tensor is defined. $T(dx^0, \cdot)$ is c times the spatial density of the energy-momentum four-vectors of the physical particles at P. Then $T^{00} = T(dx^0, dx^0) = dx^0(T(dx^0, \cdot))$ is precisely the *energy density* ρ at P as measured in this frame. It is instructive at this point, since it may seem that T is a complicated object, to note as a simple exercise that the tensor T can be determined in terms of the energy densities for inertial observers.

For $i = 1, 2, 3$, $T^{0i} = cT(dt, dx^i)$ is c times the x^i-component of the momentum density (per unit spatial volume) at P. Similarly, for $T(dx^1, \cdot)$, a spacetime box B is now determined by a rectangle R of area ΔA in the x^2x^3-plane, as well as a side of length $\Delta x^0 = c\Delta t$. Thus $\Delta V = c\,\Delta t\,\Delta A$, and so $T^{10} \approx \Delta E/(c\,\Delta t\,\Delta A)$ equals $1/c$ times the rate of energy flux (per unit area per unit time) in the x^1-direction. Note that by symmetry, $T^{10} = T^{01}$, which reflects the equivalence of mass and energy (and note how the units work out). We move on to the purely spatial components, such as $T^{11} \approx \Delta p^1/(\Delta t\,\Delta A)$; as the force is given by $\boldsymbol{F} = d\boldsymbol{p}/dt$, we see that T^{11} is the *normal stress* in the x^1-direction, that is, the force per unit area normal to the x^2x^3-direction. Another way to frame this is as the flux in the x^1-direction of the x^1-component of the momentum. The normal stress is sometimes called the *pressure* when it is independent of direction. On the other hand the component T^{12} is then a *shear*, the force per unit area on the x^2x^3-plane acting in the x^2-direction; this is a component of force *tangential* to the relevant area element, or alternately, as the flux in the x^1-direction of the x^2-component of the momentum.

As defined, the total stress-energy tensor encompasses all relevant physical interactions, so that there is no net external force; from this we can derive another key property of a stress-energy tensor, namely that it is *divergence-free*. To motivate this, we compute a component of the divergence in inertial coordinates (so that the Christoffel symbols vanish), again using the spatial index $i \in \{1, 2, 3\}$, and with the Einstein summation convention still in force. In such a coordinate system, c times the zero-component of the divergence is then $c(T^{00}_{\ ,0} + T^{i0}_{\ ,i}) = \partial\rho/\partial t + cT^{i0}_{\ ,i}$. As discussed above, $cT^{i0}\partial/\partial x^i$ is the vector field measuring the rate of flux (per unit area per unit time) of the energy. Applying the divergence theorem to a spatial region R, we obtain the time rate of change of the energy in the region as

$$\frac{\partial}{\partial t}\int_R \rho\,dx = \int_R \frac{\partial\rho}{\partial t}\,dx = \int_R cT^{00}_{\ ,0}\,dx.$$

By conservation of energy, this must balance with the rate of flux of energy *into* the region across the spatial boundary ∂R with *outward* unit normal $\boldsymbol{n}$:

$$\int_R cT^{00}_{\ ,0}\,dx = \int_{\partial R} \left\langle cT^{i0}\frac{\partial}{\partial x^i}, -\boldsymbol{n}\right\rangle d\sigma = -\int_R cT^{i0}_{\ ,i}\,dx$$

where we used the divergence theorem. Since the region R may be made arbitrarily small, we obtain the 0-component of $\mathrm{div}_\eta T = 0$. The other components may be derived similarly using conservation of momentum.

We now introduce several standard examples of the stress-energy tensor. Of course, when in vacuum (free of fields or particles), $T = 0$. The next simplest example is that of *dust*. Consider a collection of particles which are all at rest in some inertial frame $\tilde{\mathcal{O}}$. In this frame, the *rest* energy density ρ of the dust determines the stress-energy tensor:

$$T = c^{-2}\rho\,\frac{\partial}{\partial \tilde{t}} \otimes \frac{\partial}{\partial \tilde{t}} = \rho\,\frac{\partial}{\partial \tilde{x}^0} \otimes \frac{\partial}{\partial \tilde{x}^0}.$$

This can be written invariantly as $T = c^{-2}\rho\,\mathbb{U} \otimes \mathbb{U}$, where $\mathbb{U}$ is the four-velocity of the dust. Using a Lorentz transformation, one can determine the energy density in a frame $\mathcal{O}$ with respect to which the dust moves with velocity v aligned along the x-axis. Indeed, in such a frame, (1.2.4) gives

$$\mathbb{U} = \frac{\partial}{\partial \tilde{t}} = \frac{1}{\sqrt{1-(v/c)^2}}\frac{\partial}{\partial t} + \frac{v}{\sqrt{1-(v/c)^2}}\frac{\partial}{\partial x}$$
$$= \frac{c}{\sqrt{1-(v/c)^2}}\frac{\partial}{\partial x^0} + \frac{v}{\sqrt{1-(v/c)^2}}\frac{\partial}{\partial x},$$

so that

$$T = c^{-2}\rho\,\mathbb{U} \otimes \mathbb{U}$$
$$= \frac{\rho}{1-(v/c)^2}\frac{\partial}{\partial x^0} \otimes \frac{\partial}{\partial x^0}$$
$$+ \frac{\rho v/c}{1-(v/c)^2}\left(\frac{\partial}{\partial x^0} \otimes \frac{\partial}{\partial x} + \frac{\partial}{\partial x} \otimes \frac{\partial}{\partial x^0}\right) + \frac{\rho(v/c)^2}{1-(v/c)^2}\frac{\partial}{\partial x} \otimes \frac{\partial}{\partial x}. \tag{1.3.2}$$

We note that the energy density as measured in $\mathcal{O}$ is ρ divided by *two* factors $\sqrt{1-(v/c)^2}$: one for the transformation of energy and another for the length contraction, which affects the value of the measured density in $\mathcal{O}$. The other summands in (1.3.2) likewise show two factors of $\sqrt{1-(v/c)^2}$ in the denominator, for similar reasons.

To amplify this further, we may write $c^{-2}\rho\,\mathbb{U} \otimes \mathbb{U} = \mathbb{P} \otimes \mathbb{N}$, where $\mathbb{P} = m_0\mathbb{U}$ is the energy-momentum four-vector for a single dust particle (rest mass m_0) and $\mathbb{N} = n\mathbb{U}$, for n the *number density* (number of dust particles per unit volume), as measured in the rest frame of the dust. Thus $m_0 n$ is the mass density in the rest frame, and $m_0 n c^2 = \rho$ is the energy density in the rest frame. Moreover, the $\partial/\partial t$ components of $\mathbb{P}$ and $\mathbb{N}$ are $m_0/\sqrt{1-(v/c)^2}$ and $n/\sqrt{1-(v/c)^2}$, respectively, the latter being the particle density as measured in $\mathcal{O}$.

A *perfect fluid* is a continuum model specified by its four-velocity $\mathbb{U}$ and density ρ, as well as a further scalar quantity, the *pressure* p. For a perfect fluid, in a momentarily comoving inertial reference frame, there is no heat conduction ($T^{0i} = 0$) and no shear stress ($T^{ij} = 0$ for $i \neq j$); since the latter requirement must hold upon spatial rotation of such a frame, the matrix $[T^{ij}]$ must be a multiple of the identity matrix. Perfect fluids exhibit no viscosity, and the pressure p encodes the normal stress in an inertial frame momentarily comoving with the fluid, so that $T^{ij} = p\delta^{ij}$ in such a frame.

An invariant way to express the perfect fluid stress-energy tensor is then $T = c^{-2}(\rho + p)\,\mathbb{U} \otimes \mathbb{U} + p\eta^{\sharp}$ where $\eta^{\sharp}$ is the contravariant form of the metric tensor η (indices up). In a rest frame of a fluid element,

$$T = \rho \frac{\partial}{\partial x^0} \otimes \frac{\partial}{\partial x^0} + p\delta^{ij} \frac{\partial}{\partial x^i} \otimes \frac{\partial}{\partial x^j}.$$

For comparison, the dust model is an idealization in which the constituent particles have no random motion at all (they are all at rest in a certain frame), and thus there is no pressure.

Finally, as we will see in Example 2-20, the electromagnetic stress-energy tensor is given by

$$T^{\mu\nu} = \frac{1}{4\pi}\left(F^{\mu\alpha}\eta_{\alpha\beta}F^{\nu\beta} - \tfrac{1}{4}\eta^{\mu\nu}F_{\alpha\beta}F^{\alpha\beta}\right).$$

By (1.1.7) we have

$$T^{00} = \frac{1}{4\pi}\left(F^{0\alpha}\eta_{\alpha\beta}F^{0\beta} + \tfrac{1}{4}(-2|\boldsymbol{E}|^2 + 2|\boldsymbol{B}|^2)\right) = \frac{1}{8\pi}\left(|\boldsymbol{E}|^2 + |\boldsymbol{B}|^2\right), \quad (1.3.3)$$

the familiar formula for the energy density of the electromagnetic field (in cgs units), and similarly

$$T^{0i} = \frac{1}{4\pi}(\boldsymbol{E} \times \boldsymbol{B})^i. \tag{1.3.4}$$

1.3.2.1. *Brief remarks on force.* Whereas force plays a key role in Newtonian mechanics, energy and momentum play a more central role in relativistic dynamics. For a discussion of transformation rules for acceleration and force, see, for example, [96; 188; 189]. We will be content with introducing a four-vector for the force, with Newton's second law in mind, and see what it means.

If $\gamma(\tau)$ is a timelike curve parametrized by proper time τ, recall that the four-velocity is $\mathbb{U}(\tau) = \gamma'(\tau)$, and the four-acceleration is the covariant derivative $\mathbb{A}(\tau) = \gamma''(\tau)$.

For the path of a particle of rest mass m_0, with energy-momentum $\mathbb{P} = m_0\mathbb{U}$, we define the four-force (sometimes called the *Minkowski force*) as $\mathbb{F} = D\mathbb{P}/d\tau$. Thus $\mathbb{F} = m_0\mathbb{A} + (dm_0/d\tau)\,\mathbb{U}$, so that $\mathbb{F} = m_0\mathbb{A}$ if the rest mass is unchanging, which in turn is easily seen to be equivalent to the orthogonality of $\mathbb{U}$ and $\mathbb{F}$: since $\mathbb{U}(\tau) = \gamma'(\tau)$ has constant length, $\mathbb{A}$ and $\mathbb{U}$ are orthogonal, and so $c^2 dm_0/d\tau = -\langle \mathbb{F}, \mathbb{U} \rangle$.

Consider an inertial frame $\mathcal{O}$ with respect to which the particle has instantaneous velocity $\boldsymbol{v} = v^i \, \partial/\partial x^i$ (summation convention in use!), and let $v = |\boldsymbol{v}|$. Recall that $dt/d\tau = 1/\sqrt{1 - (v/c)^2} = m(v)/m_0$, and note that if $\boldsymbol{v} = 0$, then *at this instant*, $d(m(v))/d\tau = d(m(v))/dt = dm_0/dt = dm_0/d\tau$. We have $\mathbb{P} = m(v)\,\partial/\partial t + m(v)v^i\,\partial/\partial x^i = m(v)\,\partial/\partial t + \boldsymbol{p}$, and thus

$$\frac{D\mathbb{P}}{d\tau} = \frac{1}{\sqrt{1-(v/c)^2}}\frac{D\mathbb{P}}{dt} = \frac{1}{\sqrt{1-(v/c)^2}}\left(\frac{d(m(v))}{dt}\frac{\partial}{\partial t} + \frac{dp^i}{dt}\frac{\partial}{\partial x^i}\right).$$

Aside from the time dilation factor out front, the spatial part of this vector is the classical force $\boldsymbol{f} = d\boldsymbol{p}/dt$ (written lowercase to avoid confusion here), so that

$$\mathbb{F} = \frac{1}{\sqrt{1-(v/c)^2}}\left(\frac{d(m(v))}{dt}\frac{\partial}{\partial t} + \boldsymbol{f}\right).$$

Note that if $\mathcal{O}$ were momentarily comoving with the particle, we would have $\mathbb{F} - (dm_0/dt)\partial/\partial t + \boldsymbol{f}$, and that in general with respect to another frame $\widehat{\mathcal{O}}$, $\hat{\boldsymbol{f}} = d\hat{\boldsymbol{p}}/d\hat{t}$ is not the same as $\boldsymbol{f}$. Now, as noted above, $\mathbb{A}$ is orthogonal to

$$\mathbb{U} = \frac{1}{\sqrt{1-(v/c)^2}}\left(\frac{\partial}{\partial t} + v^i\frac{\partial}{\partial x^i}\right),$$

and $c^2 dm_0/d\tau = -\langle \mathbb{F}, \mathbb{U} \rangle$, so that

$$\frac{c^2}{\sqrt{1-(v/c)^2}}\frac{dm_0}{dt} = c^2\frac{dm_0}{d\tau} = -\langle \mathbb{F}, \mathbb{U} \rangle = \frac{1}{1-(v/c)^2}\left(c^2\frac{dm(v)}{dt} - \langle \boldsymbol{f}, \boldsymbol{v}\rangle\right).$$

This yields $c^2 dm(v)/dt = \langle \boldsymbol{f}, \boldsymbol{v}\rangle + \sqrt{1-(v/c)^2}\,c^2 dm_0/dt$; note that $\langle \boldsymbol{f}, \boldsymbol{v}\rangle$ is the classical measure of the rate at which the force $\boldsymbol{f}$ applied with velocity $\boldsymbol{v}$ is doing work, i.e., the (mechanical) *power* developed by the force. As $E = m(v)c^2$, we see that the measured rate of change of energy dE/dt in $\mathcal{O}$ is the sum of two terms, the first of which we might interpret as the rate of change of kinetic energy, and the second as the rate of change of heat energy, as it arises from the rate of change of the internal energy as measured in $\mathcal{O}$ (cf. [189, Chapter V]). This should not be surprising, as the energy-momentum four-vector delineates relations between energy and momentum (hence mass) in different frames.

Given the discussion of the stress-energy tensor above in terms of energy and momentum fluxes, if one measures the stress-energy fluxes for a system upon which external forces or fields act (a non-closed system, then), the divergence $(\mathrm{div}_\eta T)^\mu = T^{\mu\nu}_{\;\;;\nu}$ of the stress-energy tensor T for the system is then a vector field whose components encode the density of the net power and force applied to the system (see [162, p. 188], for instance).

Before we move on, we give one example, that of the force on a particle of charge q moving with four-velocity $\mathbb{U}$ in an electromagnetic field given by the Faraday tensor $F^{\mu\nu}$. The four-force is given by $\mathbb{F}^\mu = (q/c)\mathbb{U}_\nu F^{\mu\nu}$. By skew-symmetry, $\mathbb{F}^\mu \mathbb{U}_\mu$ vanishes, so that the force is purely mechanical and does not change the rest mass of the charged particle. We compute the components in an inertial frame with respect to which the particle has velocity $\boldsymbol{v} = v^i \, \partial/\partial x^i$. Then with $v_i = v^i$, we have

$$\mathbb{U}_\nu \, dx^\nu = \frac{1}{\sqrt{1-(v/c)^2}}(-c \, dx^0 + v_i \, dx^i),$$

so that

$$\mathbb{F} = \frac{1}{\sqrt{1-(v/c)^2}}\left(\frac{q}{c}\langle \boldsymbol{v}, \boldsymbol{E}\rangle \frac{\partial}{\partial x^0} + q\left(\boldsymbol{E} + \frac{\boldsymbol{v}}{c}\times \boldsymbol{B}\right)^i \frac{\partial}{\partial x^i}\right)$$

$$= \frac{1}{\sqrt{1-(v/c)^2}}\left(\frac{1}{c^2}\langle \boldsymbol{v}, q\boldsymbol{E}\rangle \frac{\partial}{\partial t} + q\left(\boldsymbol{E} + \frac{\boldsymbol{v}}{c}\times \boldsymbol{B}\right)\right).$$

We plainly see here the Lorentz force in the spatial components, as well as (c^{-2} times) the power developed by the electric field on the moving charge in the time component.

1.4. Some geometric aspects of Minkowski spacetime

In this section we again use inertial coordinates (x^0, x^1, x^2, x^3) for Minkowski spacetime $\mathbb{M}^4 = \mathbb{R}^4_1$, with metric $\langle \cdot, \cdot \rangle = -(dx^0)^2 + (dx^1)^2 + (dx^2)^2 + (dx^3)^2$, and analogously for $\mathbb{M}^{1+k} = \mathbb{R}^{1+k}_1$. We will blur the distinction between a point P and its coordinates $(x^\mu(P))$. We let D be the Levi-Civita connection for the Minkowski metric, so that if a vector field Y has the expression $Y^\mu \, \partial/\partial x^\mu$ (summation convention) in inertial coordinates, then since $D_X(\partial/\partial x^\mu) = 0$, we have simply

$$D_X Y = D_X(Y^\mu)\frac{\partial}{\partial x^\mu} = X[Y^\mu]\frac{\partial}{\partial x^\mu}.$$

1.4.1. *Hyperquadrics.* We consider the level sets of the function

$$F(x) = -(x^0)^2 + (x^1)^2 + (x^2)^2 + \cdots + (x^k)^2.$$

We start with the zero level set, $\Sigma = F^{-1}(0)$, which is precisely the set of all points whose position vector from the origin O is null; in other words, Σ is the lightcone from the origin O. The complement $\Sigma \setminus \{O\} = \Sigma^+ \cup \Sigma^-$ is a smooth *null* hypersurface: along $\Sigma \setminus \{O\}$, the null vector field $x^\mu \, \partial/\partial x^\mu$ is *both tangent and normal* to Σ.

We move on to discuss the other level sets Σ, which are regular hypersurfaces. Recall (or see 5.1.2) that for Y and Z tangent to Σ, if $D_Y^\Sigma Z$ is the tangential component of $D_Y Z$, then D^Σ is the Levi-Civita connection for Σ in the induced metric. We write $D_Y Z = D_Y^\Sigma Z + \mathrm{I\!I}(Y, Z)$, where $\mathrm{I\!I}$ is the *second fundamental form* of Σ.

1.4.1.1. *Hyperbolic space.* We now consider the smooth hypersurface Σ equal to one of the two components of $F^{-1}(-r^2)$, for $r > 0$, say the component where $x^0 > 0$. Along Σ, $n = r^{-1} x^\mu \, \partial/\partial x^\mu$ is a unit *timelike* normal vector field. The induced metric on the hypersurface is thus Riemannian. If $Y = Y^\mu \, \partial/\partial x^\mu$ is tangent to Σ, then

$$D_Y n = r^{-1} D_Y(x^\mu) \frac{\partial}{\partial x^\mu} = r^{-1} Y^\mu \frac{\partial}{\partial x^\mu} = r^{-1} Y.$$

From here we see that Σ is *totally umbilic*, that is, the second fundamental form is proportional to the induced metric:

$$\mathrm{I\!I}(Y, Z) = \langle D_Y Z, n \rangle \langle n, n \rangle n = \langle D_Y n, Z \rangle n = r^{-1} \langle Y, Z \rangle n.$$

We can compute its curvature via the Gauss equation (see Proposition 5-5): for X, Y, Z, W tangent to Σ,

$$\langle R^\Sigma(X, Y, Z), W \rangle$$
$$= \langle R(X, Y, Z), W \rangle - \langle \mathrm{I\!I}(X, Z), \mathrm{I\!I}(Y, W) \rangle + \langle \mathrm{I\!I}(X, W), \mathrm{I\!I}(Y, Z) \rangle. \quad (1.4.1)$$

Thus with $R = 0$ on Minkowski spacetime, we insert the formula for the second fundamental form to obtain

$$\langle R^\Sigma(X, Y, Z), W \rangle = r^{-2} \left(-\langle X, Z \rangle \langle Y, W \rangle \langle n, n \rangle + \langle X, W \rangle \langle Y, Z \rangle \langle n, n \rangle \right). \quad (1.4.2)$$

If E_1, E_2 are orthonormal and span a two-plane Π tangent to Σ, then the sectional curvature (see (0.0.1)) of Π in Σ is $K(\Pi) = \langle R^\Sigma(E_1, E_2, E_2), E_1 \rangle = -r^{-2}$. Thus Σ (for $k \geq 2$) has constant negative sectional curvature; being simply connected, Σ is therefore isometric to the hyperbolic space of curvature $-r^{-2}$. In particular, $\Sigma \subset \mathbb{M}^4$ is isometric to $\mathbb{H}^3$ when $r = 1$. In $\mathbb{M}^{1+k}$ with $k \geq 2$, each component of $F^{-1}(-1)$ is isometric to hyperbolic space $\mathbb{H}^k$.

1.4.1.2. *De Sitter spacetime.* The level set $F^{-1}(r^2) =: \mathbb{S}_1^k(r)$ for $r > 0$ is a smooth connected hypersurface which is topologically $\mathbb{R} \times \mathbb{S}^{k-1}$, so that for $k \geq 3$, $\mathbb{S}_1^k(r)$ is simply connected. Along the level set, $n = r^{-1} x^\mu \, \partial/\partial x^\mu$ is a unit *spacelike* normal vector field. The induced metric on $\mathbb{S}_1^k(r)$ is *Lorentzian*: the vector field

$$(r^2 + (x^0)^2)\frac{\partial}{\partial x^0} + x^0 x^i \frac{\partial}{\partial x^i}$$

is tangent and timelike along the submanifold. If Π is a nondegenerate two-plane tangent to $\mathbb{S}_1^k(r)$, then (0.0.1) and (1.4.2) imply $K(\Pi) = r^{-2}$; note that this holds not only for a spacelike two-plane, but also for a nondegenerate plane Π spanned by orthonormal vectors E_0 (timelike) and E_1 (spacelike). We call $\mathbb{S}_1^4(1)$ *de Sitter spacetime*.

1.4.2. *Conformal compactification of* $\mathbb{M}^4$. We now present the classical compactification of $\mathbb{M}^4$. The motivation is to topologically compactify the spacetime in a way that preserves the causal structure, and in particular preserves the nullcones. In this way, one can faithfully represent the null geometry on a compact region, whose boundary in part represents the null paths *at infinity*. This can be generalized to $\mathbb{M}^{1+k}$, and has been used to prove the existence of global solutions to quasilinear hyperbolic systems with "small" initial data (cf. [54], [51, XV.2]).

We will obtain the compactification via a conformal embedding, built using the null structure, of $\mathbb{M}^4$ into the *Einstein static universe* $\mathbb{R} \times \mathbb{S}^3$ with the Lorentzian product metric. We compute in explicit coordinates for convenience, but the construction is geometric in nature; for example, we first introduce coordinates based on affine parameters along null geodesics. Indeed, consider advanced and retarded null coordinates $v = x^0 + r$, $u = x^0 - r$, where $r^2 = (x^1)^2 + (x^2)^2 + (x^3)^2$. The Minkowski metric becomes

$$-du\, dv + \tfrac{1}{4}(v - u)^2 \mathring{g}_{\mathbb{S}^2},$$

where $du\, dv = \tfrac{1}{2}(du \otimes dv + dv \otimes du)$, and $\mathring{g}_{\mathbb{S}^2}$ is the metric on the round unit sphere. Note that the level sets of u and v are null. If one holds u and a point $\omega \in \mathbb{S}^2$ fixed, then varying $v \to +\infty$ corresponds to going forward in time along a null geodesic, to infinity. Likewise, with v fixed, $u \to -\infty$ corresponds to a path of light going to past infinity. The goal is to represent where these null paths "are" at infinity. One way to do this is to use the inverse tangent function to define new coordinates

$$T = \tan^{-1} v + \tan^{-1} u, \quad R = \tan^{-1} v - \tan^{-1} u.$$

Since $dT = (1+v^2)^{-1}\,dv + (1+u^2)^{-1}\,du$ and $dR = (1+v^2)^{-1}\,dv - (1+u^2)^{-1}\,du$, we can easily derive the Jacobian determinant

$$\frac{\partial(T, R)}{\partial(u, v)} = \frac{2}{(1+v^2)(1+u^2)} > 0.$$

Moreover,

$$-dT^2 + dR^2 = \frac{4}{(1+v^2)(1+u^2)}(-du\,dv) =: \Omega^2(-du\,dv).$$

Note that $\Omega^2 = \dfrac{4}{(1+t^2+r^2)^2 - 4t^2r^2}$ is smooth on all of $\mathbb{M}^4$, and that

$$\sin R = \sin(\tan^{-1} v)\cos(\tan^{-1} u) - \sin(\tan^{-1} u)\cos(\tan^{-1} v) = \frac{v - u}{\sqrt{1+v^2}\sqrt{1+u^2}}.$$

Thus $\sin^2 R = \frac{1}{4}\Omega^2(v - u)^2$. This implies

$$-dT^2 + dR^2 + (\sin^2 R)\,\mathring{g}_{\mathbb{S}^2} = \Omega^2\left(-du\,dv + \tfrac{1}{4}(v - u)^2\mathring{g}_{\mathbb{S}^2}\right).$$

The metric on the left is manifestly *conformal* to the Minkowski metric, and is readily identified as a Lorentzian product metric $(\mathbb{R} \times \mathbb{S}^3, -dt^2 + \mathring{g}_{\mathbb{S}^3})$, which is in fact the *Einstein static universe* (Section 2.4.2). In other words, we have produced an embedding of Minkowski spacetime into the Einstein static universe, which is not an isometry, but a *conformal isometry*. Thus it preserves the causal nature of vectors, in particular the null structure. The image of the embedding is a bounded set, since $-\pi < T < \pi$ and $0 \le R < \pi$ on Minkowski spacetime. The boundary of the set is the union of two smooth null hypersurfaces $\mathscr{J}^\pm$, "scri-plus" and "scri-minus", where "scri" is short for "script I." We note that Ω is a defining function for $\mathscr{J}^\pm$, since $\Omega = 0$ here, with $d\Omega \ne 0$.

We now briefly describe some of the features of the boundary (Figure 3). Since $T + R = 2\tan^{-1} v$, the null rays to the future end up ($v \to +\infty$) at $T + R = \pi$, which for $0 < R < \pi$ gives $\mathscr{J}^+$. Similarly for $u \to -\infty$ we get $T - R = 2\tan^{-1} u = \pi$, which gives $\mathscr{J}^-$. The null vector $\partial/\partial T \mp \partial/\partial R$ is both tangent and normal to $\mathscr{J}^\pm$. The closure of the image of Minkowski spacetime can be represented by a T-R triangle, bounded by $R = 0$ and $T \pm R = \pm\pi$. Every point in this region represents a two-sphere, except where $R = 0$ or $R = \pi$, each point of which represents a point. One can argue that timelike geodesics must start at i^- in the past, corresponding to $(T, R) = (-\pi, 0)$ and must end at i^+ corresponding to $(T, R) = (\pi, 0)$. We let i^0 be the point corresponding to $(T, R) = (0, \pi)$, which is called *spacelike infinity*. Spacelike curves with $r \to +\infty$ have a limit at i^0 in the conformal picture.

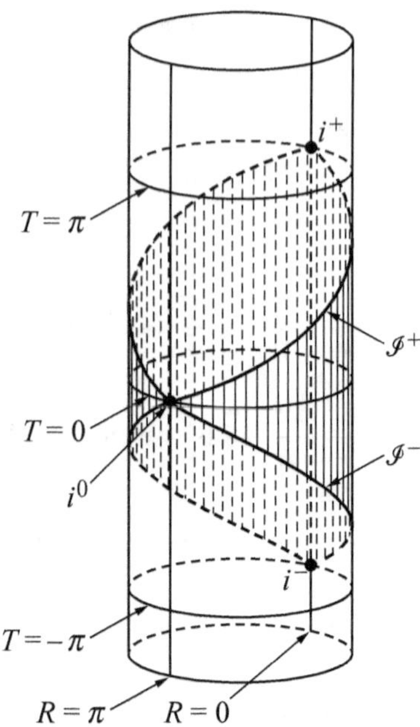

Figure 3. Conformal compactification of Minkowski spacetime.

Exercises

Exercise 1-4. a. Show that if $T : \mathbb{R}^2 \to \mathbb{R}^2$ is continuous and invertible, maps each line onto a line, and fixes the origin, then T is a linear transformation. (Hint: Use the parallelogram law for addition to get additivity of T. Infer from this and continuity that $T(q\boldsymbol{w}) = qT(\boldsymbol{w})$ for all $\boldsymbol{w} \in \mathbb{R}^2$ and for any rational q, then use continuity to extend to all $q \in \mathbb{R}$.)

b. Suppose $T : \mathbb{R}^n \to \mathbb{R}^n$ preserves Euclidean distance between points and fixes the origin. Show that T is linear, and hence is given by multiplication by an orthogonal matrix.

c. Suppose $T : \mathbb{R}^{n+1} \to \mathbb{R}^{n+1}$ preserves the Minkowski spacetime interval $\Delta s^2 = -(\Delta x^0)^2 + \sum_{k=1}^n (\Delta x^k)^2$ between points and fixes the origin. Show that T is linear, and hence is a given by multiplication by a Lorentz transformation matrix Λ.

d. Suppose that $\varphi : U \to V$ is a diffeomorphism between open, connected subsets of $\mathbb{R}^{n+1}$, and suppose that $\varphi^* \eta = \eta$, i.e., φ preserves the Minkowski metric $\eta = -(dx^0)^2 + \sum_{k=1}^n (dx^k)^2$. Show that φ is an affine linear function of the coordinates (x^μ), and thus up to translation is given by multiplication by a Lorentz matrix. (Hint: You might let $y^\mu(x) = \varphi^\mu(x)$ be the component functions of φ in inertial coordinates on V, and show these have vanishing second derivatives. The components of $\varphi^*\eta$ are given in terms of the partials of y^μ; differentiate and permute indices in a manner similar to the computation of the Christoffel symbols (which vanish in these coordinates) in terms of the metric to conclude.)

Exercise 1-5 (Doppler factor). Suppose light with wavelength λ_0 emanates from an emitter at rest in an inertial frame $\mathcal{O}$, with respect to which a detector moves at velocity v along the same line as the propagation of light; let $\widetilde{\mathcal{O}}$ be a rest frame for the detector. As measured in $\mathcal{O}$, the frequency ν_0 of the light satisfies $\lambda_0 \nu_0 = c$, and the time from the start to the end of emission of one wavelength is $\Delta t_0 = 1/\nu_0$. Let ν_1 be the frequency of the light as measured by the detector, so that $\Delta t_1 = 1/\nu_1$ is the time for the reception of one wavelength as measured in $\widetilde{\mathcal{O}}$. Derive the Doppler shift formula

$$\frac{\Delta t_0}{\Delta t_1} = \frac{\nu_1}{\nu_0} = \frac{\sqrt{1-v/c}}{\sqrt{1+v/c}},$$

and show this is consistent with the quantum mechanical formula $E = h\nu$. (Hint: This is basically recasting the derivation of the Doppler factor from (1.2.3) in the twin paradox example. Set up respective inertial coordinate systems with common axis along the direction of motion, so that one can analyze the problem in the relevant coordinates (t, x) for $\mathcal{O}$ and $(\tilde{t}, \tilde{x})$ for $\widetilde{\mathcal{O}}$. Suppose that in $\mathcal{O}$, the start and end of the emission of one wavelength of a photon of light occur at $(0, 0)$ and $(\Delta t_0, 0)$, respectively. Let the worldline of the detector be given in $\mathcal{O}$ by $x = x_0 + vt$, and let the detections of the start and end of the wavelength occur at (t_1, x_1) and (t_2, x_2), respectively, as measured in $\mathcal{O}$. Observe that $x_1 = ct_1 = x_0 + vt_1$ and $x_2 = c(t_2 - \Delta t_0) = x_0 + vt_2$. Use a Lorentz transformation to convert to $\widetilde{\mathcal{O}}$. Apply (1.3.1) to show that the ratio E_1/E_0 of photon energies at reception and emission agrees with the ratio of the frequencies.)

The following exercises represent a small sampling of basic problems from semi-Riemannian geometry. They should be a review and a warmup for some of the geometry you will see in later chapters. Please refer to the preface for curvature and index conventions. We reiterate that we use the Einstein summation convention, summing over a pair of repeated indices, one upper and one lower (the latter including such notation as $\partial/\partial x^i$).

By a connection ∇, we mean an affine connection on the tangent bundle TM of a manifold M. When $(M, g = \langle \cdot, \cdot \rangle)$ is semi-Riemannian, we use the Levi-Civita connection ∇ (metric compatible ($\nabla g = 0$) with vanishing torsion). We also assume tensor fields are (sufficiently) smooth.

Exercise 1-6. a. If ∇ and $\hat{\nabla}$ are connections on TM, show that their difference $S(X, Y) = \nabla_X Y - \hat{\nabla}_X Y$ is tensorial in both X and Y; that is, it is $C^\infty(M)$-linear in X and Y.

b. If ∇ is a connection on TM, show that $T(X, Y) = \nabla_X Y - \nabla_Y X - [X, Y]$ is tensorial in X and Y. We call T the *torsion tensor*.

Note: S and T each determine a $(1, 2)$-tensor, e.g., $(\theta, X, Y) \mapsto \theta(T(X, Y))$, where θ is a one-form. Observe that for each one-form θ, if we let $\tau^\theta(X, Y) = \theta(T(X, Y))$, then τ^θ is a two-form (alternating).

c. Prove that for a one-form α: $d\alpha(X, Y) = X[\alpha(Y)] - Y[\alpha(X)] - \alpha([X, Y])$.

d. If α is a one-form and ∇ is a connection on TM with torsion tensor T, prove that $d\alpha(X, Y) = (\nabla_X \alpha)(Y) - (\nabla_Y \alpha)(X) + \tau^\alpha(X, Y)$.

Exercise 1-7. Suppose ∇ is a connection on TM, extended to tensor bundles by the product rule. Suppose (x^i) is a local coordinate system on M, and

$$\nabla_{\frac{\partial}{\partial x^i}} \frac{\partial}{\partial x^j} = \Gamma_{ij}^k \frac{\partial}{\partial x^k}.$$

a. If we express $\nabla_{\frac{\partial}{\partial x^i}} dx^j$ as $C_{ik}^j dx^k$, find C_{ik}^j.

b. If T is a $(1, 2)$-tensor field on M, and if the components of T in the coordinate system (x^i) are T_{jk}^i, i.e.,

$$T = T_{jk}^i \frac{\partial}{\partial x^i} \otimes dx^j \otimes dx^k,$$

show the components $T_{jk;\ell}^i$ of ∇T satisfy

$$T_{jk;\ell}^i = T_{jk,\ell}^i + \Gamma_{\ell m}^i T_{jk}^m - \Gamma_{\ell j}^m T_{mk}^i - \Gamma_{\ell k}^m T_{jm}^i.$$

Exercise 1-8 (parallel transport). Let I be an interval containing the origin, and let $\gamma : I \to (M, g)$ be a smooth curve. In local coordinates we have $\gamma'(t) = \dot{\gamma}^i(t)\partial/\partial x^i$, where $\dot{\gamma}^i(t) = d\gamma^i/dt$. For a vector field W along γ, recall that the *covariant derivative* DW/dt can be expressed in local coordinates as

$$\frac{DW}{dt} = \left(\frac{dW^k}{dt} + \Gamma_{ij}^k \big|_{\gamma(t)} \dot{\gamma}^i(t) W^j(t) \right) \frac{\partial}{\partial x^k} \bigg|_{\gamma(t)}.$$

Let $P_t : T_{\gamma(0)}M \to T_{\gamma(t)}M$ be the *parallel transport* operator: for $w \in T_{\gamma(0)}M$, $P_t(w) = W(t)$, where $W(t)$ solves the linear ODE $DW/dt = 0$ along γ, with $W(0) = w$.

a. Use parallel transport to prove there exists a smooth orthonormal parallel frame field $e_1(t), \ldots, e_n(t)$ along γ.

b. If V is a smooth vector field along γ, show (perhaps using part a.) that the covariant derivative satisfies

$$\frac{DV}{dt}\bigg|_{t=0} = \frac{d}{dt}\bigg|_{t=0} P_t^{-1}(V(t)).$$

Exercise 1-9 (the Hessian). Suppose (M, g) is semi-Riemannian, and u is a smooth function on M. The *Hessian* of u is defined by $\mathrm{Hess}_g u = \nabla(du)$. It is a $(0, 2)$-tensor; in local coordinates, $(\mathrm{Hess}_g u)_{ij} = u_{;ij}$ (and recall $u_{,i} = u_{;i}$, since $du = u_{,i} dx^i$).

a. Show that $\mathrm{Hess}_g u(X, Y) = Y\big[X[u]\big] - (\nabla_Y X)[u]$, where we recall that $X[u] = du(X)$ is the directional derivative of u in the direction X.

b. Show that the Hessian is symmetric, in two ways: (i) using the identity in a., and (ii) writing out the components $u_{;ij}$. Identify the common fact needed for either proof.

c. The *Laplacian* Δ_g is the trace of the Hessian: $\Delta_g u = \mathrm{tr}_g(\mathrm{Hess}_g u) = g^{ij} u_{;ij}$. Compare the component expressions at the center point of a normal coordinate chart for (M, g) in case g is Riemannian versus when g is Lorentzian. The term *Laplacian* is often reserved for the case (M, g) is Riemannian (though sometimes defined as the negative of our definition), and in the case (M, g) is Lorentzian, the trace of the Hessian is (again, up to a sign) the *wave operator* $\Box_g$.

Exercise 1-10 (divergence and Laplacian). Consider a semi-Riemannian manifold (M, g) of dimension n with Levi-Civita connection ∇. For any vector field X, ∇X is a $(1, 1)$-tensor: $\nabla X(\theta, Y) = \theta(\nabla_Y X)$, whose components are $X^i{}_{;j}$:

$$\nabla X = X^i{}_{;j} \frac{\partial}{\partial x^i} \otimes dx^j.$$

The *divergence* of X is the contraction of ∇X; in coordinates, $\mathrm{div}_g X = X^i{}_{;i}$.

a. Explain how any $(1, 1)$-tensor, such as ∇X, can be construed at each $p \in M$ as a linear operator on $T_p M$, and conversely. What is contraction in terms of linear operator terminology?

b. Suppose ω is a volume form on $T_p M$; i.e., $\omega(e_1, \ldots, e_n) = \pm 1$ for any orthonormal frame $\{e_1, \ldots, e_n\}$ for $T_p M$. Let $\{E_1, \ldots, E_n\}$ be an orthonormal frame for $T_p M$ with dual frame $\{\theta^1, \ldots, \theta^n\}$ (i.e., $\theta^i(E_j) = \delta^i{}_j$). Show that $\theta^1 \wedge \theta^2 \wedge \cdots \wedge \theta^n$ is a volume form, and that if ω is any volume form on $T_p M$, then $\omega = \pm \theta^1 \wedge \theta^2 \wedge \cdots \wedge \theta^n$.

c. A smooth (or at least continuous) n-form defined on an open set $U \subset M$ is a *local volume form* if it is a volume form on $T_p M$ for each $p \in U$. Show that, in local coordinates, $\mathring{\omega} = \sqrt{|\det(g_{ij})|}\, dx^1 \wedge \cdots \wedge dx^n$ gives a local volume form.

Show that M is orientable if and only if it admits a continuous (smooth in fact) global volume form ω.

d. For any p-form α, consider the interior product $\iota_X(\alpha) = \alpha(X, \ldots)$, which is a $(p-1)$-form. Show that for a smooth local volume form ω, $d(\iota_X\omega) = (\operatorname{div}_g X)\,\omega$. Conclude that, in local coordinates,

$$\operatorname{div}_g X = \frac{1}{\sqrt{|\det g|}} \frac{\partial}{\partial x^i}\left(\sqrt{|\det g|}\, X^i\right).$$

What does this reduce to at the center point of a normal coordinate chart?

e. Argue that $\operatorname{tr}_g(\operatorname{Hess}_g u) = \operatorname{div}_g(\operatorname{grad}_g u)$, where $\operatorname{grad}_g u = (du)^\sharp$ is the vector dual to the one-form du under the metric g, i.e., $du(X) = \langle X, \operatorname{grad}_g u\rangle$. Use part d. to express $\operatorname{tr}_g(\operatorname{Hess}_g u)$ in local coordinates, and conclude that, in local coordinates on a Riemannian manifold (M, g),

$$\Delta_g u = \frac{1}{\sqrt{\det g}} \frac{\partial}{\partial x^i}\left(g^{ij}\sqrt{\det g}\,\frac{\partial u}{\partial x^j}\right).$$

Exercise 1-11 (Gauss's divergence theorem and Green's identity). Let (M, g) be a Riemannian manifold, possibly with nonempty (smooth) boundary ∂M. We assume either that M is compact, or that the functions or vector fields in the integrands below have compact support.

a. Assume (M, g) is oriented with global volume form ω_g. If ∂M is nonempty, give it the induced orientation with induced volume form σ_g. If ν is the outward unit normal to the boundary of M, use Exercise 1-10 to show that

$$\int_M (\operatorname{div}_g X)\omega_g = \int_{\partial M} \langle X, \nu\rangle\,\sigma_g.$$

What, if anything, changes in the case of (M, g) semi-Riemannian?

b. Show that the analogous formula to that in part a. holds even if M is not orientable, if the form ω is replaced by the volume measure dv_g (and induced measure $d\sigma_g$ on the boundary); recall that in local coordinates $dv_g = \sqrt{\det(g_{ij})}\,dx$, where dx is Lebesgue measure on $\mathbb{R}^n$.

c. For a function u, we let the *normal derivative* $du(\nu) = \partial u/\partial \nu$ be the directional derivative in the ν-direction. Prove that

$$\int_M v\,\Delta_g u\,dv_g = \int_{\partial M} v\,\frac{\partial u}{\partial \nu}\,d\sigma_g - \int_M \langle \operatorname{grad}_g u, \operatorname{grad}_g v\rangle\,dv_g, \tag{1-11a}$$

and then deduce *Green's identity*

$$\int_M (u\,\Delta_g v - v\,\Delta_g u)\,dv_g = \int_{\partial M} \left(u\,\frac{\partial v}{\partial \nu} - v\,\frac{\partial u}{\partial \nu}\right)d\sigma_g. \tag{1-11b}$$

d. Equation (1-11a) is often called *integration by parts*. Suppose u and v are supported in a compact set contained in an open coordinate neighborhood as follows: in terms of the coordinates, the boundary points correspond to $x^n = 0$ (if this is a boundary chart), and there is some $K > 0$ such that u and v are supported in $\{x \in \mathbb{R}^n : x^n \geq 0 \text{ and } |x^i| \leq K \text{ for all } 1 \leq i \leq n - 1\}$. Write out the identity (1-11a) in coordinates and verify that it corresponds to integration by parts along coordinate directions. You might consider using a dual frame to an adapted orthonormal frame along the boundary to relate dv_g and $d\sigma_g$ in coordinates.

e. Suppose $\Delta_g u = -\lambda u$ for some nontrivial (smooth) function u on a closed (compact without boundary) Riemannian manifold (M, g). Show that $\lambda \geq 0$. If $\lambda = 0$, what is u?

Exercise 1-12 (Ricci formula). Let (M, g) be semi-Riemannian.

a. Prove the vector field version of the *Ricci formula*, which essentially is the definition of the Riemann curvature tensor: $Z^i_{\;;jk} - Z^i_{\;;kj} = Z^\ell R^i_{kj\ell}$.

b. Deduce the equivalent one-form version of the Ricci formula: if α is a one-form, then $\alpha_{i;jk} - \alpha_{i;kj} = \alpha_\ell R^\ell_{jki}$.

c. Use the Ricci formula to prove the following, for smooth functions u, where $\nabla_g u := \operatorname{grad}_g u$ and $\Delta_g u = \operatorname{tr}_g(\operatorname{Hess}_g u)$ (for any signature metric):

$$\langle \nabla_g(\Delta_g u), \nabla_g u \rangle + \langle \operatorname{Hess}_g u, \operatorname{Hess}_g u \rangle + \operatorname{Ric}_g(\nabla_g u, \nabla_g u) = \tfrac{1}{2}\Delta_g(\langle \nabla_g u, \nabla_g u \rangle).$$

Exercise 1-13 (Cartan's structural equations). Let $E_1, \ldots, E_n$ be a local frame field for TM with dual frame $\theta^1, \ldots, \theta^n$, i.e., $\theta^i(E_j) = \delta^i{}_j$. Let ∇ be a connection on TM, with torsion tensor T (Exercise 1-6). Since $\nabla_X Y$ is tensorial in X, there is a matrix $(\omega_i{}^j)$ of *connection one-forms* such that $\nabla_X E_i = \omega_i{}^j(X)E_j$. Furthermore, there are two-forms τ^j with $T(X, Y) = \tau^j(X, Y)F_j = \theta^j(T(X, Y))E_j$, so that $\tau^J = \tau^{\theta^J}$ in the notation of Exercise 1-6.

a. What is $(\nabla_X \theta^j)(E_k)$ in terms of the connection one-forms? Write the one-form $\nabla_X \theta^j$ as a linear combination of $\theta^1, \ldots, \theta^n$.

b. Prove Cartan's *first structural equation*: $d\theta^j = \theta^i \wedge \omega_i{}^j + \tau^j$. If the connection is torsion-free, and if the frame $\theta^i = dx^i$ is dual to a local coordinate frame, what does this reduce to?

c. If (M, g) is Riemannian with Levi-Civita connection ∇, and if $E_1, \ldots, E_n$ is a local *orthonormal* frame field, what can you say about the matrix $(\omega_i{}^j)$?

Now suppose (M, g) is semi-Riemannian and ∇ is the Levi-Civita connection. In particular, $\tau = 0$. Let $(\Omega_i^{\,j})$ be a matrix of two-forms defined by $\Omega_i^{\,j} = \frac{1}{2} R_{k\ell i}^{j} \theta^k \wedge \theta^\ell$, so that $\Omega_i^{j}(X, Y) E_j = R(X, Y, E_i)$.

d. Prove Cartan's *second structural equation* $\Omega_i^{\,j} = d\omega_i^{\,j} - \omega_i^{\,k} \wedge \omega_k^{\,j}$.

e. Prove that $d\Omega_i^{\,j} = \omega_i^k \wedge \Omega_k^{j} - \Omega_i^k \wedge \omega_k^{j} = \omega_i^k \wedge \Omega_k^{j} - \omega_k^{j} \wedge \Omega_i^{k}$. Use this to derive the differential Bianchi identity (cf. Proposition 2-1), written in components as $R_{k\ell i;s}^{j} + R_{\ell s i;k}^{j} + R_{ski;\ell}^{j} = 0$. (Hint: If you expand this in a local coordinate frame using Christoffel symbols and set $C_{k\ell s} = (\Gamma_{ks}^m R_{m\ell i}^{j} + \Gamma_{\ell s}^m R_{kmi}^{j})$, observe that $C_{k\ell s} + C_{\ell sk} + C_{sk\ell}$ must vanish.)

Remark: This technique for computing curvature is often applied to local orthonormal frame fields. While the metric takes a simple form in such a frame, there is a trade-off, since unlike coordinate fields the Lie brackets $[E_i, E_j]$ do not vanish for general frame fields.

Exercise 1-14. Let $I \subset \mathbb{R}$ be an open interval, and let u and v be smooth *positive* functions defined on I. Consider the two-dimensional Lorentzian metric defined for $(t, r) \in \mathbb{R} \times I$ by $\bar{g} = -u(r)\, dt^2 + v(r)\, dr^2$.

a. Show that the nonzero Christoffel symbols of $\bar{g}$ in the (t, r)-coordinates are given by

$$\Gamma_{rr}^r = \frac{1}{2} \frac{v'(r)}{v(r)}, \qquad \Gamma_{tt}^r = \frac{1}{2} \frac{u'(r)}{v(r)}, \qquad \Gamma_{tr}^t = \frac{1}{2} \frac{u'(r)}{u(r)} = \Gamma_{rt}^t.$$

b. Find $\langle \nabla r, \nabla r \rangle_{\bar{g}}$ and $\mathrm{Hess}_{\bar{g}} r$.

c. Show that $\mathrm{Ric}(\bar{g}) = K \bar{g}$, where

$$K = -\frac{1}{2} \frac{u''(r)}{u(r)v(r)} + \frac{1}{4} \frac{(u'(r))^2}{(u(r))^2 v(r)} + \frac{1}{4} \frac{u'(r)v'(r)}{u(r)(v(r))^2}.$$

d. Compute and verify the Cartan structural equations (cf. Exercise 1-13) for the Lorentzian-orthonormal frame $E_1 = u(r)^{-1/2} \partial/\partial t$, $E_2 = v(r)^{-1/2} \partial/\partial r$.

Exercise 1-15 (curvature of warped products). Let B and F be manifolds of respective dimensions b and d, and let $M = B \times F$. For $p \in B$ and $q \in F$, we can identify $T_{(p,q)}M$ with $T_p B \oplus T_q F$, by *lifting* via the canonical projection maps $\pi_B : M \to B$ and $\pi_F : M \to F$ given by $\pi_B(p, q) = p$ and $\pi_F(p, q) = q$. Given a point $(p, q) \in M$ and vector $v \in T_p B$, there is a *unique* vector $V \in T_{(p,q)}M$ such that $(\pi_B)_*(V) = v$ and $(\pi_F)_*(V) = 0$, and similarly $W \in T_{(p,q)}M$ can be analogously associated to $w \in T_q F$. We thus identify vectors tangent to B or to F as tangent vectors to M, and any vector in $T_{(p,q)}M$ can be uniquely expressed as $V + W$, with V and W as above. If g_B and g_F are semi-Riemannian metrics on B and F respectively, and if $f : B \to (0, +\infty)$ is a smooth positive function, then we can consider the *warped product metric* $g = g_B + f^2 g_F$, or more precisely, $g = \pi_B^*(g_B) + (\pi_B^* f)^2 \pi_F^*(g_F)$, where $\pi_B^* f = f \circ \pi_B$. Bracket notation for the metric g will be employed. Let ∇, ∇^B and ∇^F be the associated Levi-Civita connections for g, g_B and g_F, and let R, R^B and R^F be their Riemann tensors.

In what follows, these identifications of the tangent spaces will be in force and f^{-1} will denote $1/f$. Furthermore, computations can be done either invariantly (cf. [174, Chapter 7], with the opposite curvature convention) or using adapted coordinates $(x^1, \ldots, x^b; y^1, \ldots, y^d)$, where (x^i) are local coordinates for B and (y^j) for F, with associated tangent vectors $\partial/\partial x^i$ and $\partial/\partial y^j$, regarded also as tangent vectors to M under the identifications above.

a. For X and Y tangent to B, and V and W tangent to F, show that

$$\nabla_X Y = \nabla_X^B Y,$$
$$\nabla_X V = f^{-1}(\nabla_X f)V = \nabla_V X,$$
$$\nabla_V W = -f^{-1}\langle V, W\rangle \operatorname{grad}_{g_B} f + \nabla_V^F W$$
$$= -f g_F(V, W) \operatorname{grad}_{g_B} f + \nabla_V^F W.$$

(Note: Computing Christoffel symbols in local coordinates is straightforward.)

b. Show that for any smooth function $\varphi : B \to \mathbb{R}$, the gradient $\operatorname{grad}_g(\pi_B^* \varphi)$ agrees (under the identification above) with $\operatorname{grad}_{g_B}(\varphi)$. Show that $\operatorname{Hess}_{g_B}\varphi(X, Y) = \operatorname{Hess}_g(\pi_B^*\varphi)(X, Y)$ if X and Y are vectors tangent to B. Can you give an example of a warped product metric, along with a function φ and a vector field Z tangent to M, for which $\operatorname{Hess}_{g_B}\varphi((\pi_B)_*(Z), (\pi_B)_*(Z)) \neq \operatorname{Hess}_g(\pi_B^*\varphi)(Z, Z)$?

c. Let X, Y and Z be tangent fields to B, and let U, V and W be tangent to F. Verify the following curvature formulae:

$$R(X, Y, Z) = R^B(X, Y, Z),$$
$$R(X, V, Y) = f^{-1}(\operatorname{Hess}_{g_B} f(X, Y))V,$$
$$R(X, Y, V) = 0 = R(V, W, X),$$
$$R(V, X, W) = f^{-1}\langle V, W\rangle \nabla_X^B(\operatorname{grad}_{g_B} f) = f g_F(V, W)\nabla_X^B(\operatorname{grad}_{g_B} f),$$
$$R(U, V, W) = R^F(U, V, W) - f^{-2}\langle \operatorname{grad}_{g_B} f, \operatorname{grad}_{g_B} f\rangle(\langle V, W\rangle U - \langle U, W\rangle V)$$
$$= R^F(U, V, W) - \langle \operatorname{grad}_{g_B} f, \operatorname{grad}_{g_B} f\rangle(g_F(V, W)U - g_F(U, W)V).$$

d. Let X, Y be tangent fields to B, and let V and W be tangent to F. Verify the following Ricci curvature formulae, where $\Delta_{g_B} f = \operatorname{tr}_{g_B}(\operatorname{Hess}_{g_B} f)$ (in any signature):

$$\mathrm{Ric}_g(X,Y) = \mathrm{Ric}_{g_B}(X,Y) - d \cdot f^{-1}\mathrm{Hess}_{g_B} f(X,Y),$$
$$\mathrm{Ric}_g(X,V) = 0,$$
$$\mathrm{Ric}_g(V,W) = \mathrm{Ric}_{g_F}(V,W) - \langle V,W\rangle(f^{-1}\Delta_{g_B} f + (d-1)f^{-2}\langle \mathrm{grad}_{g_B} f, \mathrm{grad}_{g_B} f\rangle)$$
$$= \mathrm{Ric}_{g_F}(V,W) - g_F(V,W)(f\Delta_{g_B} f + (d-1)\langle \mathrm{grad}_{g_B} f, \mathrm{grad}_{g_B} f\rangle).$$

Exercise 1-16 (totally geodesic submanifolds). Let $M \subset (\overline{M}, \bar{g})$ be an embedded submanifold of a semi-Riemannian manifold, with induced metric g on M. If v is tangent to M, let γ_v be the g-geodesic on M with $\gamma_v'(0) = v$, and let $\bar{\gamma}_v$ be the $\bar{g}$-geodesic on $\overline{M}$ with $\bar{\gamma}_v'(0) = v$. For convenience, we restrict the domain of these curves to a common interval I containing 0.

a. Show that the second fundamental form $\mathrm{II}(X, Y) = (\nabla_X Y)^{\perp}$ vanishes identically on M if and only if for every v tangent to M, $\gamma_v = \bar{\gamma}_v$. In this case we say M is *totally geodesic* in $\overline{M}$.

b. For semi-Riemannian manifolds (M_1, g_1) and (M_2, g_2), we consider the product $(\overline{M} = M_1 \times M_2,\ \bar{g} = g_1 \oplus g_2)$. Show that $\{p_1\} \times M_2$ and $M_1 \times \{p_2\}$, for any $p_i \in M_i$, are totally geodesic in $\overline{M}$.

Exercise 1-17 (curvature and parallel transport). The curvature tensor can be computed via parallel transport, and in fact it measures the failure of path independence of parallel transport. Some geometry texts skip this, whereas some general relativity texts point this out, to varying degrees of mathematical precision. See, for example, [207; 218].

Let g be a semi-Riemannian metric on M, and let $p \in M$. Consider a coordinate chart $\varphi : U \subset \mathbb{R}^n \to M$ centered at p, and let $B \subset U$ be a closed rectangle in a two-dimensional coordinate plane around the origin, say $(x^1, x^2) \in B$ for $\max(|x^1|, |x^2|) \le \varepsilon_0$. Given a vector $V \in T_p M$, define a vector field along the coordinate mapping from B to M as follows: for $(x^1, x^2) \in B$, let $V(x^1, x^2) \in T_{\varphi(x^1,x^2)} M$ be the vector obtained by the parallel transport of $V \in T_p M$ along the image under φ of the segment from $(0, 0)$ to $(x^1, 0)$, and then along the image of the segment from $(x^1, 0)$ to (x^1, x^2). By smooth dependence of solutions to ODE, this depends smoothly on (x^1, x^2). Similarly, let $\widetilde{V}(x^1, x^2)$ be the vector field obtained by the parallel transport of $V \in T_p M$ along the image of the segment from $(0, 0)$ to $(0, x^2)$, and then along the image of the segment from $(0, x^2)$ to (x^1, x^2). We remark that we can compute in coordinates, i.e., compute in $(U, \varphi^* g)$, which we will do without further comment, pushing forward or pulling back quantities between U and M as needed.

Given $(x^1, x^2) \in B$, let $\gamma_{(x^1,x^2)}$ be the piecewise linear path parametrizing the rectangle with vertices $(0, 0)$, $(x^1, 0)$, (x^1, x^2) and $(x^2, 0)$, oriented by this ordering, and let $\widetilde{W}_{(x^1,x^2)}$ be the vector field along $\gamma_{(x^1,x^2)}$ obtained by parallel transport of $V \in T_p M$ around $\gamma_{(x^1,x^2)}$. Then $\widetilde{W}_{(x^1,x^2)}(\xi^1, \xi^2) = V(\xi^1, \xi^2)$ for any (ξ^1, ξ^2) on either of the first two segments of $\gamma_{(x^1,x^2)}$. Let $W(x^1, x^2) \in T_p M$ be the value of $\widetilde{W}_{(x^1,x^2)}$ at the final point along the map $\gamma_{(x_1,x_2)}$; in other words, $W(x^1, x^2)$ is the vector obtained by parallel transport of $V(x^1, x^2)$ first along the line from (x^1, x^2) to $(0, x^2)$, and then along the line from $(x^2, 0)$ to $(0, 0)$.

Given V, we can choose a uniform constant for the "O"-estimates below, or likewise, given K, we have uniform "O"-estimates for all $|V|_g \leq K$.

a. Note that $V(x^1, x^2) = \widetilde{V}(x^1, x^2)$ along the axes, where $x^1 x^2 = 0$. Show that $\left| V(x^1, x^2) - \widetilde{V}(x^1, x^2) \right| = O(|x^1 x^2|)$. (You might write the parallel transport system in coordinates along the sides of the rectangle and estimate the change in the vector fields along pairs of parallel sides by estimating an appropriate integral.)

b. Note that $W = V \in T_p M$ for $x^1 x^2 = 0$. Argue that $|W - V| = O(|x^1 x^2|)$.

Remark. You can make the corresponding construction on any rectangle

$$[x^1, x^1 + \Delta x^1] \times [x^2, x^2 + \Delta x^2] \subset B,$$

starting from $V(x^1, x^2)$, say, and the analogous difference is $O(|\Delta x^1 \Delta x^2|)$.

c. Show that

$$\lim_{\substack{(x^1,x^2) \to (0,0) \\ x^1 x^2 \neq 0}} \frac{W(x^1, x^2) - V}{x^1 x^2} = R\left(\frac{\partial}{\partial x^2}, \frac{\partial}{\partial x^1}, V \right).$$

(Hint: Write the numerator above in terms of integrals along the sides of $\gamma_{(x^1,x^2)}$. You again want to estimate the integrals in parallel pairs. The answer will drop out once you argue that up to an acceptable error, you can replace the $\widetilde{W}_{(x^1,x^2)}$-term in the integrals by the appropriate V-term or $\widetilde{V}$-term.)

CHAPTER 2

The Einstein equation

The theory of special relativity incorporates a modification of Newtonian mechanics together with electromagnetism. A natural question to consider is how gravitation fits into the framework of relativity. We focus our analysis of this question along two main ideas, that of the equivalence between uniform acceleration and a uniform gravitational field, and that of the gravitational redshift, which will lead us to the Einstein equation. We begin, however, by reviewing Newton's law of gravitation.

2.1. Newtonian gravity

Newton's law of gravity can be formulated as follows. If two objects are separated by a spatial distance r, then the magnitude of the gravitational force between them is given by $F = Gm_g M_g/r^2$, where the direction of the force is along the line from one mass to the other. Here m_g and M_g are the *gravitational masses* associated to the two objects, and G is Newton's gravitational constant. If $\hat{r}$ is the unit vector from the object of mass M_g to the other object, then the force on the object of mass m_g is $F = -(Gm_g M_g/r^2)\hat{r}$. If the object of mass M_g is located at the origin, and $x \in \mathbb{R}^3$ is the position of the other object, then $r = |x|$, $\hat{r} = x/r$, and we can write the force as follows, where ∇ is the Euclidean gradient:

$$F = -\frac{Gm_g M_g}{|x|^2}\hat{r} = m_g \nabla\left(\frac{GM_g}{|x|}\right) = -m_g \nabla\Phi, \qquad (2.1.1)$$

where $\Phi(x) := -GM_g/|x|$ is the *gravitational potential* associated to the mass M_g. In analogy with *Coulomb's law* of electrostatics, namely that the electric force between *stationary* charged particles (with charges q_1 and q_2) is given by $F = q_1 q_2/r^2$ (in cgs units), m_g and M_g play the role of *gravitational charges*. Of course, there is already a notion of mass embodied in Newton's second law: in this context we write it as $F = m_i a$, and call m_i the *inertial mass*. By equating forces, we solve for the acceleration of an object of gravitational mass m_g and inertial mass m_i due to the gravitational force of an object of gravitational

47

mass M_g:

$$a = -\frac{m_g}{m_i}\nabla\Phi.$$

We could in principle use this equation to discern the ratio of the inertial to the gravitational mass for various objects. It turns out that the acceleration is the same for all bodies, and hence the mass ratio is *constant*, a result epitomized by the apocryphal experiments of Galileo dropping objects of different masses from the tower in Pisa. By adjusting G, we may assume, then, that $m_i = m_g$: the inertial and gravitational masses agree. The effect of gravity is *universal*: it accelerates all objects the same way, independent of what precisely comprises the mass. In this way gravity is decidedly different from electromagnetism.

Before we move on, we note that the potential function for Newtonian gravity satisfies a simple partial differential equation. Indeed, away from $x = 0$, the function $\Phi(x) = -GM/|x|$ is harmonic (with respect to the Euclidean metric), i.e., $\Delta\Phi = 0$, as you can easily check. Of course, $\Delta\Phi$ can be interpreted globally as a distribution, say $T = \Delta\Phi$, and we obtain the equation

$$\Delta\Phi = 4\pi GM\delta_0, \tag{2.1.2}$$

where δ_0 is the Dirac measure at the origin (the "location" of the mass M). Indeed, suppose $\psi \in C_c^\infty(\mathbb{R}^3)$ is any smooth function supported in the ball of radius r_0 around the origin. There exists a $C > 0$ such that $|(\Phi\nabla\psi)(x)| \le C/|x|$ for all $|x| \ne 0$; hence, for all $\varepsilon > 0$, we have $\left|\int_{\{|x|=\varepsilon\}}\Phi\nabla\psi\cdot\hat{\mathbf{r}}\,d\sigma\right| \le 4\pi C\varepsilon$. Then, since $1/|x| \in L^1_{\text{loc}}(\mathbb{R}^3)$, an application of Gauss's divergence theorem, along with Green's identity $\text{div}(\Phi\nabla\psi - \psi\nabla\Phi) = \Phi\Delta\psi - \psi\Delta\Phi$ (compare (1-11b)) and the vanishing of $\Delta\Phi(x)$ for $|x| \ne 0$, yields

$$\begin{aligned}
T(\psi) &:= \int_{\mathbb{R}^3}\Phi\Delta\psi\,dx = \lim_{\varepsilon\searrow 0}\int_{\{\varepsilon\le|x|\le r_0\}}\Phi\Delta\psi\,dx \\
&= -\lim_{\varepsilon\searrow 0}\int_{\{|x|=\varepsilon\}}(\Phi\nabla\psi - \psi\nabla\Phi)\cdot\hat{\mathbf{r}}\,d\sigma \\
&= \lim_{\varepsilon\searrow 0}\int_{\{|x|=\varepsilon\}}\psi\frac{GM}{|x|^2}\hat{\mathbf{r}}\cdot\hat{\mathbf{r}}\,d\sigma \\
&= 4\pi GM\lim_{\varepsilon\searrow 0}\frac{1}{4\pi\varepsilon^2}\int_{\{|x|=\varepsilon\}}\psi\,d\sigma \\
&= 4\pi GM\psi(0),
\end{aligned}$$

where the last identity follows by continuity.

In this case the matter density is $\sigma = M\delta_0$. For a more general matter distribution of density σ, the gravitational potential solves Poisson's equation

$$\Delta\Phi = 4\pi G\sigma. \tag{2.1.3}$$

If σ is compactly supported (or more generally, if σ decays sufficiently at infinity), we may choose $\Phi(x) \to 0$ as $|x| \to \infty$.

Newton's law of gravitation has a flaw, which Newton himself critiqued. Namely, the gravitational force between two masses does not appear to be effected through an intermediary, even if the masses are far apart. This "action at a distance" leads immediately to causality violation, since the motion of one mass would *instantly* affect the gravitational field everywhere else. This would mean that gravitational effects have infinite speed of propagation, which seems troublesome whether or not the discussion is framed in the context of special relativity. Einstein sought to rectify this, and in doing so to incorporate another fundamental force within a relativistic framework.

2.2. From the equivalence principle to general relativity

In Minkowski spacetime there is a preferred set of coordinate charts, corresponding in physics terminology to *inertial observers*. In such charts the metric for Minkowski spacetime has the familiar form $\eta_{\mu\nu}dx^\mu dx^\nu = -(dx^0)^2 + \delta_{ij}dx^i dx^j$, and, up to a spacetime translation, any two such charts are related by a Lorentz transformation. These charts are the analogues of Cartesian coordinate systems for Euclidean space. From the point of view of physics, coordinates from an inertial chart correspond to measurements made by an inertial observer. The principle of (special) relativity asserts that there are no preferred inertial observers, and as such the form of physical laws should take the same form in all inertial frames (sometimes called the *principle of special covariance*). An interesting physics question is whether inertial observers can exist in principle when a nontrivial gravitational field is present, which relates mathematically to whether the spacetime metric can be flat, as in Minkowski spacetime.

2.2.1. *The equivalence principle.* We can bring to bear upon the question of the existence of inertial frames a famous thought experiment of Einstein. Suppose there were an inertial frame of reference, say a small lab room isolated from other forces or fields. In such a frame, if one lets go of a ball (or rather a test particle, say), it would tend to stay at rest. Now suppose a rocket is attached to the top of the room, and then accelerates the room "upward" at a uniform rate. If one lets go of a ball now, it will "fall" toward the floor, just as it would in a *uniform* gravitational field.[3] In this case, one cannot distinguish, by making local measurements of the paths of objects upon which no forces (other than possibly gravity) act, whether the frame is non-inertial, or whether there is a uniform gravitational field present. Note that this depends on the universality of gravity: it imparts the same acceleration to all objects. This version of the *equivalence principle* basically comes down to the equality of gravitational and inertial masses: if they were different, then one could distinguish between the two situations by making appropriate measurements. The equality of the gravitational and inertial masses is thus a reflection of the equivalence of a uniformly accelerating frame with an inertial frame in which a uniform gravitational field is present. Other versions of the equivalence principle involve other laws of physics, asserting that measurements involving those laws cannot distinguish between a uniformly accelerating frame and a frame in which a uniform gravitational field is present.

We have just seen how to "create" a uniform gravitational field via acceleration. Conversely, consider now the lab room as a spaceship in orbit, or closer to home, a compartment of one of those amusement park "free fall" rides. In free fall, if one lets go of a small object (not safe to try this on the free fall ride!), its position with respect to the room is constant, because a uniform gravitational field accelerates all objects the same. The law of inertia appears to hold, and the observer in free fall will then *not* detect a uniform gravitational field. We on the earth claim to detect such a field precisely because we are *not* in free fall: the contact forces with the earth keep us from a free fall path, and thus if we drop an apple, we observe it "fall" to the earth under the force of gravity. This is equivalent to the accelerated room: we feel the contact force between the floor and our legs, and we observe objects falling toward the ground.

Even light cannot escape "gravity's" pull: if we imagine a light ray entering the room at one end moving in a straight line in an inertial frame, the path of the light is curved in the accelerated frame of reference. Einstein reasoned that by equivalence, a gravitational field should bend the paths of light rays too. Thus, were there to exist an inertial reference frame, where a nonzero gravitational field accounts for acceleration not attributed to other forces, light rays would apparently not move along straight lines in this frame. (While Newton had actually anticipated the deflection of light by a massive object, Einstein was able

---

[3] Strictly speaking, we are considering a lab frame which is limited in extent in both space and time, so that we can reliably expect a gravitational field to be roughly uniform, causing an approximately constant acceleration. In regions more extended in spacetime, the non-uniformity of the gravitational field will give rise to *tidal forces* that could be used to distinguish between uniform acceleration and a gravitational field, and in fact will be directly related to the curvature of spacetime, as we will see below.

to use his theory of gravitation to give a much more accurate prediction than that of classical physics.)

A natural question is how to proceed with these notions and their ramifications for the question of the existence of inertial frames. Before doing this, we introduce the physical phenomenon of *gravitational redshift*.

2.2.2. *Gravitational redshift.* We recall the Doppler shift formula, as seen earlier in the twin paradox example. Suppose light with wavelength λ_0 emanates from an emitter at rest in an inertial frame $\mathcal{O}$; the frequency of the light ν_0 satisfies $\lambda_0 \nu_0 = c$, and the time between the start and end of the emission of one full wavelength is $\Delta t_0 = 1/\nu_0$. The light is absorbed by a detector moving at velocity v with respect to $\mathcal{O}$, along the same line as the propagation of light. We let $\widetilde{\mathcal{O}}$ be the rest frame of the detector, and let ν_1 be the frequency of the light as measured by the detector in $\widetilde{\mathcal{O}}$, so that $\Delta t_1 = 1/\nu_1$ is the time for the absorption of one wavelength as measured in $\widetilde{\mathcal{O}}$. As shown in Exercise 1-5, we have

$$\frac{\Delta t_0}{\Delta t_1} = \frac{\nu_1}{\nu_0} = \frac{\sqrt{1 - v/c}}{\sqrt{1 + v/c}}.$$

We next show that the gravitational redshift, which is derived in concert with and provides evidence for the equivalence between uniform acceleration and a uniform gravitational field, places a roadblock in the way of the existence of an inertial observer, and thus calls into question whether one can mesh gravity with the Minkowskian geometry of special relativity. Indeed, imagine two rockets moving along the y-axis, one following the other at a fixed distance Δy, and with a uniform acceleration $a > 0$ with respect to some inertial frame $\mathcal{O}$. Suppose the frame $\mathcal{O}$ is momentarily comoving with the rockets at the instant a photon of wavelength $\lambda_0 = c/\nu_0$ is emitted from the trailing rocket to the lead rocket. In the time Δt (measured in $\mathcal{O}$) the photon travels to the lead rocket, the rockets have a change in velocity $\Delta v = a\,\Delta t$, so there will be a Doppler shift in the reported frequency of the received photon, from the difference in velocity at reception and emission. The Doppler shift can be computed using the formula recalled above, since the shift will be the same as that computed by inertial frames with relative velocity Δv (momentarily comoving with the respective rocket at emission and absorption, respectively; see Section 1.2.4). Thus if ν_1 is the frequency of the received photon (as measured by the lead rocket), then

$$\frac{\nu_1}{\nu_0} = \frac{\sqrt{1 - \Delta v/c}}{\sqrt{1 + \Delta v/c}} = \frac{1 - \Delta v/c}{\sqrt{1 - (\Delta v/c)^2}}.$$

If we arrange $a\Delta t = \Delta v \ll c$, then $\Delta y = (c - \frac{1}{2}a\Delta t)\Delta t \approx c\Delta t$, and $\nu_0/\nu_1 \approx 1 + \Delta v/c$, to first order in $\Delta v/c$. We relate this to Δy as follows:

$$\frac{\Delta\lambda}{\lambda_0} = \frac{\lambda_1 - \lambda_0}{\lambda_0} = \frac{\nu_0}{\nu_1} - 1 \approx \frac{\Delta v}{c} = \frac{a\Delta t}{c} \approx \frac{a\Delta y}{c^2}. \qquad (2.2.1)$$

With the equivalence principle in mind, we compare the above situation to that in which there is a uniform gravitational field, with gravitational potential Φ, which we assume is time-independent, and only varies in one spatial variable $y \in \mathbb{R}$. We imagine two observers stationary in the gravitational field, each of whom would measure that an object, in the absence of other forces, would experience a "downward" (i.e., in the negative y-direction) acceleration of magnitude $a = \partial\Phi/\partial y$. The equivalence principle asserts this situation should be equivalent to the observers in an accelerating "elevator," or to the accelerating rockets as in the preceding paragraph. Now suppose a photon of frequency ν_0 travels a "height" Δy in the gravitational field, from an observer at y (akin to the trailing rocket) to an observer at height $y + \Delta y$. Then using (2.2.1), we should see a relative Doppler shift of $\Delta\lambda/\lambda_0 \approx c^{-2}(\partial\Phi/\partial y)\Delta y \approx c^{-2}(\Phi(y + \Delta y) - \Phi(y))$.

This leads to an experimentally verified phenomenon about clock rates in a gravitational field. Suppose a photon is emitted from a height y and is received at height $y + \Delta y$ in the gravitational field. If the photon has wavelength λ_0 at emission, the time between the beginning and end of a wavelength (as measured by an inertial observer at emission, say) is $\Delta t_0 = c^{-1}\lambda_0$. As we have just seen, an inertial observer at reception measures the time between the beginning and end of a wavelength as $\Delta t_1 = \Delta t_0 + c^{-1}\Delta\lambda > \Delta t_0$, since the wavelength of the absorbed photon is *longer*: the frequency, and hence the energy, is lower, accounting for the gravitational potential energy via the relativistic mass-energy equivalence and the equality of inertial and gravitational masses. This means that clocks run at different rates in different places in a gravitational field; in this case the clock at $y + \Delta y$ runs faster than an identical clock at y. This was confirmed experimentally by the Pound–Rebka–Snider experiment (see [161, p. 1055–1060], for example). A similar analysis can be done involving acceleration via rotation, cf. [94, p. 24–25].

We return to the question at hand: could there be inertial reference frames (with respect to which the principle of special relativity is formulated) in the presence of a gravitational field, where acceleration unaccounted for by other forces could be interpreted to be gravitational in nature? We certainly cannot reconcile the principle of special relativity with the existence of inertial frames corresponding to two observers stationary at the heights y and $y + \Delta y$ as in the above paragraph. Two such observers are not in motion relative to each other, so should they be inertial observers, then since the spacetime paths of the beginning and tail ends of the photon wavelength should be related by a simple time-translation (the gravitational field is time-independent, and only depends on y), the Δt measurements should be the same at the two different heights. The fact that experiment shows otherwise indicates an incompatibility between gravitation and the existence of such inertial observers.

2.2.3. *Towards a geometric solution.* Einstein made an argument [85; 84] that the spacetime continuum in the presence of a gravitational field (related to acceleration via the equivalence principle) should be "non-Euclidean" (i.e., non-flat, so non-Minkowskian). As we will see, the argument might be construed as more heuristic than precise. We present it briefly for historical reasons, and hopefully the reader will get some utility from it without getting the wrong impression.

Consider a frame of reference $\mathcal{O}$ which represents uniform rotation with angular velocity $\omega > 0$ with respect to an inertial frame of reference. We pick the origin for coordinate charts adapted to the frames as the center of rotation, synchronize the clocks for the two observers at the origin of coordinates, and align the axes at $t = 0$, so that the x and y axes are in the plane of rotation. One can conceive ways to make measurements, and thus build coordinates for spacetime, adapted in some way to $\mathcal{O}$. There is not a canonical way to do this, though one might initially be inclined to think the following is such a way: relate the coordinates (t, x, y, z) in $\mathcal{O}$ to the inertial coordinates $(\mathring{t}, \mathring{x}, \mathring{y}, \mathring{z})$ of an event by the transformation (where $(x, y) = (r\cos\theta, r\sin\theta)$ for $r \geq 0$), $\mathring{t} = t$, $\mathring{z} = z$, $\mathring{x} = x\cos\omega t - y\sin\omega t = r\cos(\theta + \omega t)$, $\mathring{y} = x\sin\omega t + y\cos\omega t = r\sin(\theta + \omega t)$. For example, one might use radar (by sending and receiving signals to and from an event), using the clock at the origin to read off the time, which then agrees in both frames because there is no time dilation. One could also build coordinates for spacetime by making measurements of events using measuring devices (a grid of rods and clocks) associated to the observers at rest relative to $\mathcal{O}$. See [161; 38; 162] for further discussion of coordinates adapted to accelerated observers.

Let C_r be a circle centered at the origin, of radius r as measured in the inertial frame, with respect to which the length is $2\pi r$. An observer on C_r at rest in $\mathcal{O}$ is moving *along* the circle C_r with angular velocity ω relative to the inertial frame; of course we require $\omega r < c$. We proceed following Einstein: relative to the inertial frame, the length of a unit measuring stick in the frame $\mathcal{O}$ oriented *tangentially* along C_r is shorter than one unit, by the factor $\sqrt{1 - (\omega r/c)^2}$. Thus, the length $L(C_r)$ measured using rods adapted to $\mathcal{O}$ is $2\pi r/\sqrt{1 - (\omega r/c)^2}$, which is *more* than $2\pi r$: if $\omega r/c$ is small, then in $\mathcal{O}$, $L(C_r) \approx 2\pi r\left(1 + \frac{1}{2}(\omega r/c)^2\right)$. Another way to think of this is that the observers at rest in $\mathcal{O}$ located along C_r place non-overlapping measuring sticks along the circle to mark its circumference, and one just counts how many of these are needed to go around the perimeter to determine the length of the circle as measured in $\mathcal{O}$. On the other hand, distances perpendicular to the direction of motion agree in both frames. Thus both frames agree on the radius r. It appears that from the point of view of the frame $\mathcal{O}$, one might conclude that the spatial geometry is *curved*. Indeed, recall the classical formula for the Gauss curvature (Exercise 2-55), which applied to the above analysis would yield nonzero (negative) curvature:

$$K(p) = \lim_{r \to 0} \frac{3}{\pi} \cdot \frac{2\pi r - L(C_r)}{r^3}.$$

While Einstein cited this intriguing argument as motivation for the introduction of non-Euclidean geometry into the theory of gravitation, one must critique it in various ways. Bearing in mind the relativity of simultaneity, for instance, has the argument above really succeeded in showing there is an observer who will measure the circumference and radius of a circle to be out of step with Euclidean geometry? Or does the analysis just yield some sort of non-Euclidean geometry on the set of worldlines of rotating observers? While the spacetime interval between two *events* is invariant, one needs to consider carefully to what extent one can define and compare the spatial length and radius of a circle in the two frames, keeping in mind that the observers will not necessarily agree on simultaneity. Clock rates will vary for rotating clocks depending on the location relative to the center; the clock rates depend on r, so while the rates are the same for observers at rest in $\mathcal{O}$ on each C_r (each of which we note has a different local rest frame), we might ask to what extent the observers can agree on a set of events to be deemed the *disk* spanned by C_r. Imagine too if the rotating frame is slowly brought to rest relative to the inertial frame: what happens as it slows and the length contraction factor goes to 1? There are in principle too many measuring rods positioned around the circumference, so something would have to give. This discussion is related to the breakdown of standard notions of *rigidity* in relativity, and the thought experiment of Einstein is related to similar arguments in discussions of *Ehrenfest's paradox* concerning the fate of a rotating cylinder in relativity. For more on this topic, see [110], for example, and for Kaluza's argument relating Ehrenfest's paradox to hyperbolic geometry, see [132].

From a geometric point of view, if spacetime is a manifold equipped with a Lorentzian metric, then spacetime geometry is either Minkowskian or not. If spacetime were Minkowskian, then the events which comprise a circle at a fixed time in an inertial frame would yield a spacelike curve ξ embedded in a

Euclidean three-space (and hence a Euclidean plane), so that the length of the curve would follow the well-known formula for the circumference of a circle. The length of the curve ξ is a geometric invariant; if another frame of reference were used to build spacetime coordinates (e.g., coordinates somehow adapted to an accelerating frame), the spacelike curve ξ might not be comprised of events simultaneous in this frame, but its length would be invariant. The coordinate measurements might need to be converted using metric components to obtain truly invariant geometric distances; this is akin to using curvilinear coordinates for the Euclidean plane. If on the other hand there is a region where the spacetime geometry is curved, there is no frame from which one could build coordinates for which the metric is everywhere identical to that of Minkowski spacetime in inertial coordinates, a situation analogous to the familiar fact that one cannot make maps (charts) of the Earth's surface which are isometries (up to a constant scale factor) between the geometry on the surface and the Euclidean geometry of the planar map.

Similar comments apply to the gravitational redshift scenario above. For the observers stationary in a gravitational field with time-independent potential varying in y as above, we maintain that the paths of successive photon crests should be congruent in a coordinate chart associated to either observer. The conundrum about clock rates discussed above would persist if this observer were an inertial observer, say, using the Minkowski metric in inertial coordinates to compute spacetime intervals. On the other hand, if the metric components are not the Minkowskian components in inertial coordinates, but rather components for Minkowski spacetime in a non-inertial frame, or components for a metric for which the geometry of spacetime is curved, then it is generally expected that coordinate measurements do not give geometric invariants, and that one would need to use the metric to compute the relevant invariant, and thus give the physical quantity of interest (in this case a proper time interval).

2.2.4. *Free fall and geodesics.* Beyond the issues we have seen above regarding the obstruction to the existence of observers which are at once inertial and stationary in a gravitational field, another fundamental issue arises from the universality of gravity. An inertial frame of reference is one in which test particles upon which no net force acts move with constant velocity. Consider a region where the only forces are electromagnetic in nature: neutral particles and charged particles could in principle be distinguished by their relative motions. In an inertial frame, the neutral particles would move with constant velocity according to the law of inertia, while charged particles would move according to the Lorentz force law.

The situation is decidedly different in the context of gravitation. In a region where the only forces are due to a nontrivial gravitational field, there would be no free test particle to illustrate the law of inertia. Imagine a Newtonian inertial frame in a region far away from massive objects, in which free particles (including photons) are moving; if we now "dial up" a gravitational field (by moving a massive object nearby), then what happens? All free particles, including light, would be accelerated in the same manner away from the straight line paths in this frame. In fact, if we conceive of a frame in terms of an observer, the universality of gravity implies that the observer would also be affected by gravity.

In seeking a frame where the law of inertia holds, we have to reconcile the notion of gravitational force with the equivalence principle. Namely, observations of physical phenomena in a frame in which there is a uniform gravitational field should be equivalent to those from a frame uniformly accelerating with respect to an inertial frame. Recall that this was tied into the equivalence of inertial and gravitational masses: if these differed, you could in principle perform experiments on test particles to distinguish between the dynamic effects of a gravitational field and the basic kinematic effects of frame acceleration. In any case, local experiments (where tidal effects due to a non-uniform gravitational field are too small to detect) cannot distinguish the two situations, which suggests, then, that test particles should behave like free particles, with the local effects of the gravitational force interpreted as a fictitious force arising from an accelerated frame of reference. From this viewpoint, then, the worldlines of particles which undergo only gravitational forces, i.e., *freely falling particles*, should then give rise to a class of observers which to some extent might assume the role inertial observers played in Minkowski spacetime of special relativity.

We test this hypothesis via the gravitational redshift analysis above: what if we do the analysis instead from the perspective of two observers *falling freely* in the gravitational field, as opposed to those remaining stationary with respect to the gravitational potential? If the freely falling observers build coordinates adapted to their motion (normal coordinates, or Fermi coordinates along the worldline, as we will later employ in deriving the Einstein equation), then the physics in this frame corresponds, to good approximation and locally in spacetime, to what would be measured in an inertial frame. For example, consider the above photon with emission frequency ν_0, having traveled a height Δy. As we saw above, a stationary observer in the field measures the photon frequency ν_1 with $\nu_0/\nu_1 \approx (1 + c^{-2}a\,\Delta y)$. On the other hand, a freely falling observer which was stationary in the field when the photon was emitted will have attained a velocity $\Delta v \approx -c^{-1}a\,\Delta y$ at absorption (sign indicates relative direction), and so will measure the photon to have a Doppler shift given by

$$\frac{\nu_2}{\nu_1} = \frac{\sqrt{1 - \Delta v/c}}{\sqrt{1 + \Delta v/c}} \approx 1 - \frac{\Delta v}{c} \approx 1 + c^{-2} a \Delta y.$$

We see that ν_0 and ν_2 *agree* to leading order in $|\Delta v|/c$. Thus while spacetime is not Minkowskian, there are certain observers that correspond, approximately and locally in spacetime, to inertial frames.

In light of all this, one might dispense with the notion of gravitational force and acceleration due to gravity, and instead assert that a gravitational field manifests itself through a family of timelike paths that represent *freely falling* test particles (null paths for photons), whose trajectories are determined by and encode gravitational effects, but upon which no other forces act. We stress that while this assertion identifies a class of observers, this class is not determined *a priori*, so such observers will not be preferred for the formulation of physical laws, in accordance with Einstein's principle of general relativity (see Section 2.3).

Geodesics are paths that have zero *covariant acceleration* (with respect to an affine connection ∇), and as such are analogues of straight lines for a curved space. Einstein asserted that objects that are experiencing no other force except possibly that of gravity should move along timelike geodesics, while light should propagate along null geodesics. In any coordinate chart for spacetime, the covariant acceleration can be computed, and if it vanishes in one chart, it vanishes in all charts. In this way, the paths of light rays in vacuum obey a rule that takes the same form in *every* frame of reference, consistent with the principle of general relativity, and providing a simple example of the *principle of general covariance*: the laws of physics should in principle be diffeomorphism-invariant, and so should be able to be formulated in a coordinate-independent way.

Of course, the *coordinate* acceleration along a geodesic may not vanish, as the equation for a geodesic $\gamma(t)$ with coordinates $x^\mu(t)$ can be expressed as

$$\frac{d^2 x^\mu}{dt^2} + \Gamma^\mu_{\nu\sigma}\Big|_{\gamma(t)} \frac{dx^\nu}{dt} \frac{dx^\sigma}{dt} = 0,$$

where $\nabla_{\partial/\partial x^\nu} \partial/\partial x^\sigma = \Gamma^\mu_{\nu\sigma} \partial/\partial x^\mu$; if we use the Levi-Civita connection for a metric, we have the formula $\Gamma^\mu_{\nu\sigma} = \frac{1}{2} g^{\mu\rho}(g_{\rho\sigma,\nu} + g_{\nu\rho,\sigma} - g_{\nu\sigma,\rho})$ for the Christoffel symbols. As broached earlier, then, gravitational force may be locally akin to a fictitious force in an accelerating frame in Euclidean space. By choosing local coordinates adapted to a freely falling observer, one can arrange for the Christoffel symbols $\Gamma^\mu_{\nu\sigma}$ to vanish at an event (or even along a worldline). However, unlike in Euclidean space or Minkowski spacetime, in the presence of curvature, there is no coordinate system in which the Christoffel symbols can be made to vanish identically, and as we will soon see, spacetime curvature produces tidal effects ascribed to a gravitational field.

The universality of gravity has led to its incorporation into the structure of spacetime at a fundamental level, determining inertial properties (motion of test particles which experience no non-gravitational forces) through the geometry of spacetime. Furthermore, as experiments continue to confirm that signals in vacuum cannot travel at speeds faster than the speed of light, gravity plays a distinguished role in determining the causal structure of spacetime through its effect on paths of light rays. For a concrete geometric consequence, consider a spacetime modeled on a Lorentzian manifold, and assert that light rays move along null paths. The collection of lightcones determines the conformal class of the spacetime metric g (and recall for comparison, the conformal compactification of Minkowski spacetime as presented in the first chapter indeed preserves the lightcone structure). Indeed, if X is timelike and $Y \neq 0$ is spacelike at a point p, then $g(X + aY, X + aY) = a^2 g(Y, Y) + 2ag(X, Y) + g(X, X)$, a quadratic polynomial in a with a positive leading coefficient and a negative constant term. Hence there are exactly two real roots of this quadratic, which in principle we can glean from the lightcone at p. The product of these roots gives $g(X, X)/g(Y, Y)$. If V and W are any tangent vectors at p, then

$$g(V, W) = \tfrac{1}{2}\big(g(V + W, V + W) - g(V, V) - g(W, W)\big).$$

Knowing the lightcone at p, then, allows us to find the ratio $g(V, W)/g(X, X)$, since any of the terms on the right of the preceding equation, if nonzero, can be gleaned in ratio with either $g(X, X)$ or $g(Y, Y)$.

The question remains how to connect the geometry of spacetime to the distribution of matter and energy, whose motion should be in part determined by the curvature of spacetime and whose gravitational effects should in turn influence the geometry of spacetime. An answer lies in the Einstein equation, to which we now turn.

2.3. The Einstein equation

We begin with a quote from Einstein [83, p. 113]: "The laws of physics must be of such a nature that they apply to systems of reference in any kind of motion." This *principle of general relativity* puts all frames of reference on an equal footing, in contrast to the privileged inertial frames of reference of special relativity. Together with the *equivalence principle*, that an accelerating frame ought to be locally (and approximately) equivalent from the point of view of physics to a frame in which there is a uniform gravitational field, we see that a theory obeying general relativity ought to naturally be in part a theory of gravitation, whereby the global inertial frames of special relativity are replaced by local inertial frames adapted to freely falling observers. Such a theory incorporating gravity is consistent with and gives impetus for the assertion that no coordinate system should be preferred for the formulation of physical laws. This is echoed in a more geometric way by the *principle of general covariance*, which asserts that the laws of physics should be invariant under diffeomorphisms, so that it should be possible to formulate their equations in a coordinate-independent manner. In particular, the laws of physics should have the same form for all frames of reference (compare the formulation in [218, p. 57]). While equations of tensorial physical laws in Minkowski spacetime can be cast in generally covariant form, we have argued that other spacetime metrics should arise when there is a nontrivial gravitational field. We emphasize that this can be the case *even in vacuum regions of spacetime*, possibly corresponding to the gravitational field outside a compact massive object, say.

One seeks to relate the spacetime geometry to the distribution of matter fields and energy within spacetime: the result is the Einstein equation, which will be discussed at length below. That the theory should incorporate some features of special relativity suggests we need to strengthen the equivalence principle, and assert that *spacetime geometry should be given by a Lorentzian metric* (recall that our signature for Lorentzian metrics is $(-, +, +, \ldots, +)$); in normal coordinates at a point in spacetime, the laws of physics will take the same form as they would in special relativity, and thus locally, the laws are approximately of the same form as in special relativity. Said another way, local experiments in a small region of spacetime cannot detect a gravitational field which is roughly uniform, and could be ascribed to a uniformly accelerating frame of reference, with the results being in accordance with special relativity. In larger regions of spacetime, non-uniformities of the gravitational field can be detected by measuring tidal forces. As a final guide to the Einstein equation, we have a *correspondence principle*: the equations governing gravity should yield Newton's law of gravitation when the gravitational field is sufficiently weak.

A few words are in order before we move on. We will not go into a discussion of *Mach's principle* and its influence on Einstein's development of a theory of gravitation and cosmology (cf. [85; 161; 171]), other than to say that some of the spirit of Mach's ideas may be present in the way spacetime geometry, from which we discern the timelike geodesics which represent inertial motion, *interacts* with matter and energy, through the forthcoming Einstein equation; in other words, the inertia of a test particle is determined in relation to the rest of matter and energy of spacetime. We also will not spend any more time delineating differences and relationships between the various principles at the foundation of the theory of gravitation. It may indeed appear that we have been reciting the same themes over and over. There is a whole literature on the foundational underpinnings of the theory, stemming from writings of Einstein and his contemporaries, continuing right up to the present, from physicists, mathematicians, as well as philosophers and historians of science; see, e.g., [169; 170].

Einstein searched for a way to relate the geometry of spacetime to the energy-momentum distribution of matter and fields it within it. The gravitational field itself is encoded in the metric of spacetime. Einstein sought to equate the stress-energy tensor T describing the energy-momentum densities of the fields and matter to some tensor created from a Lorentzian metric g. In this quest he faced some restrictions. The tensor T is symmetric and divergence-free in special relativity, and therefore these properties should hold at the center of a normal coordinate chart (locally inertial frame), in accordance with the equivalence principle; the divergence of T must then vanish any coordinate system. We note that it is a *mathematical* statement whether the divergence of a tensor vanishes (which can be computed in any coordinate system), whereas the *physics* dictates whether the stress-energy properties can be encoded by a tensor, which then behaves in accordance with general covariance. Einstein originally tried to equate T with the Ricci tensor, up to a constant scalar multiple, but that was doomed to fail in general, because of the contracted Bianchi identity (Corollary 2-2). Nowadays this is a well-known fact covered in introductory graduate geometry courses, but as it is so important, we give the proof here, after recalling a few facts and conventions about the curvature.

Throughout the remainder of the chapter, indices will run over all components, except where we specifically note otherwise. Recall that we often employ angle brackets $\langle\,\cdot\,,\,\cdot\,\rangle$ for the metric. For convenience, we recall our various curvature conventions here. We take $R(X, Y, Z) = \nabla_X \nabla_Y Z - \nabla_Y \nabla_X Z - \nabla_{[X,Y]} Z$ to be the curvature tensor, with index convention

$$R^\ell_{ijk}\frac{\partial}{\partial x^\ell} = R\left(\frac{\partial}{\partial x^i}, \frac{\partial}{\partial x^j}, \frac{\partial}{\partial x^k}\right),$$

in which R is a $(1, 3)$-tensor, while

$$R_{ijk\ell} = g_{\ell m} R^m_{ijk} = \left\langle R\left(\frac{\partial}{\partial x^i}, \frac{\partial}{\partial x^j}, \frac{\partial}{\partial x^k}\right), \frac{\partial}{\partial x^\ell}\right\rangle$$

gives the components of the corresponding $(0, 4)$-tensor. $R(X, Y, Z)$ is alternating in (X, Y) and enjoys *symmetry-by-pairs*: $\langle R(V, W, Y), Z\rangle = \langle R(Y, Z, V), W\rangle$. Thus we have the component identities $R_{k\ell ij} = R_{ijk\ell} = -R_{jik\ell} = -R_{ij\ell k}$. The *Ricci tensor* $\mathrm{Ric}(g)$ is a symmetric $(0, 2)$-tensor with components $R_{jk} = R^i_{ijk} =$

$g^{i\ell}R_{ijk\ell} = R_{kj}$, and the metric trace of the Ricci tensor is the scalar curvature $R(g) = g^{ij}R_{ij}$.

Proposition 2-1 (Bianchi identities). *In a semi-Riemannian manifold (M, g), with Levi-Civita connection ∇, the curvature tensor satisfies an algebraic Bianchi identity,*

$$R(X, Y, Z) + R(Y, Z, X) + R(Z, X, Y) = 0,$$

for all vectors X, Y, and Z. The curvature tensor also satisfies a differential Bianchi identity: for all vectors $X, Y, Z, V,$ and W,

$$\langle(\nabla_X R)(V, W, Y), Z\rangle + \langle(\nabla_Y R)(V, W, Z), X\rangle + \langle(\nabla_Z R)(V, W, X), Y\rangle = 0.$$

By symmetry-by-pairs, this is equivalent to

$$\langle(\nabla_X R)(V, W, Y), Z\rangle + \langle(\nabla_V R)(W, X, Y), Z\rangle + \langle(\nabla_W R)(X, V, Y), Z\rangle = 0.$$

Proof. Applying $R(X, Y, Z) = \nabla_X \nabla_Y Z - \nabla_Y \nabla_X Z - \nabla_{[X,Y]}Z$, we can rearrange terms to obtain

$$R(X, Y, Z) + R(Y, Z, X) + R(Z, X, Y)$$
$$= \nabla_X(\nabla_Y Z - \nabla_Z Y) + \nabla_Y(\nabla_Z X - \nabla_X Z) + \nabla_Z(\nabla_X Y - \nabla_Y X)$$
$$\quad - \nabla_{[X,Y]}Z - \nabla_{[Y,Z]}X - \nabla_{[Z,X]}Y$$
$$= \nabla_X[Y, Z] + \nabla_Y[Z, X] + \nabla_Z[X, Y] - \nabla_{[X,Y]}Z - \nabla_{[Y,Z]}X - \nabla_{[Z,X]}Y$$
$$= [X, [Y, Z]] + [Y, [Z, X]] + [Z, [X, Y]] = 0.$$

On the last line we used the Jacobi identity, and in the line above that we used the torsion-free property of the Levi-Civita connection,

By symmetry-by-pairs, it remains only to prove the second differential Bianchi identity above, for which it suffices to verify on a coordinate frame $\left\{\frac{\partial}{\partial x^i}\right\}$. In fact, we use normal coordinates at a point $p \in M$, so that $g_{ij}(p) = \pm\delta_{ij}$, and $\nabla_{\frac{\partial}{\partial x^i}}\frac{\partial}{\partial x^j}\big|_p = \Gamma_{ij}^k\big|_p\frac{\partial}{\partial x^k}\big|_p = 0$. Since $\left[\frac{\partial}{\partial x^i}, \frac{\partial}{\partial x^j}\right] = 0$, we have at the point p, where the Christoffel symbols vanish (and hence a component of a covariant derivative at the point agrees with the corresponding partial derivative):

$$R_{ijk\ell;m}(p) = \frac{\partial}{\partial x^m}\bigg|_p \left\langle R\left(\frac{\partial}{\partial x^i}, \frac{\partial}{\partial x^j}, \frac{\partial}{\partial x^k}\right), \frac{\partial}{\partial x^\ell}\right\rangle$$
$$= \left\langle\nabla_{\frac{\partial}{\partial x^m}}\left(\nabla_{\frac{\partial}{\partial x^i}}\nabla_{\frac{\partial}{\partial x^j}}\frac{\partial}{\partial x^k} - \nabla_{\frac{\partial}{\partial x^j}}\nabla_{\frac{\partial}{\partial x^i}}\frac{\partial}{\partial x^k}\right), \frac{\partial}{\partial x^\ell}\right\rangle\bigg|_p.$$

By combining terms in pairs and using that $\nabla_{\frac{\partial}{\partial x^i}} \frac{\partial}{\partial x^j}\Big|_p = 0$ for all i, j, we find

$$R_{ijk\ell;m}(p) + R_{jmk\ell;i}(p) + R_{mik\ell;j}(p)$$

$$= \left\langle \nabla_{\frac{\partial}{\partial x^m}} \left(\nabla_{\frac{\partial}{\partial x^i}} \nabla_{\frac{\partial}{\partial x^j}} \frac{\partial}{\partial x^k} - \nabla_{\frac{\partial}{\partial x^j}} \nabla_{\frac{\partial}{\partial x^i}} \frac{\partial}{\partial x^k} \right), \frac{\partial}{\partial x^\ell} \right\rangle\Big|_p$$

$$+ \left\langle \nabla_{\frac{\partial}{\partial x^i}} \left(\nabla_{\frac{\partial}{\partial x^j}} \nabla_{\frac{\partial}{\partial x^m}} \frac{\partial}{\partial x^k} - \nabla_{\frac{\partial}{\partial x^m}} \nabla_{\frac{\partial}{\partial x^j}} \frac{\partial}{\partial x^k} \right), \frac{\partial}{\partial x^\ell} \right\rangle\Big|_p$$

$$+ \left\langle \nabla_{\frac{\partial}{\partial x^j}} \left(\nabla_{\frac{\partial}{\partial x^m}} \nabla_{\frac{\partial}{\partial x^i}} \frac{\partial}{\partial x^k} - \nabla_{\frac{\partial}{\partial x^i}} \nabla_{\frac{\partial}{\partial x^m}} \frac{\partial}{\partial x^k} \right), \frac{\partial}{\partial x^\ell} \right\rangle\Big|_p$$

$$= \left\langle R\left(\frac{\partial}{\partial x^j}, \frac{\partial}{\partial x^m}, \nabla_{\frac{\partial}{\partial x^i}} \frac{\partial}{\partial x^k}\right) + R\left(\frac{\partial}{\partial x^m}, \frac{\partial}{\partial x^i}, \nabla_{\frac{\partial}{\partial x^j}} \frac{\partial}{\partial x^k}\right) + R\left(\frac{\partial}{\partial x^i}, \frac{\partial}{\partial x^j}, \nabla_{\frac{\partial}{\partial x^m}} \frac{\partial}{\partial x^k}\right), \frac{\partial}{\partial x^\ell} \right\rangle\Big|_p$$

$$= 0. \qquad \square$$

Corollary 2-2. *If (M, g) is a semi-Riemannian manifold with scalar curvature $R(g)$, then*

$$2 \operatorname{div}_g \operatorname{Ric}(g) = dR(g).$$

Proof. We use the differential Bianchi identity, along with symmetries of the curvature tensor, and the fact that $\nabla g = 0$, so that $g^{ij}_{\ ;k} = 0$ for all i, j, k:

$$\begin{aligned}
dR(g)_i &= (g^{j\ell} g^{km} R_{kj\ell m})_{;i} \\
&= g^{j\ell} g^{km}(-R_{kjmi;\ell} - R_{kji\ell;m}) \\
&= g^{j\ell} g^{km}(R_{jkmi;\ell} + R_{jki\ell;m}) \\
&= g^{j\ell} R_{ji;\ell} + g^{km} R_{ki;m} \\
&= 2(\operatorname{div}_g \operatorname{Ric}(g))_i. \qquad \square
\end{aligned}$$

With this corollary in mind, we introduce the *Einstein tensor*

$$G_\Lambda = \operatorname{Ric}(g) - \tfrac{1}{2}R(g)\, g + \Lambda g, \qquad (2.3.1)$$

where Λ is a constant, called the *cosmological constant*. Note that sometimes the Einstein tensor refers only to the case $\Lambda = 0$ above, i.e., $G = \operatorname{Ric}(g) - \tfrac{1}{2}R(g)\, g$. The Einstein tensor is divergence-free, as is any constant scalar multiple, and thus provides a candidate for the stress-energy tensor of spacetime. In fact, it is known that up to scalar multiple, G_Λ is the *only* divergence-free symmetric tensor whose coordinate expression is a function of the components $g_{\mu\nu}$ of the metric tensor, along with their first and second partial derivatives. This result was known to Cartan and Weyl in the special case that the tensor is quasilinear, and the more general result was proved by Lovelock [148, p. 322].

From this result, if the Einstein equation should be as simple as possible, and thus be second-order in the metric components, then it must take the form

$$G_\Lambda = \kappa T \tag{2.3.2}$$

for some constant κ that will be determined by the Newtonian limit, as we now show.

2.3.1. *The Newtonian limit.* Consider a spacetime metric g that is close to the Minkowski metric η, in the sense that there are coordinates in which $g_{\mu\nu} = \eta_{\mu\nu} + h_{\mu\nu}$, where $h_{\mu\nu}$ and its derivatives can be taken to be "small", and $\eta_{\mu\nu}$ takes the standard inertial form. We assume that $g_{\mu\nu,0} = 0$ (or at least $g_{\mu\nu,0} \sim 0$; see below), so that with $x^0 = ct$, the field is (approximately) time-independent in these coordinates. We let i and j run over spatial indices ($i, j \neq 0$). Let the trajectory of a slowly moving particle be modeled by a geodesic with coordinates $x^\mu(\tau)$, parametrized by proper time τ, so that $|dx^i/d\tau| \ll c\,dt/d\tau \approx c$. We will expand to first order in h (and derivatives of h) and $c^{-1}dx^i/d\tau$, and denote expressions that are equal up to terms quadratic in these quantities (with bounded coefficients) using "$\sim$". Thus $dt/d\tau \sim 1$, so that $c^{-1}dx^i/dt \sim c^{-1}dx^i/d\tau$. Since $g^{\mu\nu} = \eta^{\mu\nu} + O(h)$, we have

$$\Gamma^\mu_{00} = \tfrac{1}{2}g^{\mu\nu}(g_{\nu0,0} + g_{0\nu,0} - g_{00,\nu}) \sim -\tfrac{1}{2}\eta^{\mu\nu}h_{00,\nu}. \tag{2.3.3}$$

Since $\Gamma^\mu_{\rho\sigma} = \tfrac{1}{2}g^{\mu\nu}(h_{\nu\sigma,\rho} + h_{\rho\nu,\sigma} - h_{\rho\sigma,\nu})$, the geodesic equation becomes

$$0 = c^{-2}\left(\frac{d^2x^\mu}{d\tau^2} + \Gamma^\mu_{\rho\sigma}\frac{dx^\rho}{d\tau}\frac{dx^\sigma}{d\tau}\right)$$

$$\sim c^{-2}\left(\frac{d^2x^\mu}{d\tau^2} + \Gamma^\mu_{00}\left(\frac{dx^0}{d\tau}\right)^2\right) \sim c^{-2}\left(\frac{d^2x^\mu}{d\tau^2} + c^2\Gamma^\mu_{00}\right).$$

The time component gives, using (2.3.3),

$$\frac{d^2t}{d\tau^2} = c^{-2}\frac{d^2x^0}{d\tau^2} \sim -\Gamma^0_{00} \sim 0.$$

For the spatial components we have

$$c^{-2}\frac{d^2x^i}{d\tau^2} = c^{-2}\frac{d}{d\tau}\left(\frac{dx^i}{dt}\frac{dt}{d\tau}\right) = c^{-2}\left(\frac{d^2x^i}{dt^2}\left(\frac{dt}{d\tau}\right)^2 + \frac{dx^i}{dt}\frac{d^2t}{d\tau^2}\right) \sim c^{-2}\frac{d^2x^i}{dt^2},$$

so that

$$c^{-2}\frac{d^2x^i}{dt^2} \sim c^{-2}\frac{d^2x^i}{d\tau^2} \sim -\Gamma^i_{00} = \tfrac{1}{2}h_{00,i}.$$

If we let $\Phi = -\tfrac{1}{2}c^2h_{00}$, so that $g_{00} = -(1 + 2\Phi c^{-2})$, we recover the Newtonian relation between acceleration and the gradient of the gravitational potential

(2.1.1). We remark that this analysis can be interpreted in the case where g *is* the Minkowski metric, and $g_{\mu\nu} = \eta_{\mu\nu} + h_{\mu\nu}$ gives the components of the Minkowski metric in a (weakly) accelerating coordinate system, consistent with the equivalence principle.

We now determine κ. To do this, we consider the stress-energy for a dust model. For a perfect fluid, the pressure becomes important due to high random motion of the particles, and we are assuming our particles are slowly moving. So we are just considering the dust particles at rest in a given frame, without any pressure forces between them, each with four-velocity $\mathbb{U}$. Hence $T^{\mu\nu} = c^{-2}\rho\, \mathbb{U}^\mu \mathbb{U}^\nu$. We again consider $g_{\mu\nu} = \eta_{\mu\nu} + h_{\mu\nu}$, where $g_{\mu\nu,0} = 0$ (or at least $g_{\mu\nu,0} \sim 0$). We expand to first order in h (and its derivatives), $\mathbb{U}^i$ and ρ. Since $g_{\mu\nu}\mathbb{U}^\mu \mathbb{U}^\nu = -c^2$, we have $\operatorname{tr}_g T = g_{\mu\nu}T^{\mu\nu} = -\rho$ and $g_{00}\mathbb{U}^0 \mathbb{U}^0 \sim -c^2$, as well as

$$T_{00} = g_{\mu 0}g_{\nu 0}T^{\mu\nu} \sim g_{00}g_{00}T^{00} = (-1+h_{00})^2 c^{-2}\rho\,\mathbb{U}^0\mathbb{U}^0 \sim \rho.$$

Starting from the Einstein equation (2.3.2) with $\Lambda = 0$, $\operatorname{Ric}(g) - \frac{1}{2}R(g)\,g = \kappa T$, and taking a trace yields $R(g) = \kappa\rho$; evaluating the $(0,0)$-component of the Einstein equation then gives $R_{00} + \frac{1}{2}\kappa\rho \sim \kappa\rho$, or $R_{00} \sim \frac{1}{2}\kappa\rho$. We can also compute R_{00} in terms of Christoffel symbols: from equation (2-8a) on page 72 we have

$$\begin{aligned}
R^i{}_{j00} &= \Gamma^i{}_{00,j} - \Gamma^i{}_{j0,0} + \Gamma^i{}_{j\mu}\Gamma^\mu{}_{00} - \Gamma^i{}_{0\mu}\Gamma^\mu{}_{j0}\\
&\sim \Gamma^i{}_{00,j} = \left(\tfrac{1}{2}g^{i\mu}(g_{\mu 0,0} + g_{0\mu,0} - g_{00,\mu})\right)_{,j}\\
&\sim -\tfrac{1}{2}\eta^{i\mu}g_{00,\mu j} = -\tfrac{1}{2}\delta^{i\mu}g_{00,\mu j}
\end{aligned}$$

and so

$$\tfrac{1}{2}\kappa\rho = R_{00} = R^i{}_{i00} \sim -\tfrac{1}{2}\Delta(h_{00}) = c^{-2}\Delta\Phi.$$

To compare with the Newtonian limit, we convert the energy density to mass density, $\sigma = c^{-2}\rho$, to obtain $\Delta\Phi = \frac{1}{2}\kappa c^4\sigma$. Thus from (2.1.3) we get (in spacetime dimension four) $\frac{1}{2}\kappa c^4 = 4\pi G$, or

$$\kappa = \frac{8\pi G}{c^4}. \tag{2.3.4}$$

2.3.2. *Energy conditions*. Without the imposition of additional conditions on T, the Einstein equation does not impose any restrictions on a metric g, since the Einstein tensor is always symmetric and divergence-free. We note here some conditions often imposed on T based on physically reasonable energy considerations.

We rewrite the Einstein equation (2.3.2) as follows. From (2.3.1) and (2.3.4) we have $\operatorname{Ric}(g) - \frac{1}{2}R(g)\,g + \Lambda g = \kappa T$; take the trace to obtain (in spacetime dimension four) $-R(g) + 4\Lambda = \kappa \operatorname{tr}_g T$, so

$$\operatorname{Ric}(g) = \kappa\left(T - \tfrac{1}{2}\left(\operatorname{tr}_g T\right)g\right) + \Lambda g. \tag{2.3.5}$$

In the vacuum case $(T = 0)$ the Einstein equation becomes $\operatorname{Ric}(g) = \Lambda g$; a metric satisfying an equation of this form is an *Einstein metric*. The vacuum Einstein equation commonly refers to $\operatorname{Ric}(g) = 0$, which holds when $T = 0$ and $\Lambda = 0$.

We now introduce several conditions coming from physical notions that are sometimes imposed on T, some of which will appear in later chapters. The *weak energy condition* is that $T(\xi, \xi) \geq 0$ for all timelike ξ. If $c = 1$, say, then unit timelike vectors $\mathbb{U}$ correspond to (instantaneous) physical observers, and $T(\mathbb{U}, \mathbb{U})$ is the energy density as measured by such an observer. The *strong energy condition* is that for all unit timelike $\mathbb{U}$, $T(\mathbb{U}, \mathbb{U}) \geq -\tfrac{1}{2}\operatorname{tr}_g T$. From (2.3.5), we see this is equivalent when $\Lambda = 0$ to the *timelike convergence condition* $\operatorname{Ric}(\xi, \xi) \geq 0$ for all timelike ξ; replacing timelike ξ by null ξ, we have the *null energy condition*. In a time-oriented Lorentz manifold (i.e., if the manifold admits a smooth timelike vector field that can be used to give a smooth assignment of a future timecone in the tangent space at each point), we define the *dominant energy condition* that for all future-directed timelike ξ, the vector given by $-T^a_{\ b}\xi^b$ is future-directed causal, or in other words, for all future-directed timelike (causal) ξ and χ, we have $T(\xi, \chi) \geq 0$. The dominant energy condition clearly implies the weak energy condition.

2.3.3. *The Einstein equation in Fermi coordinates along a timelike geodesic.*
Consider the motion of particles under gravitational force with potential Φ in the Newtonian framework. We consider a family of paths $\xi(t, s)$ with coordinates $x^k(t, s)$, where s parametrizes the family of paths by, say, their initial position s along an axis. Newton's law becomes

$$\frac{\partial^2 x^k}{\partial t^2} = -\frac{\partial \Phi}{\partial x^k}(\xi(t, s)).$$

We now consider the variation vector

$$V = \frac{\partial \xi}{\partial s} = \frac{\partial x^k}{\partial s}\frac{\partial}{\partial x^k}$$

in the direction across nearby paths. It satisfies the equation ($1 \leq i, j, k \leq 3$)

$$\left(\frac{D^2 V}{dt^2}\right)^k = \frac{\partial^2}{\partial t^2}\left(\frac{\partial x^k}{\partial s}\right) = -\left(\frac{\partial^2 \Psi}{\partial x^j \partial x^k}\right)\frac{\partial x^j}{\partial s} = -\left(\frac{\partial^2 \Phi}{\partial x^j \partial x^k}\right)V^j. \tag{2.3.6}$$

This equation describes the relative motion of particles moving on nearby paths under the force of gravity. The relative motion is sometimes described in terms of *tidal forces* from non-uniformities in the gravitational field, which is consistent with what we see in the equation. The matrix governing this behavior is minus the Hessian of Φ, so its trace is $-\Delta\Phi = -4\pi G\sigma$, where σ is the mass density.

We can use this formulation as a motivation for Einstein's equation. In general relativity, the paths of observers in free fall (i.e., subject only to gravitation) are given by timelike geodesics, which are timelike paths with vanishing covariant acceleration. By the equivalence principle, experiments cannot distinguish a uniform gravitational field from uniform acceleration. It is as if locally, gravitational field effects can be either created via an accelerating frame, or effectively cancelled out (at least approximately near a given point in spacetime) by using a suitable ("freely falling") reference frame; we emphasize that we are not claiming that *any* gravitational field can be precisely generated by accelerating some inertial frame.

To be precise, if $\gamma(t)$ is a geodesic, the geodesic equation in coordinates is

$$\frac{d^2\gamma^\mu}{dt^2} = -\Gamma^\mu_{\nu\sigma}\Big|_{\gamma(t)} \frac{d\gamma^\nu}{dt}\frac{\gamma^\sigma}{dt}.$$

In normal coordinates at $p = \gamma(0)$ (or in Fermi coordinates along γ, as we discuss below), these equations reduce to $\frac{d^2\gamma^\mu}{dt^2}\big|_0 = 0$, which is analogous to the Newtonian equation with vanishing gravitational field. Unlike the components of the curvature tensor, for example, the Christoffel symbols do not form the components of a tensor field, and can be transformed away at a point by a coordinate change. At such a point in such coordinates, covariant derivatives reduce to partial derivatives, and tensor equations for physical laws take their special relativistic form as in inertial coordinates in Minkowski spacetime.

We now derive the equation governing the behavior of a family of nearby geodesics. Let $f(t, s)$ be a two-parameter map, and let $D/\partial s$ and $D/\partial t$ be the covariant derivatives along the s- and t-curves under the map f. Suppose that for each s, $\gamma_s(t) = f(t, s)$ is a geodesic. The vector field $V = \partial f/\partial s$ along f is the *variation field* for the family of geodesics. If we focus on $\gamma = \gamma_0$ and consider $V(t)$ along γ, then $V(t)$ satisfies the differential equation (2.3.7), the *Jacobi equation*.

Proposition 2-3. *Consider a family of geodesics $f(t, s)$ as above, with variation field V. Then, along $\gamma = f(\cdot, 0)$,*

$$\frac{D^2V}{dt^2} = R(\gamma'(t), V(t), \gamma'(t)). \tag{2.3.7}$$

Proof. If V is a smooth vector field defined in a neighborhood of a curve γ, then $DV/dt = \nabla_{\gamma'(t)}V$. For example, V might be a local coordinate vector field, or possibly $V(t)$ is defined along an immersed curve $\gamma(t)$, so that V can be locally extended and the identity holds (compare to the case $\gamma(t) = p_0$ is constant, while $V(t) \in T_{p_0} M$ has a nonzero derivative).

With this in mind, we find (using the symmetry of the Christoffel symbols at the last step, and with the summation convention running over all indices)

$$
\begin{aligned}
\frac{D}{\partial t} \frac{\partial f}{\partial s} &= \frac{D}{\partial t} \left(\frac{\partial f^k}{\partial s} \frac{\partial}{\partial x^k} \right) \\
&= \frac{\partial^2 f^k}{\partial t \, \partial s} \frac{\partial}{\partial x^k} + \frac{\partial f^k}{\partial s} \nabla_{\frac{\partial f}{\partial t}} \frac{\partial}{\partial x^k} \\
&= \frac{\partial^2 f^k}{\partial t \, \partial s} \frac{\partial}{\partial x^k} + \frac{\partial f^k}{\partial s} \frac{\partial f^\ell}{\partial t} \nabla_{\frac{\partial}{\partial x^\ell}} \frac{\partial}{\partial x^k} \\
&= \left(\frac{\partial^2 f^m}{\partial t \, \partial s} + \frac{\partial f^k}{\partial s} \frac{\partial f^\ell}{\partial t} \Gamma^m_{\ell k} \right) \frac{\partial}{\partial x^m} \\
&= \frac{D}{\partial s} \frac{\partial f}{\partial t}.
\end{aligned}
\tag{2.3.8}
$$

As we did not use the geodesic equation, this identity holds for general $f(t, s)$.

For a smooth vector field $W(t, s)$ along the map f, we have similarly,

$$
\begin{aligned}
\frac{D}{\partial t} \frac{DW}{\partial s} &= \frac{D}{\partial t} \left(\frac{\partial W^k}{\partial s} \frac{\partial}{\partial x^k} + W^k \nabla_{\frac{\partial f}{\partial s}} \frac{\partial}{\partial x^k} \right) \\
&= \frac{D}{\partial t} \left(\frac{\partial W^k}{\partial s} \frac{\partial}{\partial x^k} + W^k \frac{\partial f^\ell}{\partial s} \nabla_{\frac{\partial}{\partial x^\ell}} \frac{\partial}{\partial x^k} \right) \\
&= \frac{\partial^2 W^k}{\partial t \, \partial s} \frac{\partial}{\partial x^k} + \frac{\partial W^k}{\partial s} \nabla_{\frac{\partial f}{\partial t}} \frac{\partial}{\partial x^k} + \frac{\partial W^k}{\partial t} \nabla_{\frac{\partial f}{\partial s}} \frac{\partial}{\partial x^k} + W^k \frac{\partial^2 f^\ell}{\partial t \, \partial s} \nabla_{\frac{\partial}{\partial x^\ell}} \frac{\partial}{\partial x^k} \\
&\quad + W^k \frac{\partial f^\ell}{\partial s} \frac{\partial f^j}{\partial t} \nabla_{\frac{\partial}{\partial x^j}} \nabla_{\frac{\partial}{\partial x^\ell}} \frac{\partial}{\partial x^k}.
\end{aligned}
$$

Thus we see that

$$
\frac{D}{\partial t} \frac{DW}{\partial s} - \frac{D}{\partial s} \frac{DW}{\partial t} = W^k \frac{\partial f^\ell}{\partial s} \frac{\partial f^j}{\partial t} \left(\nabla_{\frac{\partial}{\partial x^j}} \nabla_{\frac{\partial}{\partial x^\ell}} \frac{\partial}{\partial x^k} - \nabla_{\frac{\partial}{\partial x^\ell}} \nabla_{\frac{\partial}{\partial x^j}} \frac{\partial}{\partial x^k} \right).
$$

Since $R(X, Y, Z) = \nabla_X \nabla_Y Z - \nabla_Y \nabla_X Z - \nabla_{[X,Y]} Z$ (from the definition of curvature) and since $[\partial/\partial x^i, \partial/\partial x^j] = 0$, we get

$$
\begin{aligned}
\frac{D}{\partial t} \frac{DW}{\partial s} - \frac{D}{\partial s} \frac{DW}{\partial t} &= W^k \frac{\partial f^\ell}{\partial s} \frac{\partial f^j}{\partial t} R\left(\frac{\partial}{\partial x^j}, \frac{\partial}{\partial x^\ell}, \frac{\partial}{\partial x^k} \right) \\
&= R\left(\frac{\partial f}{\partial t}, \frac{\partial f}{\partial s}, W \right).
\end{aligned}
\tag{2.3.9}
$$

Specializing to $W = \partial f / \partial t$, and with $\gamma'(t) = \partial f / \partial t$ along $s = 0$, we have

$$\frac{D^2 V}{\partial t^2} = \frac{D}{\partial t} \frac{D}{\partial t} \frac{\partial f}{\partial s} = \frac{D}{\partial t} \frac{D}{\partial s} \frac{\partial f}{\partial t}$$

$$= \frac{D}{\partial s} \frac{D}{\partial t} \frac{\partial f}{\partial t} + R(\gamma'(t), V(t), \gamma'(t))$$

$$= R(\gamma'(t), V(t), \gamma'(t));$$

in the last step we used that $D\gamma'(t)/\partial t = 0$ since γ is a geodesic.

We note that when f is an embedding (or an immersion, which is an embedding upon suitably restricting the domain of f), so that vector fields along f can be extended, we can see the above much more simply as follows. Since in this case we can use $D/\partial t = \nabla_{\gamma'(t)}$ and $D/\partial s = \nabla_{\partial f / \partial s} = \nabla_{V(t)}$, and since $0 = df\left([\partial/\partial s, \partial/\partial t]\right) = [\partial f/\partial s, \partial f/\partial t]$, we again have, using that $\gamma'(t) = \partial f/\partial t$ along $s = 0$,

$$\frac{D^2 V}{\partial t^2} = \frac{D}{\partial t} \frac{D}{\partial t} \frac{\partial f}{\partial s} = \frac{D}{\partial t} \frac{D}{\partial s} \frac{\partial f}{\partial t} = \nabla_{\frac{\partial f}{\partial t}} \nabla_{\frac{\partial f}{\partial s}} \frac{\partial f}{\partial t}$$

$$= \nabla_{\frac{\partial f}{\partial s}} \nabla_{\frac{\partial f}{\partial t}} \frac{\partial f}{\partial t} + R(\gamma'(t), V(t), \gamma'(t))$$

$$= R(\gamma'(t), V(t), \gamma'(t)). \qquad \square$$

We compare this to (2.3.6) to infer that the tidal acceleration relative to $\mathbb{U} = \gamma'(t)$, i.e., the acceleration $D^2(V\Delta s)/dt^2$, where $V\Delta s$ is approximately the displacement from the reference observer $\gamma(t)$ to a nearby observer, is given by $R(\mathbb{U}, V\Delta s, \mathbb{U})$, which must be akin to the (tidal) force (per unit mass). We see that the analogue of the matrix $\left(\dfrac{\partial^2 \Phi}{\partial x^j \partial x^k} \right)$ is the matrix

$$\left(R^k_{\ j\mu\nu} \frac{d\gamma^\mu}{dt} \frac{d\gamma^\nu}{dt} \right)$$

(note the index swap to match the sign), so that the analogue of $\Delta\Phi$ is obtained by tracing over j and k to obtain $\mathrm{Ric}(\gamma'(t), \gamma'(t))$, where t is proper time. Since $\mathrm{Ric}(\gamma'(t), \gamma'(t)) = 0$ for all timelike $\gamma'(t)$ implies that the Ricci curvature must vanish, we see that the Newtonian law of gravitation in vacuum is analogous to the vacuum Einstein field equation $\mathrm{Ric}(g) = 0$.

To make a further link to Newtonian theory, we now construct coordinates along a timelike geodesic adapted to the geodesic and the geometry. We use Greek letters for spacetime indices and Roman letters for spatial indices. Consider a timelike geodesic $\gamma(\tau)$, parametrized by proper time τ, with $\gamma(0) = p$. Choose an orthonormal frame $\{e_1, e_2, e_3\}$ for the orthogonal complement of $\gamma'(0)$ in $T_p M$. We parallel translate these vectors along γ to produce an orthonormal frame $\{e_1(\tau), e_2(\tau), e_3(\tau)\}$ for the orthogonal complement of $\gamma'(\tau)$ in $T_{\gamma(\tau)}M$. We define coordinates in a neighborhood of the geodesic by the map

$$\varphi(\tau, x) = \exp_{\gamma(\tau)}(x^i e_i(\tau)).$$

Since

$$d\varphi\left(\frac{\partial}{\partial \tau}\Big|_{(\tau,0)}\right) = \gamma'(\tau) \quad \text{and} \quad d\varphi\left(\frac{\partial}{\partial x^i}\Big|_{(\tau,0)}\right) = e_i(\tau),$$

φ defines a coordinate system, called *Fermi coordinates*, in a neighborhood of γ. For index purposes, we rescale the time coordinate to $x^0 = c\tau$, and refer to the coordinates x^μ as Fermi coordinates. Note that the coordinates of $\gamma(\tau) = \varphi(\tau, 0)$ are $\gamma^0(\tau) = c\tau$, while $\gamma^j(\tau) = 0$, and we let $\Gamma_{\nu\sigma}^\mu(c\tau, x) = \Gamma_{\nu\sigma}^\mu|_{\varphi(\tau,x)}$.

It is clear by construction that the metric components along γ in this coordinate system agree with the components of the Minkowski metric in inertial coordinates. We now establish a lemma regarding the behavior of the Christoffel symbols along γ.

Lemma 2-4. *For all* $\mu, \nu, \sigma \in \{0, 1, 2, 3\}$, $\Gamma_{\nu\sigma}^\mu(c\tau, 0) = 0$.

Proof. For any (τ, b), with $b = (b^1, b^2, b^3)$, consider the curve defined via the exponential map as $\beta(s) = \exp_{\gamma(\tau)}(sb^i e_i(\tau)) = \varphi(\tau, sb)$. Let $b^0 = 0$. Then $\beta^0(s) = c\tau$ and $\beta^k(s) = sb^k$ for $k = 1, 2, 3$. By definition, β is a geodesic. Since $d^2\beta^\mu/ds^2 = 0$ for $\mu = 0, 1, 2, 3$, we have

$$0 = \Gamma_{\nu\sigma}^\mu(c\tau, sb)\frac{d\beta^\nu}{ds}\frac{d\beta^\sigma}{ds} = \Gamma_{\nu\sigma}^\mu(c\tau, sb)b^\nu b^\sigma = \Gamma_{ij}^\mu(c\tau, sb)b^i b^j.$$

Since at $s = 0$, we can consider β defined by an arbitrary $b \in \mathbb{R}^3$, we have $\Gamma_{ij}^\mu(c\tau, 0) = 0$ for $0 \le \mu \le 3$ and $1 \le i, j \le 3$. Similarly, the geodesic equation for γ yields $\Gamma_{00}^\mu(c\tau, 0) = 0$ for $0 \le \mu \le 3$.

To get the other Christoffel symbols, we use the equations for parallel transport. Along γ, the x^ℓ-coordinate vector is e_ℓ for $\ell = 1, 2, 3$, so the components of e_ℓ along γ are $e_\ell^\mu(\tau) = \delta_\ell^\mu$. The parallel transport equations are then given by

$$0 = \frac{d}{d\tau}(e_\ell^\mu(\tau)) + \Gamma_{\nu\sigma}^\mu(c\tau, 0)e_\ell^\nu(\tau)\frac{d\gamma^\sigma}{d\tau} = \Gamma_{\nu\sigma}^\mu(c\tau, 0)\delta_\ell^\nu\frac{d\gamma^\sigma}{d\tau} = c\Gamma_{\ell 0}^\mu(c\tau, 0),$$

where we used the fact that $\gamma^0(\tau) = c\tau$ and $\gamma^j(\tau) = 0$ for $j = 1, 2, 3$. $\square$

Lemma 2-5. *Along* γ, *we have* $R_{\nu 00}^\mu = \partial\Gamma_{00}^\mu/\partial x^\nu$, *so that* $R_{j00}^k = -\frac{1}{2}g_{00,kj}$ *for* $j, k \in \{1, 2, 3\}$, *while* $R_{j00}^0 = 0$.

Proof. Along γ the Christoffel symbols vanish and hence $\partial\Gamma^{\mu}_{\nu\sigma}/\partial\tau = 0$, so that $R^{\mu}_{000} = 0 = \partial\Gamma^{\mu}_{00}/\partial x^0$ along γ, while for $j = 1, 2, 3$,

$$
\begin{aligned}
R^{\mu}_{j00}\frac{\partial}{\partial x^{\mu}} &= \nabla_{e_j(\tau)}\nabla_{\frac{\partial}{\partial x^0}}\frac{\partial}{\partial x^0} - \nabla_{c^{-1}\gamma'(\tau)}\nabla_{\frac{\partial}{\partial x^j}}\frac{\partial}{\partial x^0} \\
&= \nabla_{e_j(\tau)}\left(\Gamma^{\mu}_{00}\frac{\partial}{\partial x^{\mu}}\right) - c^{-1}\nabla_{\gamma'(\tau)}\left(\Gamma^{\mu}_{j0}\frac{\partial}{\partial x^{\mu}}\right) \\
&= \frac{\partial\Gamma^{\mu}_{00}}{\partial x^j}\frac{\partial}{\partial x^{\mu}}.
\end{aligned}
$$

Moreover, since the Christoffel symbols vanish along γ, so do the first partials of $g_{\mu\nu}$ and $g^{\mu\nu}$. Thus, along γ, we have

$$
\frac{\partial\Gamma^{k}_{00}}{\partial x^j} = \frac{\partial}{\partial x^j}\left(\tfrac{1}{2}g^{k\sigma}(2g_{0\sigma,0} - g_{00,\sigma})\right) = -\tfrac{1}{2}\delta^{km}g_{00,mj} = -\tfrac{1}{2}g_{00,kj}.
$$

Similarly, $\partial\Gamma^{0}_{00}/\partial x^j = -\tfrac{1}{2}g_{00,0j} = 0$. $\qquad\qquad\square$

Now, as we noted above, the analogue of the matrix $\left(\dfrac{\partial^2\Phi}{\partial x^j\partial x^k}\right)$ is

$$
c^2 R^{k}_{j00} = -\tfrac{1}{2}c^2 g_{00,jk}.
$$

Thus the analogue of the gravitational potential Φ is $-\tfrac{1}{2}c^2 g_{00}$, which is equivalent to what we had earlier (since Φ is defined only up to an additive constant). The analogue of its Laplacian $\Delta\Phi$ is then $\mathrm{Ric}(\gamma'(\tau), \gamma'(\tau)) = c^2 R_{00}$, as we had before. Our analysis leads us again to propose that the Ricci curvature should be related to the matter density. Of course, the mass-energy density is not an invariant object, but the stress-energy tensor is. We can argue as in our earlier analysis how to get from here to the Einstein equation.

2.3.4. *Variational formulation.* We now consider a Lagrangian variational formulation for the Einstein equation, as first derived by Hilbert around the time Einstein proposed the equation to model gravitation. Consider the *Einstein–Hilbert action*, or total scalar curvature functional, $\mathcal{R}(g) = \int_M R(g)\,dv_g$ (or the related action $\mathcal{S}(g) = \int_M \frac{1}{2\kappa}R(g)\,dv_g$; cf. (2.3.15)), where, in local coordinates, $dv_g = \sqrt{|\det(g_{ij})|}\,dx$. (The definition holds for semi-Riemannian metrics g, and thus includes both the Lorentzian and Riemannian cases.) We assume that M is compact, or more generally that $R(g) \in L^1(M, dv_g)$. We want to compute the first variation of $\mathcal{R}$ and the associated Euler–Lagrange equation.

We will need *Cramer's rule*. If $A = (A_{ij})$ is an $n \times n$ matrix, we let M_{ij} be the determinant of the $(n-1) \times (n-1)$ *minor* obtained by deleting row i and column j of A. The determinant of A is given by column or row expansion:

$$\det A = \sum_{i=1}^{n} (-1)^{i+j} M_{ij} A_{ij} = \sum_{j=1}^{n} (-1)^{i+j} M_{ij} A_{ij}.$$

The $n \times n$ matrix with entries $\mathrm{cof}(A)_{ij} = (-1)^{i+j} M_{ij}$ is known as the *cofactor matrix*, and its transpose, $A^{\mathrm{adj}} = (\mathrm{cof}(A))^T$, is called the *Cramer's rule adjoint*, so $A^{\mathrm{adj}}_{ij} = (-1)^{i+j} M_{ji} = \mathrm{cof}(A)_{ji}$. Thus for any $j \in \{1, 2, \ldots, n\}$ we have $\det A = (A^{\mathrm{adj}} \cdot A)_{jj}$. For $i \neq j$, we have $(A^{\mathrm{adj}} \cdot A)_{ij} = \sum_{k=1}^{n} (-1)^{i+k} M_{ki} A_{kj} = 0$: indeed, we can interpret this sum as the determinant of the matrix $\tilde{A}$ obtained by replacing column i of A by column j of A. The minors M_{ki} are obtained by crossing out column i, so M_{ki} are the same for A and $\tilde{A}$. But $\det \tilde{A} = 0$ since $\tilde{A}$ has two equal columns. In summary we arrive at *Cramer's rule*: If I_n is the $n \times n$ identity matrix, then $A^{\mathrm{adj}} \cdot A = (\det A) I_n$. If A is invertible, then

$$A^{-1} = \frac{1}{\det A} A^{\mathrm{adj}}.$$

Now we turn to variational formulae.

Lemma 2-6. *If $A(t)$ is a smooth path of $n \times n$ matrices,*

$$\frac{d}{dt} \det A(t) = \mathrm{tr}(A^{\mathrm{adj}}(t) \cdot A'(t)).$$

If $A(t)$ is invertible,

$$\frac{d}{dt} \log |\det A(t)| = \mathrm{tr}(A^{-1}(t) \cdot A'(t))$$

and

$$\frac{d}{dt} \sqrt{|\det A(t)|} = \tfrac{1}{2} \sqrt{|\det A(t)|} \, \mathrm{tr}(A^{-1}(t) \cdot A'(t)). \qquad (2.3.10)$$

Proof. If we consider $\det A$ as a function of $(A_{ij}) \in \mathbb{R}^{n^2}$, then

$$\frac{\partial \det A}{\partial A_{ij}} = (-1)^{i+j} M_{ij} = A^{\mathrm{adj}}_{ji}.$$

By the chain rule,

$$\frac{d}{dt} \det A(t) = \sum_{i,j=1}^{n} \frac{\partial \det A}{\partial A_{ij}} \cdot \frac{\partial A_{ij}}{\partial t}.$$

Together with the preceding equation, this establishes the lemma. $\square$

We need one more variational formula, for the scalar curvature, to compute the Euler–Lagrange equation for the Einstein–Hilbert action. For a symmetric $(0, 2)$-tensor field h, at any point, $g_t := g + th$ is a metric for $|t|$ small enough on any compact subset, or on all of M if h has compact support. We define $L_g h := \frac{d}{dt}\big|_{t=0} R(g + th)$, which may be computed locally, so we may assume in the computation that $|t|$ is small enough that g_t is indeed a metric.

Lemma 2-7. *For a symmetric* $(0, 2)$-*tensor field* h,

$$L_g h := -\Delta_g (\mathrm{tr}_g h) + \mathrm{div}_g \, \mathrm{div}_g \, h - h \cdot \mathrm{Ric}(g) \qquad (2.3.11)$$

where $h \cdot \mathrm{Ric}(g) = g^{ik} g^{j\ell} h_{ij} R_{k\ell}$.

The proof can be carried out in a straightforward computation. For instance, one might compute in normal coordinates at a point, and it might be useful to employ the fact that if $\Gamma_{ij}^k (t)$ are the Christoffel symbols (which do not comprise components for a local tensor field) for $g_t = g + th$ in a coordinate chart, then $\delta\Gamma_{ij}^k := \frac{d}{dt}\big|_{t=0}\Gamma_{ij}^k (t)$ in fact *do* give the components for a local *tensor* field; this follows from the fact that the *difference* of two connections is tensorial, so that if ∇^g is the Levi-Civita connection for g, then $S(X, Y) := \nabla_X^{g_t} Y - \nabla_X^g Y$ is tensorial (Exercise 1-6).

Exercise 2-8. Prove Lemma 2-7. You might start by deriving the local coordinate formula for the curvature,

$$R_{kij}^\ell = \Gamma_{ij,k}^\ell - \Gamma_{jk,i}^\ell + \Gamma_{km}^\ell \Gamma_{ij}^m - \Gamma_{jk}^m \Gamma_{im}^\ell, \qquad (2\text{-}8\text{a})$$

and thus for the Ricci and scalar curvatures:

$$R(g) = g^{ij} R_{ij} = g^{ij}\left(\Gamma_{ij,k}^k - \Gamma_{ik,j}^k + \Gamma_{km}^k \Gamma_{ij}^m - \Gamma_{jk}^m \Gamma_{im}^k\right). \qquad (2\text{-}8\text{b})$$

You might then argue that $\delta\Gamma_{ij}^k = \frac{1}{2} g^{km}(h_{mj;i} + h_{im;j} - h_{ij;m})$, and observe that the variation of the Ricci tensor is given by $\frac{d}{dt}\big|_{t=0} R_{ij} = (\delta\Gamma)_{ij;k}^k - (\delta\Gamma)_{ik;j}^k$.

We now derive the Euler–Lagrange equation for the Einstein–Hilbert action. We will vary the metric g in the direction of a symmetric $(0, 2)$-tensor h. We will take h to be compactly supported, so we can make sense out of the variation for a given h even in case $R(g)$ fails to be integrable, by integrating only over the support of h, where the metric g is actually changing.

Theorem 2-9. *The first variation of the Einstein–Hilbert action* (*total scalar curvature functional*) $\mathcal{R}$ *is given by*

$$\frac{d}{dt}\bigg|_{t=0} \mathcal{R}(g + th) = -\int_M h \cdot \left(\mathrm{Ric}(g) - \tfrac{1}{2} R(g) g\right) dv_g$$

for all compactly supported tensors h (*vanishing near the boundary* ∂M *if* ∂M *is nonempty*). *Thus the Euler–Lagrange equation is* $\mathrm{Ric}(g) - \frac{1}{2}R(g)g = 0$. *This equation is satisfied on all two-dimensional manifolds* (M, g). *For* $n = \dim M \geq 3$, *the Euler–Lagrange equation is equivalent to* $\mathrm{Ric}(g) = 0$.

Proof. In local coordinates x, we have $dv_g = \sqrt{|\det(g_{ij})|}\,dx$ (which we can treat as a measure or locally as an n-form). Hence, from Lemma 2-7, the symmetry of g (or h), and the fact that $g^{ij}h_{ij} = \mathrm{tr}_g h$, we have, for $g_t = g + th$,

$$\left.\frac{d}{dt}\right|_{t=0} dv_{g_t} = \tfrac{1}{2}g^{ij}h_{ij}dv_g = \tfrac{1}{2}(\mathrm{tr}_g h)\,dv_g. \tag{2.3.12}$$

Integrating by parts and observing that boundary terms vanish by the choice of h, we obtain

$$\left.\frac{d}{dt}\right|_{t=0}\mathcal{R}(g+th)$$

$$= \int_M L_g h\,dv_g + \int_M R(g)\cdot\tfrac{1}{2}(\mathrm{tr}_g h)\,dv_g$$

$$= \int_M \left((-\Delta_g(\mathrm{tr}_g h) + \mathrm{div}_g\,\mathrm{div}_g h - h\cdot\mathrm{Ric}(g)) + \tfrac{1}{2}R(g)g\cdot h\right)dv_g$$

$$= -\int_M h\cdot\left(\mathrm{Ric}(g) - \tfrac{1}{2}R(g)g\right)dv_g.$$

If M is closed, we let $h = \mathrm{Ric}(g) - \tfrac{1}{2}R(g)g$ to finish the proof. In any case, the preceding equation holds for all h in a dense subset of $L^2(M, dv_g)$, so we see that we must have $\mathrm{Ric}(g) - \tfrac{1}{2}R(g)g = 0$.

If M is two-dimensional and $\{e_1, e_2\}$ is an orthonormal basis of T_pM, say $\langle e_1, e_2\rangle = 0$, $\langle e_1, e_1\rangle = \epsilon_1 = \pm 1$, and $\langle e_2, e_2\rangle = \epsilon_2 = 1$, then

$$\begin{aligned}
\mathrm{Ric}_g(e_1, e_j) &= \epsilon_1\langle R(e_1, e_1, e_j), e_1\rangle + \epsilon_2\langle R(e_2, e_1, e_j), e_2\rangle \\
&= \langle R(e_2, e_1, e_1), e_2\rangle\delta_{1j} \\
&= \epsilon_1\langle R(e_2, e_1, e_1), e_2\rangle g(e_1, e_j), \\
\mathrm{Ric}_g(e_2, e_j) &= \epsilon_1\langle R(e_2, e_1, e_1), e_2\rangle\delta_{2j} \\
&= \epsilon_1\langle R(e_2, e_1, e_1), e_2\rangle g(e_2, e_j).
\end{aligned}$$

Hence the desired equation follows from

$$R(g) = \epsilon_1\mathrm{Ric}_g(e_1, e_1) + \epsilon_2\mathrm{Ric}_g(e_2, e_2) = 2\epsilon_1\langle R(e_2, e_1, e_1), e_2\rangle.$$

If M has higher dimension, we trace the Euler–Lagrange equation to obtain $R(g) - \tfrac{1}{2}(\dim M)R(g) = 0$, or $R(g) = 0$. Thus $\mathrm{Ric}(g) = 0$ in this case; the converse is trivial. $\square$

Definition 2-10. A semi-Riemannian manifold (M^n, g) is called an *Einstein manifold* if $\mathrm{Ric}(g) = Cg$ for some constant C. In this case the scalar curvature is constant: $R(g) = nC$.

Exercise 2-11. Consider an Einstein semi-Riemannian manifold (M^n, g). In the case $n = 2$, the sectional curvature is clearly constant. Prove this as well for $n = 3$. (Hint: consider using an orthonormal frame $\{E_1, E_2, E_3\}$ for $T_p M$, mindful of the signature $\epsilon_i = \langle E_i, E_i \rangle$.)

Lemma 2-12. *Consider (M^n, g) semi-Riemannian with $n \geq 3$. If $\mathrm{Ric}(g) = fg$ for some function f, then f is (locally) constant.*

Exercise 2-13. Prove the lemma. You might recall the Bianchi identity; compare Corollary 2-2.

For any constant Λ, we can also consider $\mathcal{R}_\Lambda(g) = \int_M (R(g) - 2\Lambda)\, dv_g$. The formula (2.3.12) for the variation of the volume element dv_g easily gives this:

Corollary 2-14. *For all compactly supported tensors h (vanishing near ∂M if ∂M is nonempty), we have*

$$\frac{d}{dt}\Big|_{t=0} \mathcal{R}_\Lambda(g + th) = -\int_M h \cdot \left(\mathrm{Ric}(g) - \tfrac{1}{2}R(g)g + \Lambda g\right) dv_g.$$

Thus the Euler–Lagrange equation for $\mathcal{R}_\Lambda$ is

$$\mathrm{Ric}(g) - \tfrac{1}{2}R(g)g + \Lambda g = 0.$$

Suppose that g solves the Euler–Lagrange equation. Then $\Lambda = 0$ if M has dimension 2. For $\dim M = n \geq 3$, we have

$$R(g) = \frac{2n}{n-2}\Lambda,$$

that is, the metric g is Einstein, $\mathrm{Ric}(g) = Cg$, with $C = \dfrac{2\Lambda}{n-2}$.

Definition 2-15. The *Einstein tensor* is given by $G = G(g) = \mathrm{Ric}(g) - \tfrac{1}{2}R(g)g$. For Λ constant, we also define the related tensor $G_\Lambda(g) = \mathrm{Ric}(g) - \tfrac{1}{2}R(g)g + \Lambda g$. The *vacuum Einstein equation* (with cosmological constant Λ) is given by $G_\Lambda = 0$. If $\Lambda = 0$ and $n > 2$, this is equivalent to $\mathrm{Ric}(g) = 0$.

2.3.4.1. *Lagrangian formulation with matter fields.* We now discuss how to include matter fields into the variational formulation. We assume the matter fields are given in terms of a collection Ψ of tensor fields (including scalars, vectors, etc.), which we assume are *independent* of the metric g and are governed by an action $\int_M \mathcal{L}_m(g, \Psi)\, dv_g$. Note that $\mathcal{L}_m$ can also depend on derivatives of the fields in Ψ. If h is a symmetric $(0, 2)$-tensor and Φ is a collection of tensor fields representing a direction of variation of Ψ, we write

$$\frac{d}{dt}\Big|_{t=0} \mathcal{L}_m(g+th, \Psi) =: (D_1\mathcal{L}_m)_{(g,\Psi)}(h, 0),$$

$$\frac{d}{dt}\Big|_{t=0} \mathcal{L}_m(g, \Psi+t\Phi) =: (D_2\mathcal{L}_m)_{(g,\Psi)}(0, \Phi).$$

We remark that it is sometimes useful, especially when varying the metric, to express the action though the Lagrangian density, $\mathcal{L}_m \cdot \sqrt{|\det g|}$.

The fields Ψ must satisfy certain equations for the action to be stationary at (g, Ψ): for any direction of variation Φ (compactly supported and vanishing near the boundary, if nonempty) of the fields, we must have

$$0 = \frac{d}{dt}\Big|_{t=0} \int_M \mathcal{L}_m(g, \Psi+t\Phi)\,dv_g = \int_M (D_2\mathcal{L}_m)_{(g,\Psi)}(0, \Phi)\,dv_g. \qquad (2.3.13)$$

From here we can derive field equations (sometimes called *equations of motion*) for the matter fields. Instead of introducing new notation, we illustrate with examples.

Example 2-16. Consider $g - \eta$, the Minkowski metric on $\mathbb{M}^{1+k} = \mathbb{R}_1^{1+k}$, and work with global inertial coordinates (t, x^i). Then for any smooth function ψ,

$$\eta(d\psi, d\psi) = -c^{-2}\left(\frac{\partial\psi}{\partial t}\right)^2 + \sum_{i=1}^{k}\left(\frac{\partial\psi}{\partial x^i}\right)^2.$$

If the action integrand is $\mathcal{L}_m = -\frac{1}{2}\eta(d\psi, d\psi)$, then for any smooth compactly supported φ, the field equation is obtained from

$$0 = \frac{d}{dt}\Big|_{t=0} \int_{\mathbb{M}} -\tfrac{1}{2}\eta(d\psi + t\,d\varphi, d\psi + t\,d\varphi)\,dt\,dx^1 \cdots dx^k$$

$$= \int_{\mathbb{M}} -\eta(d\psi, d\varphi)\,dt\,dx^1 \cdots dx^k = \int_{\mathbb{M}} \varphi\,\Box\psi\,dt\,dx^1 \cdots dx^k$$

where

$$\Box - -c^{-2}\left(\frac{\partial}{\partial t}\right)^2 + \sum_{i=1}^{k}\left(\frac{\partial}{\partial x^i}\right)^2$$

is the *wave operator* in Minkowski spacetime. Since this must hold for all appropriate φ, the field equation is the *wave equation* $\Box\psi = 0$.

Exercise 2-17. If we let $\mathcal{L}_m = -\frac{1}{2}\eta(d\psi, d\psi) - V(\psi)$, where V is a smooth function of one variable, show that the resulting field equation is $\Box\psi = V'(\psi)$. In case $V(\psi) = \frac{1}{2}m^2\psi^2$, the resulting equation $\Box\psi - m^2\psi = 0$ is called the *Klein–Gordon equation* (with $c = 1$, $\hbar = 1$).

We remark that the Lagrangian density in the Einstein–Hilbert action for the gravitational field involves second-order derivatives of the metric, as does the corresponding Euler–Lagrange equation (the Einstein equation), whereas for the preceding two examples, the Lagrangian density is first-order in ψ, while the Euler–Lagrange equation is second-order in ψ.

Example 2-18. For the source-free electromagnetic field, one has

$$\mathcal{L}_m = -\frac{1}{16\pi} F_{ab} F_{cd} g^{ac} g^{bd},$$

where $F_{ab} = (dA)_{ab}$ for a one-form A, the *potential*. If φ is a direction of variation of the potential, then the stationarity of the action requires (we use the antisymmetry of F and Exercise 1-6)

$$
\begin{aligned}
0 &= \int_M -\frac{1}{8\pi} \langle d\varphi, F \rangle_g \, dv_g = \int_M -\frac{1}{8\pi} (\varphi_{b;a} - \varphi_{a;b}) F_{cd} g^{ac} g^{bd} \, dv_g \\
&= \int_M \frac{1}{4\pi} \varphi_{a;b} F_{cd} g^{ac} g^{bd} \, dv_g \\
&= \int_M -\frac{1}{4\pi} \varphi_a (F_{cd;b} g^{bd}) g^{ac} \, dv_g.
\end{aligned}
$$

In other words, we obtain the equation $\mathrm{div}_g F = 0$, where div_g is the *spacetime* divergence $F_{cd;b} g^{bd}$, which forms part of Maxwell's equations. The other part comes from the fact that F_{ab} is a closed two-form.

We now define the *stress-energy tensor T* corresponding to matter fields given by a Lagrangian as above, via the following equation identifying T as a symmetric $(0, 2)$-tensor via an integral pairing:

$$
\begin{aligned}
\frac{d}{dt}\bigg|_{t=0} \int_M \mathcal{L}_m(g + th, \Psi) \, dv_{g+th} & \\
&= \int_M \left((D_1 \mathcal{L}_m)_{(g,\Psi)}(h, 0) + \mathcal{L}_m(g, \Psi) \cdot \tfrac{1}{2} \mathrm{tr}_g h \right) dv_g \\
&=: \frac{1}{2} \int_M \langle h, T \rangle_g \, dv_g.
\end{aligned}
\tag{2.3.14}
$$

Example 2-19. Consider a Klein–Gordon field $\mathcal{L}_m = -\tfrac{1}{2}\eta(d\psi, d\psi) - \tfrac{1}{2}m^2\psi^2$ at the Minkowski metric on $\mathbb{M}^{1+k}$. The stress tensor T satisfies

$$
\begin{aligned}
\int_{\mathbb{M}} \eta(h, T) \, dv_\eta & \\
&= 2 \int_{\mathbb{M}} \left(\tfrac{1}{2}\eta^{ij} h_{js} \eta^{s\ell} \psi_{;i}\psi_{;\ell} - \left(\tfrac{1}{2}\eta(d\psi, d\psi) + \tfrac{1}{2}m^2\psi^2 \right) \tfrac{1}{2} h_{js} \eta^{js} \right) dv_\eta.
\end{aligned}
$$

Thus we see $T_{ab} = \psi_{;a}\psi_{;b} - \frac{1}{2}(\eta(d\psi, d\psi) + m^2\psi^2)\eta_{ab}$. There is an analogue on any spacetime (M, g): $T_{ab} = \psi_{;a}\psi_{;b} - \frac{1}{2}(\psi_{;c}\psi_{;d}g^{cd} + m^2\psi^2)g_{ab}$.

Example 2-20. For the source-free Maxwell field $\mathcal{L}_m = -\dfrac{1}{16\pi}F_{ab}F_{cd}g^{ac}g^{bd}$, the method above yields

$$\int_M \langle h, T\rangle_g \, dv_g$$

$$= \int_M -\frac{1}{8\pi}F_{ab}F_{cd}\left(-g^{ai}h_{ij}g^{jc}g^{bd} - g^{ac}g^{bi}h_{ij}g^{jd} + \frac{1}{2}h_{ij}g^{ij}g^{ac}g^{bd}\right)dv_g$$

$$= \int_M \frac{1}{8\pi}h_{ij}F_{ab}F_{cd}\left(g^{ai}g^{jc}g^{bd} + g^{ac}g^{bi}g^{jd} - \frac{1}{2}g^{ac}g^{bd}g^{ij}\right)dv_g$$

$$= \int_M h_{ij}\frac{1}{4\pi}F_{ab}F_{cd}\left(g^{bd}g^{ai}g^{jc} - \frac{1}{4}g^{ac}g^{bd}g^{ij}\right)dv_g$$

where in the last step we used the antisymmetry of F_{ab}. Thus

$$T_{ab} = \frac{1}{4\pi}\left(F_{ac}F_{bd}g^{cd} - \frac{1}{4}F_{ij}F_{s\ell}g^{is}g^{j\ell}g_{ab}\right).$$

One can derive the Einstein equation starting from the ansatz that the action for the combination of gravitation with the fields is simply the sum of a constant multiple of $\mathcal{R}_\Lambda$ with the action for the matter. In particular, we are assuming that the fields do not themselves appear in the Lagrangian for the gravitational field. With κ a positive constant as above (where $\kappa = 8\pi G/c^4$ in spacetime dimension four by the Newtonian limit), we consider the action

$$S(g, \Psi) = \int_M \left(\frac{1}{2\kappa}(R(g) - 2\Lambda) + \mathcal{L}_m(g, \Psi)\right)dv_g. \qquad (2.3.15)$$

An important consideration in the earlier derivation of the Einstein tensor is that the stress-energy tensor is divergence-free. This can be derived from the variational definition above, along with *diffeomorphism invariance* of the action. The Einstein–Hilbert action $\mathcal{R}_\Lambda$ is clearly diffeomorphism invariant: if ϕ is a diffeomorphism of M, then $\mathcal{R}_\Lambda(g) = \mathcal{R}_\Lambda(\phi^*g)$. Thus the diffeomorphism invariance of $S(g, \Psi)$ is equivalent to the condition that the action for the matter, $\mathcal{R}_m(g, \Psi) = \int_M \mathcal{L}_m(g, \Psi)\, dv_g$, is diffeomorphism invariant.

Now, suppose that X is a smooth vector field on M, which generates a local one-parameter subgroup of diffeomorphisms ϕ_t. Recall that $\frac{d}{dt}\big|_{t=0}\phi_t^*g = L_Xg$, where $(L_Xg)_{ij} = X_{i;j} + X_{j;i}$.

Exercise 2-21. Suppose X is compactly supported away from the boundary, and let $h = L_Xg$. Show directly from the formula for the variation of the Einstein–Hilbert action and the Bianchi identities that $\frac{d}{dt}\big|_{t=0}\mathcal{R}(g + th) = 0$.

For X compactly supported away from the boundary with one-parameter group of diffeomorphisms ϕ_t, we then have by using the preceding exercise, along with (2.3.13) and (2.3.14), the symmetry of T and the diffeomorphism invariance of the action,

$$
\begin{aligned}
0 &= \frac{d}{dt}\Big|_{t=0} \int_M \mathcal{L}_m(\phi_t^* g, \phi_t^* \Psi)\, dv_{\phi_t^* g} \\
&= \frac{1}{2}\int_M \langle L_X g, T\rangle_g\, dv_g = \int_M X_{i;j} T^{ij}\, dv_g = -\int_M X_i T^{ij}_{\;;j}\, dv_g.
\end{aligned}
$$

Since this holds for any compactly supported X, we see that T must be divergence-free as desired.

2.3.4.2. *Related variational problems.* Let M be a closed (compact without boundary) manifold of dimension $n \geq 3$. We want to characterize critical points of $\mathcal{R}$ constrained to metrics of fixed volume, say $\operatorname{Vol} g = 1$. We let $\mathcal{M}$ be the set of (smooth) metrics, and let $\mathcal{M}_1 \subset \mathcal{M}$ be the subset of unit volume metrics. Note that if $g \in \mathcal{M}$, and if $C > 0$ is a constant, then $dv_{(Cg)} = C^{n/2}dv_g$, and so $\operatorname{Vol}(Cg) = C^{n/2}\operatorname{Vol} g$. Thus $(\operatorname{Vol} g)^{-2/n}g \in \mathcal{M}_1$. We also note that $R(Cg) = C^{-1}R(g)$.

We can consider $\mathcal{R}$ on $\mathcal{M}_1$, or by rescaling $g \in \mathcal{M}$ to $\bar{g} = (\operatorname{Vol} g)^{-2/n}g \in \mathcal{M}_1$, we can consider the functional

$$
\overline{\mathcal{R}}(g) = \mathcal{R}(\bar{g}) = \frac{\mathcal{R}(g)}{(\operatorname{Vol} g)^{1-\frac{2}{n}}}.
$$

Proposition 2-22. *A metric $g \in \mathcal{M}$ is critical for $\overline{\mathcal{R}}$ if and only if g is Einstein. Thus $g \in \mathcal{M}_1$ is critical for $\mathcal{R}$ restricted to $\mathcal{M}_1$ if and only if g is Einstein.*

Proof. Let $g_t = g + th$, and let $\bar{g}_t = (\operatorname{Vol} g_t)^{-2/n}g_t \in \mathcal{M}_1$. Then $\overline{\mathcal{R}}(g_t) = \mathcal{R}(\bar{g}_t) = \mathcal{R}(g_t)/(\operatorname{Vol} g_t)^{1-\frac{2}{n}}$. We compute, using $\frac{d}{dt}\big|_{t=0} \operatorname{Vol} g_t = \int_M \frac{1}{2}(\operatorname{tr}_g h)\, dv_g$,

$$
\begin{aligned}
\frac{d}{dt}\Big|_{t=0}&\mathcal{R}(\bar{g}_t) \\
&= -\frac{\int_M h \cdot (\operatorname{Ric}(g) - \frac{1}{2}R(g)g)\, dv_g}{(\operatorname{Vol} g)^{1-\frac{2}{n}}} - \left(1 - \frac{2}{n}\right)\frac{\mathcal{R}(g)}{(\operatorname{Vol} g)^{2-\frac{2}{n}}}\int_M \frac{1}{2}(\operatorname{tr}_g h)\, dv_g \\
&= -\frac{1}{(\operatorname{Vol} g)^{1-\frac{2}{n}}}\int_M h \cdot \left(\operatorname{Ric}(g) - \frac{1}{2}R(g)g + \frac{n-2}{2n}(\operatorname{Vol} g)^{-1}\mathcal{R}(g)g\right) dv_g.
\end{aligned}
$$

For this to vanish for all symmetric $(0, 2)$-tensors h, we must have

$$
\operatorname{Ric}(g) - \frac{1}{2}R(g)g + \frac{n-2}{2n}(\operatorname{Vol} g)^{-1}\mathcal{R}(g)g = 0.
$$

Taking the trace yields

$$\frac{n-2}{2}\big(R(g) - (\text{Vol } g)^{-1}\mathcal{R}(g)\big) = 0.$$

Hence $R(g) = (\text{Vol } g)^{-1}\mathcal{R}(g)$, a constant. From here, we clearly have

$$\text{Ric}(g) = \frac{R(g)}{n}g,$$

and $R(g)$ is constant. $\qquad\square$

We note that while the unconstrained problem has critical points characterized by $\text{Ric}(g) = 0$, the constrained problem has a more general critical point equation, which can be interpreted as a Lagrange multiplier condition. If we interpret the gradient (in an L^2 integral sense) of $\mathcal{R}$ as $-\big(\text{Ric}(g) - \frac{1}{2}R(g)g\big)$, the gradient of the volume functional would be $\frac{1}{2}g$, and a Lagrange multiplier condition would be of the form $-(\text{Ric}(g) - \frac{1}{2}R(g)g) = \lambda \cdot \frac{1}{2}g$.

Remark 2-23. Einstein metrics arise as stationary points of *normalized Ricci flow*. From (2.3.12), we see that a solution $g = g_t$ to the normalized Ricci flow

$$\frac{\partial g}{\partial t} = -2\,\text{Ric}(g) + \frac{2}{n}\frac{\mathcal{R}(g)}{\text{Vol } g}g = -2\left(\text{Ric}(g) - \frac{1}{n}\frac{\mathcal{R}(g)}{\text{Vol } g}g\right)$$

has volume $\text{Vol } g_t$ constant.

Exercise 2-24. Suppose $g = g_t$ is a one-parameter family of metrics satisfying the Ricci flow $\partial g/\partial t = -2\text{Ric}(g)$ on a closed manifold M. Let $V_t = \text{Vol } g_t$, and let $\tilde{g} = \tilde{g}_t$ be the rescaled metric $\tilde{g}_t = V_t^{-2/n}g_t$. Note that $\tilde{g}$ has unit volume. Let $\tilde{t} = \tilde{t}(t)$ be a rescaled time function determined by $d\tilde{t}/dt = V_t^{-2/n}$, with $\tilde{t}(0) = 0$. Show that $\tilde{g}$ solves the normalized Ricci flow with respect to $\tilde{t}$, i.e., $\partial\tilde{g}/\partial\tilde{t} = -2\big(\text{Ric}(\tilde{g}) - \frac{1}{n}R(\tilde{g})\tilde{g}\big)$.

We could further constrain the variation so that the metrics under consideration are not only of unit volume, but *pointwise conformal* to g, that is, expressible as $f \cdot g$, for $f > 0$ a smooth function on M. One can partition $\mathcal{M}$ into equivalence classes $[g]$ of pointwise conformal metrics. As an immediate corollary of the proof of Proposition 2-22 we have the following.

Proposition 2-25. *A metric g is critical for $\overline{\mathcal{R}}$ amongst variations $g_t \in [g]$ if and only if $R(g)$ is constant. A metric $g \in \mathcal{M}_1$ is critical for $\mathcal{R}$ amongst variations $g_t \in \mathcal{M}_1 \cap [g]$ if and only if $R(g)$ is constant.*

Proof. Let $g_t = f_t g$, $t \in I$, be smooth in $I \times M$, with $f_0 = 1$. Then f_t is smooth, and $h = dg_t/dt = (df_t/dt)g$. Let $\psi = df_t/dt|_{t=0}$, which can be any smooth function on M, since we could let $f_t = 1 + t\psi$. Applying the argument in the proof of Proposition 2-22, with $h = \psi g$, we have

$$\frac{d}{dt}\bigg|_{t=0} \overline{\mathcal{R}}(g_t)$$

$$= -\frac{1}{(\operatorname{Vol} g)^{1-\frac{2}{n}}} \int_M \psi g \cdot \left(\operatorname{Ric}(g) - \tfrac{1}{2}R(g)g + \frac{n-2}{2n}(\operatorname{Vol} g)^{-1}\mathcal{R}(g)g\right) dv_g$$

$$= \frac{1}{(\operatorname{Vol} g)^{1-\frac{2}{n}}} \cdot \frac{n-2}{n} \int_M \psi \left(R(g) - (\operatorname{Vol} g)^{-1}\mathcal{R}(g)\right) dv_g.$$

Since this must vanish for all ψ, we must have $R(g) = (\operatorname{Vol} g)^{-1}\mathcal{R}(g)$, and conversely. $\qquad\qquad\qquad\qquad\qquad\qquad\qquad\qquad\qquad\qquad\qquad\qquad\qquad\square$

2.3.5. *The Gauss–Bonnet theorem for closed surfaces.* Before we return to general relativity, we discuss how the analysis of the Einstein–Hilbert action yields the global Gauss–Bonnet theorem.

Theorem 2-26. *Suppose that (Σ, g) is an orientable, closed Riemannian surface with Euler characteristic $\chi(\Sigma)$. If $K = K(g)$ is the Gauss (sectional) curvature, then*

$$\int_\Sigma K \, dv_g = 2\pi \chi(\Sigma).$$

Proof. Any connected such surface Σ is topologically a sphere, or a torus, or a torus with some number γ of handles attached (genus γ). A sphere has Euler characteristic 2, as can be seen using a tetrahedral triangulation. It is also not too hard to triangulate a torus and compute its Euler characteristic to be 0. Higher-genus surfaces are obtained by doing a connected sum to a torus. We can view this as removing a triangle from a surface and a torus, and gluing along the edges of the triangle. Suppose the original surface had Euler characteristic χ, and we know the torus has Euler characteristic 0. So we know the total alternating sum of vertices, edges and faces at the start for both surfaces is χ. In the process of adding a handle, we lose two faces, three edges and three vertices. Thus adding a handle brings the Euler characteristic down by 2. Thus $\chi(\Sigma) = 2 - 2\gamma$, $\gamma = 0, 1, 2, \ldots$.

As we proved earlier, the Einstein–Hilbert action $\mathcal{R}(g)$ is critical at *every* metric g on a two-dimensional manifold. This means that $\mathcal{R}$ is *constant* on the space $\mathcal{M}$ of metrics: any two metrics $g_1, g_2 \in \mathcal{M}$ can be connected by a linear path $g_t = (1-t)g_1 + tg_2$, $0 \le t \le 1$, and we have seen that $\mathcal{R}(g_t)$ is constant in t. Since $R(g) = 2K(g)$, we see that $\int_\Sigma K(g) \, dv_g$ is independent of g, and thus it suffices to compute it at any single chosen metric.

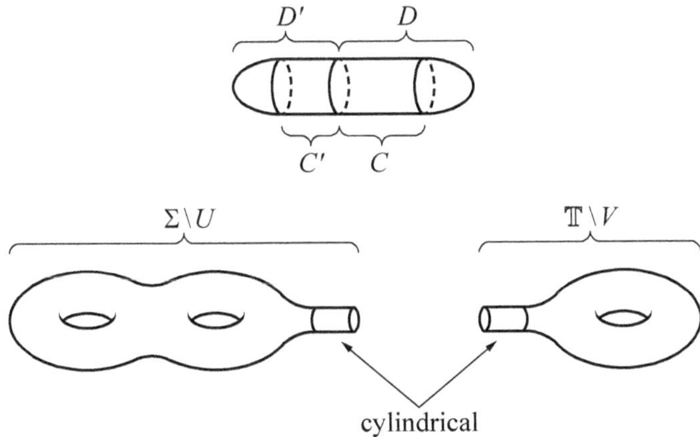

Figure 4. Connected sum construction for the proof of the Gauss–Bonnet theorem.

For a sphere, we take the metric of the unit (round) sphere, $K = 1$ and Vol $g = 4\pi$. The Euler characteristic of the sphere is 2, as mentioned.

For a torus, we have Euler characteristic 0, and we can use a flat metric on the torus to compute $\mathcal{R}(g)$.

To progress to the higher-genus case, we use an auxiliary construction. By smoothly capping off an end of a circular cylinder, we obtain a surface D which is topologically a closed disk, with a cylindrical collar neighborhood of its boundary, which is a circular geodesic on the cylinder. Reflect the surface D across the plane of the geodesic circle to produce a congruent surface D', and let $S = D \cup D'$, which is smooth with spherical topology, and induced metric g. Thus

$$\int_D K(g)\, dv_g = \frac{1}{2} \int_S K(g)\, dv_g = 2\pi. \tag{2.3.16}$$

Observe that (2.3.16) is independent of the precise geometry of the smooth cap, as well as the size of the cylindrical portion (C in Figure 4), where $K = 0$.

Given a closed surface Σ, let Σ' be the connected sum of Σ and a torus $\mathbb{T}$. We can readily put a metric g on Σ and g_0 on $\mathbb{T}$, so that there are neighborhoods $U \subset \Sigma$ and $V \subset \mathbb{T}$, each isometric to the interior of the surface D as above, and whose complements $\Sigma \setminus U$ and $\mathbb{T} \setminus V$ each possess a cylindrical collar neighborhood of their boundary circle. We can represent Σ' as the union $(\Sigma \setminus U) \cup (\mathbb{T} \setminus V)$ along the common boundary circle, and take the metric g' coinciding with the metrics on each piece, noting that g' is smooth by construction. If the Gauss–Bonnet theorem holds for Σ, then by (2.3.16)

$$\int_{\Sigma'} K(g')\, dv'_g = \int_\Sigma K(g)\, dv_g + \int_\mathbb{T} K(g_0)\, dv_{g_0} - 4\pi = 2\pi(\chi(\Sigma) - 2).$$

Since adding a handle contributes to a decrease in the Euler characteristic by 2, the Gauss–Bonnet theorem follows by induction. $\qquad\square$

The interested reader can readily extend the Gauss–Bonnet theorem to the case where Σ is closed but nonorientable, by applying the preceding to the orientation double covering for any nonorientable component of Σ.

2.4. Spacetime examples

In this section we consider some examples of spacetimes and the form of the Einstein equation they satisfy.

2.4.1. *Constant curvature spacetimes.* Let $\mathbb{R}^n_\nu$, for $0 \leq \nu \leq n$, denote the semi-Riemannian manifold $\mathbb{R}^n$ with the metric $\langle \partial/\partial x^i, \partial/\partial x^j \rangle = \epsilon_i \delta_{ij}$, where $\epsilon_i = -1$ for $1 \leq i \leq \nu$ and $\epsilon_i = 1$ for $\nu + 1 \leq i \leq n$. We define the following manifolds as level sets of certain quadratic polynomials, with the metric induced from the indicated inclusions: for $n \geq 2$ and $r > 0$, we let

$$\mathbb{S}^n_1(r) = \{x \in \mathbb{R}^{n+1} : -(x^0)^2 + (x^1)^2 + \cdots + (x^n)^2 = r^2\} \subset \mathbb{R}^{n+1}_1 = \mathbb{M}^{1+n},$$

$$\mathbb{H}^n_1(r) = \{x \in \mathbb{R}^{n+1} : -(x^0)^2 - (x^1)^2 + (x^2)^2 + \cdots + (x^n)^2 = -r^2\} \subset \mathbb{R}^{n+1}_2.$$

It is easy to see that $\mathbb{S}^n_1(r)$ is diffeomorphic to $\mathbb{R} \times \mathbb{S}^{n-1}$, which is simply connected for $n \geq 3$, while $\mathbb{H}^n_1(r)$ is diffeomorphic to $\mathbb{S}^1 \times \mathbb{R}^{n-1}$. We let $\widetilde{\mathbb{S}}^2_1(r)$ and $\widetilde{\mathbb{H}}^n_1(r)$ be the universal covering manifolds with the associated covering metrics.

Proposition 2-27 [174, p. 228–229]. *A complete, simply connected n-dimensional Lorentzian manifold of constant sectional curvature C is isometric to one of the following:*

- $\mathbb{S}^n_1(r)$ $\quad (n \geq 3, C = r^{-2})$,
- $\widetilde{\mathbb{S}}^2_1(r)$ $\quad (n = 2, C = r^{-2})$,
- $\mathbb{R}^n_1$ $\qquad (C = 0)$,
- $\widetilde{\mathbb{H}}^n_1(r)$ $\quad (C = -r^{-2})$.

$\mathbb{S}^4_1(r)$ is *de Sitter* spacetime of curvature $C = r^{-2}$, $\mathbb{H}^4_1(r)$ is *anti-de Sitter* spacetime of curvature $C = -r^{-2}$, while the universal cover $\widetilde{\mathbb{H}}^4_1(r)$ is *universal anti-de Sitter* spacetime.

Constant curvature manifolds are Einstein. Indeed, in a constant curvature Lorentz four-manifold $(\mathcal{S}^4, g)$, we have $\mathrm{Ric}(g) = \frac{1}{4}R(g)\,g$, where $R(g)$ is constant. Then $\left(\mathrm{Ric}(g) - \frac{1}{2}R(g)\,g\right) + \frac{1}{4}R(g)\,g = 0$. One can interpret this as Einstein's equation with $T = 0$ and $\Lambda = \frac{1}{4}R(g)$. Thus $\mathrm{Ric}(g) = \Lambda g$, and in the examples above, $\Lambda > 0$ corresponds to de Sitter spacetime, while $\Lambda < 0$ corresponds to anti-de Sitter spacetime.

2.4.2. *The Einstein static universe.* Consider the space $\mathbb{R} \times \mathbb{S}^3$ with the product metric $g = -c^2 dt^2 + \mathring{g}_{\mathbb{S}^3}$, where $\mathring{g}_{\mathbb{S}^3}$ is the unit round metric on the three-sphere.

Exercise 2-28. Show that the Ricci curvature of this metric is $\mathrm{Ric}(g) = 2\mathring{g}_{\mathbb{S}^3}$.

From this we see that $R(g) = 6$, so the Einstein tensor is $\mathrm{Ric}(g) - \frac{1}{2}R(g)g = 3c^2 dt^2 - \mathring{g}_{\mathbb{S}^3}$, and $G_\Lambda(g) = (3-\Lambda)c^2 dt^2 + (\Lambda - 1)\mathring{g}_{\mathbb{S}^3}$. We identify this with the stress-energy tensor of a perfect fluid, with fluid velocity $\mathbb{U} = \partial/\partial t$, density ρ and pressure p. If we write T as a $(0, 2)$-tensor, we have $T = (\rho + p)c^2 dt^2 + pg = \rho c^2 dt^2 + p\mathring{g}_{\mathbb{S}^3}$. The Einstein equation $G_\Lambda(g) = \kappa T$ is then equivalent to

$$3 - \Lambda = \kappa\rho, \quad \Lambda - 1 = \kappa p.$$

For $p \geq 0$, we must have $\Lambda \geq 1$. Note that $\rho + p = 2/\kappa$ and $\kappa = 8\pi G/c^4$.

2.4.3. *Friedmann–Lemaître–Robinson–Walker (FLRW) spacetimes.* Let I be an interval and Σ a three-manifold. We consider a warped product metric of the form $g = -c^2 dt^2 + (cf(t))^2 g_0$, where g_0 is a Riemannian metric of constant curvature k_0 on Σ and f is a positive function on I. Such metrics arise in cosmological models which incorporate the observation that the universe seems to be everywhere isotropic (with respect to a class of observers, which may be on the galactic scale), which in turn implies spatial homogeneity of the manifolds orthogonal to the trajectories of such observers (spaces of simultaneity). See [42; 174; 218] for more discussion of how isotropy and spatial homogeneity are translated into a metric of the above form.

We let $g = \langle \cdot, \cdot \rangle$, and for the rest of this section, we let $\mathbb{U} = \partial/\partial t$. Further, let X, Y and Z denote tangent vectors to $\Sigma_t = \{t\} \times \Sigma$ for $t \in I$, so that $g_0(X, Y) = (cf(t))^{-2}\langle X, Y \rangle$.

Exercise 2-29. Verify the following formulae for the Riemann tensor of g (cf. Exercise 1-15):

$$R(X, Y, Z) = c^{-2}\left(\left(\frac{f'(t)}{f(t)}\right)^2 + \frac{k_0}{(f(t))^2}\right)(\langle Y, Z \rangle X - \langle X, Z \rangle Y),$$

$$R(X, \mathbb{U}, \mathbb{U}) = -\frac{f''(t)}{f(t)}X,$$

$$R(X, Y, \mathbb{U}) = 0,$$

$$R(X, \mathbb{U}, Y) = -c^{-2}\frac{f''(t)}{f(t)}\langle X, Y \rangle \mathbb{U}.$$

We can now readily find the Ricci and scalar curvatures of g:

$$\mathrm{Ric}(\mathbb{U}, \mathbb{U}) = -3\frac{f''(t)}{f(t)} = 3\frac{f''(t)}{f(t)}\langle \mathbb{U}, \mathbb{U}\rangle c^{-2},$$

$$\mathrm{Ric}(\mathbb{U}, X) = 0,$$

$$\mathrm{Ric}(X, Y) = \left(2\left(\frac{f'(t)}{f(t)}\right)^2 + 2\frac{k_0}{(f(t))^2} + \frac{f''(t)}{f(t)}\right)\langle X, Y\rangle c^{-2},$$

$$R(g) = 6\left(\left(\frac{f'(t)}{f(t)}\right)^2 + \frac{k_0}{(f(t))^2} + \frac{f''(t)}{f(t)}\right)c^{-2}.$$

The stress-energy tensor that corresponds to this metric can be found using the Einstein equation:

$$T = \frac{c^4}{8\pi G}\left(\mathrm{Ric}(g) - \frac{1}{2}R(g)\,g + \Lambda g\right).$$

Exercise 2-30. Verify the following formulas.

$$T(\mathbb{U}, X) = 0,$$

$$T(X, Y) = -\frac{c^2}{8\pi G}\left(\left(\frac{f'(t)}{f(t)}\right)^2 + \frac{k_0}{(f(t))^2} + 2\frac{f''(t)}{f(t)} - \Lambda c^2\right)\langle X, Y\rangle,$$

$$T(\mathbb{U}, \mathbb{U}) = \frac{c^4}{8\pi G}\left(3\left(\frac{f'(t)}{f(t)}\right)^2 + 3\frac{k_0}{(f(t))^2} - \Lambda c^2\right).$$

We can try to identify this with a perfect fluid, which has stress tensor $T = c^{-2}(\rho + p)\mathbb{U}^\flat \otimes \mathbb{U}^\flat + pg$. For instance, $T(\mathbb{U}, \mathbb{U}) = c^2\rho$, $T(\mathbb{U}, X) = 0$ and $T(X, Y) = p\langle X, Y\rangle$. We can identify the stress-energy tensor of the warped product as that of a perfect fluid, with

$$\rho = \frac{c^2}{8\pi G}\left(3\left(\frac{f'(t)}{f(t)}\right)^2 + 3\frac{k_0}{(f(t))^2} - \Lambda c^2\right)$$

$$= \frac{c^4}{8\pi G}\left(3\left(\frac{f'(t)}{cf(t)}\right)^2 + 3\frac{k_0}{(cf(t))^2} - \Lambda\right),$$

$$p = -\frac{c^2}{8\pi G}\left(\left(\frac{f'(t)}{f(t)}\right)^2 + \frac{k_0}{(f(t))^2} + 2\frac{f''(t)}{f(t)} - \Lambda c^2\right)$$

$$= \frac{c^4}{8\pi G}\left(-\left(\frac{f'(t)}{cf(t)}\right)^2 - \frac{k_0}{(cf(t))^2} - 2\frac{f''(t)}{c^2 f(t)} + \Lambda\right).$$

Note that $(4\pi G/c^4)(\rho + 3p) - \Lambda = -3f''(t)/f(t)$. It is also easy to derive the following equation from the above: $\rho'(t) = -3(\rho(t) + p(t))f'(t)/f(t)$;

it is basically the vanishing of the t-component of $\mathrm{div}_g T$, and so expresses conservation of energy.

When $f(t)$ is constant, for example in the Einstein static universe, the geometry of the spatial slices is not dynamic. In this case, if $\Lambda = 0$, then ρ and p must have opposite signs, which is not appealing in terms of standard physics.

The Friedmann cosmological models are the cases of the above that correspond to dust models, so that $p = 0$, with $f'(t_0)/f(t_0) > 0$ for some time $t_0 \in I$. This would model situations when the energy density dominates pressure, as might be the present situation in the universe (but not, say, near the Big Bang). When $p = 0$, we obtain the first-order linear equation $\rho'(t) + 3\rho(t) f'(t)/f(t) = 0$, so by integrating we obtain $\rho f^3 =: \mathfrak{m}$, a constant. Substituting into the equation for ρ above, we get the *Friedmann equation*

$$\frac{8\pi G\mathfrak{m}}{3c^2} \cdot \frac{1}{f(t)} = (f'(t))^2 + k_0 - \tfrac{1}{3}\Lambda c^2 (f(t))^2.$$

Note that there is a critical value of Λ for which one can achieve a static model (constant $f(t)$) when $p = 0$, given by $\Lambda c^2 - k_0^3 (4\pi G\mathfrak{m}/c^2)^{-2}$.

We let $\Lambda = 0$ and $A = 8\pi G\mathfrak{m}/(3c^2)$, so that the Friedmann equation becomes

$$\frac{A}{f(t)} = (f'(t))^2 + k_0.$$

One can solve this for each possible sign of k_0. For $k_0 = 0$, we get, assuming $f(t) > 0$ and $f'(t) \neq 0$, $\sqrt{f(t)}f'(t) = \pm\sqrt{A}$, so if we let $f(0) = 0$, then $f(t) = Ct^{2/3}$, where $C > 0$ is constant. This describes a universe that expands from an initial "Big Bang" singularity (note that $f'(t) \to +\infty$ as $t \searrow 0$). If $k_0 = 1$, the solution graph can be written in parametrized form as $t = \tfrac{1}{2}A(u - \sin u)$, $f = \tfrac{1}{2}A(1 - \cos u)$, which describes a cycloid. The geometry here expands from $f(0) = 0$ to a maximum value, then recollapses at $t = \pi A$ ($u = 2\pi$), the "Big Crunch". Similarly, if $k_0 = -1$ and $\rho > 0$, then $(f'(t))^2 = 1 + A/f(t) > 1$, so that $f(t)$ keeps growing without bound, and the spatial slices expand in size over time. We note that you can parametrize the solution graph as $t = \tfrac{1}{2}A(\sinh u - u)$, $f = \tfrac{1}{2}A(\cosh u - 1)$.

2.4.4. *Schwarzschild spacetime.* We consider the vacuum Einstein equation $\mathrm{Ric}(g) = 0$. This is a nonlinear system of second-order partial differential equations for the metric components. One can reduce this to ordinary differential equations by imposing an ansatz of spherical symmetry, and argue (see the proof of Birkhoff's theorem in [112], [161, Section 32.2], or [218, Section 6.1]) that in spherical symmetry, the $(n+1)$-dimensional spacetime metric has the form

$$g = -u(r)dt^2 + v(r)dr^2 + r^2 \mathring{g}_{\mathbb{S}^{n-1}},$$

where $\mathring{g}_{\mathbb{S}^{n-1}}$ is the unit round metric on the sphere. (We take $n \geq 3$, since any Ricci-flat surface or three-manifold is flat; see Exercise 2-11.) So a spherically symmetric metric is also *static* (see Section 2.4.5, and compare [218, Chapter 6; 174, Chapter 12]), where the vector field $\partial/\partial t$ is a Killing field, timelike for $u(r) > 0$, which is orthogonal to the orbits of the isometry group.

We impose the Einstein equation $\mathrm{Ric}(g) = 0$ to determine g. We will sketch this here, referring to the works just cited, as well as, e.g., [182, Chapter 3]. We will use results from Exercises 1-14 and 1-15, namely the formulas for the curvature of a warped product, with the tr-plane as the base B with metric $g_B = -u(r)dt^2 + v(r)dr^2$ and with fiber F the round sphere $(\mathbb{S}^{n-1}, \mathring{g}_{\mathbb{S}^{n-1}})$, with warping function r.

We apply Exercises 1-14 and 1-15 to find $\mathrm{Ric}_{\mathring{g}}(X, Y)$ for vectors X and Y tangent to the base, which yields

$$\left(K\left(-u(r)dt^2 + v(r)dr^2\right) - \frac{n-1}{r} \cdot \left(-\frac{1}{2}\frac{u'(r)}{v(r)}dt^2 - \frac{1}{2}\frac{v'(r)}{v(r)}dr^2\right) \right)(X, Y)$$

where

$$K = -\frac{1}{2}\frac{u''(r)}{u(r)v(r)} + \frac{1}{4}\frac{(u'(r))^2}{(u(r))^2 v(r)} + \frac{1}{4}\frac{u'(r)v'(r)}{u(r)(v(r))^2}.$$

For this to vanish for all X and Y is equivalent to

$$-Ku(r) + \frac{n-1}{2r}\frac{u'(r)}{v(r)} = 0 = Kv(r) + \frac{n-1}{2r}\frac{v'(r)}{v(r)}. \tag{2.4.1}$$

Thus a necessary condition for $\mathrm{Ric}(g) = 0$ is that $\dfrac{u'(r)}{u(r)} = -\dfrac{v'(r)}{v(r)}$, i.e., $u(r)v(r)$ is *constant*.

Exercise 2-31. Use the formula for K, as well as $u(r)v(r) = C$, to show that the condition (2.4.1) above reduces to $ru''(r) + (n-1)u'(r) = 0$. Show the general solution is given by $u(r) = c_1 + c_2/r^{n-2}$, where c_1 and c_2 are constants.

Thus we can achieve the Ricci-flat condition along base directions with $u(r) = c_1 + c_2/r^{n-2}$ and $v(r) = C/u(r)$. From Exercise 1-15, it follows that we just have to consider the fiber directions, for which we note $\mathrm{Ric}(g_F) = (n-1)\mathring{g}_{\mathbb{S}^{n-1}}$. Furthermore, as in Exercise 1-15, the meaning of $\langle \nabla r, \nabla r \rangle = (v(r))^{-1}$ is the same computed in g_B or g. Moreover from Exercise 1-14 we have

$$\mathrm{Hess}_{g_B} r = -\frac{1}{2}\frac{u'(r)}{v(r)}dt^2 - \frac{1}{2}\frac{v'(r)}{v(r)}dr^2,$$

so

$$\Delta_{g_B} r = \mathrm{tr}_{g_B}(\mathrm{Hess}_{g_B} r) = \frac{1}{2}\left(\frac{u'(r)}{u(r)v(r)} - \frac{v'(r)}{(v(r))^2}\right).$$

Therefore, for vectors V and W tangent to the fiber, we find $\mathrm{Ric}_g(V, W)$ is precisely

$$\mathring{g}_{\mathbb{S}^{n-1}}(V, W)\left((n-2) - \left(\frac{r}{2}\left(\frac{u'(r)}{u(r)v(r)} - \frac{v'(r)}{(v(r))^2}\right) + \frac{n-2}{v(r)}\right)\right).$$

Exercise 2-32. Show that under the condition $u(r)v(r) = C$, the vanishing of $\mathrm{Ric}_g(V, W)$ for all V and W tangent to the fiber is tantamount to

$$ru'(r) + (n-2)u(r) = C(n-2).$$

Solve this to obtain $u(r) = C + c_3/r^{n-2}$ for some constant c_3.

Together with the result of Exercise 2-31, we see that $\mathrm{Ric}(g) = 0$ boils down to (letting α satisfy $c_3 = -\alpha C$)

$$u(r) = C + \frac{c_3}{r^{n-2}} = C\left(1 - \frac{\alpha}{r^{n-2}}\right) \quad \text{and} \quad v(r) = \left(1 - \frac{\alpha}{r^{n-2}}\right)^{-1},$$

for $C > 0$ and α arbitrary constants. We remark that we can rescale the time variable by a constant factor (changing time units effectively adjusts the numerical value of the speed of light c), to normalize the constant value of $u(r)v(r)$. If we take $C = c^2$, we can write the Schwarzschild spacetime metric $\bar{g}_S$, for any constant α, as

$$\bar{g}_S = -\left(1 - \frac{\alpha}{r^{n-2}}\right)c^2 dt^2 + \left(1 - \frac{\alpha}{r^{n-2}}\right)^{-1} dr^2 + r^2 \mathring{g}_{\mathbb{S}^{n-1}}.$$

Note that for large r, $\bar{g}_S \approx -c^2 dt^2 + g_{\mathbb{E}^n} = -(dx^0)^2 + \delta_{ij} dx^i dx^j$, the Minkowski metric. So not only must the spacetime be static, which follows directly from the ansatz of spherical symmetry, but in the far field it must approach Minkowski spacetime, a consequence of the symmetry together with the Ricci-flat condition from the vacuum Einstein equation.

We want to identify α as proportional to the *mass m* of the spacetime, eventually settling on units for which $\alpha = 2m$. We motivate this in dimension four, i.e., $n = 3$, and we restrict the discussion to this dimension until further notice. The Schwarzschild metric can represent the gravitational field in the exterior of a non-rotating spherically symmetric massive body, and in the weak field regime (large r), the effect of the gravitational field on test particles (ascertained by finding the geodesics in the Schwarzschild metric) is roughly that of a Newtonian gravitational field for a point mass m. In fact, note that if we write $(\bar{g}_S)_{\mu\nu} = \eta_{\mu\nu} + h_{\mu\nu}$, with $h_{00} = 2Gm/(c^2 r)$ and $dx^0 = c\, dt$, we can as on p. 63

identify a gravitational potential $\Phi = -\frac{1}{2}c^2 h_{00} = -Gm/r$, in agreement with Newtonian theory for the field of a body of mass m. The spacetime so obtained is then

$$\bar{g}_S = -\left(1 - \frac{2Gm}{c^2 r}\right) c^2 \, dt^2 + \left(1 - \frac{2Gm}{c^2 r}\right)^{-1} dr^2 + r^2 \mathring{g}_{\mathbb{S}^2}.$$

For simplicity, we take units for which $G = 1$ and $c = 1$, so that

$$\bar{g}_S = -\left(1 - \frac{2m}{r}\right) dt^2 + \left(1 - \frac{2m}{r}\right)^{-1} dr^2 + r^2 \mathring{g}_{\mathbb{S}^2}. \tag{2.4.2}$$

For $m = 0$ we recover Minkowski spacetime. For $m > 0$, there are metric components that are singular at $r = 2m$ and at $r = 0$, so that we have spacetime regions with $0 < r < 2m$ and $r > 2m$, respectively. It turns out the *geometry* does not become singular as $r \to 2m$, and in fact it can be extended across $r = 2m$, as we will investigate and make precise in the next section (2.4.4.1). However, as $r \searrow 0$, there is a curvature blowup, as we conclude in Remark 2-35; see also [161, p. 822].

Exercise 2-33. Show that the sectional curvature K of the metric

$$-\left(1 - \frac{2m}{r}\right) dt^2 + \left(1 - \frac{2m}{r}\right)^{-1} dr^2$$

is $K = 2m/r^3$. (ii) Let $I \subset (0, +\infty)$ be an open interval, and consider a (Lorentzian or Riemannian) metric on $I \times \mathbb{S}^2$ of the form $g = h(r) dr^2 + r^2 \mathring{g}_{\mathbb{S}^2}$. Show that $\mathrm{Ric}_g(\partial/\partial r, \partial/\partial r) = h'(r)/(rh(r))$. Use this to show that, if v is the unit vector $v = \sqrt{|1 - 2m/r|}\, \partial/\partial r$, then $\mathrm{Ric}_{g_S}(v, v) = -2m/r^3$ for the induced metric $g_S = (1 - 2m/r)^{-1} dr^2 + r^2 \mathring{g}_{\mathbb{S}^2}$ $(0 < r \neq 2m)$ on a constant t-slice in Schwarzschild. Note that for $m > 0$, the curvatures computed here have a finite limit as $r \to 2m$.

Exercise 2-34. Find the Christoffel symbols Γ^j_{i0} of the Schwarzschild metric $\bar{g}_S$, where i and j index spherical spatial coordinates and 0 is the t-index. Conclude that the second fundamental form II of a constant t-slice of Schwarzschild vanishes.

Remark 2-35. While the spacetime Schwarzschild metric $\bar{g}_S$ has vanishing Ricci curvature, the *Riemannian Schwarzschild metric* g_S on a constant t-slice does not. The *scalar* curvature of g_S does vanish, as we will see later in the section (you can also just compute it directly). The norm of the ambient curvature tensor for $\bar{g}_S$ blows up as $r \searrow 0$, by direct calculation, or by using the Gauss equation (Proposition 5-5) together with $\mathrm{II} = 0$ and $\mathrm{Ric}_{g_S}(v, v) = -2m/r^3$ for a constant t-slice.

For $m < 0$, the metric g_S on a constant t-slice is defined for all $r > 0$ and is Riemannian. There are radial geodesics for this metric g_S, which are then easily seen to be spacelike geodesics for $\bar{g}_S$, since $\amalg = 0$. If $r_0 > 0$, then for $m < 0$, $\int_0^{r_0}(1 - 2m/r)^{-1/2}\,dr < +\infty$. Thus a radial geodesic going from r_0 toward $r = 0$ has finite length, so that both $(\mathbb{R}^3\setminus\{0\}, g_S)$ and $(\mathbb{R}\times(\mathbb{R}^3\setminus\{0\}), \bar{g}_S)$ are geodesically incomplete. Neither metric can be smoothly extended in order to extend the geodesic, due to the curvature blowup as $r \searrow 0$. As we will see below, for $m > 0$, while the geometry of the spacetime metric $\bar{g}_S$ is well-behaved near $r = 2m$, the metric is causally geodesically incomplete: there are timelike geodesics which approach $r = 0$ in finite proper time (as well as null geodesics which approach $r = 0$ at finite affine parameter), and along which there is curvature blowup.

We collect a few formulas that will be useful here and in the next section, and make one more observation before moving on.

Exercise 2-36. By direct calculation, either computing the Christoffel symbols $\Gamma^\mu_{\rho\sigma} = \frac{1}{2}g^{\mu\nu}(g_{\nu\sigma,\rho} + g_{\rho\nu,\sigma} - g_{\rho\sigma,\nu})$ or by using the Cartan equations (Exercise 1-13), show that in the Schwarzschild metric $\bar{g}_S$ we have

$$\nabla_{\frac{\partial}{\partial t}}\frac{\partial}{\partial t} = \frac{m}{r^2}\left(1 - \frac{2m}{r}\right)\frac{\partial}{\partial r},$$

$$\nabla_{\frac{\partial}{\partial t}}\frac{\partial}{\partial r} = \frac{m}{r^2}\left(1 - \frac{2m}{r}\right)^{-1}\frac{\partial}{\partial t} = \nabla_{\frac{\partial}{\partial r}}\frac{\partial}{\partial t},$$

$$\nabla_{\frac{\partial}{\partial r}}\frac{\partial}{\partial r} = -\frac{m}{r^2}\left(1 - \frac{2m}{r}\right)^{-1}\frac{\partial}{\partial r}.$$

Consider the unit timelike vector field $\mathbb{U} = (1 - 2m/r)^{-1/2}\,\partial/\partial t$, for $r > 2m$. Integral curves $\gamma(\tau)$ of this vector field are *Schwarzschild observers*. In these coordinates, their spatial positions are fixed. From the preceding exercise, we note that $D\gamma'(\tau)/d\tau = \nabla_{\mathbb{U}}\mathbb{U} = (m/r^2)\,\partial/\partial r$. Thus, for $m \neq 0$, such observers are not in free fall. In fact, as we have seen earlier (cf. Section 1.3.2.1), if m_0 is the rest mass of the observer, one might interpret $f = (m_0 m/r^2)\,\partial/\partial r$ as the force on the observer. As the observer has fixed spatial coordinates in this chart, one might interpret this as the force required to oppose the effect of gravitation, to keep the observer from geodesic motion (free fall); moreover, $\partial/\partial r$ is approximately a unit vector if r is large, so the interpretation is consistent with Newtonian gravity (with $G = 1$), at least in the far field (weak field limit).

2.4.4.1. *Kruskal extension.* We have seen that the null structure plays a key role in the geometry of spacetime, so we study the behavior of null geodesics near $r = 2m > 0$. Fix a point ω_0 on the sphere, and consider a curve $\gamma(s)$ which is given in coordinates by $(t(s), r(s), \omega_0)$. Thus $\gamma'(s) = t'(s)\partial/\partial t + r'(s)\partial/\partial r$.

Therefore, using Exercise 2-36, we obtain

$$\gamma''(s) = \left(t''(s) + 2t'(s)r'(s)\frac{m}{r^2}\left(1 - \frac{2m}{r}\right)^{-1} \right)\frac{\partial}{\partial t}$$
$$+ \left(r''(s) + (t'(s))^2\frac{m}{r^2}\left(1 - \frac{2m}{r}\right) - (r'(s))^2\frac{m}{r^2}\left(1 - \frac{2m}{r}\right)^{-1} \right)\frac{\partial}{\partial r}.$$

We observe that the $\frac{\partial}{\partial t}$-component of the geodesic equation $\gamma''(s) = 0$ is first-order linear in $t'(s)$, and in fact can be written $\frac{d}{ds}\left(t'(s)(1 - 2m/r)\right) = 0$, which reduces to the condition that $t'(s)(1 - 2m/r)$ be constant.

Note that γ is null if and only if $(t'(s))^2(1 - 2m/r)^2 = (r'(s))^2$, in which case

$$\gamma''(s) = \left(t''(s) + 2t'(s)r'(s)\frac{m}{r^2}\left(1 - \frac{2m}{r}\right)^{-1} \right)\frac{\partial}{\partial t} + r''(s)\frac{\partial}{\partial r}.$$

If we want $\gamma(s)$ to be a null geodesic, we must require $r''(s) = 0$, i.e., $r'(s)$ is a constant, which we can assume to equal 1 by rescaling the affine parameter s. We then see that

$$t'(s) = \pm r'(s)\left(1 - \frac{2m}{r}\right)^{-1} = \pm\frac{r}{r - 2m} = \pm\left(1 + \frac{2m}{r - 2m}\right)$$

from the null condition, which we can integrate to get $t = C \pm (r + 2m \log|r - 2m|)$, for some constant C. We thus obtain the solutions to the null geodesic equation for $\gamma(s)$, with $r(s)$ linear in s. Note that for $m > 0$, as $r(s) \to 2m^\pm$, $|t(s)| \to \infty$: so, in the tr-plane, these null geodesics are asymptotic to $r = 2m$ *over a bounded affine parameter interval*; compare this to the case $m < 0$, for which $t(s)$ remains bounded as $r(s) \searrow 0$, again over a bounded affine parameter interval.

We can analyze this in terms of lightcones. We again fix a point on the sphere, and study the lightcones as $r \to 2m$. Null vectors of the form $a\,\partial/\partial t + b\,\partial/\partial r$ must satisfy $a/b = \pm(1 - 2m/r)^{-1}$, which approaches infinity as $r \to 2m$. As we saw above, $a/b = dt/dr$ along a null geodesic path. Thus the lightcones seem to be pinching in these coordinates.

To see how the geometry is not ill-behaved at $r = 2m > 0$, we seek coordinates in which the lightcones are better behaved. They are suggested by the null geodesics, whose form is $t = \pm r_*$ (up to an additive constant), where $r_* :=$ $r + 2m \log|r/(2m) - 1|$ is known as the *Regge–Wheeler* coordinate. Note that $dr_*/dr = (1 - 2m/r)^{-1}$. If we use this to define new coordinates via incoming and outgoing null directions, we have, say, $u = t - r_*$ and $v = t + r_*$, so that $du = dt - (1 - 2m/r)^{-1}dr$, $dv = dt + (1 - 2m/r)^{-1}dr$, and thus

$$-\left(1 - \frac{2m}{r}\right)dt^2 + \left(1 - \frac{2m}{r}\right)^{-1}dr^2 = -\left(1 - \frac{2m}{r}\right)du\,dv.$$

The metric is now adapted to the structure of null geodesics, but it still has singular components at $r = 2m$. Note that $r_* = \frac{1}{2}(v - u)$, so that

$$e^{\frac{v-u}{4m}} = |F(r)|, \quad \text{with } F(r) := e^{\frac{r}{2m}}\left(\frac{r}{2m} - 1\right).$$

It is a simple exercise to show that F is a diffeomorphism between $(0, +\infty)$ and $(-1, +\infty)$, mapping $(0, 2m)$ to $(-1, 0)$ and $(2m, +\infty)$ to $(0, +\infty)$.

With this in mind, we let $U = e^{-\frac{u}{4m}}$ and $V = e^{\frac{v}{4m}}$, so that for $r > 2m$, $UV = F(r)$. The portion of the Schwarzschild spacetime for $r > 2m$ corresponds, with r_* taking on all values in $\mathbb{R}$, to $\{(U, V) : U > 0, \ V > 0\}$. Moreover, we compute, since $dU = -\frac{1}{4m} e^{-\frac{u}{4m}} du$ and $dV = \frac{1}{4m} e^{\frac{v}{4m}} dv$, that

$$-\left(1 - \frac{2m}{r}\right)dt^2 + \left(1 - \frac{2m}{r}\right)^{-1} dr^2 = -\left(1 - \frac{2m}{r}\right) du\, dv$$

$$= -16m^2 \frac{1 - \frac{2m}{r}}{e^{\frac{r}{2m}}\left(\frac{r}{2m} - 1\right)} dU\, dV$$

$$= \frac{32m^3}{r} e^{-\frac{r}{2m}} dU\, dV.$$

Let F^{-1} be the inverse function for F. We can construe $(U, V) \mapsto F^{-1}(UV)$ as a smooth function on the set $\{(U, V) : UV > -1\}$, for which we note that $\{(U, V) : F^{-1}(UV) > 2m\} = \{(U, V) : UV > 0\}$, one component of which we just used above to map to $r > 2m$ in Schwarzschild. Likewise we have $\{(U, V) : 0 < F^{-1}(UV) < 2m\} = \{(U, V) : -1 < UV < 0\}$, one component of which we will map to the region $0 < r < 2m$ in Schwarzschild, namely $\{(U, V) : UV > -1, \ U < 0, \ V > 0\}$. Note that as r increases in the interval $(0, 2m)$, r_* decreases over the interval $(-\infty, 0)$; if we now let $U = -e^{-\frac{u}{4m}}$ and $V = e^{\frac{v}{4m}}$, we again have in this region $UV = F(r)$, and also again we compute with $dU = \frac{1}{4m} e^{-\frac{u}{4m}} du$ and $dV = \frac{1}{4m} e^{\frac{v}{4m}} dv$, that

$$\left(1 - \frac{2m}{r}\right)dt^2 + \left(1 - \frac{2m}{r}\right)^{-1} dr^2 = -\left(1 - \frac{2m}{r}\right) du\, dv = \frac{32m^3}{r} e^{-\frac{r}{2m}} dU\, dV.$$

Again, $r = F^{-1}(UV)$ can be construed as a smooth function of (U, V) in $\{(U, V) : UV > -1\}$, taking on all values in $(0, +\infty)$: $r = 2m$ does not correspond to a singularity of this metric. We have the original two disconnected regions $r > 2m$ and $0 < r < 2m$ embedded isometrically into a smooth manifold, connected along the axis $U = 0$, $V > 0$. Notice that the level sets of r, which correspond to varying t, are hyperbolae in the UV-plane, with one exception: $r = 2m$ corresponds to the axes, $UV = 0$. In any case, we see that there is not a singularity at $r = 2m$ in the spacetime metric, which smoothly extends over this set, not only connecting $r > 2m$ with $0 < r < 2m$, but also giving rise to two other related regions as well. They are topologically connected to each other, but not all regions are causally connected, as will soon be clear.

If we want to take the metric out of null form, we can let $T = \frac{1}{2}(V - U)$, and $R = \frac{1}{2}(V + U)$, and on $\{(T, R) : R^2 - T^2 > -1\}$, we have

$$-\left(1 - \frac{2m}{r}\right)dt^2 + \left(1 - \frac{2m}{r}\right)^{-1}dr^2 = \frac{32m^3}{r}e^{-\frac{r}{2m}}\,dU\,dV$$

$$= \frac{32m^3}{r}e^{-\frac{r}{2m}}(-dT^2 + dR^2),$$

where $r = F^{-1}(R^2 - T^2)$; the level sets of r are hyperbolae, and $r = 2m$ now corresponds to $R = \pm T$, which is clearly a null hypersurface. The lightcones behave uniformly in these coordinates, which of course is how the coordinates were constructed.

The original region $r > 2m$, corresponds to $R + T = V > 0$ and $R - T = U > 0$, i.e., $R > |T|$, and in this region, we can relate the coordinates to the original (t, r) coordinates, as follows:

$$T = \tfrac{1}{2}(V - U) = \tfrac{1}{2}\left(e^{\frac{t+r_*}{4m}} - e^{-\frac{t-r_*}{4m}}\right) = e^{\frac{r_*}{4m}}\sinh\frac{t}{4m} = \left(\frac{r}{2m} - 1\right)^{\frac{1}{2}}e^{\frac{r}{4m}}\sinh\frac{t}{4m},$$

$$R = \tfrac{1}{2}(V + U) = \left(\frac{r}{2m} - 1\right)^{\frac{1}{2}}e^{\frac{r}{4m}}\cosh\frac{t}{4m}.$$

The region $0 < r < 2m$ corresponds to $R + T > 0$, $R - T < 0$, i.e., $T > |R|$, with $R^2 - T^2 > -1$, with coordinates related to the (t, r) coordinates by (watch the log term in r_*)

$$T = \tfrac{1}{2}(V - U) = \tfrac{1}{2}\left(e^{\frac{t+r_*}{4m}} + e^{-\frac{t-r_*}{4m}}\right) = e^{\frac{r_*}{4m}}\cosh\frac{t}{4m} = \left(1 - \frac{r}{2m}\right)^{\frac{1}{2}}e^{\frac{r}{4m}}\cosh\frac{t}{4m},$$

$$R = \tfrac{1}{2}(V + U) = \left(1 - \frac{r}{2m}\right)^{\frac{1}{2}}e^{\frac{r}{4m}}\sinh\frac{t}{4m}.$$

We note several things about the causal structure (Figure 5). Spacetime has a time orientation given by $\partial/\partial T$. Because the lightcones in the (T, R)-plane are at $45°$ from the T-axis, we see that no causal curve from the region E_2 given by $R < -|T|$ can enter the region E_1, where $R > |T|$ (the original $r > 2m$ region). Moreover, any ingoing (to the future) radial null geodesic starting from E_1 will cross $R = |T|$ ($r = 2m$), entering into the region B_1 (the original $0 < r < 2m$ region), where $T > |R|$, $R^2 - T^2 > -1$, and will inevitably "reach" $r = 0$ ($R^2 - T^2 = -1$) at a finite affine parameter. In fact, *any* causal curve emanating from B_1 stays within this region to the future, and must reach $r = 0$ at finite proper time (or affine parameter). This is a simple consequence of the structure of lightcones in the (T, R)-plane, and recall that a null geodesic has

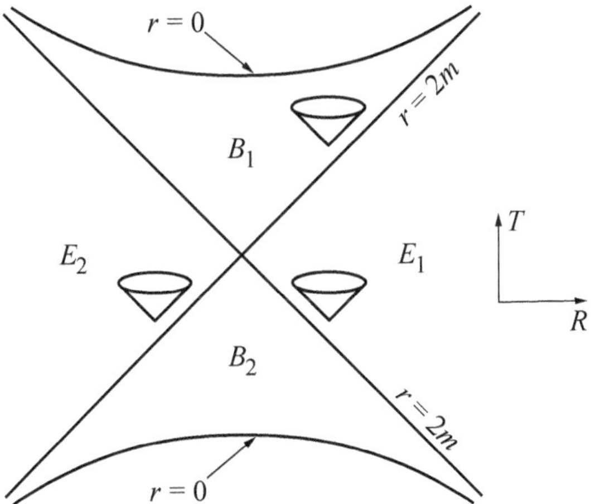

Figure 5. Schwarzschild black hole: Kruskal extension.

affine parameter s with $r'(s)$ constant. The final region B_2 is dual to B_1, and is given by $T < -|R|$, $R^2 - T^2 > -1$; future-directed causal curves starting from B_2 can enter E_1, say, but no future-directed causal curve can go from E_1 to B_2. Interpreting causal curves in terms of signals, one can interpret B_1 as a *black hole* and B_2 as a *white hole*.

2.4.4.2. *Isotropic coordinates.* The spatial slice at constant t in the Schwarzschild metric is actually conformally flat. To see this, we perform a monotone increasing change in the radial variable, from r to $\tilde{r}$. We have

$$\left(1 - \frac{2m}{r}\right)^{-1} dr^2 + r^2 \mathring{g}_{\mathbb{S}^2} = \left(\left(1 - \frac{2m}{r}\right)^{-\frac{1}{2}} \frac{dr}{d\tilde{r}}\right)^2 d\tilde{r}^2 + \left(\frac{r}{\tilde{r}}\right)^2 \tilde{r}^2 \mathring{g}_{\mathbb{S}^2}.$$

We would like the expression in braces to equal $r/\tilde{r}$. This gives

$$\int \frac{d\tilde{r}}{\tilde{r}} = \int \frac{dr}{\sqrt{r^2 - 2mr}} - \int \frac{dr}{\sqrt{(r-m)^2 - m^2}}.$$

We can integrate this with the substitution $r - m = m \cosh w$ to get $w = \log \tilde{r} + \tilde{C}$, or $e^w = C\tilde{r}$. Thus

$$r = m + m \cosh w = m + \frac{m}{2}\left(C\tilde{r} + \frac{1}{C\tilde{r}}\right),$$

and so $\dfrac{r}{\tilde{r}} = \dfrac{mC}{2} + \dfrac{m}{\tilde{r}} + \dfrac{m}{2C\tilde{r}^2}$. If we let $C = 2/m$, we get

$$\frac{r}{\tilde{r}} = \left(1 + \frac{m}{2\tilde{r}}\right)^2.$$

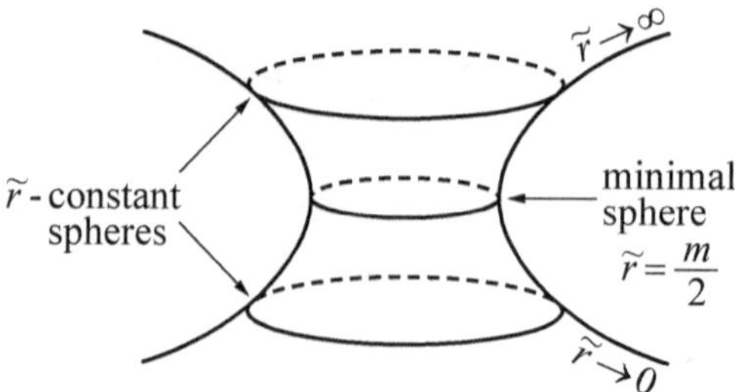

Figure 6. Riemannian Schwarzschild metric of positive mass.

Thus the Schwarzschild metric can also be written

$$\bar{g}_S = -\frac{\left(1 - \frac{1}{2}m/\tilde{r}\right)^2}{\left(1 + \frac{1}{2}m/\tilde{r}\right)^2}\, dt^2 + \left(1 + \frac{m}{2\tilde{r}}\right)^4 (d\tilde{r}^2 + \tilde{r}^2 \mathring{g}_{\mathbb{S}^2}). \qquad (2.4.3)$$

Of course the Euclidean metric $g_{\mathbb{E}^3}$ on $\mathbb{R}^3$ is written in standard spherical coordinates with $\tilde{r} = |x|$, $x \in \mathbb{R}^3$, as $g_{\mathbb{E}^3} = d\tilde{r}^2 + \tilde{r}^2 \mathring{g}_{\mathbb{S}^2}$, so we see that the constant time slice is conformally Euclidean:

$$g_S = \left(1 + \frac{m}{2|x|}\right)^4 g_{\mathbb{E}^3}.$$

For $m > 0$, the conformal metric is defined for all $x \neq 0$. For $m < 0$, $r \searrow 0$ corresponds to $\tilde{r} \searrow -\frac{m}{2}$. The manifold $(\mathbb{R}^3 \setminus \{x : |x| \leq -\frac{m}{2}\}, g_S)$ is incomplete, by the same argument given earlier, and cannot be extended over $|x| = -\frac{m}{2}$: the g_S-areas of the spheres $\{|x| = r\}$ tend to 0 as $r \searrow -\frac{m}{2}$ (Exercise 2-49), and radial geodesics from $r = r_0 > -\frac{m}{2}$ have finite length as $r \searrow -\frac{m}{2}$, but the curvature blowup precludes being able to add a point to complete the metric, as is done in the case $m = 0$ for the Euclidean metric.

For $m > 0$, the values $r > 2m$ correspond to $\tilde{r} > \frac{m}{2}$, and the set where $0 < \tilde{r} \leq \frac{m}{2}$ is in the extended Schwarzschild spacetime developed above. The set where $t = 0$ and $\tilde{r} = \frac{m}{2}$ is a round two-sphere, which is actually totally geodesic inside the three-slice (Figure 6). We note that as $\tilde{r} = |x| \to \infty$, g_S approaches the Euclidean metric. What is not immediately obvious is that the same can be said of the geometry as $\tilde{r} \searrow 0$, so that the metric is complete. This follows from the following exercise.

Exercise 2-37. There is an isometric inversion of $(\mathbb{R}^3 \setminus \{0\}, g_S)$ through the two-sphere $\tilde{r} = \frac{m}{2}$, given by $\tilde{r} \to \left(\frac{m}{2}\right)^2/\tilde{r}$, which fixes this two-sphere, and maps the set where $\tilde{r} > \frac{m}{2}$ isometrically onto the set where $0 < \tilde{r} < \frac{m}{2}$. Conclude that the fixed two-sphere is totally geodesic.

A similar change of radial coordinate can be made on the Schwarzschild metric in any dimension. Indeed from $g_S = (1 - \alpha/r^{n-2})^{-1} dr^2 + r^2 \mathring{g}_{\mathbb{S}^{n-1}}$ (as above, we may let $\alpha = 2m$), we see as above the desired change would satisfy

$$\left(1 - \frac{\alpha}{r^{n-2}}\right)^{-\frac{1}{2}} \frac{dr}{d\tilde{r}} = \frac{r}{\tilde{r}}. \tag{2.4.4}$$

Exercise 2-38. Suppose $k > 0$ is a constant and

$$r = (k\tilde{r})\left(1 + \frac{\alpha}{4(k\tilde{r})^{n-2}}\right)^{\frac{2}{n-2}}.$$

Show that r satisfies (2.4.4). If we let $k = 1$, then

$$g_S = \left(1 + \frac{\alpha}{4\tilde{r}^{n-2}}\right)^{\frac{4}{n-2}} (d\tilde{r}^2 + \tilde{r}^2 \mathring{g}_{\mathbb{S}^{n-1}}).$$

Write the spacetime Schwarzschild metric $\bar{g}_S$ in the form $\bar{g}_S = -(f(\tilde{r}))^2 dt^2 + g_S$. (Answer: $f(\tilde{r}) = (1 - \frac{1}{4}\alpha/\tilde{r}^{n-2})/(1 + \frac{1}{4}\alpha/\tilde{r}^{n-2})$.)

Remark 2-39. We saw that the Schwarzschild family $\bar{g}_S$ is obtained by imposing a rotationally symmetric ansatz for a Lorentzian metric, together with the vanishing of the Ricci curvature (the vacuum Einstein equation). Likewise, one obtains a metric for a spatial slice in Schwarzschild by imposing a rotationally symmetric ansatz for a Riemannian metric and solving for vanishing *scalar* curvature. To see this, let $I \subset \mathbb{R}$ be an interval, and consider the manifold $M = I \times \mathbb{S}^{n-1}$, on which we prescribe a rotationally symmetric metric $g = ds^2 + H(s)\mathring{g}_{\mathbb{S}^{n-1}}$, for $s \in I$ and some smooth positive function H on I. It is easy to see that for fixed $\omega_0 \in \mathbb{S}^{n-1}$, the path $\gamma(s) = (s, \omega_0)$ is a unit-speed geodesic orthogonal to the family of spheres $\{s\} \times \mathbb{S}^{n-1}$, $s \in I$. We let $\rho = \sqrt{H}$, a smooth positive function called the *area radius*, since the geometry of $\{s\} \times \mathbb{S}^{n-1}$ in (M, g) (and thus in particular its area $A(s)$) is precisely that of the Euclidean sphere of radius $\rho(s)$. We remark that we can use the area radius as the radial coordinate on the set where $\rho'(s) \neq 0$, equivalently where $A'(s) \neq 0$. Once we impose vanishing scalar curvature, there must be points where $\rho'(s) \neq 0$, since $\rho(s)$ cannot be constant on any interval: the simple metric product of an interval and a sphere has positive scalar curvature. That said, we will use a different reparametrization.

Suppose $s = s(r)$ defines a smooth function with $s'(r) \neq 0$, giving a bijection $s : J \to I$, where $J \subset \mathbb{R}$ is thus an interval. We can change coordinates on M and write the metric as

$$g = (s'(r))^2 dr^2 + H(s)\mathring{g}_{\mathbb{S}^{n-1}} = (s'(r))^2 \left(dr^2 + \frac{H(s)}{(s'(r))^2} \mathring{g}_{\mathbb{S}^{n-1}}\right).$$

We define $r = r(s)$ to be a function satisfying $dr/ds = r(s)/\sqrt{H(s)}$, i.e., $r = r_0 e^{F(s)}$ where $dF/ds = 1/\sqrt{H(s)}$; we take $r_0 > 0$. Thus $r : I \to r(I) =: J$ is a smooth bijection, with smooth inverse $s = s(r)$, for which we have $g = (s'(r))^2 (dr^2 + r^2 \mathring{g}_{\mathbb{S}^{n-1}}) = (s'(r))^2 g_{\mathbb{E}^n}$. Thus g is conformally Euclidean. As $n \geq 3$, we can write this in a standard form given in Exercise 2-51 as $g = u^{4/(n-2)} g_{\mathbb{E}^n}$, where in this case $u = u(r)$ and $r = |x|$. Applying (2-51a) from that exercise, we see that the vanishing of the scalar curvature amounts to $\Delta u = 0$, which for a function of $r = |x|$ alone yields $u = c_1 + c_2/r^{n-2}$; cf. (7.1.3). If $c_2 = 0$, the metric g is flat; more generally, by a radial rescaling, it is easy to see that g is isometric to the Riemannian Schwarzschild metric g_S of mass $2c_1 c_2$. See Exercise 7-36.

We make one final remark here. Let $I = (0, +\infty)$. For $m > 0$, we consider the Riemannian Schwarzschild metric g_S as a metric on $I \times \mathbb{S}^{n-1}$. This metric admits an isometric $\mathbb{Z}_2$-action, given by radial inversion (cf. Exercises 2-37 and 2-49) along I, together with the antipodal map on the sphere $\mathbb{S}^{n-1}$. The Schwarzschild metric then induces a complete Riemannian metric with zero scalar curvature on the quotient space, a smooth noncompact manifold diffeomorphic to $\mathbb{RP}^n$ with a point removed, known as the $\mathbb{RP}^n$-*geon*. The geometry can still be construed as rotationally symmetric, with one asymptotically flat end instead of two.

2.4.5. *Static vacuum metrics*. We briefly consider spacetime metrics on $\mathcal{S} = I \times M$ of the form $\bar{g} = -f^2 \, dt^2 + g$, where $I \subset \mathbb{R}$ is an open interval, $f : M \to \mathbb{R}$, and g is a Riemannian metric on M. For $\bar{g}$ to be a Lorentzian metric, we assume that f is nowhere-vanishing on M. In this case, $\partial/\partial t$ is a timelike Killing vector field orthogonal to the time slices $M_{t_0} := \{t_0\} \times M$ for $t_0 \in I$. We recall the formulae for $\mathrm{Ric}(\bar{g})$, from Exercise 1-15.

Lemma 2-40. *Let X and Y be (lifts of) vector fields tangent to the base M. Then*

$$\mathrm{Ric}_{\bar{g}}(X, Y) = \mathrm{Ric}_g(X, Y) - f^{-1}\mathrm{Hess}_g f(X, Y),$$
$$\mathrm{Ric}_{\bar{g}}(X, \partial/\partial t) = 0,$$
$$\mathrm{Ric}_{\bar{g}}(\partial/\partial t, \partial/\partial t) = f \Delta_g f.$$

Thus the condition that $\bar{g}$ be Ricci-flat is given by the *static vacuum equations* $\mathrm{Ric}(g) = f^{-1}\mathrm{Hess}_g f$ and $\Delta_g f = 0$. Note that from here we get $R(g) = 0$. We can relate the static equations to scalar curvature deformation, by recalling the linearization $L_g h$ of the scalar curvature from (2.3.11), written in local coordinates as

$$L_g h = -g^{k\ell} g^{ij} h_{ij;k\ell} + g^{ik} g^{j\ell} h_{ij;k\ell} - g^{ik} g^{j\ell} h_{ij} R_{k\ell}.$$

For h compactly supported, we can integrate by parts to get $\int_M f L_g h \, dv_g = \int_M h \cdot L_g^* f \, dv_g$, where

$$L_g^* f = -(\Delta_g f) g + \mathrm{Hess}_g f - f \, \mathrm{Ric}(g). \tag{2.4.5}$$

Thus, if $\bar{g} = -f^2 \, dt^2 + g$ is a Ricci-flat Lorentzian metric, then $L_g^* f = 0$.

Example 2-41. Consider the Riemannian Schwarzschild metric

$$g_S = \left(1 - \frac{2m}{r}\right)^{-1} dr^2 + r^2 \mathring{g}_{\mathbb{S}^2}$$

induced on a constant time slice from the spacetime metric $\bar{g}_S$ in (2.4.2). From the form of $\bar{g}_S$, we see that $f = \sqrt{1 - 2m/|x|}$ is a solution of $L_{g_S}^* f = 0$. If we change to isotropic coordinates, $g_S = (1 + \frac{1}{2}m/|x|)^4 g_{\mathbb{E}^3}$, then using the form (2.4.3) of the Schwarzschild metric $\bar{g}_S$ in these coordinates, we see that

$$f = \frac{1 - \frac{1}{2}m/|x|}{1 + \frac{1}{2}m/|x|}$$

is a solution to $L_{g_S}^* f = 0$.

Now consider the converse: suppose $f : M \to \mathbb{R}$ is a smooth function with $L_g^* f = 0$. Such a function is called a *static potential*. By taking the trace of the equation, we obtain $\Delta_g f = -\frac{1}{n-1} R(g) f$, with $n = \dim M \geq 2$. Thus we can rewrite the system $L_g^* f = 0$ as $f \, \mathrm{Ric}(g) = \mathrm{Hess}_g f + \frac{1}{n-1} R(g) f g$. On the open set where $f \neq 0$, then, we see from Lemma 2-40 that $\mathrm{Ric}(\bar{g}) = \frac{1}{n-1} R(g) \bar{g}$. By applying this to the Schwarzschild metric $\bar{g}_S$, we could conclude $R(g_S) = 0$, i.e., the Riemannian Schwarzschild metric g_S has vanishing scalar curvature.

Exercise 2-42. For $(M, g) = (\mathbb{R}^n, g_{\mathbb{E}^n})$, show that the solution space to $L_{g_{\mathbb{E}^n}}^* f = 0$ is the span of the constants and coordinate functions x^i, $i = 1, \ldots, n$. In the case of a flat torus $(\mathbb{T}^n, g_{\mathbb{F}})$, the only solutions to $L_{g_{\mathbb{F}}}^* f = 0$ are the constant functions. Derive this from knowing the solutions on Euclidean space, or from the general fact that $L_g^* f = 0$ and $R(g) = 0$ implies f is harmonic.

Exercise 2-43. In the case of a round sphere $(\mathbb{S}^n, \mathring{g})$, which we consider embedded as the unit sphere about the origin in $\mathbb{R}^{n+1}$, the solutions to $L_{\mathring{g}}^* f = 0$ are the restrictions of the coordinates $x^1, \ldots, x^{n+1}$ to the sphere. Show that these functions solve $L_{\mathring{g}}^* f = 0$, and apply the next proposition to conclude these span the space of all solutions. These functions are eigenfunctions for $\Delta_{\mathring{g}}$, with $\lambda = n$; cf. Remark 7-27. (Hint: Compute $\mathrm{Hess}_{\mathring{g}}(x^i)$ by computing $\langle D_Y(\mathrm{grad}_{\mathring{g}} x^i), Z \rangle$ for Y and Z tangent to the sphere, where D is the Levi-Civita connection for the Euclidean metric; use the fact that $\mathrm{grad}_{\mathring{g}} x^i$ is the tangential component of

$\mathrm{grad}_{g_{\mathbb{E}}} x^i = e_i = \partial/\partial x^i$, and along the unit sphere centered at the origin, the position vector x is a unit normal.)

Proposition 2-44. *Let (M, g) be a connected Riemannian manifold with $n = \dim M \geq 2$. The dimension of the space of solutions to $L_g^* f = 0$ is at most $(n+1)$. For any $f : M \to \mathbb{R}$ which is a nontrivial solution to $L_g^* f = 0$, the level set $\Sigma := \{p \in M : f(p) = 0\}$ is either empty or a smooth totally geodesic hypersurface, along any component of which $|df|_g^2$ is a nonzero constant.*

Proof. We first show that the set $\{f = 0\}$ is a regular level set for f. Suppose $L_g^* f = 0$, and assume $\{f = 0\}$ is nonempty. Let $p \in M$ with $f(p) = 0$. We claim that since f is nontrivial, df_p cannot vanish; equivalently, if $df_p = 0$, then f would be trivial. To see this, observe that the system $L_g^* f = 0$ can be rewritten as $\mathrm{Hess}_g f + f\left(\frac{1}{n-1} R(g) g - \mathrm{Ric}(g)\right) = 0$. If $\gamma(t)$ is a unit-speed geodesic emanating from $p = \gamma(0)$, and if we let $h(t) = f(\gamma(t))$, then since $D\gamma'/dt = 0$, we have $h''(t) = \mathrm{Hess}_g f(\gamma'(t), \gamma'(t))$, so that $h(t)$ satisfies the second-order linear ODE

$$h''(t) + \left(\tfrac{R(g)}{n-1} - \mathrm{Ric}_g(\gamma'(t), \gamma'(t))\right) h(t) = 0.$$

By the conditions satisfied by f and p, $h(0) = 0$ and $h'(0) = df_p(\gamma'(0)) = 0$, so that $h(t)$ vanishes identically. Since the geodesic was arbitrary, we have that $f = 0$ in a neighborhood of p.

If (M, g) were complete, the argument above would suffice. More generally, let S be the interior of $\{f = 0\}$, which we have just argued is nonempty. If q is in the closure of S, there is a point $s \in S$ close enough to q so that q is contained in a geodesic ball B around s. Since $s \in S$, $f(s) = 0$ and $df_s = 0$, from which we conclude as above that $f = 0$ on B; in particular, $q \in B \subset S$. Thus $S \subset M$ is nonempty, open (by definition) and closed. By connectedness, $S = M$.

The dimension bound follows from the fact that by the ODE argument, a solution is determined by the values of $f(p)$ and df_p at a single point.

That $\Sigma = \{f = 0\}$ is totally geodesic in case it is a hypersurface (i.e., nonempty and not all of M) follows directly from the fact that along Σ, $L_g^* f = 0$ reduces to $\mathrm{Hess}_g f = 0$, and thus the gradient of f, which is normal to Σ, has constant nonzero length along any component of Σ. $\qquad\square$

We summarize our discussion with a corollary (using Lemma 2-12, cf. Exercise 2-52), and we emphasize that a nontrivial solution to $L_g^* f = 0$ gives rise to a *static vacuum metric* $\bar{g} = -f^2 dt^2 + g$, possibly with nonzero Λ.

Corollary 2-45. *Let (M^n, g), $n \geq 2$, be a connected Riemannian manifold which admits a nontrivial solution f of $L_g^* f = 0$, and let $\bar{g} = -f^2 dt^2 + g$ on $\mathbb{R} \times \{f \neq 0\}$. Then the scalar curvature $R(g)$ is constant, and $\bar{g}$ is an Einstein metric with $\mathrm{Ric}(\bar{g}) = \frac{R(g)}{n-1} \bar{g}$.*

Exercise 2-46. What changes, if anything, if at the beginning of this section, we allowed (M, g) to be semi-Riemannian, and/or if we defined $\bar{g} = f^2 \, dt^2 + g$?

2.4.6. *The Kerr metric.* The Kerr metrics form a family of axisymmetric Ricci-flat metrics, which includes as a subset the Schwarzschild metrics, with their full spherical symmetry. Such metrics model the exterior of a rotating gravitational object. We will write down the metric in a certain coordinate system, the Boyer–Lindquist coordinates. Let r, ϕ and θ be spherical coordinates on three-space. We will use the mathematical convention that θ is the polar angle, whereas ϕ is the angle between the position vector and the z-axis: in most physics books, the angle notation is reversed, so be careful when comparing. In any case, the metric appears somewhat complicated, and depends on two parameters, m and a. For simplicity, we use units where $G = 1$ and $c = 1$; in general, we can replace "m" with "Gm/c^2" and "dt" with "$c\,dt$" in what follows to obtain the formula for general units. See [161, p. 36] for more on geometrized units.

Let $\Delta = r^2 - 2mr + a^2$ and $\rho^2 = r^2 + a^2 \cos^2 \phi$. The Kerr metric is given by

$$g = -\left(1 - \frac{2mr}{\rho^2}\right) dt^2 - a\frac{2mr \sin^2 \phi}{\rho^2}(dt \otimes d\theta + d\theta \otimes dt)$$

$$+ \frac{(r^2 + a^2)^2 - a^2 \Delta \sin^2 \phi}{\rho^2} \sin^2 \phi \, d\theta^2 + \frac{\rho^2}{\Delta} dr^2 + \rho^2 \, d\phi^2. \quad (2.4.6)$$

This spacetime metric is *stationary*: $\partial/\partial t$ is a timelike Killing vector, but unlike the Schwarzschild situation, if $a \neq 0$, this Killing field is not orthogonal to the constant t-slices. The axisymmetry is apparent, as $\partial/\partial\theta$ is also a Killing vector. One can define suitably the angular momentum (see (7.2.6)) and show that it is $J = am$ in the z-direction. When $a = 0$, the metric reduces to the Schwarzschild metric.

General relativity predicts that a rotating massive object will cause the spacetime around it to be "warped". One such manifestation of this is the *frame dragging effect*, which we illustrate in the Kerr metric. We use the following lemma saying that a symmetry gives rise to a conserved quantity along geodesics.

Lemma 2-47. *If X is a Killing field on (M, g), and if γ is a geodesic, then $\langle X, \gamma' \rangle$ is constant.*

Proof. Since $\gamma(\tau)$ has constant speed, $\langle \nabla_X \gamma', \gamma' \rangle = 0$. By the geodesic equation, then, $d\langle X, \gamma' \rangle/d\tau = \langle \nabla_{\gamma'} X, \gamma' \rangle = g_{ij}(\gamma^k)' X^i_{;k}(\gamma^j)' = X_{j;k}(\gamma^k)'(\gamma^j)' = X_{k;j}(\gamma^k)'(\gamma^j)'$. We see this must vanish by applying the Killing equation $X_{j;k} + X_{k;j} = 0$. $\square$

Consider in any spacetime a timelike geodesic parametrized by proper time τ, with velocity vector $\mathbb{U}(\tau)$. In the Kerr metric, $X = \partial/\partial\theta$ is a Killing vector, so that by Lemma 2-47, $\mathbb{U}_\theta = \langle \mathbb{U}, \partial/\partial\theta \rangle$ is conserved along the geodesic. One can see this directly from the geodesic equations $\mathbb{U}^\alpha \mathbb{U}_{\alpha;\beta} = 0$, which can be written

$$\frac{d\mathbb{U}_\beta}{d\tau} = \Gamma^\gamma_{\alpha\beta}\mathbb{U}^\alpha\mathbb{U}_\gamma = \tfrac{1}{2}g^{\gamma\nu}(g_{\nu\beta,\alpha} + g_{\alpha\nu,\beta} - g_{\alpha\beta,\nu})\mathbb{U}^\alpha\mathbb{U}_\gamma$$

$$= \tfrac{1}{2}(g_{\nu\beta,\alpha} + g_{\alpha\nu,\beta} - g_{\alpha\beta,\nu})\mathbb{U}^\alpha\mathbb{U}^\nu = \tfrac{1}{2}g_{\alpha\nu,\beta}\mathbb{U}^\alpha\mathbb{U}^\nu.$$

Since the metric is independent of θ, then $\mathbb{U}_\theta$ is conserved. If the geodesic represents the path of a particle of rest mass m_0, then $\mathbb{P}_\theta = m_0\mathbb{U}_\theta$ is a component of the momentum one-form which is conserved.

Now consider a geodesic in Kerr with $\mathbb{U}_\theta = \langle \mathbb{U}, \partial/\partial\theta \rangle = 0$. Although $\mathbb{U}$ remains orthogonal to $\partial/\partial\theta$, the fact that $g_{t\theta} \neq 0$ means that $\mathbb{U}^\theta$ will be nonzero. In fact, we have $\mathbb{U}^t = dt/d\tau \neq 0$ and $\mathbb{U}^\theta = d\theta/d\tau$. Thus

$$\frac{d\theta}{dt} = \frac{\mathbb{U}^\theta}{\mathbb{U}^t} = \frac{g^{\theta\alpha}\mathbb{U}_\alpha}{g^{t\alpha}\mathbb{U}_\alpha} = \frac{g^{\theta t}}{g^{tt}} = -\frac{g_{\theta t}}{g_{\theta\theta}} = \frac{2mra}{(r^2+a^2)^2 - a^2\Delta\sin^2\phi}.$$

Although in the metric the trajectory is orthogonal to the direction of the symmetry given by the Killing field $\partial/\partial\theta$, the velocity has a nontrivial component in the θ-direction. Thus a freely falling object seems to pick up some coordinate angular momentum in the direction of rotation of the massive object, say, generating this gravitational field (metric). For tests of the effect of the rotation of a massive object on spacetime, see the results of the Gravity Probe B experiment [87].

Exercises

Exercise 2-48. Consider the Minkowski plane with inertial coordinates $x^0 = ct$ and x^1. Let $g > 0$ be a constant, and introduce a set of coordinates (ξ^0, ξ^1) with $\xi^1 > -c^2/g$, such that the change of coordinates is given by

$$x^0 = \left(\xi^1 + \frac{c^2}{g}\right)\sinh\frac{g\xi^0}{c^2}, \qquad x^1 = \left(\xi^1 + \frac{c^2}{g}\right)\cosh\frac{g\xi^0}{c^2}.$$

It is a straightforward matter to express (ξ^0, ξ^1) as smooth functions of the inertial coordinates.

a. Sketch a grid of level sets of ξ^0 and ξ^1 in the x^0x^1-plane, and indicate the region covered by these coordinates. Show that the ξ^0 and ξ^1 coordinate curves are orthogonal for the Minkowski metric, and that the metric spacing of the grid points along $\xi^0 = b$ is independent of b. To do this, find the components $\bar{g}_{\mu\nu}$ for the Minkowski metric $-(dx^0)^2 + (dx^1)^2 = \bar{g}_{\mu\nu}d\xi^\mu d\xi^\nu$, and confirm that it retains a static form in these coordinates.

b. Find the proper acceleration along the worldlines of constant ξ^1. Show that for $|g\xi^0/c^2|$ small, the motion along any such worldline approximates that of uniform Newtonian acceleration in the inertial coordinates.

c. Interpreting the metric in the (ξ^0, ξ^1) coordinates for $|g\xi^1/c^2|$ small in terms of a Newtonian limit, estimate the Newtonian gravitational potential Φ, and show it is consistent with that of a uniform gravitational field.

d. Let C_ξ be the worldline $\xi^1 = \xi$. Suppose that light is emitted along C_ξ with *coordinate* frequency ν, i.e. with $\Delta\xi^0 = c/\nu$ between successive wave crests. Argue that the light will have the same coordinate frequency of absorption as measured along $C_{\tilde\xi}$ for any $\tilde\xi$. Find the relative change in *proper* frequency from emission along C_ξ to absorption at a detector carried along C_0, conclude that the change in energy per unit *mass* is $c^2(\sqrt{-\bar{g}_{00}(\xi)} - 1)$, and relate this to Φ.

Exercise 2-49 (geometry of the Riemannian Schwarzschild metric). Define

$$
M - \begin{cases} \{x \in \mathbb{R}^3 : x \neq 0\} & \text{if } m > 0, \\ \mathbb{R}^3 & \text{if } m = 0, \\ \{x \in \mathbb{R}^3 : |x| > -\tfrac{m}{2}\} & \text{if } m < 0. \end{cases}
$$

On M, we let $g_S = \left(1 + \frac{m}{2|x|}\right)^4 g_{\mathbb{E}^3}$, so $(g_S)_{ij} = \left(1 + \frac{m}{2|x|}\right)^4 \delta_{ij}$ in Cartesian coordinates, while in spherical coordinates we find $g_S = (1 + \frac{m}{2r})^4(dr^2 + r^2 \mathring{g}_{\mathbb{S}^2})$ with $r = |x|$. For each $r > 0$, let $A(r)$ be the g_S-area of $S_r := \{|x| = r\}$, and for $m > 0$, let $\Sigma = S_{m/2}$ (cf. Exercise 2-37).

a. Calculate $A(r)$. Observe that $\lim_{r \searrow -\frac{m}{2}} A(r) = 0$ for $m < 0$, while if $m > 0$, $A(r)$ has a global minimum at $r = \frac{m}{2}$, and that $A(r) = A\left(\frac{m^2}{4r}\right)$, which is consistent with Exercise 2-37. Solve for m in terms of the area A_Σ of Σ.

b. For $m > 0$, show that Σ is totally geodesic in M by direct computation in coordinates (as opposed to using the inversion isometry, cf. Exercise 2-37).

c. Let $m > 0$. Find an isometric embedding of (M, g_S) into $\mathbb{E}^4$, identified in Cartesian coordinates (x, y, z, w) with $(\mathbb{R}^4, dx^2 + dy^2 + dz^2 + dw^2)$. It might be easiest to use the first set of coordinates we introduced for the Schwarzschild metric, for which $g_S = \left(1 - \frac{2m}{r}\right)^{-1} dr^2 + r^2 \mathring{g}_{\mathbb{S}^2}$, $r > 2m$. For $\omega \in \mathbb{S}^2$, look for an embedding of the form $r\omega \mapsto (r\omega, \xi(r)) \in \mathbb{R}^4$, for $r > 2m$ and $\xi'(r) > 0$. This corresponds to half of (M, g_S). The map you get will then extend by reflection to the other half. Use this to sketch a representative picture of the Riemannian Schwarzschild metric. (If you let $w = \xi(r)$, with $\lim_{r \searrow 2m} \xi(r) = 0$, the solution set in the wr-plane is (half of) a parabola. The image of the embedding in $\mathbb{R}^4$ is called *Flamm's paraboloid*.)

d. When $m < 0$ the above argument breaks down. Instead, look for an isometric embedding into Minkowski spacetime $\mathbb{M}^4$, which is identified with $\mathbb{R}^4$ with the metric $dx^2 + dy^2 + dx^2 - dw^2$.

e. Generalize the above to higher dimensions: consider the higher-dimensional Riemannian Schwarzschild metric g_S, with $\alpha = 2m > 0$, given in isotropic coordinates in Exercise 2-38. Find the value of r that minimizes the area $A(r)$ of $S_r = \{x : |x| = r\}$. $\left(\text{Answer: } r^{n-2} = \frac{m}{2}\right)$. Find an isometric inversion through the corresponding minimal sphere Σ. Show this sphere is totally geodesic, and express m in terms of the area of Σ. Can you embed $(\mathbb{R}^n \setminus \{0\}, g_S)$ with $m > 0$ isometrically in Euclidean space $(\mathbb{R}^{n+1}, g_{\mathbb{E}^{n+1}})$?

Exercise 2-50. In Euclidean space, the spheres minimize surface area for a given enclosed volume V. In fact if a closed surface of area A encloses a volume V, the *isoperimetric inequality* in three dimensions is $V \le A^{3/2}/(6\sqrt{\pi})$.

Let $m > 0$. In his Ph.D. thesis [25] (see also [29]), Hubert Bray showed that the spheres $S_r = \{x : |x| = r\}$ in the Riemannian Schwarzschild metric $\left(\mathbb{R}^3 \setminus \{0\}, \left(1 + \frac{1}{2}m/|x|\right)^4 g_{\mathbb{E}^3}\right)$ are isoperimetric in the homology class of $\Sigma = S_{m/2}$. In other words, amongst all surfaces homologous to the minimal surface Σ and enclosing a certain volume V with Σ, the one with smallest area is the sphere S_r of the correct r value to enclose volume V.

a. Show that the volume $V(r)$ enclosed by Σ and S_r ($r \ge m/2$) has the expansion

$$V(r) = \frac{4\pi r^3}{3}\left(1 + \frac{9m}{2r} + O(mr^{-2})\right).$$

b. Conclude that the volume V enclosed by Σ and the sphere S_r of area A has the expansion

$$V(A) = \frac{A^{3/2}}{6\sqrt{\pi}}\left(1 + \frac{(3\sqrt{\pi})m}{\sqrt{A}} + O(mA^{-1})\right).$$

Exercise 2-51 (conformal deformation of scalar curvature). Suppose (M, g) is a semi-Riemannian manifold of dimension n, and $\tilde{g} = e^{\varphi} g$ where φ is a smooth function on M, and let $u > 0$ be a smooth function on M.

a. Show that, with $\Delta_g \varphi = \mathrm{tr}_g(\mathrm{Hess}_g \varphi)$ in any signature,

$$R(\tilde{g}) = e^{-\varphi}\left(R(g) - (n-1)\Delta_g \varphi - \tfrac{1}{4}(n-1)(n-2)\langle \nabla_g \varphi, \nabla_g \varphi \rangle_g\right).$$

b. In case $n \ge 3$, show that there is a unique choice of $q \ne 0$ such that if we write $e^{\varphi} = u^q$ for $u > 0$, the resulting formula for $R(\tilde{g})$ does not contain any $\langle \nabla_g u, \nabla_g u \rangle_g$-term. In fact, for this to be the case, we must have $q = \frac{4}{n-2}$, and with $\tilde{g} = u^{4/(n-2)}g$, for $u > 0$, also

$$R(\tilde{g}) = u^{-\frac{n+2}{n-2}}\left(R(g)u - 4\frac{(n-1)}{(n-2)}\Delta_g u\right) = -(c(n))^{-1}u^{-\frac{n+2}{n-2}}\mathcal{L}_g u, \quad \text{(2-51a)}$$

where $c(n) = \frac{n-2}{4(n-1)}$ and $\mathcal{L}_g u = \Delta_g u - c(n)R(g)u$ defines the *conformal Laplacian* $\mathcal{L}_g$.

c. Suppose (M, g) is a closed (compact without boundary) Riemannian manifold. Show that the total scalar curvature of $\tilde{g} = u^{4/(n-2)}g$ is given by

$$\int_M R(\tilde{g})\, dv_{\tilde{g}} = c(n)^{-1}\int_M \left(|\nabla_g u|_g^2 + c(n)R(g)u^2\right)dv_g.$$

(Hint: Show that $dv_{\tilde{g}} = u^{2n/(n-2)}\, dv_g$.)

d. For $\tilde{g} = u^{4/(n-2)}g$ and any smooth f, prove that $\mathcal{L}_g f = u^{(n+2)/(n-2)}\mathcal{L}_{\tilde{g}}(f/u)$.

Exercise 2-52. Suppose (M, g) is a Riemannian manifold.

a. Use the Bianchi identity and the Ricci formula to show that $\operatorname{div}_g L_g^* f = -\frac{1}{2}f\, dR(g)$. Use this and Proposition 2-44 to give another proof of Corollary 2-45.

b. Find the kernel of L_g^* if (M, g) is closed and has negative scalar curvature. (Hint: Exercise 1-11.)

c. Observe that if $\operatorname{Ric}(g) = 0$, then L_g^* has nontrivial kernel. Recall an example of a Ricci-flat metric where the kernel of L_g^* has dimension greater than one, and an example of a metric with zero scalar curvature and which is not Ricci-flat, but for which L_g^* is nontrivial. What can you say about the kernel of L_g^* if (M, g) is closed with zero scalar curvature?

d. Consider the metric $g = (n-2)^{-1}\mathring{g}_{\mathbb{S}^1} \oplus \mathring{g}_{\mathbb{S}^{n-1}}$ on $\mathbb{S}^1 \times \mathbb{S}^{n-1}$. Show that $f(t, \omega) = \sin t$ is in the kernel of L_g^*.

Exercise 2-53 (kernel of $L_{g_S}^*$). Let g_S be a Riemannian Schwarzschild metric of nonzero mass m. Show that there is a one-dimensional kernel for $L_{g_S}^*$. Outline: Observe that $L_{g_S}^* f = 0$ implies $\operatorname{Hess}_{g_S} f = f\operatorname{Ric}(g_S)$. (Recall from Remark 2-35 that whereas $\operatorname{Ric}(\bar{g}_S) = 0$, $\operatorname{Ric}(g_S) \neq 0$). In three dimensions, write this out in coordinates for which $g_S = (1 - 2m/r)^{-1}dr^2 + r^2(d\varphi^2 + \sin^2\varphi\, d\theta^2)$. Show that $\partial_\theta f = 0$ and $\partial_\varphi f = 0$, and then solve the remaining ODE for f. Compare your answer with Example 2-41; for $m > 0$, note where this potential vanishes (cf. Exercise 2-49). Generalize the argument to higher dimensions.

Exercise 2-54 (metric expansion in normal coordinates). Suppose ∇ is the Levi-Civita connection for a metric $g = \langle \cdot, \cdot \rangle$ on M. Suppose that $\gamma(t)$ is a unit-speed geodesic, and that $J(t)$ is a Jacobi field along γ: $J''(t) = R(\gamma'(t), J(t), \gamma'(t))$ (cf. Proposition 2-3).

a. If R is the Riemann curvature tensor, show that

$$J'''(t) = (\nabla_{\gamma'(t)}R)(\gamma'(t), J(t), \gamma'(t)) + R(\gamma'(t), J'(t), \gamma'(t)).$$

b. Suppose $J(0) = 0$. Let $\chi(t) = \langle J(t), J(t)\rangle$. Derive the fourth-order Taylor expansion

$$\chi(t) = \langle J'(0), J'(0)\rangle t^2 - \tfrac{1}{3}\langle R(\gamma'(0), J'(0), J'(0)), \gamma'(0)\rangle t^4 + O(t^5).$$

c. Suppose (M, g) is a Riemannian manifold and $p \in M$. Show that in normal coordinates centered at p (so $x^i(p) = 0$)

$$g_{ij}(x) = \delta_{ij} - \tfrac{1}{3}R_{kij\ell}x^k x^\ell + O(|x|^3).$$

(Hint: In normal coordinates (x^i) centered at p, consider a unit speed radial geodesic $\gamma(t)$ and the vector field $W(t) = tW^i \partial/\partial x^i$ along γ, where the W^i are constants. (Note that the summation convention applies here.) Show that $W(t)$ is a Jacobi field along γ with $W'(0) = W^i(\partial/\partial x^i)|_p$. One way to do this is to build a variation $\Gamma(s, t)$ of $\gamma = \Gamma(0, \cdot)$ through geodesics. In any case, you might first observe that in normal coordinates, the curve β with components $\beta^i(t) = tV^i$ is a geodesic with $\beta(0) = p$ and $\beta'(0) = V^i(\partial/\partial x^i)|_p$.)

Exercise 2-55 (geometric formula for Gaussian curvature). Let (M, g) be a surface with a Riemannian metric g. Consider an orthonormal basis $\{e_1, e_2\}$ of T_pM. Note that the Gauss curvature at p is just $K(p) = \langle R(e_1, e_2, e_2), e_1\rangle$, where R is the Riemann tensor. Consider a normal neighborhood of radius $a > 0$ about p, with normal coordinates (x, y) built using the orthonormal basis $\{e_1, e_2\}$ of T_pM: $(x, y) \mapsto \exp_p(xe_1 + ye_2)$. Define geodesic polar coordinates by $(r, \theta) \mapsto f(r, \theta) = \exp_p(r\cos\theta\, e_1 + r\sin\theta\, e_2)$. Note that the change of coordinates map is just $(r, \theta) \mapsto (x, y) = (r\cos\theta, r\sin\theta)$, which shows that the map f, which is clearly smooth for $r < a$ and all θ, is a diffeomorphism for $0 < r < a$ and $\theta \in I$, where I in any open interval of length at most 2π. Note that by Gauss's lemma, the metric components in geodesic polar coordinates are $g_{rr} = 1$, $g_{r\theta} = 0$ and $g_{\theta\theta} = |\partial f/\partial\theta|_g^2$. Since the radial curves of constant θ on M are geodesics, by Proposition 2-3, for any θ_0, $J(r) = (\partial f/\partial\theta)(r, \theta_0)$ is a Jacobi field along the radial geodesic $r \mapsto \gamma(r) = f(r, \theta_0)$.

a. For any θ_0, show that $J'(0) \perp \gamma'(0)$ and $|J'(0)|_g = 1$. (Equation (2.3.8) might be useful.)

b. Use part b. of Exercise 2-54 to show that $g_{\theta\theta}(r, \theta) = r^2 - \tfrac{1}{3}K(p)r^4 + \mathcal{E}(r, \theta)$, where $\mathcal{E}(r, \theta) = O(r^5)$ uniformly in θ, i.e., $\mathcal{E}(r, \theta) \leq Cr^5$, where C can be chosen independent of r (small) and θ. Use the Taylor expansion $(1 + x)^\alpha = 1 + \alpha x + O(x^2)$ to derive $\sqrt{g_{\theta\theta}}(r, \theta) = r - \tfrac{1}{3!}K(p)r^3 + O(r^4)$.

c. Let $L(r)$ be the length of a geodesic circle of radius r about p, and let $A(r)$ be the area enclosed by this circle, both computed using the metric g. Show that

$$\lim_{r \searrow 0} \frac{3}{\pi} \frac{2\pi r - L(r)}{r^3} = K(p) = \lim_{r \searrow 0} \frac{12}{\pi} \frac{\pi r^2 - A(r)}{r^4}.$$

d. Let $\mathbb{D} = \{(x, y) : x^2 + y^2 < 1\}$ be the unit disk in the plane, with polar coordinates (ρ, θ), and consider the hyperbolic metric

$$g_\mathbb{H} = \frac{4}{(1 - (x^2 + y^2))^2}(dx^2 + dy^2) = \frac{4}{(1 - \rho^2)^2}(d\rho^2 + \rho^2 d\theta^2).$$

By solving the differential equation

$$\frac{2d\rho}{1 - \rho^2} = dr,$$

show how to rewrite the hyperbolic metric as $g_\mathbb{H} = dr^2 + \sinh^2 r\, d\theta^2$. Use this along with the formulas above to show $K = -1$ at the origin of coordinates (of course, $K = -1$ everywhere).

Exercise 2-56 (volume expansion of geodesic balls). In this problem you will generalize the relation of curvature and area from part c. of Exercise 2-55 to higher dimensions, showing how the scalar curvature measures the top-order deviation of the volume of small geodesic balls from that of Euclidean geometry.

a. Suppose $(V, \langle\,,\,\rangle)$ is an n-dimensional real inner product space. Suppose that $T : V \to V$ is a self-adjoint linear operator. If $d\sigma$ is the Euclidean area measure, $\mathbb{B}^n$ is the unit ball, and $\mathbb{S}^{n-1} = \{x \in V : \langle x, x \rangle = 1\} \subset V$ is the unit sphere in V, then if Vol is the Euclidean volume,

$$\int_{\mathbb{S}^{n-1}} \langle T(x), x \rangle\, d\sigma = \text{trace}(T)\; \text{Vol}\,\mathbb{B}^n.$$

b. If (M, g) is Riemannian and $p \in M$, let $B_r(p) \subset M$ be the geodesic ball of radius $r > 0$ (for sufficiently small r). Then

$$\text{Vol}_g(B_r(p)) = \text{Vol}(\mathbb{B}^n)\, r^n \left(1 - \frac{R(g)|_p}{6(n+2)}\, r^2 + O(r^3)\right).$$

(Hint: You may wish to use $\det(I + tA) = 1 + t\,\text{trace}(A) + O(t^2)$, along with Exercise 2-54.)

CHAPTER 3

Basics of Lorentzian causality

In this chapter we introduce some basic concepts of Lorentzian causality needed in the discussion of the Penrose singularity theorem in the next chapter. For a more comprehensive treatment of the causal structure of a spacetime, readers are referred to [179; 112; 174; 218; 59; 62; 101].

3.1. Preliminaries from Lorentzian geometry

Let $\mathcal{S}$ be a connected, smooth Lorentz manifold of dimension $n + 1$. Let $\langle \cdot, \cdot \rangle$ denote the Lorentzian metric on $\mathcal{S}$ which has signature $(-, +, \ldots, +)$. We classify tangent vectors v to $\mathcal{S}$ as follows:

$$v \text{ is } \begin{cases} \text{timelike} & \text{if } \langle v, v \rangle < 0, \\ \text{null} & \text{if } \langle v, v \rangle = 0, \text{ and } v \neq 0, \\ \text{spacelike} & \text{if } \langle v, v \rangle > 0, \text{ or } v = 0. \end{cases}$$

3.1.1. *Time-orientability.* Given $p \in \mathcal{S}$, the set of timelike vectors $v \in T_p\mathcal{S}$ is the disjoint union of two open cones, known as *timecones* at p. While there is no intrinsic choice of one of these timecones at each point in $\mathcal{S}$ to serve as the designation of the *future* direction, for the purpose of doing this in a continuous manner, we introduce the concept of time-orientability. Let $\mathcal{C}$ be the set of all timecones on $\mathcal{S}$. We say that a map $\phi : \mathcal{S} \to \mathcal{C}$ is a *time-orientation* on $\mathcal{S}$ if $\phi(p) \subset T_p\mathcal{S}$ and $\phi(p)$ depends smoothly on p in the sense that, for all $p \in \mathcal{S}$, there exists an open neighborhood U of p and a smooth timelike vector field X on U such that $X(q) \in \phi(q)$, for all $q \in U$. If $\mathcal{S}$ admits such a time-orientation ϕ, we say $\mathcal{S}$ is *time-orientable.* In this case, one can designate $\phi(p)$ as the *future* timecone at p, and the other cone as the *past* timecone at p. A time-orientable $\mathcal{S}$ with designated future (and past) timecones is called *time-oriented.*

Using a partition of unity, one obtains:

Lemma 3-1 [174, Lemma 5.32]. *$\mathcal{S}$ is time-orientable if and only if there exists a smooth timelike vector field on $\mathcal{S}$.*

Lemma 3-1 reveals a sufficient (and indeed necessary) topological condition for a differentiable manifold to admit a time-orientable Lorentzian metric:

Proposition 3-2 [174, Proposition 5.37]. *Let M be a smooth manifold. If M is noncompact or if M is compact and has Euler number $\chi(M) = 0$, then there exists a smooth time-orientable Lorentzian metric on M.*

Proof. The given topological assumption implies that M has a smooth, nowhere vanishing vector field X (see [217]). Let g be any smooth Riemannian metric on M. Since X is nowhere vanishing, one can normalize X to obtain a g-unit vector field n. Let ω be the 1-form dual to n with respect to g. Then $\bar{g} = g - 2\omega \otimes \omega$ is time-orientable Lorentzian metric on M. $\qquad\square$

Remark 3-3. By Lemma 3-11 below, all physical spacetimes are noncompact.

3.1.2. Causal curves. Suppose $\mathcal{S}$ is time-oriented. A timelike or null tangent vector v is called *future-directed* if v lies in the closure of the future timecone; *past-directed* is defined similarly. Given a differentiable curve $\sigma : [a, b] \to \mathcal{S}$, we say

$$\sigma \text{ is } \begin{cases} \text{timelike} & \text{if } \sigma'(t) \text{ is timelike for all } t, \\ \text{null} & \text{if } \sigma'(t) \text{ is null for all } t, \\ \text{causal} & \text{if } \sigma'(t) \text{ is either timelike or null for all } t. \end{cases}$$

A timelike, null, or causal curve σ is called *future-directed* if $\sigma'(t)$ is future-directed. Past-directed curves are defined in a similar way.

These notions extend naturally to piecewise differentiable curves σ by requiring that, when σ is differentiable on $[a, b]$ and $[b, c]$, the vectors $\sigma'(b^-)$ and $\sigma'(b^+)$ point into the closure of the same timecone.

From now on this chapter, we will work within the class of piecewise differentiable causal curves. We state a basic result regarding the deformation of causal curves:

Theorem 3-4 [174, Theorem 10.51]. *Given a causal curve σ from p to q in $\mathcal{S}$, there exists a timelike curve from p to q which is arbitrarily near σ unless σ is a null geodesic (when suitably parametrized) along which there are no conjugate points of p before q.*

We refer readers to Section 4.1 for the definition of a conjugate point.

3.1.3. Convex open sets. For the later purpose of understanding the causality relations locally on $\mathcal{S}$, we recall the concept of convex open sets in $\mathcal{S}$.

Given $p \in \mathcal{S}$, an open neighborhood U of p is called a *normal neighborhood* of p provided $U = \exp_p(\tilde{U})$, where $\exp_p(\cdot)$ denotes the exponential map at p and $\tilde{U} \subset T_p\mathcal{S}$ is a starshaped open set containing 0 such that $\exp_p : \tilde{U} \to U$ is a diffeomorphism (see [174, p. 71]).

An open set U in S is said to be *convex* if U is a normal neighborhood of each $p \in U$. Thus, given any two points p, q in a convex U, there exists a unique geodesic segment $\sigma_{pq} : [0, 1] \to S$ from p to q which lies entirely in U. Every point in S has a convex open neighborhood [174, Proposition 5.7].

3.2. Causality relations

Henceforth in this chapter, S denotes a *spacetime*, i.e., a connected, time-oriented Lorentz manifold. The *causality relations* $\ll$ and $<$ on S are defined as follows.

Definition 3-5. For $p, q \in S$,

- $p \ll q$ means there exists a future-directed timelike curve from p to q;

- $p < q$ means there exists a future-directed causal curve from p to q.

By Theorem 3-4, one has:

Lemma 3-6. *If $p < q$ and $q \ll r$, or if $p \ll q$ and $q < r$, then $p \ll r$.*

Given $p \in S$, the *timelike future* and *causal future* of p, denoted by $I^+(p)$ and $J^+(p)$ respectively, are defined by

$$I^+(p) = \{q \in S : p \ll q\} \quad \text{and} \quad J^+(p) = \{q \in S : p \le q\}.$$

Here, the notation $p \le q$ means either $p = q$ or $p < q$. The *timelike past* and *causal past* of p, denoted by $I^-(p)$ and $J^-(p)$ respectively, are defined in a time-dual manner.

Naturally one wonders if $I^+(p)$ is open in S. To answer this question, it is convenient to restrict the causality relations to an open set U of S. Given $p \in U$, the *timelike future of p within U*, denoted by $I^+(p, U)$, consists of all points q in U for which there exists a future-directed timelike curve within U from p to q. Similarly, one defines $J^+(p, U)$.

When U is *convex*, the causality relations in U are as simple as those of the Minkowski spacetime.

Lemma 3-7 (see [174, p. 403]). *Let U be a convex open set in S, and let $p, q \in U$.*

- *$q \in I^+(p, U)$ if and only if the unique geodesic segment σ_{pq} connecting p to q within U is future-directed timelike. Consequently, $I^+(p, U)$ is open in U (hence open in S).*

- *If $q \ne p$, then $q \in J^+(p, U)$ if and only if the unique geodesic segment σ_{pq} in U is future-directed causal. Consequently, $J^+(p, U)$ is the closure of $I^+(p, U)$ in U.*

Lemma 3-7 implies that $\ll$ is indeed an open relation on S.

Proposition 3-8. *Let $p, q \in \mathcal{S}$. If $p \ll q$, then there are open neighborhoods U and V of p and q respectively, such that $\tilde{p} \ll \tilde{q}$ for all $\tilde{p} \in U$ and $\tilde{q} \in V$.*

Proof. Let $\sigma : [0, 1] \to \mathcal{S}$ be a future-directed timelike curve from p to q. Let W_q be a convex open neighborhood of $q = \sigma(1)$. Choose $\epsilon > 0$ small such that $\sigma(t) \in W_q$ for all $t \in [1 - \epsilon, 1]$. Then $q \in I^+(\sigma(1 - \epsilon), W_q)$. Similarly, one can choose a convex open neighborhood W_p of $p = \sigma(0)$ and a small $\delta > 0$ such that $p \in I^-(\sigma(\delta), W_p)$. By Lemma 3-6, the open sets $I^+(\sigma(1 - \epsilon), W_q)$ and $I^-(\sigma(\delta), W_p)$ have the required properties of V and U respectively. $\qquad\square$

Given a subset $A \subset \mathcal{S}$, the *timelike future* and *causal future* of A are defined respectively by

$$I^+(A) = \bigcup_{p \in A} I^+(p) \quad \text{and} \quad J^+(A) = \bigcup_{p \in A} J^+(p).$$

It follows from Proposition 3-8 that $I^+(A)$ is an open set in $\mathcal{S}$.

We now turn to the structure of $J^+(A)$. In general, $J^+(A)$ need not be the closure of $I^+(A)$; in fact, $J^+(A)$ may not even be closed, as illustrated by the Minkowski spacetime with one point removed. However, we do have this:

Lemma 3-9. *Let $A \subset \mathcal{S}$.*

 (i) *int $J^+(A) = I^+(A)$. Here int $J^+(A)$ denotes the interior of $J^+(A)$.*

 (ii) *$J^+(A) \subset \overline{I^+(A)}$ with equality if and only if $J^+(A)$ is a closed set.*

Proof. (i) It suffices to show int $J^+(A) \subset I^+(A)$. Let $q \in$ int $J^+(A)$, then there is an open neighborhood U of q such that $U \subset J^+(A)$. Take any $q^- \in I^-(q, U)$, then $q^- \in J^+(A)$. By Lemma 3-6, $q \in I^+(A)$.

The proof of (ii) is left as an exercise. $\qquad\square$

As a direct corollary of Theorem 3-4, the set $J^+(A) \setminus (A \cup I^+(A))$ has an interesting geometric structure.

Proposition 3-10. *Given $A \subset \mathcal{S}$, suppose $q \in J^+(A) \setminus (A \cup I^+(A))$. Let σ be a future-directed causal curve ending at q and starting from some $p \in A$. Then σ (when suitably parametrized) is a future-directed null geodesic along which there are no conjugate points of p before q; moreover σ does not intersect $I^+(A)$.*

3.3. Causality conditions

In a physical spacetime, one does not expect observers to be able to travel to their own past. A spacetime $\mathcal{S}$ is said to satisfy the *chronology condition* if it contains no closed timelike curves.

Lemma 3-11. *A compact spacetime does not satisfy the chronology condition.*

Proof. Suppose S is compact. Since $I^+(p)$ is open for all p, there exist a finite number of points $p_1, \ldots, p_m$ such that $S \subset \bigcup_{i=1}^{m} I^+(p_i)$. If $p_m \in I^+(p_m)$, then there is a closed timelike curve through p_m. If $p_m \notin I^+(p_m)$, then $p_m \in I^+(p_j)$ for some $j \le m - 1$, hence $S \subset \bigcup_{i=1}^{m-1} I^+(p_i)$. Repeating this argument, one concludes that S must contain a closed timelike curve. $\qquad\square$

A spacetime S is said to satisfy the *causality condition* if it contains no closed causal curves. Despite being stronger than the chronology condition, the causality condition itself in many cases is not well suited for doing analysis on S because it does not rule out existence of "almost closed" causal curves. For this reason, the following condition is often imposed.

Definition 3-12. The *strong causality condition* is said to hold at a point $p \in S$ provided that given any open neighborhood U of p there exists an open neighborhood $V \subset U$ of p such that every causal curve $\sigma : [0, 1] \to S$ with $\alpha(0) \in V$ and $\alpha(1) \in V$ lies entirely in U.

A spacetime S is said to be *strongly causal* if the strong causality condition holds at every $p \in S$.

The following lemma gives a good illustration of the implication of the strong causality condition.

Lemma 3-13. *Suppose $K \subset S$ is a compact set and the strong causality condition holds at every point in K. Let $\sigma : [0, b) \to S$, where $b \le \infty$, be a future-directed causal curve such that $\sigma(0) \in K$. If σ is future-inextendible, i.e., if $\lim_{t \nearrow b} \sigma(t)$ does not exist, then σ eventually leaves K: that is, there exists a $T \in (0, b)$ such that $\sigma(t) \notin K$, for all $t > T$.*

Proof. If the conclusion is false, there exists an increasing sequence $\{t_i\} \subset (0, b)$ such that $\sigma(t_i) \in K$ and $\lim_{i \to \infty} t_i = b$. As K is compact, passing to a subsequence, one may assume $\lim_{i \to \infty} \sigma(t_i) = p$ for some $p \in K$. Applying the strong causality condition at p, one can show $\lim_{t \nearrow b} \sigma(t) = p$, which is a contradiction. $\qquad\square$

The (strong) causality condition is used to define a globally hyperbolic spacetime.

Definition 3-14. A spacetime S is said to be *globally hyperbolic* if

(1) S is strongly causal, and

(2) the sets $J^+(p) \cap J^-(q)$ are compact for all $p, q \in S$.

Remark 3-15. In [21], it was shown that the causality condition together with condition (2) in fact implies the strong causality condition.

A direct consequence of condition (2) in Definition 3-14 is the closedness of $J^{\pm}(p)$.

Proposition 3-16. *Suppose a spacetime $\mathcal{S}$ satisfies condition 2 above, that is, $J^+(p) \cap J^-(q)$ is compact for all $p, q \in \mathcal{S}$. Then $J^+(A)$ and $J^-(A)$ are closed for any compact subset $A \subset \mathcal{S}$.*

Proof. First consider the case where $A = \{p\}$. Let $\{q_k\} \subset J^+(p)$ and assume $\lim_{k \to \infty} q_k = q$. Take $q^+ \in I^+(q)$, then $q_k \in I^-(q^+) \subset J^-(q^+)$ for large k. Hence, $q_k \in J^+(p) \cap J^-(q^+)$ which is compact by assumption. Therefore, $q = \lim_{k \to \infty} q_k \in J^+(p)$. This shows $J^+(p)$ (and $J^-(p)$) are always closed.

Next suppose A is compact. Suppose $\{q_k\} \subset J^+(A)$ and $\{p_k\} \subset A$ such that $q_k \in J^+(p_k)$ and $\lim_{k \to \infty} q_k = q$. As A is compact, passing to a subsequence, one may assume $\lim_{k \to \infty} p_k = p \in A$. Now take a sequence $\{p_m^-\} \subset I^-(p)$ such that $\lim_{m \to \infty} p_m^- = p$. For each fixed m, when k is sufficiently large, $p_k \in I^+(p_m^-)$ which implies $q_k \in J^+(p_m^-)$. Since $J^+(p_m^-)$ is closed, one has $q \in J^+(p_m^-)$ or equivalently $p_m^- \in J^-(q)$. Since $J^-(q)$ is closed, one concludes $p \in J^-(q)$ and hence $q \in J^+(p) \subset J^+(A)$. This shows $J^+(A)$ is closed. $\square$

Remark 3-17. The proof of Proposition 3-16 indeed shows that $\leq$ is a closed relation on $\mathcal{S}$ if the spacetime $\mathcal{S}$ satisfies condition (2) in Definition 3-14.

3.4. Achronal sets

We now discuss properties of certain subsets in a spacetime $\mathcal{S}$. A subset $A \subset \mathcal{S}$ is said to be *achronal* if $A \cap I^+(A) = \varnothing$, i.e., if the relation $p \ll q$ never holds for $p, q \in A$.

Since $\ll$ is an open relation (Proposition 3-8), one has:

Lemma 3-18. *If A is achronal, so is its closure $\bar{A}$.*

An important example of an achronal set is the topological boundary of the timelike future (and past) of any given set.

Lemma 3-19. *Let $A \subset \mathcal{S}$.*

(i) *If $p \in \partial I^+(A)$, then $I^+(p) \subset I^+(A)$, and $I^-(p) \subset \mathcal{S} \setminus \overline{I^+(A)}$.*

(ii) *$\partial I^+(A)$ is achronal.*

Proof. (i) Take $q \in I^+(p)$. Let $\{p_k\} \subset I^+(A)$ such that $\lim_{k \to \infty} p_k = p$. As $I^-(q)$ is an open set containing p, $p_k \in I^-(q)$ for large k. Hence, $q \in I^+(A)$. This proves $I^+(p) \subset I^+(A)$. Similarly, one shows $I^-(p) \subset \mathcal{S} \setminus \overline{I^+(A)}$.

Statement (ii) follows from (i) and fact that $I^+(A)$ is open. $\square$

Because of Lemma 3-19(ii), a subset $B \subset \mathcal{S}$ is said to be an *achronal boundary* if $B = \partial I^+(A)$ or $B = \partial I^-(A)$ for some $A \subset \mathcal{S}$. A main result in this section is that *an achronal boundary, if nonempty, is always a C^0 hypersurface.* To prove this, one needs the concept of an *edge point*.

Definition 3-20. Given an achronal set A, a point $p \in \bar{A}$ is called an *edge point* of A if every open neighborhood U of p contains a timelike curve from $I^-(p, U)$ to $I^+(p, U)$ that does not meet A.

Remark 3-21. By definition, a point $p \in \bar{A}$ is *not* an edge point of A if there exists an open neighborhood V of p such that every timelike curve, which is contained in V and is from $I^-(p, V)$ to $I^+(p, V)$, must meet A.

We will denote by $\mathrm{edge}(A)$ the set of all edge points of an achronal set A. By Lemma 3-18, if $p \in \bar{A} \setminus A$, no timelike curve passing p meets A. Hence

$$\bar{A} \setminus A \subset \mathrm{edge}(A) \subset \bar{A}.$$

In particular, A being edgeless ($\mathrm{edge}(A) = \varnothing$) implies that A is closed.

Lemma 3-22. *Given any $A \subset \mathcal{S}$, the achronal boundary $\partial I^+(A)$ is edgeless.*

Proof. Taking any $p \in \partial I^+(A)$, it follows from Lemma 3-19(i) that any timelike curve from $I^-(p)$ to $I^+(p)$ must meet $\partial I^+(A)$. Hence, $\partial I^+(A)$ is edgeless by Remark 3-21. $\qquad\square$

We now state a basic structural result.

Proposition 3-23. *Suppose A is achronal. Then $A \setminus \mathrm{edge}(A)$, if nonempty, is a C^0 hypersurface in $\mathcal{S}$.*

Proof. Suppose $p \in A \setminus \mathrm{edge}(A)$; we want to show that there exists an open neighborhood U of p such that $U \cap A$ is homeomorphic to an open set in $\mathbb{R}^n$.

Let $\{e_0, e_1, \ldots, e_n\}$ be an orthonormal basis of $T_p\mathcal{S}$ such that e_0 is future-directed timelike. Using this basis, one identifies $T_p\mathcal{S}$ with $\mathbb{R}^{n,1}$, with coordinates $(t, x_1, \ldots, x_n)$. For any $\delta > 0$, define $\tilde{W}_\delta = \{(t, x) . |t| < \delta, |x| < \delta\}$, where $x = (x_1, \ldots, x_n) \in \mathbb{R}^n$ and $|x| = \sqrt{x_1^2 + \cdots + x_n^2}$. When δ is small, the exponential map $\exp_p(\cdot)$ is a diffeomorphism from $\tilde{W}_\delta$ onto its image. Denote this image by W_δ.

Since $p \notin \mathrm{edge}(A)$, there exists an open neighborhood V of p such that every timelike curve, which is contained in V and is from $I^-(p, V)$ to $I^+(p, V)$, must meet A. Fix a small $\delta > 0$ such that $W_{2\delta} \subset V$. Given any $x \in D_\delta = \{x \in \mathbb{R}^n : |x| < \delta\}$, consider the curve $\gamma_x(t) = \exp_p(te_0 + x_1e_1 + \cdots + x_ne_n)$, $|t| < 2\delta$. By choosing δ small, one can assume γ_x is timelike for all $x \in D_\delta$. Now one restricts attention within W_δ. As t increases within $(-\delta, \delta)$, γ_x is from $I^-(p, W_\delta) \subset I^-(p, V)$ to

$I^+(p, W_\delta) \subset I^+(p, V)$. Hence, γ_x must meet A. Moreover, γ_x only meets A once since A is achronal. Therefore, $\exp_p^{-1}(W_\delta \cap A)$ is the graph over D_δ of some function $f : D_\delta \to (-\delta, \delta)$.

To complete the proof, one needs to show f is continuous. If f were not continuous at some point $x_* \in D_\delta$, there would exist an $\epsilon > 0$ and a sequence $\{y_k\} \subset D_\delta$ such that $y_k \to x_*$ as $k \to \infty$ and $|t_k - t_*| > \epsilon$, where $t_k = f(y_k)$ and $t_* = f(x_*)$. Let $\tilde{\sigma}_k \subset T_p\mathcal{S}$ be the line segment that connects (x_*, t_*) to (y_k, t_k). Let σ_k be the image of $\tilde{\sigma}_k$ under the exponential map $\exp_p(\cdot)$. Since the curve segment γ_{x_*}, $t \in [-\delta, \delta]$, is timelike, one sees that σ_k is also timelike for large k. This contradicts the fact that A is achronal. Hence, f is continuous on D_δ. $\square$

The next corollary follows directly from Lemma 3-19(ii), Lemma 3-22 and Proposition 3-23.

Corollary 3-24. *Given any $A \subset \mathcal{S}$, the achronal boundary $\partial I^+(A)$, if nonempty, is a C^0 hypersurface in $\mathcal{S}$.*

3.5. Cauchy hypersurfaces

We give two equivalent definitions of a *Cauchy hypersurface* in a spacetime $\mathcal{S}$. Recall that a causal curve $\sigma : (a, b) \to \mathcal{S}$ is *inextendible* if $\lim_{t \searrow a} \sigma(t)$ and $\lim_{t \nearrow b} \sigma(t)$ do not exist.

Definition I. A subset $\Sigma \subset \mathcal{S}$ is said to be a *Cauchy hypersurface* if Σ is met exactly once by every inextendible timelike curve in $\mathcal{S}$.

Definition II. A subset $\Sigma \subset \mathcal{S}$ is said to be a *Cauchy hypersurface* if Σ is achronal and is met by every inextendible causal curve in $\mathcal{S}$.

It is evident that Definition II implies Definition I. The following lemmas may be used to verify the reverse direction.

Lemma 3-25. *Let Σ be a Cauchy hypersurface according to Definition I.*

(i) *$\mathcal{S}$ is the disjoint union of Σ, $I^+(\Sigma)$ and $I^-(\Sigma)$.*

(ii) *Σ is a closed, edgeless, achronal set. Hence Σ is a C^0 hypersurface.*

Proof. The proof of (i) is left as an exercise. Given any $p \in \Sigma$, the fact that any timelike curve passing p meets instantly both $I^-(\Sigma)$ and $I^+(\Sigma)$ shows $\Sigma \subset \partial I^+(\Sigma)$ and $\Sigma \subset \partial I^-(\Sigma)$. This combined with (i) then implies $\Sigma = \partial I^+(\Sigma) = \partial I^-(\Sigma)$. (ii) now follows from Lemma 3-19, Lemma 3-22 and Proposition 3-23. $\square$

Lemma 3-26. *Let $C \subset \mathcal{S}$ be a closed subset. Let $\beta : [0, b) \to \mathcal{S}$ be a past-directed, past-inextendible causal curve, i.e., $\lim_{t \nearrow b} \beta(t)$ does not exist. Suppose β does not meet C.*

(i) *Given any $p_0 \in I^+(\beta(0), \mathcal{S} \setminus C)$, there is a past-inextendible timelike curve starting at p_0 that does not meet C.*

(ii) *If β is not a null geodesic free of conjugate points of $\beta(0)$, there is a past-inextendible timelike curve starting at $\beta(0)$ that does not meet C.*

Proof. (i) By reparametrization, one may assume $b = \infty$. Let $\gamma \subset \mathcal{S} \setminus C$ be a past-directed timelike curve from p_0 to $\beta(0)$. Applying Theorem 3-4 to the union of γ and $\beta|_{[0,1]}$, one knows there exists a past-directed timelike curve γ_1 contained in $\mathcal{S} \setminus C$, connecting p_0 and $\beta(1)$. Let p_1 be a point on γ_1 that is close to $\beta(1)$. Similarly, one can find a past-directed timelike curve γ_2 contained in $\mathcal{S} \setminus C$, connecting p_1 and $\beta(2)$. Let p_2 be a point on γ_2 that is close to $\beta(2)$. Repeating this argument, one obtains a sequence of points $p_1, p_2, \ldots$ in $\mathcal{S} \setminus C$ and a sequence of past-directed timelike curves $\gamma_1, \gamma_2, \ldots$ contained in $\mathcal{S} \setminus C$ such that γ_k connects p_{k-1} and p_k; moreover, with respect to any metric $d(\cdot, \cdot)$ for the topology on $\mathcal{S}$, $\{p_k\}$ can be arranged so that $d(p_k, \beta(k)) < 1/k$. In particular, $\lim_{k \to \infty} p_k$ does not exist since $\lim_{k \to \infty} \beta(k)$ does not exist. Now let $\tilde{\gamma}$ be the past-directed timelike curve which is the union of $\gamma_1, \gamma_2, \ldots$, then $\tilde{\gamma}$ is past-inextendible and is contained in $\mathcal{S} \setminus C$.

(ii) Suppose $b = \infty$. Choose $T < \infty$ such that $\beta|_{[0,T]}$ is not a null geodesic that has no conjugate points of $\beta(0)$ before $\beta(T)$. Applying Theorem 3-4 to $\beta|_{[0,T]}$, one knows there exists a past-directed timelike curve $\sigma : [0, 1] \to \mathcal{S} \setminus C$, connecting $\beta(0)$ and $\beta(T)$. Let α be the union of $\sigma|_{[\frac{1}{2},1]}$ and $\beta|_{[T,\infty)}$, then α is a past-inextendible causal curve that does not meet C. Since $\beta(0) \in I^+\left(\sigma\left(\frac{1}{2}\right), \mathcal{S} \setminus C\right)$, (ii) now follows from (i). $\square$

Proof that Definition I implies Definition II. Let $\Sigma \subset \mathcal{S}$ satisfy Definition I. Then Σ is achronal since no timelike curves meet Σ more than once. Let $\sigma : (-\infty, \infty) \to S$ be an inextendible causal curve. If $\sigma(0) \in \Sigma$, then σ meets Σ. Suppose $\sigma(0) \notin \Sigma$, then $\sigma(0) \in I^+(\Sigma) \cup I^-(\Sigma)$ by Lemma 3-25(i). Without loss of generality, one can assume $\sigma(0) \in I^+(\Sigma)$. By reversing the time direction of σ, one may also assume σ is past-directed. Now suppose $\sigma|_{[0,\infty)}$ does not meet Σ. Since Σ is closed, by Lemma 3-26 (i), for a fixed $p \in I^+(\sigma(0)) \subset I^+(\Sigma)$, there exists a past-inextendible timelike curve α starting at p which does not meet Σ. On the other hand, let η be any future-inextendible timelike curve starting at p, then η stays in $I^+(\Sigma)$. Putting together η and $-\alpha$, one obtains an inextendible timelike curve that never meets Σ, contradicting Definition I. Therefore, σ must meet Σ. $\square$

Definition I suggests using a timelike vector field to study the topology of a spacetime admitting a Cauchy hypersurface.

Let $\mathcal{T}$ be a smooth timelike vector field on $\mathcal{S}$ (the existence of such a $\mathcal{T}$ is guaranteed by the time-orientability of $\mathcal{S}$). Given any $p \in \mathcal{S}$, let $\gamma_p(t)$ be the maximal integral curve of $\mathcal{T}$ with $\gamma_p(0) = p$. It is easily seen that γ_p is inextendible. Consider the map $F : (p, t) \mapsto \gamma_p(t)$. The domain of F is the open subset of $\mathcal{S} \times \mathbb{R}$ given by $\mathcal{U} = \bigcup_{p \in \mathcal{S}} (\{p\} \times (a_p, b_p))$, where (a_p, b_p) is the maximal interval on which $\gamma_p(\cdot)$ is defined. If $\Sigma \subset \mathcal{S}$ is a Cauchy hypersurface, one has the continuous map

$$\Phi_\Sigma : \mathcal{U} \cap (\Sigma \times \mathbb{R}) \to \mathcal{S},$$

where $\Phi_\Sigma(x, t) = \gamma_x(t)$ for all $x \in \Sigma$. It is a good exercise to check that Definition I implies Φ_Σ is one-to-one and onto. Since $\mathcal{U} \cap (\Sigma \times \mathbb{R})$ and $\mathcal{S}$ are topological manifolds, Φ_Σ is a homeomorphism by the theorem on invariance of domain (for this result see [217] and [165], for example). Denote by π_Σ the natural projection from $\Sigma \times \mathbb{R}$ to Σ. Composing Φ_Σ^{-1} with π_Σ, one obtains a map

$$\psi_\Sigma = \pi_\Sigma \circ \Phi_\Sigma^{-1} : \mathcal{S} \to \Sigma,$$

which is a continuous open map and satisfies the property $\psi_\Sigma(p) = p$ for all $p \in \Sigma$. In particular, this shows that Σ is connected, since $\mathcal{S}$ is connected.

Proposition 3-27. *Let $\mathcal{S}$ be a spacetime that has a Cauchy hypersurface Σ.*

(i) *Σ is homeomorphic to any other Cauchy hypersurface in $\mathcal{S}$.*

(ii) *Suppose $A \subset \mathcal{S}$ is achronal. If A is a compact C^0 hypersurface, then A and Σ are homeomorphic. Consequently, Σ must be compact.*

Proof. (i) Let Σ' be a second Cauchy hypersurface in $\mathcal{S}$. For a fixed timelike vector field $\mathcal{T}$, consider the maps ψ_Σ and $\psi_{\Sigma'}$ defined as above. For any $p \in \Sigma$ and $p' \in \Sigma'$, it is evident that $\psi_\Sigma(\psi_{\Sigma'}(p)) = p$ and $\psi_{\Sigma'}(\psi_\Sigma(p')) = p'$. Hence, Σ and Σ' are homeomorphic.

(ii) Consider the restriction of ψ_Σ to A. The fact A is achronal shows that $\psi_\Sigma|_A$ is one-to-one. Since A is a C^0 hypersurface, by invariance of domain, $\psi_\Sigma|_A$ is an open map. On the other hand, A being compact shows that $\psi_\Sigma(A)$ is compact, hence closed in Σ. Therefore, since Σ is connected, $\psi_\Sigma(A) = \Sigma$ and $\psi_\Sigma|_A$ is a homeomorphism between A and Σ. $\qquad\square$

Remark 3-28. The compact achronal C^0 hypersurface A in Proposition 3-27(ii) is indeed a Cauchy hypersurface itself. We refer readers to Proposition 4.8 in [101] for a discussion of the proof.

3.6. Domains of dependence

Definition 3-29. Given an achronal set $A \subset \mathcal{S}$, the *future and past domains of dependence* of A, $D^+(A)$ and $D^-(A)$, are defined as follows:

$$D^+(A) = \{q : \text{every past-inextendible causal curve from } q \text{ meets } A\},$$

$$D^-(A) = \{p : \text{every future-inextendible causal curve from } p \text{ meets } A\}.$$

The union of $D^+(A)$ and $D^-(A)$, denoted by $D(A)$, is called the *domain of dependence* of A.

Remark 3-30. $D^+(A)$ and $D^-(A)$ are also known as the *future* and *past Cauchy developments* of A respectively. Similarly, $D(A) = D^+(A) \cup D^-(A)$ is also called the *Cauchy development* of A.

Remark 3-31. It follows from Definition II on p. 114 that an achronal set Σ is a Cauchy hypersurface if and only if $D(\Sigma) = \mathcal{S}$.

The following facts relating $D^{\pm}(A)$, $I^{\pm}(A)$ and $J^{\pm}(A)$ are easily checked:

$$A \subset D^+(A) \subset A \cup I^+(A) \subset J^+(A),$$
$$A \subset \partial D^+(A),$$
$$D^+(A) \cap I^-(A) = \varnothing,$$
$$D^+(A) \cap D^-(A) = A,$$
$$D(A) \cap I^+(A) = D^+(A) \setminus A.$$

A basic feature about $D^+(A)$ is that information outside $D^+(A)$ traveling into $D^+(A)$ must first pass through A.

Lemma 3-32. *Suppose* $\sigma : [0, 1] \to \mathcal{S}$ *is a past-directed causal curve with* $\sigma(0) \in D^+(A)$ *and* $\sigma(1) \notin D^+(A)$, *then* $\sigma(t) \in A$ *for some* $t \in [0, 1)$.

Proof. Since $\sigma(1) \notin D^+(A)$, there is a past-inextendible causal curve β from $\sigma(1)$ that does not meet A. The union $\sigma \cup \beta$ is a past-inextendible causal curve starting at $\sigma(0) \in D^+(A)$. Hence, it must meet A somewhere on $\sigma|_{[0,1)}$. $\qquad\square$

One of the important aspects of $D(A) = D^+(A) \cup D^-(A)$ is that its interior int $D(A)$, if nonempty, is an open set with appealing properties.

Lemma 3-33. *Suppose* $q \in \text{int } D(A)$. *If* $q \in D^+(A)$, *then any past-inextendible causal curve must meet* $I^-(A)$; *similarly, if* $q \in D^-(A)$, *any future-inextendible causal curve must meet* $I^+(A)$.

Proof. It suffices to consider the case $q \in D^+(A)$. Let $\beta : [0, b) \to \mathcal{S}$ be a past-inextendible causal curve with $\beta(0) = q$. The fact $q \in \text{int } D(A)$ implies that there is a nearby point $p_0 \in I^+(q) \cap D(A)$. Let γ be a past-directed timelike curve from

p_0 to $\beta(0)$. Repeating the construction in the proof of Lemma 3-26(i), one obtains a past-inextendible timelike curve $\tilde{\gamma}$ starting from p_0 and a sequence of points $\{p_k\}$ on $\tilde{\gamma}$ such that $\beta(k) \ll p_k$, for all $k \geq 1$. Since $p_0 \in I^+(A) \cap D(A) \subset D^+(A)$, $\tilde{\gamma}$ meet A somewhere. Therefore, $\beta(k) \in I^-(A)$ for large k. $\qquad\square$

Remark 3-34. It follows from Lemma 3-33 that if $p \in \operatorname{int} D(A)$, then every inextendible causal curve through p must meet both $I^-(A)$ and $I^+(A)$.

Lemma 3-35. *The causality condition holds on* $\operatorname{int} D(A)$, *i.e., no causal loop meets* $\operatorname{int} D(A)$.

Proof. Suppose α is a causal loop passing some $p \in \operatorname{int} D(A)$. Traveling along α infinitely many times, one gets an inextendible causal curve $\tilde{\alpha}$. By Lemma 3-33, $\tilde{\alpha}$ meets both $I^+(A)$ and $I^-(A)$. On the other hand, $\tilde{\alpha}$ meets A. This contradicts the achronality of A. $\qquad\square$

Lemma 3-36. *If* $p, q \in \operatorname{int} D(A)$ *and* $p \leq q$, *then* $J^+(p) \cap J^-(q) \subset \operatorname{int} D(A)$.

Proof. When $q = p$, $J^+(q) \cap J^-(q) = \{q\}$ by Lemma 3-35. Hence it suffices to assume $p < q$. There are a few cases to consider:

<u>Case 1</u>: $q, p \in D^+(A) \setminus A = D(A) \cap I^+(A)$. In this case, points that are close to q are still in $D(A) \cap I^+(A)$. Let $q^+ \in I^+(q)$ be chosen such that $q^+ \in D(A) \cap I^+(A)$. Consider the open set $U = I^-(q^+) \cap I^+(A)$. Given any past-directed causal curve $\alpha : [0, 1] \to \mathcal{S}$ from q to p, one has $\alpha(t) \in U$, for all $t \in [0, 1]$.

We proceed to show that $U \subset D^+(A)$. Suppose $y \in U$. Let $\sigma : [0, 1] \to \mathcal{S}$ be a past-directed timelike curve from q^+ to y. If $y \notin D^+(A)$, Lemma 3-32 implies $\sigma(s) \in A$ for some $s \in [0, 1)$. Since $\sigma(1) = y \in I^+(A)$, this contradicts the achronality of A. Hence, $U \subset D^+(A)$.

<u>Case 2</u>: $q \in D^+(A) \setminus A = D(A) \cap I^+(A)$ and $p \in D^-(A) \setminus A = D(A) \cap I^-(A)$. In this case, choose $p^- \in I^-(p)$ and $q^+ \in I^+(q)$ respectively such that $p^- \in D(A) \cap I^-(A)$ and $q^+ \in D(A) \cap I^+(A)$. Any future-directed causal curve α from p to q now is contained in the open set $V = I^+(p^-) \cap I^-(q^+)$.

We conclude by showing that $V \subset D(A)$. Suppose $x \in V$. Let $\gamma : [0, 1] \to \mathcal{S}$ and $\tau : [0, 1] \to \mathcal{S}$ be past-directed timelike curves from q^+ to x and from x to p^- respectively. If $x \notin D(A) = D^+(A) \cup D^-(A)$, then Lemma 3-32 implies $\gamma(s) \in A$ for some $s \in [0, 1)$ and $\tau(t) \in A$ for some $t \in (0, 1]$. Again, this contradicts the achronality of A. Hence, $V \subset D(A)$.

<u>Case 3</u>: $q \in D^+(A) \setminus A = D(A) \cap I^+(A)$ and $p \in A$. The proof of this case is identical to that of Case 2.

<u>Case 4</u>: $q, p \in A$. Again, this case can be proved in the same way as Case 2.

Any remaining case is dual to one of the cases above by reversing the time orientation on S. This completes the proof. $\qquad\square$

Results stronger than Lemmas 3-35 and 3-36 indeed hold on int $D(A)$. Interested readers are referred to [174, Theorem 14.38] for a complete proof of the following theorem.

Theorem 3-37. *Let $A \subset S$ be an achronal set. Then* int $D(A)$*, if nonempty, satisfies the following properties:*

(i) *The strong causality condition holds at every $p \in$ int $D(A)$.*

(ii) *Given $p, q \in$ int $D(A)$, if $p \leq q$, then $J^+(p) \cap J^-(q)$ is compact and is contained in* int $D(A)$.

Corollary 3-38. *If a spacetime S has a Cauchy hypersurface, then S is globally hyperbolic.*

This follows from Theorem 3-37 and Remark 3-31. The corollary's converse is also true; see [104].

3.7. Cauchy horizons

We end this chapter with a brief introduction to Cauchy horizons. Although this concept is not needed in the proof of the Penrose singularity theorem, it arises naturally when an achronal set A is not a Cauchy hypersurface,

Definition 3-39. Suppose $A \subset S$ is achronal. Its *future Cauchy horizon $H^+(A)$* is defined as

$$H^+(A) = \{p \in \overline{D^+(A)} : I^+(p) \cap D^+(A) = \varnothing\}.$$

The *past Cauchy horizon $H^-(A)$* is defined dually. The *Cauchy horizon* of A is $H(A) = H^+(A) \cup H^-(A)$.

By definition, $H^+(A) = \overline{D^+(A)} \setminus I^-(D^+(A))$. Therefore, $H^+(A)$ is closed. Moreover, $I^+(p) \cap D^+(A) = \varnothing$ implies $I^+(p) \cap \overline{D^+(A)} = \varnothing$ as $I^+(p)$ is open. Hence, $H^+(A)$ is achronal. By Proposition 3-23, $H^+(A) \setminus \text{edge}(H^+(A))$, if nonempty, is always a C^0 hypersurface.

Lemma 3-40. *If A is a closed achronal set, then*

$$\overline{D^+(A)} = \{q : \text{every past inextendible timelike curve from } q \text{ meets } A\}.$$

Consequently, $\overline{D^+(A)} \subset A \cup I^+(A)$.

Proof. Given $q \in \overline{D^+(A)}$, suppose there is a past-inextendible timelike curve $\sigma : [0, b) \to \mathcal{S}$ with $\sigma(0) = q$ such that σ does not meet A. As A is closed, $\mathcal{S} \setminus A$ is open. Let $p = \sigma(s)$ for some $s \in (0, b)$, then $q \in I^+(p, \mathcal{S} \setminus A)$. Hence, there exists $\tilde{q} \in D^+(A) \cap I^+(p, \mathcal{S} \setminus A)$. Note that $\sigma : [s, b) \to \mathcal{S}$ is a past-inextendible timelike curve starting at p that does not meet A. By Lemma 3-26, there exists a past-inextendible timelike curve β starting at $\tilde{q}$ that does not meet A, contradicting the fact that $\tilde{q} \in D^+(A)$.

Next, suppose q is a point such that every past-inextendible timelike curve from q meets A. Suppose $q \in \mathcal{S} \setminus \overline{D^+(A)}$. Take $q^- \in I^-(q, \mathcal{S} \setminus \overline{D^+(A)})$. Since $q^- \notin D^+(A)$, there is a past-inextendible causal curve starting from q^- that does not meet A. By Lemma 3-26, there exists a past-inextendbile timelike curve starting at q that does not meet A, which is a contradiction. $\qquad\square$

Proposition 3-41. *If A is a closed achronal set, then $\partial D^+(A) = A \cup H^+(A)$.*

Proof. It suffices to show $\partial D^+(A) \subset A \cup H^+(A)$. To do so, suppose $p \in \overline{D^+(A)} \setminus (A \cup H^+(A))$. The fact $p \in \overline{D^+(A)} \setminus A$ implies $p \in I^+(A)$ by Lemma 3-40. The fact $p \in \overline{D^+(A)} \setminus H^+(A)$ implies there exists $q \in I^+(p) \cap D^+(A)$. Consider the open set $U = I^-(q) \cap I^+(A)$, which contains p. Given any $y \in U$, there exists a past-directed timelike curve σ from q to y. If $y \notin D^+(A)$, Lemma 3-32 implies σ must meet A before y, contradicting the fact that $y \in I^+(A)$. Therefore $y \in D^+(A)$ and hence $U \subset \operatorname{int} D^+(A)$. In particular, p must be an interior point of $D^+(A)$. This shows $\partial D^+(A) \subset A \cup H^+(A)$. $\qquad\square$

Exercises

We assume that $(\mathcal{S}, \langle \cdot, \cdot \rangle)$ is a connected, time-oriented Lorentzian manifold.

Exercise 3-42. Let $\alpha : [0, 1] \to \mathcal{S}$ be a smooth causal curve segment. Let $\varepsilon > 0$ and $x : (-\varepsilon, \varepsilon) \times [0, 1] \to \mathcal{S}$ be a smooth variation of α, i.e., $x(0, t) = \alpha(t)$ for all $t \in [0, 1]$. Let $V = (\partial x / \partial s)|_{s=0}$ be the *variation vector field* along α, with covariant derivative $V'(t)$ along α. Suppose for all $t \in [0, 1]$, $\langle V'(t), \alpha'(t) \rangle < 0$. Show that there is $\varepsilon_0 > 0$ small enough so that for all $0 < s < \varepsilon_0$, the curve $x_s : [0, 1] \to \mathcal{S}$ with $x_s(t) = x(s, t)$ is timelike. Where could the strict inequality be relaxed to $\langle V'(t), \alpha'(t) \rangle \leq 0$?

Exercise 3-43. Suppose $p, q \in \mathcal{S}$, and $\gamma : [0, 1] \to \mathcal{S}$ is a smooth causal curve from p to q, with $\gamma'(0)$ or $\gamma'(1)$ timelike. Show that γ can be deformed to a nearby timelike curve with a fixed endpoint variation $x : (-\varepsilon, \varepsilon) \times [0, 1] \to \mathcal{S}$. (Hint: Start by considering the vector field along γ obtained by parallel transport of a timelike velocity vector at an endpoint.)

Remark. A more general result holds [174, Proposition 10.46]: If γ is a piecewise smooth causal curve from p to q which is not a smooth null geodesic, then γ can be deformed to a nearby timelike curve with a fixed endpoint variation.

Exercise 3-44. Suppose $p \in \mathcal{S}$ and $\widetilde{U} \subset T_p\mathcal{S}$ and $U \subset \mathcal{S}$ are open sets for which the exponential map $\exp_p : \widetilde{U} \to U$ is a diffeomorphism. Suppose $\tilde{\gamma} : [0, 1] \to \widetilde{U} \subset T_p\mathcal{S}$ is a piecewise smooth curve with $\tilde{\gamma}(0) = 0 \in T_p\mathcal{S}$. If $\gamma = \exp_p \circ \tilde{\gamma}$ is a future-pointing timelike curve in $\mathcal{S}$, prove that $\tilde{\gamma}((0, 1])$ lies in the future timecone of $T_p\mathcal{S}$. (Hint: Gauss lemma.)

Exercise 3-45. Suppose A is an achronal set that contains no edge points. Show that $I(A) := I^-(A) \cup A \cup I^+(A)$ is an open set.

Exercise 3-46. Suppose $\mathcal{S}$ has the property that for all $p \in \mathcal{S}$, $J^+(p)$ and $J^-(p)$ are closed sets. Show that if K is compact, then $J^+(K)$ is closed.

Exercise 3-47. Suppose $\mathcal{S}$ has a noncompact Cauchy hypersurface. Prove that for any $A \subset \mathcal{S}$, the topological boundary $\partial J^+(A)$, if nonempty, is noncompact. (Hint: Compare $\partial J^+(A)$ and $\partial I^+(A)$.)

CHAPTER 4

The Penrose singularity theorem

This chapter presents the classical Penrose singularity theorem [177]. The main ingredients in the proof concern, on the one hand, the causal structure of a globally hyperbolic spacetime, discussed in the previous chapter, and on the other, differential geometry techniques involving Jacobi fields together with the Riccati and Raychaudhuri equations.

This chapter is organized as follows. In Section 4.1, we review the concept of Jacobi fields and focal points of a spacelike submanifold. In Section 4.2, we give a geometric formulation of the Riccati and Raychaudhuri equations along causal geodesics. In Section 4.3, we state and prove the Penrose singularity theorem.

Throughout this chapter, $\mathcal{S}$ denotes an $(n+1)$-dimensional spacetime.

4.1. Jacobi fields and focal points

In general relativity, a freely falling observer is represented by a future-directed timelike geodesic. A family of such nearby observers can thus be described by a map $F : (-\epsilon, \epsilon) \times (a, b) \to \mathcal{S}$ such that for each $|s| < \epsilon$, the curve $\gamma_s(\cdot) = F(s, \cdot)$ is a (timelike) geodesic. The position of these observers relative to a given γ_s is measured by the tangent vector $V = \partial F / \partial s$ along γ_s. Differentiating (or linearizing) the geodesic equation $\nabla_{\gamma'_s} \gamma'_s = 0$, one obtains

$$
\begin{aligned}
0 = \nabla_V(\nabla_{\gamma'_s} \gamma'_s) &= \nabla_V(\nabla_{\gamma'_s} \gamma'_s) - \nabla_{\gamma'_s}(\nabla_V \gamma'_s) + \nabla_{\gamma'_s}(\nabla_{\gamma'_s} V) \\
&= V'' + R(V, \gamma'_s)\gamma'_s,
\end{aligned}
\tag{4.1.1}
$$

where ∇ denotes covariant differentiation in $\mathcal{S}$, $V'' = \nabla_{\gamma'_s}(\nabla_{\gamma'_s} V)$ and

$$
R(X, Y)Z = \nabla_X \nabla_Y Z - \nabla_Y \nabla_X Z - \nabla_{[X,Y]} Z.
$$

Thinking in the Newtonian regime, one tends to interpret $-R(V, \gamma'_s)\gamma'_s$ as a certain "force" acting on V. Thus the map $V \mapsto -R(V, \gamma'_s)\gamma'_s$ is often known as the *tidal force* operator.

A submanifold $P \subset \mathcal{S}$ is called *spacelike* if every tangent vector to P is spacelike.

123

We let $\mathrm{II}(\,\cdot\,,\cdot\,): T_p P \times T_p P \to (T_p P)^\perp$ be the second fundamental form of P at $p \in P$, i.e., $\mathrm{II}(v, w) = (\nabla_v w)^\perp$, where $\perp$ denotes orthogonal projection to $(T_p P)^\perp$, the orthogonal complement of $T_p P$ (see Lemma 5-3).

Suppose the observers $\{\gamma_s\}$ above originate from a given spacelike submanifold P, i.e. $\gamma_s(0) \in P$ and $\gamma_s'(0) \perp P$, for all s. Then $V(0) \in T_{\gamma_s(0)} P$, and $V'(0)$ satisfies (apply (2.3.8) to the map F)

$$\langle V'(0), w \rangle = \langle \nabla_{\gamma_s'(0)} V, w \rangle = \langle \nabla_V \gamma_s'(0), w \rangle = -\langle \gamma_s'(0), \mathrm{II}(V(0), w) \rangle, \quad (4.1.2)$$

for all $w \in T_{\gamma_s(0)} P$.

Equations (4.1.1) and (4.1.2) suggest the following definitions. Given a geodesic $\gamma : [0, b] \to \mathcal{S}$, a vector field V along γ is called a *Jacobi field* if

$$V'' + R(V, \gamma')\gamma' = 0. \quad (4.1.3)$$

When γ is orthogonal to a spacelike submanifold P, that is, if $\gamma(0) \in P$ and $\gamma'(0) \perp P$, a point $\gamma(t)$, $t > 0$, is called a *focal point* of P along γ, provided there exists a nontrivial Jacobi field V along γ such that $V(t) = 0$ and

$$\begin{aligned} &V(0) \in T_{\gamma(0)} P, \\ &\langle V'(0), w \rangle = -\langle \gamma'(0), \mathrm{II}(V(0), w) \rangle \quad \text{for all } w \in T_{\gamma(0)} P. \end{aligned} \quad (4.1.4)$$

Heuristically, if $q = \gamma(t)$ is a focal point of P along a timelike γ, there exist nearby freely falling observers that start from P and (almost) meet γ at q. When P consists of a single point, focal points of P are conjugate points of $\gamma(0)$.

The following basic result concerning the deformation of causal curves is needed in the proof of the Penrose singularity theorem:

Theorem 4-1 [174, Theorem 10.51]. *Let P be a spacelike submanifold in a spacetime $\mathcal{S}$. Given $p \in P$ and $q \in \mathcal{S}$, let α be a causal curve from p to q. There exists a timelike curve from p to q which is arbitrarily near α unless α is a null geodesic (when suitably parametrized) orthogonal to P along which there are no focal points of P before q.*

4.2. Riccati and Raychaudhuri equations

In this section, we present a geometric description of the Riccati and Raychaudhuri equations along causal geodesics.

4.2.1. *Geometric Riccati equation.* Let $P \subset \mathcal{S}$ be a spacelike submanifold. Consider a causal geodesic

$$\gamma : [0, L) \to \mathcal{S}, \text{ with } \gamma(0) = p \in P \text{ and } \gamma'(0) \perp P.$$

The set $\tilde{\mathcal{V}}$ of Jacobi fields along γ satisfying the initial condition (4.1.4) is a vector space of dimension $n + 1$. Consider the subspace $\mathcal{V} \subset \tilde{\mathcal{V}}$ given by

$$\mathcal{V} = \{J \in \tilde{\mathcal{V}} : J \perp \gamma' \text{ along } \gamma\}.$$

Besides (4.1.4), elements in $\mathcal{V}$ are characterized by an additional initial condition $J'(0) \perp \gamma'(0)$. Hence, $\mathcal{V}$ has dimension n.

In what follows, it is always assumed that $\gamma(t)$ is *not* a focal point of P along γ, for all $t \in (0, L)$. At each $\gamma(t)$, there is a well-defined map $B(t) : \mathcal{V} \to \gamma'(t)^{\perp}$, given by $B(t)(J) = J(t)$. Here $\gamma'(t)^{\perp} = \{v \in T_{\gamma(t)}\mathcal{S} : \langle v, \gamma'(t)\rangle = 0\}$. Since $\gamma(t)$ is not a focal point, $B(t)$ is a linear isomorphism and $B(t)^{-1}(v)$ is the unique Jacobi field $J \in \mathcal{V}$ such that $J(t) = v$. The map $A(t) : \gamma'(t)^{\perp} \to \gamma'(t)^{\perp}$ defined by

$$A(t)(v) = [B(t)^{-1}(v)]'(t), \tag{4.2.1}$$

where the prime in (4.2.1) denotes covariant differentiation by $\gamma'(t)$, satisfies (from the definition)

$$A(t)(J(t)) = J'(t) \quad \text{for all } J \in \mathcal{V}. \tag{4.2.2}$$

$A(t)$ is therefore a smooth section of the vector bundle over γ with fiber $\text{End}(\gamma'(t)^{\perp}, \gamma'(t)^{\perp})$, the space of linear maps from $\gamma'(t)^{\perp}$ to itself.

Now let $A'(t) = \nabla_{\gamma'(t)} A(t)$. By definition,

$$A'(t)(W(t)) = \nabla_{\gamma'(t)}[A(t)(W(t))] - A(t)(\nabla_{\gamma'(t)} W(t))$$

for any smooth vector field $W(t)$ along γ such that $W(t) \in \gamma'(t)^{\perp}$. Taking $W(t) = J(t) \in \mathcal{V}$ and applying (4.2.2), we have

$$\begin{aligned} A'(t)(J(t)) &= \nabla_{\gamma'(t)}[A(t)(J(t))] - A(t)(\nabla_{\gamma'(t)} J(t)) \\ &= J''(t)) - A(t)(J'(t)) \\ &= -R(J(t), \gamma'(t))\gamma'(t) - A(t) \circ A(t)(J(t)). \end{aligned} \tag{4.2.3}$$

The following proposition follows directly from (4.2.3).

Proposition 4-2. *The linear maps* $A(t) : \gamma'(t)^{\perp} \to \gamma'(t)^{\perp}$ *defined in* (4.2.1) *satisfy a Riccati equation of the form*

$$A'(t)(\cdot) = -R(\cdot, \gamma')\gamma' - A(t) \circ A(t)(\cdot). \tag{4.2.4}$$

With respect to the spacetime metric $\langle \cdot, \cdot \rangle$, $A(t)$ satisfies a further relation:

Proposition 4-3. *For all* $J_1, J_2 \in \mathcal{V}$,

$$\langle A(t)(J_1(t)), J_2(t)\rangle = \langle J_1(t), A(t)(J_2(t))\rangle. \tag{4.2.5}$$

Proof. By (4.1.4),

$$\lim_{t \searrow 0} \langle A(t)(J_i(t)), J_j(t) \rangle = \langle J_i'(0), J_j(0) \rangle = -\langle \gamma'(0), \mathrm{III}(J_i(0), J_j(0) \rangle$$

for $i, j \in \{1, 2\}$. Hence,

$$\lim_{t \searrow 0} \langle A(t)(J_1(t)), J_2(t) \rangle = \lim_{t \searrow 0} \langle J_1(t), A(t)(J_2(t)) \rangle. \tag{4.2.6}$$

On the other hand, (4.1.3) implies

$$\frac{d}{dt} \big[\langle A(t)(J_1(t)), J_2(t) \rangle - \langle J_1(t), A(t)(J_2(t)) \rangle \big] = 0. \tag{4.2.7}$$

Equation (4.2.5) follows from (4.2.6) and (4.2.7). $\qquad\qquad\square$

4.2.2. Raychaudhuri equation along timelike geodesics. If $\gamma'(t)$ is timelike, the restriction of $\langle \cdot, \cdot \rangle$ to $\gamma'(t)^\perp$ is positive definite. In this case, Proposition 4-3 shows $A(t)$ is self-adjoint with respect to $\langle \cdot, \cdot \rangle$. Let $h(t) : \gamma'(t)^\perp \times \gamma'(t)^\perp \to \mathbb{R}$ be the associated symmetric bilinear form: $h(t)(v, w) = \langle A(t)(v), w \rangle$. Define

$$\theta(t) = \mathrm{tr}_{\gamma'(t)^\perp} A(t) = \mathrm{tr}_{\gamma'(t)^\perp} h(t), \tag{4.2.8}$$

where $\mathrm{tr}_{\gamma'(t)^\perp}(\cdot)$ denotes the trace on $\gamma'(t)^\perp$. It is easily checked that

$$\frac{d}{dt} \mathrm{tr}_{\gamma'(t)^\perp} A(t) = \mathrm{tr}_{\gamma'(t)^\perp} A'(t). \tag{4.2.9}$$

With Proposition 4-2, this shows that $\theta(t)$ satisfies the following Raychaudhuri equation.

Proposition 4-4. *When $\gamma(t)$ is timelike, $\theta(t)$ defined in (4.2.8) obeys*

$$\theta'(t) = -\mathrm{Ric}(\gamma', \gamma') - \frac{1}{n}\theta(t)^2 - |\mathring{h}(t)|^2. \tag{4.2.10}$$

Here $\mathrm{Ric}(\cdot, \cdot)$ denotes the Ricci curvature of $\mathcal{S}$, $\mathring{h}(t)$ is the traceless part of $h(t)$ and $|\cdot|$ is the norm taken on $\gamma'(t)^\perp$.

Proof. Taking the trace of (4.2.4) and using (4.2.9), one has

$$\theta'(t) = -\mathrm{Ric}(\gamma', \gamma') - \mathrm{tr}_{\gamma'(t)^\perp}(A(t) \circ A(t)),$$

where $\mathrm{tr}_{\gamma'(t)^\perp}(A(t) \circ A(t)) = |h(t)|^2 = \frac{1}{n}\theta(t)^2 + |\mathring{h}(t)|^2$. This proves (4.2.10). $\square$

The physical meaning of $\theta(t)$ can be seen as follows. Let $J_1, \ldots, J_n$ be n Jacobi fields along γ such that $\{J_1, \ldots, J_n\}$ forms a basis of $\mathcal{V}$. For $t > 0$, let

$g_{ij}(t) = \langle J_i(t), J_j(t) \rangle$ and let $(g^{ij}(t))_{n \times n}$ be the inverse matrix of $(g_{ij}(t))_{n \times n}$. By definition and Proposition 4-3,

$$
\begin{aligned}
\theta(t) &= \sum_{1 \le i,j \le n} g^{ij}(t) \langle J_i'(t)), J_j(t) \rangle \\
&= \frac{1}{2} \sum_{1 \le i,j \le n} g^{ij}(t) \frac{d}{dt} \langle J_i(t), J_j(t) \rangle \\
&= \frac{1}{\sqrt{\det(g_{ij}(t))}} \frac{d}{dt} \sqrt{\det(g_{ij}(t))}. \qquad (4.2.11)
\end{aligned}
$$

Heuristically, (4.2.11) suggests that $\theta(t)$ represents the rate of the expansion of the spatial world $\gamma'(t)^\perp$ as measured by the (nearby) observers $J_1, \ldots, J_n$.

When P is a spacelike hypersurface, $\lim_{t \searrow 0} \theta(t)$ has an explicit geometric meaning.

Proposition 4-5. *Suppose $P \subset \mathcal{S}$ is a spacelike hypersurface. Then*

$$
\lim_{t \searrow 0} \theta(t) = -\langle \gamma'(0), \vec{H} \rangle,
$$

where $\vec{H}$ is the mean curvature vector of P at $\gamma(0)$.

Proof. Let $\{e_1, \ldots, e_n\} \subset T_{\gamma(0)} P$ be an orthonormal frame. By (4.1.4), there exists $\{J_1, \ldots, J_n\} \subset \mathcal{V}$ satisfying $J_i(0) = e_i$ and

$$
\langle J_i'(0), w \rangle = -\langle \gamma'(0), \mathrm{I\!I}(e_i, w) \rangle, \quad \text{for all } w \in T_{\gamma(0)} P. \qquad (4.2.12)
$$

Clearly, $\{J_1, \ldots, J_n\}$ forms a basis of $\mathcal{V}$. Let $g_{ij}(t)$ and $g^{ij}(t)$ be given as in (4.2.11). Since $g^{ij}(0) = \delta_{ij}$, by (4.2.11) and (4.2.12),

$$
\begin{aligned}
\lim_{t \searrow 0} \theta(t) &= \sum_{1 \le i,j \le n} \delta^{ij} \langle J_i'(0)), J_j(0) \rangle = -\sum_{i=1}^{n} \langle \gamma'(0), \mathrm{I\!I}(e_i, e_i) \rangle \\
&= -\langle \gamma'(0), \vec{H} \rangle. \qquad \qquad \qquad \qquad \qquad \square
\end{aligned}
$$

4.2.3. Raychaudhuri equation along null geodesics.

When $\gamma'(t)$ is null, the restriction of $\langle \cdot, \cdot \rangle$ to $\gamma'(t)^\perp$ is degenerate, since $\gamma'(t) \in \gamma'(t)^\perp$. The next two lemmas are left as exercises.

Lemma 4-6. *If $\gamma'(t)$ is null, every $v \in \gamma'(t)^\perp$ which is not a scalar multiple of $\gamma'(t)$ is spacelike.*

Lemma 4-7. $\mathrm{Ric}(\gamma'(t), \gamma'(t)) = \sum_{i=1}^{n-1} \langle R(e_i, \gamma'(t))\gamma'(t), e_i \rangle$ *for any collection of vectors $\{e_1, \ldots, e_{n-1}\} \subset \gamma'(t)^\perp$ satisfying $\langle e_i, e_j \rangle = \delta_{ij}$.*

Lemma 4-6 suggests one should define an equivalence relation $\sim$ in $\gamma'(t)^{\perp}$ as follows: given $v, w \in \gamma'(t)^{\perp}$,

$$v \sim w \ \text{ if and only if } (v - w) \parallel \gamma'(t).$$

The spacetime metric $\langle \cdot, \cdot \rangle$ descends to a positive definite metric $\langle \cdot, \cdot \rangle_{\sim}$ on the quotient space $\gamma'(t)^{\perp}/\sim$, where

$$\langle [v], [w] \rangle_{\sim} := \langle v, w \rangle.$$

Here $[v], [w]$ denote the equivalence classes containing v, w. The following facts hold about $\gamma'^{\perp}/\sim$:

(a) The tidal force operator $-R(\cdot, \gamma'(t))\gamma'(t) : \gamma'(t)^{\perp} \to \gamma'(t)^{\perp}$ descends to a linear transformation $-\tilde{R}(\cdot)$ on $\gamma'(t)^{\perp}/\sim$, since $R(\gamma', \gamma')\gamma' = 0$. Here $\tilde{R}(t)([v]) := [R(v, \gamma'(t))\gamma'(t)]$.

(b) The vector field $t\gamma'(t)$ is an element in $\mathcal{V}$, hence $A(t)(\gamma'(t)) = t^{-1}\gamma'(t)$. Therefore, $A(t)$ descends to a linear transformation $\tilde{A}(t)$ on $\gamma'(t)^{\perp}/\sim$, where $\tilde{A}(t)([v]) := [A(t)(v)]$. By Proposition 4-3,

$$\tilde{A}(t) : \gamma'(t)^{\perp}/\sim \ \to \ \gamma'(t)^{\perp}/\sim$$

is self-adjoint with respect to $\langle \cdot, \cdot \rangle_{\sim}$.

(c) $\gamma'^{\perp}/\sim$ is an $(n-1)$-dimensional vector bundle over γ, on which there is a connection $\tilde{\nabla}$ induced from ∇. Precisely, $\tilde{\nabla}_{\gamma'(t)}[V] = [\nabla_{\gamma'(t)} V]$ where V is any smooth vector field along γ with $V(t) \in \gamma'(t)^{\perp}$. Define

$$\tilde{A}'(t)([V]) := \tilde{\nabla}_{\gamma'(t)}[\tilde{A}(t)([V])] - \tilde{A}(t)(\tilde{\nabla}_{\gamma'(t)}[V]).$$

By Proposition 4-2, $\tilde{A}'(t)$ satisfies

$$\tilde{A}'(t)(\cdot) = -\tilde{R}(t)(\cdot) - \tilde{A}(t) \circ \tilde{A}(t)(\cdot). \tag{4.2.13}$$

As in the previous case, consider the associated symmetric bilinear form $\tilde{h}(t) :$ $\left(\gamma'(t)^{\perp}/\sim\right) \times \left(\gamma'(t)^{\perp}/\sim\right) \to \mathbb{R}$ where $\tilde{h}(t)([v], [w]) = \langle \tilde{A}(t)([v]), [w] \rangle_{\sim}$. Define

$$\tilde{\theta}(t) = \operatorname{tr}_{\gamma'(t)^{\perp}/\sim} \tilde{A}(t) = \operatorname{tr}_{\gamma'(t)^{\perp}/\sim} \tilde{h}(t), \tag{4.2.14}$$

where $\operatorname{tr}_{\gamma'(t)^{\perp}/\sim}(\cdot)$ is the trace on $\gamma'(t)^{\perp}/\sim$. Similar to (4.2.9), one has

$$\frac{d}{dt}\tilde{\theta}(t) = \operatorname{tr}_{\gamma'(t)^{\perp}/\sim} \tilde{A}'(t). \tag{4.2.15}$$

The following Raychaudhuri equation for $\tilde{\theta}(t)$ follows from Lemma 4-7, (4.2.15) and taking the trace of (4.2.13).

Proposition 4-8. *When $\gamma(t)$ is null, $\tilde{\theta}(t)$ defined in (4.2.14) obeys*

$$\tilde{\theta}'(t) = -\mathrm{Ric}(\gamma', \gamma') - \frac{1}{n-1}\tilde{\theta}(t)^2 - \left|\mathring{\tilde{h}}(t)\right|_{\sim}^2.$$

Here $\mathring{\tilde{h}}(t)$ is the traceless part of $\tilde{h}(t)$ and $|\cdot|_{\sim}$ is the norm taken on $\gamma'(t)^{\perp}/\sim$.

Similarly to Proposition 4-5, when P is a codimension-2 spacelike submanifold, $\lim_{\searrow 0}\tilde{\theta}(t)$ has an explicit geometric meaning.

Proposition 4-9. *Suppose P is a codimension-2 spacelike submanifold in $\mathcal{S}$. Then*

$$\lim_{t \searrow 0} \tilde{\theta}(t) = -\langle \gamma'(0), \vec{H}\rangle,$$

where $\vec{H}$ is the mean curvature vector of P at $\gamma(0)$.

Proof. Let $\{e_1, \ldots, e_{n-1}\} \subset T_{\gamma(0)} P$ be an orthonormal frame. By (4.1.4), there exists $\{J_1, \ldots, J_{n-1}\} \subset \mathcal{V}$ satisfying $J_i(0) = e_i$ and

$$\langle J_i'(0), w\rangle = -\langle \gamma'(0), \mathrm{II}(e_i, w)\rangle, \quad \text{for all } w \in T_{\gamma(0)} P. \qquad (4.2.16)$$

Clearly, $J_1, \ldots, J_{n-1}$ together with $J_n(t) = t\gamma'(t)$ form a basis of $\mathcal{V}$. Hence, $\{[J_1(t)], \ldots, [J_{n-1}(t)]\}$ is a basis of $\gamma'(t)^{\perp}/\sim$. For $\alpha, \beta \leq n - 1$, let

$$g_{\alpha\beta}(t) = \langle [J_\alpha(t)], [J_\beta(t)]\rangle_{\sim}$$

and let $(g^{\alpha\beta}(t))_{(n-1)\times(n-1)}$ be the inverse matrix of $(g_{\alpha\beta}(t))_{(n-1)\times(n-1)}$. Applying (4.2.16) and the fact that $g_{\alpha\beta}(0) = \delta_{\alpha\beta}$, one concludes that

$$\lim_{t \searrow 0} \tilde{\theta}(t) = \lim_{t \searrow 0} \sum_{1 \leq \alpha, \beta \leq n-1} g^{\alpha\beta}(t) \langle J_\alpha'(t), J_\beta(t)\rangle$$

$$= -\sum_{\alpha=1}^{n-1} \langle \gamma'(0), \mathrm{II}(e_\alpha, e_\alpha)\rangle = -\langle \gamma'(0), \vec{H}\rangle. \qquad \square$$

4.3. Proof of Penrose's singularity theorem

In general relativity, the problem of how to define a singularity is rather difficult (see [208; 103], for example). The approach adopted in the Penrose singularity theorem is to diagnose the existence of singular behavior of spacetimes in terms of the incompleteness of future null geodesics.

A future-directed causal geodesic $\beta : [0, b) \to \mathcal{S}$ is called *incomplete* if $b < \infty$, where $[0, b)$ is the maximum interval on which β exists.

Definition 4-10. A spacetime $\mathcal{S}$ is said to be *future null geodesically incomplete* if it contains an incomplete, future-directed null geodesic.

A key notion in the Penrose singularity theorem is that of a *trapped surface* (in $3 + 1$ dimensions). In general, a closed (i.e., compact without boundary) codimension-2 spacelike submanifold $\Sigma \subset \mathcal{S}$ is said to be (future) *trapped* if

$$\langle v, \vec{H} \rangle > 0 \qquad (4.3.1)$$

for all future-directed null vectors v that are normal to Σ. Here $\vec{H}$ denotes the mean curvature vector of Σ in $\mathcal{S}$.

Theorem 4-11 (Penrose singularity theorem [177]). *Let $\mathcal{S}$ be a globally hyperbolic spacetime with a noncompact Cauchy hypersurface satisfying the null energy condition, i.e.* $\mathrm{Ric}(v, v) \geq 0$, *for all null vectors v. If $\mathcal{S}$ contains a closed, trapped, codimension-2 spacelike submanifold Σ, then $\mathcal{S}$ is future null geodesically incomplete.*

Proof. We argue by contradiction. Suppose that $\mathcal{S}$ is not future null geodesically incomplete, then every future-directed null geodesic in $\mathcal{S}$ is defined on $[0, \infty)$. Given any $p \in \Sigma$, let $(T_p\Sigma)^\perp = \{v \in T_p\mathcal{S} \mid v \perp T_p\Sigma\}$. Let $\exp_p(\cdot)$ be the exponential map on $\mathcal{S}$ at p. Since Σ is compact, there exists a finite open covering $\{U_i\}_{1 \leq i \leq m}$ of Σ such that, for $1 \leq i \leq m$,

- the closure $\overline{U}_i$ is compact and is contained in some open set $V_i \subset \Sigma$, and

- on V_i, there exist two linearly independent future-directed null vector fields, which we denote by K_i^+ and K_i^-.

It follows from (4.3.1) that there exists a constant $\delta > 0$ such that

$$\langle K_i^\pm(q), \vec{H}(q) \rangle > \delta \qquad (4.3.2)$$

for all $q \in \overline{U}_i$ and $i \in \{1, \ldots, m\}$.

Now fix $i \in \{1, \ldots, m\}$. For each $p \in U_i$, consider the future-directed null geodesics $\gamma_p^+(t) = \exp_p(t K_i^+(p))$ and $\gamma_p^-(t) = \exp_p(t K_i^-(p))$. By assumption, γ_p^+ and γ_p^- are defined on $[0, \infty)$. Suppose Σ has no focal point along γ_p^+ on the interval $(0, L]$ for some $L > 0$. Let $\tilde{\theta}(t)$ be the quantity defined in Section 4.2.3 with $\gamma(t)$ replaced by $\gamma_p^+(t)$. By Propositions 4-8 and 4-9, (4.3.2) and the null energy condition, $\tilde{\theta}(t)$ satisfies

$$\tilde{\theta}'(t) \leq -\frac{1}{n-1}\tilde{\theta}(t)^2 \quad \text{for all } t \in (0, L] \qquad (4.3.3)$$

and

$$\lim_{t \searrow 0} \tilde{\theta}(t) = -\langle K_i^+(p), \vec{H}(p) \rangle < -\delta. \qquad (4.3.4)$$

It follows from (4.3.3), (4.3.4) and elementary calculus that $L < \frac{n-1}{\delta}$. Hence, we conclude that Σ must have a focal point along γ_p^+ in $\left[0, \frac{n-1}{\delta}\right]$. Similarly, Σ must also have a focal point along γ_p^- in $\left[0, \frac{n-1}{\delta}\right]$.

Next we consider the achronal boundary $\partial I^+(\Sigma)$. Since S is globally hyperbolic, $J^+(\Sigma)$ is closed by Proposition 3-16. This implies $J^+(\Sigma) = \overline{I^+(\Sigma)}$, by Lemma 3-9. Therefore, $\partial I^+(\Sigma) = J^+(\Sigma) \setminus I^+(\Sigma)$. Suppose $q \in \partial I^+(\Sigma) \setminus \Sigma$. By Theorem 4-1, there is a future-directed null geodesic β emanating from some $p \in \Sigma$ and ending at q such that β is orthogonal to Σ at p and Σ has no focal points along β before q. Let $i \in \{1, \ldots, m\}$ such that $p \in U_i$. Then $\beta'(0)$ is a positive constant multiple of either $K_i^+(p)$ or $K_i^-(p)$. Reparametrizing β, we can assume that $\beta = \gamma_p^+$ or $\beta = \gamma_p^-$. Since Σ has no focal point along β before q, q must lie in the set $W_i = W_i^+ \cup W_i^-$, where

$$
\begin{aligned}
W_i^+ &:= \left\{\exp_p(t K_i^+(p)) : p \in \overline{U}_i, \ t \in \left[0, \tfrac{n-1}{\delta}\right]\right\}, \\
W_i^- &:= \left\{\exp_p(t K_i^-(p)) : p \in \overline{U}_i, \ t \in \left[0, \tfrac{n-1}{\delta}\right]\right\}.
\end{aligned}
\tag{4.3.5}
$$

Since q is arbitrary, we have $\partial I^+(\Sigma) \subset W$, where $W = \bigcup_{j=1}^m W_j$ and each W_j is defined like W_i just above. Since W is compact and $\partial I^+(\Sigma)$ is closed, $\partial I^+(\Sigma)$ must also be compact.

Now, by Lemma 3-19(ii) and Corollary 3-24, $\partial I^+(\Sigma)$ is an achronal, C^0 hypersurface. Since $\partial I^+(\Sigma)$ is also compact, Proposition 3-27(ii) implies that any Cauchy hypersurface in S is homeomorphic to $\partial I^+(\Sigma)$, therefore must be compact. This contradicts the assumption that S has a noncompact Cauchy hypersurface, completing the proof. $\qquad\square$

CHAPTER 5

The Einstein constraint equations

5.1. Introduction

Many physical models admit an initial value formulation. In Newtonian mechanics, for instance, if we suppose the force is a function of the positions and velocities of the various particles under observation, then Newton's second law gives a system of ordinary differential equations which in principle will yield the evolution of the system once the initial positions and velocities of the particles are specified. The wave and heat equations also admit initial value problems. Maxwell's equations likewise admit an initial value formulation, but unlike the previous examples, where the initial configuration is essentially unconstrained, one cannot arbitrarily prescribe the electric and magnetic field at $t = 0$, say, and hope to solve Maxwell's equations. The reason is that parts of Maxwell's equations do not involve time derivatives: the spatial divergence of B vanishes, as does that of E (in regions with vanishing charge distribution). These two divergence constraints place restrictions on the vector fields one can use to prescribe initial data. As it turns out, these are the only constraints.

In this chapter, we discuss an initial value formulation for the vacuum Einstein equation. A vacuum *initial data set* will be given geometrically as a manifold Σ, endowed with a Riemannian metric g and symmetric $(0, 2)$-tensor K. That Σ embeds as a hypersurface in a Lorentzian manifold $(M, \bar{g})$ satisfying the vacuum Einstein equation, with induced metric g and second fundamental form K, imposes constraints on g and K. These conditions on g and K, which govern the space of allowable initial data sets for the vacuum Einstein equation, comprise the *Einstein constraint equations*, the study of solutions to which forms an interesting and rich subject for geometric analysis.

5.1.1. *Initial value formulation for Maxwell's equations.* We briefly discuss the initial value problem for the source-free Maxwell equations on Minkowski spacetime $\mathbb{M}^4$. For simplicity we take $c = 1$, so $x^0 = t$. We have seen that the Maxwell equations can be written in terms of the Faraday tensor F, an antisymmetric $(2, 0)$-tensor with corresponding two-form $F^\flat$. In inertial coordinates we write

$$F^\flat = (E_1\,dx^1 + E_2\,dx^2 + E_3\,dx^3) \wedge dx^0 + (B_1\,dx^2 \wedge dx^3 + B_2\,dx^3 \wedge dx^1 + B_3\,dx^1 \wedge dx^2).$$

Maxwell's equations can then be written

$$\mathrm{div}_\eta\, F = 0 \quad \text{and} \quad dF^\flat = 0.$$

From the second equation and the Poincaré lemma (see, e.g., [141]), we can write $F^\flat = dA$ for a one-form $A = A_\mu dx^\mu$, well-defined up to a gauge function ϕ: $F^\flat = d(A + d\phi)$ for any smooth ϕ. If $\vec{\nabla}$ is the spatial (Euclidean) gradient operator, and if we let

$$A = A_1\frac{\partial}{\partial x^1} + A_2\frac{\partial}{\partial x^2} + A_3\frac{\partial}{\partial x^3},$$

then we can make the identifications

$$B = \vec{\nabla} \times A, \quad E = \vec{\nabla}A_0 - \frac{\partial A}{\partial t}.$$

Note that the spatial divergence vanishes automatically: $\vec{\nabla} \cdot B = 0$. The spatial divergence of E is given by

$$\vec{\nabla} \cdot E = \Delta A_0 - \vec{\nabla} \cdot \frac{\partial A}{\partial t},$$

where Δ is the Euclidean Laplacian on $\mathbb{R}^3$. A simple calculation shows that $(\mathrm{div}_\eta\, F)^0 = F^{0\nu}_{\ ;\nu} = \vec{\nabla} \cdot E$, so we see that while Maxwell's equations are second order in A, this component has only one *time* derivative of A in it. In terms of formulating an evolution problem, this component of Maxwell's equations, equivalent to the vanishing of the spatial divergence of E, must be satisfied by the initial data. This imposes a constraint on the data.

    To formulate a second-order initial value problem for A, we can use the gauge freedom in A. The *Lorenz*[4] *gauge condition* imposes precisely that

$$0 = \mathrm{div}_\eta\, A = -\frac{\partial A_0}{\partial t} + \vec{\nabla} \cdot A.$$

Under the Lorenz gauge condition, Maxwell's equations are equivalent to

$$\Box A_\mu = -\frac{\partial^2 A_\mu}{\partial t^2} + \Delta A_\mu = 0.$$

Exercise 5-1. Verify this last claim. Use that $F_{\mu\nu} = \dfrac{\partial A_\nu}{\partial x^\mu} - \dfrac{\partial A_\mu}{\partial x^\nu}$.

---

[4]Note the spelling; this is not the eponym of Lorentz transformations!

We may proceed as follows. We specify initial data A_μ and $\partial A_\mu/\partial t$ at $t = 0$, satisfying the constraint

$$\vec{\nabla} \cdot \boldsymbol{E} = \Delta A_0 - \vec{\nabla} \cdot \frac{\partial \boldsymbol{A}}{\partial t} = 0. \tag{5.1.1}$$

We may also assume that the Lorenz gauge condition holds at $t = 0$. Indeed, given initial data $\mathring{A}_\mu|_{t=0}$ and $(\partial\mathring{A}_\mu/\partial t)|_{t=0}$ satisfying the constraint, we can arrange the Lorenz gauge condition at $t = 0$ by adding $d\phi$ to $\mathring{A}$, for a function ϕ depending only on (x^1, x^2, x^3); this is a gauge transformation, and does not affect the field F. In particular, it does not change $\boldsymbol{E}$, which remains divergence-free. Finding ϕ involves solving a Poisson equation $\Delta\phi = (\partial\mathring{A}_0/\partial t - \vec{\nabla}\cdot\mathring{\boldsymbol{A}})|_{t=0}$ on $\mathbb{R}^3$, where we note the right-hand side can be computed in terms of the initial data. We assume we are working in function spaces where we can solve this equation (for instance, where the fields decay sufficiently near infinity), and similar comments apply in the remainder of this section. We note that this gauge transformation $A = \mathring{A} + d\phi$ does not change the time derivative at $t = 0$.

We now solve the wave equation $\Box A_\mu = 0$. We will have produced a solution to Maxwell's equations, *provided* we can show that the Lorenz gauge condition, which we have satisfied at $t = 0$, is propagated in time. How do we do this? We have not yet incorporated the constraint. In fact, the condition (5.1.1) at $t = 0$, together with the wave equation $\Box A_0 = 0$, yields at $t = 0$

$$\frac{\partial^2 A_0}{\partial t^2} = \Delta A_0 = \sum_{\ell=1}^{3} \frac{\partial^2 A_\ell}{\partial t \, \partial x^\ell},$$

which is just $(\partial \operatorname{div}_\eta A/\partial t)|_{t=0} = 0$. Now, the wave equation for each A_μ also implies $\Box(\operatorname{div}_\eta A) = 0$. In other words, we see that $\operatorname{div}_\eta A$ satisfies the wave equation, and has *vanishing* initial data. Thus $\operatorname{div}_\eta A = 0$ for all t, and the gauge condition propagates in time. In summary, the gauge term (which we arranged to vanish at $t = 0$) is propagated with the wave equation, which, together with the constraint $\nabla \cdot \boldsymbol{E} = 0$, implies the vanishing of the first derivative of the gauge term.

We could also formulate the initial value problem for the source-free Maxwell's equations in terms of $\boldsymbol{E}$ and $\boldsymbol{B}$, with the constraints that the spatial divergences vanish. We will solve for A using the wave equation as above, where we take the initial data for A as follows: $A_0 = 0$, and $\boldsymbol{A}$ is chosen so that $\vec{\nabla} \times \boldsymbol{A} = \boldsymbol{B}$ (this is possible since $\vec{\nabla} \cdot \boldsymbol{B} = 0$ at $t = 0$). We also set $(\partial\boldsymbol{A}/\partial t)|_{t=0} = -\boldsymbol{E}$, and to arrange the Lorenz gauge condition initially, we take $\partial A_0/\partial t = \vec{\nabla} \cdot \boldsymbol{A}$ at $t = 0$. We can now solve for A as above to produce solutions to the Maxwell equations with given initial electric and magnetic fields.

We could also have proceeded as follows. The source-free Maxwell's equations imply vector wave equations for E and B, i.e., $\Box E = 0 = \Box B$. We can thus solve for E in time using the initial values for E (divergence-free at $t = 0$) as well as imposing the Maxwell equation $\partial E/\partial t = \vec{\nabla} \times B$ at $t = 0$. The vector wave equation for E implies $\vec{\nabla} \cdot E$ solves the wave equation. Moreover, $\partial(\vec{\nabla} \cdot E)/\partial t = \vec{\nabla} \cdot (\vec{\nabla} \times B) = 0$ at $t = 0$, and thus the constraint $\vec{\nabla} \cdot E = 0$ propagates in time, giving us one of Maxwell's equations. We solve for A at $t = 0$ so that $B = \vec{\nabla} \times A$, and we take $A_0 = 0$. The evolution of A is then given by $\partial A/\partial t = -E$, and we then define $B = \vec{\nabla} \times A$. This clearly gives the Maxwell equation $\partial B/\partial t = -\vec{\nabla} \times E$, from which we see that

$$\frac{\partial(\vec{\nabla} \cdot B)}{\partial t} = -\vec{\nabla} \cdot (\vec{\nabla} \times E) = 0,$$

so that $\vec{\nabla} \cdot B = 0$ propagates in time as well. We need now only show that $\partial E/\partial t = \vec{\nabla} \times B$, which holds at $t = 0$, also propagates. We have

$$\frac{\partial}{\partial t}\left(\frac{\partial E}{\partial t} - \vec{\nabla} \times B\right) = \frac{\partial^2 E}{\partial t^2} - \vec{\nabla} \times \left(\vec{\nabla} \times \frac{\partial A}{\partial t}\right) = \frac{\partial^2 E}{\partial t^2} + \vec{\nabla} \times (\vec{\nabla} \times E)$$

$$= \frac{\partial^2 E}{\partial t^2} + \vec{\nabla}(\vec{\nabla} \cdot E) - \Delta E = 0,$$

by the wave equation for E and the propagation of the constraint $\vec{\nabla} \cdot E = 0$. From this along with the definition $E = -\partial A/\partial t$, we see that $\vec{\nabla} \cdot A$ is constant in time. We can replace A by $A + \vec{\nabla}\phi$ for a suitable time-independent ϕ to arrange $\vec{\nabla} \cdot A = 0$ at $t = 0$, and hence for all time. The vanishing of the divergence of A is called the *Coulomb gauge*. Since we took $A_0 = 0$ here, this also corresponds to the Lorenz gauge in this special situation.

5.1.2. The Gauss and Codazzi equations. We will consider submanifolds Σ of a semi-Riemannian manifold $(M, \bar{g})$. If $\bar{g}$ is Riemannian, the metric induces a Riemannian metric g on Σ. This is not necessarily the case in the semi-Riemannian setting, as in the next example.

Example 5-2. Let

$$\Sigma^+ = \{(x^0, x^1, x^2, x^3) : 0 < x^0 = \sqrt{(x^1)^2 + (x^2)^2 + (x^3)^2}\} \subset \mathbb{M}^4$$

denote the forward lightcone minus the origin. Consider a point p in this submanifold at which $x^2 = 0 = x^3$, and $x^1 > 0$, so that $x^0 = x^1$ at p. The tangent space $T_p\Sigma^+$ is spanned by the vectors $(\partial/\partial x^0)|_p + (\partial/\partial x^1)|_p$, $(\partial/\partial x^2)|_p$, $(\partial/\partial x^3)|_p$. Note that the first of these is orthogonal to all of $T_p\Sigma^+$, and so the Minkowski metric does not induce a metric on Σ^+. Σ^+ is a *null* hypersurface.

Let Σ^k be a submanifold of $(M, \bar{g})$ upon which $\bar{g}$ induces a semi-Riemannian metric g. We let $\bar{g}(X, Y) = \langle X, Y \rangle$. Since at each $p \in \Sigma$ the metric is non-degenerate on $T_p\Sigma$, we have that $\bar{g}$ is also nondegenerate on $(T_p\Sigma)^\perp$, and $T_pM = T_p\Sigma \oplus (T_p\Sigma)^\perp$ (cf. [174, Chapter 2]). Thus we can decompose the tangent bundle of M along Σ into the direct sum of the tangent and normal bundles of Σ: $TM = T\Sigma \oplus N\Sigma$. It is not hard to show that the Levi-Civita connection ∇^Σ from the induced metric on Σ satisfies, for smooth vector fields X and Y tangent to Σ,

$$\nabla_X Y = \nabla_X^\Sigma Y + \mathrm{II}(X, Y)$$

where $\nabla_X^\Sigma Y$ is tangent to Σ, and $\mathrm{II}(X, Y)$ is normal to Σ.

Lemma 5-3. *For any vector fields X, Y tangent to Σ, the induced Levi-Civita connection $\nabla_X^\Sigma Y$ is the tangential projection of $\nabla_X Y$, while $\mathrm{II}(X, Y)$ is tensorial in X and Y, and is symmetric.*

Proof. We define $\nabla_X^\Sigma Y$ as the tangential component of $\nabla_X Y$, and we show it satisfies the defining properties of the Levi-Civita connection. Clearly ∇^Σ is torsion-free, since $\nabla_X Y - \nabla_Y X = [X, Y]$, which is tangent to Σ; this latter equation also implies that $\mathrm{II}(X, Y) = \mathrm{II}(Y, X)$. Clearly $\nabla_X^\Sigma Y$ is C^∞-linear in X, and $\mathbb{R}$-linear in Y, since $\nabla_X Y$ has these properties. Since $\nabla_X(fY) = X[f]Y + f\nabla_X Y$, and $X[f]Y$ is tangent to Σ, by taking projections we get the corresponding equation for ∇^Σ. Finally, the preceding equation also implies that $\mathrm{II}(X, Y)$ is C^∞-linear in Y, and hence by symmetry in X as well. $\qquad\square$

Definition 5-4. The tensor II is the (vector-valued) *second fundamental form*. The *mean curvature vector field* $\boldsymbol{H}$ (or $\vec{H}$) is $\boldsymbol{H} = \mathrm{tr}_g \mathrm{II} = \sum_{i=1}^k \epsilon_i \mathrm{II}(E_i, E_i)$, where $\{E_1, \ldots, E_k\}$ is an orthonormal basis of $T_p\Sigma$ with $\epsilon_i = \langle E_i, E_i \rangle$. The *scalar-valued second fundamental form* with respect to the unit normal vector n is given by $K(X, Y) = \langle \mathrm{II}(X, Y), n \rangle$, and the respective mean curvature H is the trace: $H = \mathrm{tr}_g K = \langle \boldsymbol{H}, n \rangle$.

If n denotes a local smooth unit normal field, $K(X, Y) = \langle \mathrm{II}(X, Y), n \rangle = \langle -\nabla_X n, Y \rangle$. While K and II depend on a choice of n, we emphasize that II and $\boldsymbol{H}$ do not. We note that K is sometimes defined with the opposite sign, so we define $\overline{K}(X, Y) = \langle \nabla_X n, Y \rangle = -K(X, Y)$, and $\overline{H} = \mathrm{tr}_g \overline{K}$. Moreover, in the hypersurface case, sometimes the scalar-valued second fundamental form is defined as $\hat{K}$, where $\hat{K}(X, Y)n = \mathrm{II}(X, Y) = \langle n, n \rangle K(X, Y)n$. If we let $\epsilon = \langle n, n \rangle = \pm 1$, then we see $\hat{K} = \epsilon K$, so that with $\hat{H} = \mathrm{tr}_g \hat{K} = \epsilon H$, we have $\boldsymbol{H} = \hat{H}n = \epsilon Hn$. In case (M, g) is Riemannian we of course have $\hat{K} = K = -\overline{K}$, whereas in the case $(M, \bar{g})$ is Lorentzian and (Σ, g) is a Riemannian hypersurface, then $\epsilon = \langle n, n \rangle = -1$, and $\hat{K} = \overline{K} = -K$, i.e., $\hat{K}(X, Y) = \langle \nabla_X n, Y \rangle$.

We begin by reviewing the proof of the Gauss equation, which relates the curvature of the submanifold to the ambient curvature and the second fundamental form.

Proposition 5-5 (the Gauss equation). *For any $X, Y, Z, W \in T_p\Sigma$, we have*

$$\langle R^\Sigma(X, Y, Z), W \rangle$$
$$= \langle R(X, Y, Z), W \rangle - \langle \mathrm{II}(X, Z), \mathrm{II}(Y, W) \rangle + \langle \mathrm{II}(X, W), \mathrm{II}(Y, Z) \rangle. \quad (5.1.2)$$

Proof. If N is normal to Σ, then since X and Y are tangential, it follows that $\nabla_X \langle Y, N \rangle = 0$, which is equivalent to $\langle \nabla_X Y, N \rangle = -\langle Y, \nabla_X N \rangle$. Moreover, by decomposing $\nabla_X Y$ into tangential and normal components, we obtain $\langle \nabla_X Y, W \rangle = \langle \nabla_X^\Sigma Y, W \rangle$, while $\langle \nabla_X Y, N \rangle = \langle \mathrm{II}(X, Y), N \rangle$. Using these we compute as follows:

$$\langle R(X, Y, Z), W \rangle = \langle \nabla_X \nabla_Y Z - \nabla_Y \nabla_X Z - \nabla_{[X,Y]} Z, W \rangle$$
$$= \langle \nabla_X(\nabla_Y^\Sigma Z + \mathrm{II}(Y, Z)) - \nabla_Y(\nabla_X^\Sigma Z + \mathrm{II}(X, Z)) - \nabla_{[X,Y]}^\Sigma Z, W \rangle$$
$$= \langle \nabla_X^\Sigma \nabla_Y^\Sigma Z + \mathrm{II}(X, \nabla_Y^\Sigma Z) + \nabla_X(\mathrm{II}(Y, Z))$$
$$- \nabla_Y^\Sigma \nabla_X^\Sigma Z - \mathrm{II}(Y, \nabla_X^\Sigma Z) - \nabla_Y(\mathrm{II}(X, Z)) - \nabla_{[X,Y]}^\Sigma Z, W \rangle$$
$$= \langle R^\Sigma(X, Y, Z), W \rangle - \langle \mathrm{II}(Y, Z), \nabla_X W \rangle + \langle \mathrm{II}(X, Z), \nabla_Y W \rangle$$
$$= \langle R^\Sigma(X, Y, Z), W \rangle - \langle \mathrm{II}(Y, Z), \mathrm{II}(X, W) \rangle + \langle \mathrm{II}(X, Z), \mathrm{II}(Y, W) \rangle. \quad \square$$

In the hypersurface case we can write the Gauss equation as

$$\langle R^\Sigma(X, Y, Z), W \rangle = \langle R(X, Y, Z), W \rangle - \langle n, n \rangle \langle K(X, Z), K(Y, W) \rangle$$
$$+ \langle n, n \rangle \langle K(X, W), K(Y, Z) \rangle$$
$$= \langle R(X, Y, Z), W \rangle - \langle n, n \rangle \langle \hat{K}(X, Z), \hat{K}(Y, W) \rangle$$
$$+ \langle n, n \rangle \langle \hat{K}(X, W), \hat{K}(Y, Z) \rangle. \quad (5.1.3)$$

To derive the Einstein constraint equations, we will use the Einstein equation, together with the Gauss equation, and the Codazzi equation, which we present now. We first define the normal connection $\nabla^\perp$ in the normal bundle $N\Sigma$ as follows: for V tangent to Σ and Z normal along Σ, we define $\nabla_V^\perp Z$ to be the normal component of $\nabla_V Z$. We can use this connection (and impose a product rule) to differentiate tensors with values in the normal bundle, in particular the second fundamental form: for V, X and Y tangent to Σ,

$$(\nabla_V \mathrm{II})(X, Y) := \nabla_V^\perp(\mathrm{II}(X, Y)) - \mathrm{II}(\nabla_V^\Sigma X, Y) - \mathrm{II}(X, \nabla_V^\Sigma Y). \quad (5.1.4)$$

For X, Y and Z tangent to Σ, let $R^\perp(X, Y, Z)$ be the normal component of $R(X, Y, Z)$.

Proposition 5-6 (the Codazzi equation). *For X, Y and Z tangent to Σ,*

$$R^\perp(X, Y, Z) = (\nabla_X \text{II})(Y, Z) - (\nabla_Y \text{II})(X, Z). \tag{5.1.5}$$

Proof. As in the proof of the Gauss equation, we decompose the curvature tensor:

$$
\begin{aligned}
R(X, Y, Z) &= \nabla_X \nabla_Y Z - \nabla_Y \nabla_X Z - \nabla_{[X,Y]} Z \\
&= \nabla_X(\nabla_Y^\Sigma Z + \text{II}(Y, Z)) - \nabla_Y(\nabla_X^\Sigma Z + \text{II}(X, Z)) - \nabla_{[X,Y]} Z.
\end{aligned}
$$

Using $[X, Y] = \nabla_X^\Sigma Y - \nabla_Y^\Sigma X$ and taking the normal component we have

$$
\begin{aligned}
R^\perp(X, Y, Z) &= \text{II}(X, \nabla_Y^\Sigma Z) + \nabla_X^\perp(\text{II}(Y, Z)) \\
&\quad - \text{II}(Y, \nabla_X^\Sigma Z) - \nabla_Y^\perp(\text{II}(X, Z)) - \text{II}([X, Y], Z) \\
&= \text{II}(X, \nabla_Y^\Sigma Z) + \nabla_X^\perp(\text{II}(Y, Z)) \\
&\quad - \text{II}(Y, \nabla_X^\Sigma Z) - \nabla_Y^\perp(\text{II}(X, Z)) - \text{II}(\nabla_X^\Sigma Y, Z) + \text{II}(\nabla_Y^\Sigma X, Z) \\
&= (\nabla_X \text{II})(Y, Z) - (\nabla_Y \text{II})(X, Z). \qquad \square
\end{aligned}
$$

Before we move on, we record the Gauss and Codazzi equations in index form in local coordinates, for the hypersurface case. Recall the convention $R_{ij\ell m} = g_{ms} R_{ij\ell}^s$. In the following, the indices refer to components tangential to Σ, while the "n" index refers to the vector n inserted in the corresponding slot in the tensor. The Gauss equation is easily seen to be

$$
\begin{aligned}
R_{ij\ell m}^\Sigma &= R_{ij\ell m} + \langle n, n \rangle(K_{im} K_{j\ell} - K_{i\ell} K_{jm}) \\
&= R_{ij\ell m} + \langle n, n \rangle(\hat{K}_{im} \hat{K}_{j\ell} - \hat{K}_{i\ell} \hat{K}_{jm}), \tag{5.1.6}
\end{aligned}
$$

while the Codazzi equation takes the form

$$R_{ij\ell n} = K_{j\ell;i} - K_{i\ell;j} = \langle n, n \rangle(\hat{K}_{j\ell;i} - \hat{K}_{i\ell;j}). \tag{5.1.7}$$

This readily follows from (5.1.4), $\text{II}(X, Y) = \langle n, n \rangle K(X, Y)n$ and $\langle \nabla_X n, n \rangle = 0$.

5.2. The Einstein constraint equations

Suppose $(M, \bar{g})$ is Lorentzian, and $\Sigma \subset M$ is a k-dimensional ($k \geq 2$) spacelike hypersurface, i.e., the induced metric g on Σ is Riemannian. Let n be a (smooth, local) timelike unit vector field to Σ (see Figure 7). We let T be a symmetric $(0, 2)$-tensor and assume that $(M, \bar{g})$ satisfies the Einstein equation $G_\Lambda(\bar{g}) = \kappa T$ (note that κ is not the dimension k of Σ). Then $\text{div}_{\bar{g}} T = 0$. We let $\mathbb{J} = (-T(n, \cdot))^\sharp$, so that $\mathbb{J}^\nu = -T^{\mu\nu} n_\mu$ (recall that summed-over Greek indices range from 0 to k). We write $\mathbb{J} = \rho n + J^\sharp$, where $J^\sharp$ is tangent to Σ. Then $\rho = T(n, n) = T_{00}$,

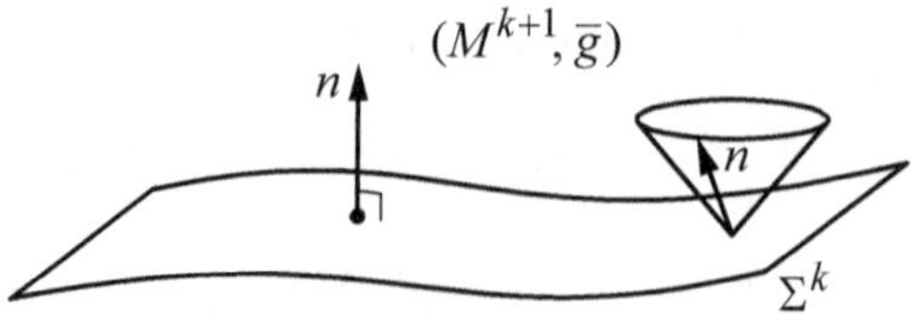

Figure 7. A spacelike hypersurface in a spacetime.

and if $E_1, \ldots, E_k$, is a local frame for $T\Sigma$ and we let $E_0 = n$ to complete the indexing, we have $J^\sharp = \sum_{i=1}^k J^i E_i$, where, for $j \geq 1$,

$$-T_{\mu j} n^\mu = -T_{0j} = -T(n, E_j) = \langle \mathbb{J}, E_j \rangle = \sum_{i=1}^k J^i g_{ij} = J_j. \tag{5.2.1}$$

Note that $J^i = \sum_{j=1}^k g^{ij} J_j = -T^{i\mu} n_\mu = T^{i0}$, $i \geq 1$. Then ρ is the energy density of the matter fields as measured by the observer with four-velocity cn, and J is (c times) the corresponding momentum density one-form. If n is future-pointing, then the dominant energy condition ($\mathbb{J}$ is future-pointing causal) implies $\rho \geq |J|_g = \sqrt{\sum_{i=1}^k J^i J_i}$.

We now come to the *Einstein constraint equations*, analogues of the divergence constraint on the initial data for the Maxwell equations we saw earlier. We recall that in spacetime dimension four, $\kappa = 8\pi G/c^4$.

Theorem 5-7 (Einstein constraint equations). *Let $(M, \bar{g})$ be Lorentzian. The following system of equations must hold on a Riemannian hypersurface $\Sigma \subset M$, where the Einstein equation $\mathrm{Ric}(\bar{g}) - \frac{1}{2}R(\bar{g})\bar{g} + \Lambda\bar{g} = \kappa T$ holds on M:*

$$R(g) - |K|_g^2 + (\mathrm{tr}_g K)^2 = 2\kappa\rho + 2\Lambda, \tag{5.2.2}$$

$$\mathrm{div}_g K - d(\mathrm{tr}_g K) = \kappa J. \tag{5.2.3}$$

Equation (5.2.2) is known as the *Hamiltonian constraint*, and (5.2.3) is the *momentum constraint*; see Sections 5.3.3.2 and 7.2.1 for further connection to the energy and momenta for gravitational systems. When we insert $\rho = 0$ and $J = 0$ into the constraints, we obtain the *vacuum constraint equations*.

Proof. Let $E_1, \ldots, E_k$ be a smooth local orthonormal frame field for Σ. We first apply the Gauss equation:

$$\sum_{i,j=1}^k \langle R(E_i, E_j, E_j), E_i \rangle$$

$$= \sum_{i,j=1}^k \left(\langle R^\Sigma(E_i, E_j, E_j), E_i \rangle - \langle \mathbb{II}(E_j, E_j), \mathbb{II}(E_i, E_i) \rangle + \langle \mathbb{II}(E_i, E_j), \mathbb{II}(E_i, E_j) \rangle \right)$$

$$= R(g) - |K|_g^2 + H^2. \tag{5.2.4}$$

Because $\langle n, n \rangle = -1$, we have for $i \geq 1$,

$$\mathrm{Ric}_{\bar{g}}(E_i, E_i) = -\langle R(n, E_i, E_i), n \rangle + \sum_{j=1}^{k} \langle R(E_i, E_j, E_j), E_i \rangle,$$

so that with the Einstein tensor $G = \mathrm{Ric}(\bar{g}) - \frac{1}{2} R(\bar{g}) \bar{g}$,

$$\sum_{i,j=1}^{k} \langle R(E_i, E_j, E_j), E_i \rangle = \mathrm{Ric}_{\bar{g}}(n, n) + \sum_{i=1}^{k} \mathrm{Ric}_{\bar{g}}(E_i, E_i)$$

$$= R(\bar{g}) + 2\mathrm{Ric}_{\bar{g}}(n, n) = 2G(n, n) \qquad (5.2.5)$$

$$= 2(-\Lambda \bar{g} + \kappa T)(n, n) = 2\Lambda + 2\kappa\rho.$$

Together with (5.2.4), we conclude (5.2.2).

For (5.2.3), we apply the Codazzi equation (5.1.7), along with the fact that $R_{ninn} = 0 = \bar{g}_{in}$ for any $i \in \{1, \ldots, k\}$, to obtain

$$G_{in} = R_{in} = \sum_{j=1}^{k} \langle R(E_j, E_i, n), E_j \rangle = -\sum_{j=1}^{k} R_{jijn}$$

$$= -\sum_{j=1}^{k} (K_{ij;j} - K_{jj;i}) = -\left(\mathrm{div}_g K - d(\mathrm{tr}_g K) \right)_i. \qquad (5.2.6)$$

By the Einstein equation and (5.2.1), we find $G_{in} = -\kappa J_i$, from which we conclude (5.2.3) as desired. $\qquad\qquad\square$

To summarize, we have from (5.2.4), (5.2.5) and (5.2.6) the following identities for the Einstein tensor, which show that these components of G can be expressed purely in terms of g and K (note the covariant derivatives are with respect to g):

$$2G_{nn} = R(g) - |K|_g^2 + (\mathrm{tr}_g K)^2, \qquad (5.2.7)$$

$$G_{ni} = -(K_{i;j}^j - K_{j;i}^j) = (\hat{K}_{i;j}^j - \hat{K}_{j;i}^j). \qquad (5.2.8)$$

Remark 5-8. The momentum constraint (5.2.3) appears on [218, p. 266] and some other works with a sign difference where the second fundamental form has the opposite sign to ours. In Chapter 8 of this volume, the momentum constraint takes the same form as (5.2.3), with both the second fundamental form and the one-form J having the opposite sign to what we have taken here.

Example 5-9. In Minkowski spacetime $\mathbb{M}^{1+k}$, the metric $\bar{g} = -dt^2 + \sum_{j=1}^{k} (dx^j)^2$ induces on the hypersurface $\Sigma = \{t = 0\}$ the Euclidean metric, and the second fundamental form vanishes. On the other hand, if we consider the hypersurface $\Sigma = \{(t, x) : -t^2 + |x|^2 = -1, \, t > 0\}$, as we saw in Section 1.4, the induced metric is the hyperbolic metric of constant curvature -1, and the second fundamental form (with respect to n future-pointing) is given by $K = -g$. The vacuum constraint equations ($\Lambda = 0$) are easily verified in each case.

Example 5-10. Let $n \geq 3$. Recall from Section 2.4.1 anti-de Sitter spacetime,
$\mathbb{H}_1^n(1) = \{x \in \mathbb{R}^{n+1} : -(x^0)^2 - (x^1)^2 + (x^2)^2 + \cdots + (x^n)^2 = -1\} \subset \mathbb{R}_2^{n+1}$.
The metric induced on $\{x \in \mathbb{R}^{n+1} : x^0 = 0\}$ is that of Minkowski spacetime
$\mathbb{M}^n$. The metric induced on the set $\Sigma = \{x \in \mathbb{H}_1^n(1) : x^0 = 0, \ x^1 > 0\}$, on
the other hand, is Riemannian; it is the unit hyperbolic metric, since Σ can be
isometrically identified with the upper unit hyperboloid in $\mathbb{R}_1^n$. However, Σ is
totally geodesic in $\mathbb{H}_1^n(1)$, i.e., its second fundamental form in $\mathbb{H}_1^n(1)$ vanishes;
this is easy to see, since $\partial/\partial x^0$ is a unit normal field to Σ which is tangent to
$\mathbb{H}_1^n(1)$ along Σ. One readily checks that the vacuum constraint equations *with
nonzero* $\Lambda = -\frac{1}{2}(n-1)(n-2)$ hold (Σ is isometric to $\mathbb{H}^{n-1}$ here).

   We can in turn consider universal anti-de Sitter spacetime $\widetilde{\mathbb{H}}_1^n(1)$, diffeomorphic
to $\mathbb{R}^n$, with covering map

$$\widetilde{\mathbb{H}}_1^n(1) \approx \mathbb{R} \times \mathbb{R}^{n-1} \ni (t, \xi) \mapsto (\sqrt{1+|\xi|^2}\,\cos t, \sqrt{1+|\xi|^2}\,\sin t, \xi) \in \mathbb{H}_1^n(1).$$

The covering map restricts to a diffeomorphism of each constant t-slice $\Sigma_t = \{t\} \times \mathbb{R}^{n-1}$ in $\widetilde{\mathbb{H}}_1^n(1)$ to the image of Σ under an isometry of $\mathbb{R}_2^{n+1}$ (a rotation of
the $x^0 x^1$-plane, fixing $x^2, \ldots, x^n$). Thus Σ_t is isometric to the unit hyperbolic
space, and is also totally geodesic.

Exercise 5-11. In an FLRW spacetime $(I \times \Sigma, -dt^2 + (f(t))^2 g_0)$, where g_0 has
constant curvature $k_0 = -1$, show that the second fundamental form of $\{t\} \times \Sigma$
with respect to $n = \partial/\partial t$ is given by $K = -f(t)f'(t)g_0$, and verify explicitly
the Einstein constraint equations in spacetime dimension four; cf. Section 2.4.3.

Example 5-12. Recall the exterior Schwarzschild spacetime metric

$$\bar{g}_S = -\left(1 - \frac{2m}{r}\right)dt^2 + \left(1 - \frac{2m}{r}\right)^{-1}dr^2 + r^2 \mathring{g}_{\mathbb{S}^2} = -\left(1 - \frac{2m}{r}\right)dt^2 + g_S.$$

The $t = 0$ slice Σ has induced metric g_S, and is easily seen to be totally geodesic,
since $\partial/\partial t$ is parallel along Σ. The vacuum Einstein constraints ($\Lambda = 0$) reduce
to a scalar curvature constraint $R(g_S) = 0$.

   Note that the constraints come from imposing $G_\Lambda(\bar{g})(n, \cdot) = \kappa T(n, \cdot)$ along
a spacelike hypersurface. As a simple exercise, one can see that if this equation
holds in the *normal* direction along *every* spacelike hypersurface, then the Einstein
equation $G_\Lambda(\bar{g}) = \kappa T$ holds. For example, if $(G_\Lambda)_{nn}$ vanishes along *every*
spacelike hypersurface of a spacetime, then the vacuum field equation $G_\Lambda(\bar{g}) = 0$
must hold, which is equivalent to $\mathrm{Ric}(\bar{g}) = \frac{2\Lambda}{k-1}\bar{g}$ in spacetime dimension $k+1 \geq 3$.

As K involves one time derivative of the metric (see (5.3.4) below), the expressions of G_{nn} and G_{ni} in (5.2.7)–(5.2.8) do not involve second *time* derivatives of the metric. Thus in formulating the Einstein equation as a second-order evolution problem for the spacetime metric, as discussed in the next section, the values of $G_{n\mu}$ (and hence $(G_\Lambda)_{n\mu}$) can be expressed in terms of the initial data set $(\Sigma, g, K; \rho, J)$ (where for the non-vacuum case, we augment the geometric initial data (g, K) with the energy-momentum density (ρ, J) along Σ). Thus, for instance, a valid initial data set (Σ, g, K) for the vacuum Einstein equation is *constrained* by the condition that these components must vanish along Σ: $(G_\Lambda)_{nn} = 0$, $(G_\Lambda)_{ni} = 0$ along Σ, i.e., (5.2.2)–(5.2.3) must hold with $(\rho, J) = (0, 0)$.

In the non-vacuum case, an initial data set may include initial data for the physical fields being modeled, from which ρ and J can be computed, and moreover the constraints may be augmented by equations that these fields must also satisfy, as in the following example.

Example 5-13. We consider the coupled Einstein–Maxwell equations in spacetime dimension four, initial data for which will be taken to be $(\Sigma, g, K; E, B)$, though we might instead encode the electromagnetic fields on Σ using a potential A, as seen in Section 5.1.1. As in (1.3.3)–(1.3.4), we have

$$\rho = \frac{1}{8\pi}(|E|_g^2 + |B|_g^2) \quad \text{and} \quad J^i = \frac{1}{4\pi}(E \times_g B)^i.$$

Along with using these in the constraint equations (5.2.2)–(5.2.3), we also impose constraints on the divergence of E and B, namely in the absence of charge, $\mathrm{div}_g E$ and $\mathrm{div}_g B$ vanish.

Any spacelike hypersurface in any Lorentzian manifold gives rise to a solution of the constraints, defining T so that the Einstein equation holds for a given Λ. Similarly, given (Σ, g, K), one can simply *define* ρ and J by (5.2.2)–(5.2.3) to get a solution $(\Sigma, g, K; \rho, J)$ of the constraint equations. With this in mind, we often impose some form for T, such as $T = 0$ (vacuum), or at least impose an energy condition (cf. Section 2.3.2) on T, such as the dominant energy condition $\rho \geq |J|_g$. Under such restrictions, the equations (5.2.2)–(5.2.3) do constrain g and K on Σ^k $(k \geq 2)$ in some way. For instance, if $\Lambda = 0$, the dominant energy condition can be written

$$\tfrac{1}{2}\big(R(g) - |K|_g^2 + (\mathrm{tr}_g K)^2\big) \geq \big|\mathrm{div}_g K - d(\mathrm{tr}_g K)\big|_g.$$

That said, even the vacuum constraints form an *underdetermined* system. In the case of spacetime dimension four, the constraints form a system of four equations for twelve independent functions in g and K (which are symmetric tensors). Even when one accounts for diffeomorphism gauge equivalence, the system is still underdetermined, and indeed there are lots of solutions to the vacuum constraints.

The *time-symmetric*, or *Riemannian*, constraints are the case $K = 0$. Then (5.2.2) reduces to $R(g) = 2\kappa\rho + 2\Lambda$, so that if $\Lambda = 0$, the scalar curvature $R(g)$ is proportional to the energy density, and $R(g) \geq 0$ if and only if $\rho \geq 0$ (which holds under the weak, and hence under the dominant, energy condition). The *maximal* case is $H = 0$, so that $R(g) = |K|_g^2 + 2\kappa\rho + 2\Lambda \geq 2\kappa\rho + 2\Lambda$. In the vacuum case ($T = 0$), the time-symmetric constraints reduce to $R(g) = 2\Lambda$ (constant scalar curvature), and in the maximal case we have $R(g) \geq 2\Lambda$; we often consider $\Lambda = 0$, which highlights the condition of zero or nonnegative scalar curvature of (Σ, g).

The *constraints operator* Φ is defined by

$$\Phi(g, K) = \big(R(g) - |K|_g^2 + (\mathrm{tr}_g K)^2, \, \mathrm{div}_g K - d(\mathrm{tr}_g K)\big).$$

As we will see in Sections 5.3.3.2 and 7.2.1, we may sometimes want to rewrite the constraints in terms of the *momentum tensor* π, which is algebraically equivalent ($k \geq 2$) to K via $\pi_{ij} = K_{ij} - (\mathrm{tr}_g K)g_{ij}$. It is easy to see that

$$R(g) - |K|_g^2 + (\mathrm{tr}_g K)^2 = R(g) - |\pi|_g^2 + \frac{1}{k-1}(\mathrm{tr}_g \pi)^2$$

$$\mathrm{div}_g K - d(\mathrm{tr}_g K) = \mathrm{div}_g \pi.$$

When we use π instead of K, we may abuse notation and write the constraints operator as $\Phi(g, \pi) = \big(R(g) - |\pi|_g^2 + \frac{1}{k-1}(\mathrm{tr}_g \pi)^2, \, \mathrm{div}_g \pi\big)$. While this abuse of notation should not cause confusion, we note that below we might use the same notation for the operator where π is treated as a $(2, 0)$-tensor, in which case $\mathrm{div}_g \pi$ is a vector field. In the literature, the constraints operator may appear as the above operator except with a factor of $\pm\frac{1}{2}$ in the first component (or a factor of ± 2 on the second component); see, e.g., (5.2.7) or Section 5.3.3.2.

5.3. The initial value formulation for the vacuum Einstein equation

In this section we discuss aspects of the analysis and geometry of the initial value formulation for $\mathrm{Ric}(\bar{g}) = 0$, or more generally $G_\Lambda(\bar{g}) = 0$. We will follow the approach of the foundational work of Choquet-Bruhat [49]. The purpose of this section is to illustrate the ideas; to be mathematically precise, we should specify function spaces and state carefully the partial differential equations results that are in play. We will not do this, but refer the reader to [51; 190; 218].

5.3.1. *Einstein's equation in Lorentz-harmonic gauge.* In order to formulate
the Einstein equation in terms of an initial value problem for a nonlinear wave
equation, we express the Einstein equations in terms of a partial differential
equation *along with* a gauge condition, as we did for Maxwell's equations above.
The gauge choice we use is the choice of coordinates. In fact, we will use
harmonic coordinates, or in the Lorentzian case, *wave coordinates* x^α, which
are coordinates so that $\lambda^\alpha := \Box_{\bar g} x^\alpha = 0$, where we recall that for a Lorentzian
metric $\bar g$, we let $\Box_{\bar g} u = \mathrm{tr}_{\bar g}(\mathrm{Hess}_{\bar g} u)$. On any Lorentzian manifold we can locally
set up wave coordinates: given any local coordinates y^μ, we solve the Cauchy
problem for the linear wave equations $\Box_{\bar g} x^\mu = 0$ with initial conditions $x^\mu = y^\mu$
and $\nabla_n x^\mu = \nabla_n y^\mu$ imposed on a level surface of y^0 ($\partial/\partial y^0$ is timelike), where
n is the unit normal to the level surface.

Let $\Gamma^\kappa_{\nu\beta} = \frac{1}{2}\bar g^{\kappa\mu}(\bar g_{\nu\mu,\beta} + \bar g_{\mu\beta,\nu} - \bar g_{\nu\beta,\mu})$ be the Christoffel symbols for $\bar g$ in a
coordinate system. It is easy to see that

$$\lambda^\alpha = \Box_{\bar g} x^\alpha = \bar g^{\mu\nu} x^\alpha_{;\mu\nu} = \bar g^{\mu\nu}(-\Gamma^\kappa_{\mu\nu} x^\alpha_{,\kappa}) = -\bar g^{\mu\nu}\Gamma^\alpha_{\mu\nu}.$$

In what follows we will write "$A \sim B$" to mean $A - B$ is a function of the
components $\bar g_{\mu\nu}$ and $\bar g_{\mu\nu,\kappa}$. In particular, if $A \sim B$, then $A - B$ does not depend
on second derivatives of the metric components. For example,

$$-(\bar g_{\alpha\nu}\lambda^\alpha_{,\beta} + \bar g_{\alpha\beta}\lambda^\alpha_{,\nu}) \sim \bar g_{\alpha\nu}\bar g^{\kappa\mu}\Gamma^\alpha_{\kappa\mu,\beta} + \bar g_{\alpha\beta}\bar g^{\kappa\mu}\Gamma^\alpha_{\kappa\mu,\nu}.$$

Exercise 5-14. Show that

$$\frac{1}{2}(\bar g_{\alpha\nu}\lambda^\alpha_{,\beta} + \bar g_{\alpha\beta}\lambda^\alpha_{,\nu}) \sim -\frac{1}{2}\bar g^{\kappa\mu}(\bar g_{\kappa\nu,\mu\beta} + \bar g_{\beta\mu,\kappa\nu} - \bar g_{\kappa\mu,\nu\beta}). \qquad (5\text{-}14a)$$

From (2-8b) (p. 72), the components of the Ricci curvature of $\bar g$ are given by

$$R_{\mu\nu} = \Gamma^\kappa_{\mu\nu,\kappa} - \Gamma^\kappa_{\mu\kappa,\nu} + \Gamma^\kappa_{\kappa\gamma}\Gamma^\gamma_{\mu\nu} - \Gamma^\kappa_{\nu\gamma}\Gamma^\gamma_{\mu\kappa} \sim \Gamma^\kappa_{\mu\nu,\kappa} - \Gamma^\kappa_{\mu\kappa,\nu}.$$

Moreover,

$$\Gamma^\kappa_{\mu\nu,\kappa} - \Gamma^\kappa_{\mu\kappa,\nu} \sim \frac{1}{2}\bar g^{\kappa\gamma}\big((g_{\mu\gamma,\nu\kappa} + g_{\nu\gamma,\mu\kappa} - g_{\mu\nu,\gamma\kappa}) - (\bar g_{\mu\gamma,\kappa\nu} + \bar g_{\kappa\gamma,\mu\nu} - \bar g_{\mu\kappa,\gamma\nu})\big)$$

$$= \frac{1}{2}\bar g^{\kappa\gamma}(\bar g_{\nu\gamma,\mu\kappa} - \bar g_{\mu\nu,\gamma\kappa} - \bar g_{\kappa\gamma,\mu\nu} + \bar g_{\mu\kappa,\gamma\nu}).$$

From this and (5-14a) we see that

$$R^H_{\mu\nu} := R_{\mu\nu} + \frac{1}{2}(\bar g_{\alpha\mu}\lambda^\alpha_{,\nu} + \bar g_{\alpha\nu}\lambda^\alpha_{,\mu}) \sim -\frac{1}{2}\bar g^{\kappa\gamma}\bar g_{\mu\nu,\kappa\gamma}.$$

By adding to the Ricci curvature operator the Lie derivative of the metric with
respect to a suitable vector field, we obtain the *reduced Ricci curvature* $R^H_{\mu\nu}$,
whose leading term plainly constitutes a nonlinear wave operator. (Recall the
Lie derivative formula $(L_X \bar g)_{\mu\nu} = \bar g_{\alpha\mu} X^\alpha_{;\nu} + \bar g_{\alpha\nu} X^\alpha_{;\mu}$.)

We let $R^H = \bar{g}^{\mu\nu} R^H_{\mu\nu}$ and $(G^H_\Lambda(\bar{g}))_{\mu\nu} = R^H_{\mu\nu} - \frac{1}{2} R^H \bar{g}_{\mu\nu} + \Lambda \bar{g}_{\mu\nu}$, and we consider the *reduced Einstein equation* $G^H_\Lambda(\bar{g}) = 0$, which can be formulated as a system of quasilinear wave equations:

$$-\tfrac{1}{2} \bar{g}^{\alpha\beta} \bar{g}_{\mu\nu,\alpha\beta} + \Psi_{\mu\nu}((\bar{g}_{\gamma\theta}), (\bar{g}_{\gamma\theta,\kappa})) = 0. \tag{5.3.1}$$

A solution to (5.3.1) will solve the Einstein equation if we can arrange $\lambda^\alpha = 0$ for all α. From the seminal work of Leray, along with a rescaling argument (see the references, e.g., [218, Chapter 10]), we can solve a system like (5.3.1) for small time.

The gauge condition above is not the only one that can be used, and moreover gauge choices can be expressed in a more geometric framework, in terms of background metrics (or connections), and the tension field for harmonic maps. See [17; 190, Chapter 14; 51, Chapter 7]. Furthermore, the same degeneracy and issue of gauge for the Ricci operator arises in Riemannian geometry, for instance in the study of the Ricci flow; cf. [73; 74].

5.3.2. The Einstein constraints and the propagation of the gauge condition.

Suppose we are given a solution (Σ, g, K) of the Einstein constraint equations for $G_\Lambda(\bar{g}) = 0$. We now incorporate this solution into initial data for (5.3.1). Choose local coordinates x^i $(i \geq 1)$ on $V \subset \Sigma$; we will obtain a solution of (5.3.1) on the product $I \times U$, where I is an interval around 0 with coordinate $x^0 = ct$, and U is a compactly contained open subset of V. We prescribe initial values of $\bar{g}_{\mu\nu}$ and $\bar{g}_{\mu\nu,0}$ at $t = 0$ as follows: for $i, j \geq 1$, let $\bar{g}_{ij} = g_{ij}$, and $\bar{g}_{ij,0} = -2K_{ij}$; let $\bar{g}_{00} = -1$, and for $j \geq 1$, let $\bar{g}_{0j} = 0$. For a spacetime metric $\bar{g}$ with these conditions, $\partial/\partial x^0$ is a unit normal field to $\{0\} \times U$, and so the components of the second fundamental form of U are indeed given by

$$K_{ij} = \bar{g}\left(\nabla_{\frac{\partial}{\partial x^i}} \frac{\partial}{\partial x^j}, \frac{\partial}{\partial x^0} \right) = -\tfrac{1}{2} \bar{g}_{ij,0},$$

as desired. For $\mu \geq 0$, we will choose the initial values of $\bar{g}_{0\mu,0}$ to arrange the gauge condition, as we explain. In this section, the Einstein summation convention will be in force for Greek indices, which as usual run over all values.

Now using the formula (2.3.10) derived earlier for the derivative of a determinant, we see that

$$\lambda^\alpha = \frac{1}{\sqrt{|\det \bar{g}|}} \left(\sqrt{|\det \bar{g}|}\, \bar{g}^{\beta\gamma} x^\alpha_{,\gamma} \right)_{,\beta} = \bar{g}^{\beta\alpha}_{\ \ ,\beta} + \tfrac{1}{2} \bar{g}^{\beta\alpha} \bar{g}^{\rho\sigma} \bar{g}_{\rho\sigma,\beta}.$$

At $t = 0$ we get

$$\lambda^0 = \bar{g}^{00}_{\ \ ,0} + \tfrac{1}{2} \bar{g}^{00} \bar{g}^{\rho\sigma} \bar{g}_{\rho\sigma,0} = -\tfrac{1}{2} \bar{g}_{00,0} - \tfrac{1}{2} \sum_{i,j\geq 1} \bar{g}^{ij} \bar{g}_{ij,0}, \tag{5.3.2}$$

and, for $i \geq 1$,

$$\lambda^i = \bar{g}^{0i}_{,0} + \sum_{j \geq 1}\left(\bar{g}^{ji}_{,j} + \tfrac{1}{2}\bar{g}^{ji}\bar{g}^{\rho\sigma}\bar{g}_{\rho\sigma,j}\right)$$

$$= \sum_{j \geq 1}\bar{g}_{0j,0}\bar{g}^{ji} + \sum_{j \geq 1}\left(\bar{g}^{ji}_{,j} + \tfrac{1}{2}\bar{g}^{ji}\bar{g}^{\rho\sigma}\bar{g}_{\rho\sigma,j}\right). \tag{5.3.3}$$

The final summation in (5.3.3) only involves spatial derivatives, so we can solve for $\bar{g}_{0j,0}$ at $t = 0$ ($j \geq 1$) to arrange $\lambda^i|_{t=0} = 0$ for $i \geq 1$. We can also clearly use (5.3.2) to determine $\bar{g}_{00,0}$ at $t = 0$ in order that $\lambda^0|_{t=0} = 0$ as well.

Now that we have specified all the Cauchy data, we can use standard theory for nonlinear wave equations (see [51; 190; 218]) to obtain a solution $\bar{g}_{\mu\nu}$ to (5.3.1), which is a Lorentzian metric on the product of U with an interval about 0, and the induced geometry on the slice $\{0\} \times U$ is precisely (U, g, K). The question now is how to guarantee that $\lambda^\alpha = 0$ propagates in time, so that $\bar{g}$ solves the Einstein equation. As we will see, a homogeneous linear wave equation for λ^α is a consequence of the Bianchi identities, while the Einstein constraints will show that the initial time derivative of λ^α vanishes. Together with the preceding paragraph, this will allow us to conclude that the gauge conditions that we have arranged at $t = 0$ propagate in time.

We begin with a simple exercise.

Exercise 5-15. Assuming $G^H_\Lambda(\bar{g}) = 0$, show that

$$(G_\Lambda(\bar{g}))_{\mu\nu} = R_{\mu\nu} - \tfrac{1}{2}R(\bar{g})\bar{g}_{\mu\nu} + \Lambda\bar{g}_{\mu\nu} = -\tfrac{1}{2}\bar{g}_{\alpha\mu}\lambda^\alpha_{,\nu} - \tfrac{1}{2}\bar{g}_{\alpha\nu}\lambda^\alpha_{,\mu} + \tfrac{1}{2}\bar{g}_{\mu\nu}\lambda^\alpha_{,\alpha}.$$

The vacuum constraints that are satisfied by the induced geometry (U, g, K) are precisely $(G_\Lambda(\bar{g}))_{\mu\nu}n^\nu = 0$ for $\mu \geq 0$, or in our set up, $(G_\Lambda(\bar{g}))_{\mu 0} = 0$ at $t = 0$; cf. (5.2.7)–(5.2.8). Note that by our arrangement of the gauge condition at $t = 0$, $\lambda^\mu_{,i} = 0$ for $i \geq 1$ and all μ. The component of the vacuum constraint for $\mu = 0$ (at $t = 0$) is just

$$0 = (G_\Lambda(\bar{g}))_{00} = -\tfrac{1}{2}\bar{g}_{\alpha 0}\lambda^\alpha_{,0} - \tfrac{1}{2}\bar{g}_{\alpha 0}\lambda^\alpha_{,0} + \tfrac{1}{2}\bar{g}_{00}\lambda^\alpha_{,\alpha} = \tfrac{1}{2}\lambda^0_{,0}.$$

For $i \geq 1$, we have at $t = 0$ (again using the vanishing of the spatial derivatives of λ^μ, and the initial condition on the metric)

$$0 = (G_\Lambda(\bar{g}))_{i0} = -\tfrac{1}{2}\bar{g}_{\alpha i}\lambda^\alpha_{,0} - \tfrac{1}{2}\bar{g}_{\alpha 0}\lambda^\alpha_{,i} + \tfrac{1}{2}\bar{g}_{i0}\lambda^\alpha_{,\alpha} = -\tfrac{1}{2}\sum_{j \geq 1}g_{ji}\lambda^j_{,0}.$$

Since this is true for $i \geq 1$ and the matrix (g_{ij}) is invertible, $\lambda^i_{,0}$ must vanish as well at $t = 0$.

Exercise 5-16. Use the preceding exercise to show that for a solution of (5.3.1), the vanishing of the divergence of $G_\Lambda(\bar{g})$ is equivalent to

$$0 = -\tfrac{1}{2}\bar{g}_{\alpha\nu}\,\Box_{\bar{g}}\,\lambda^\alpha + B^\theta_{\nu\gamma}\left((\bar{g}_{\rho\sigma}),(\bar{g}_{\rho\sigma,\beta})\right)\lambda^\gamma_{,\theta}.$$

The functions $B^\theta_{\nu\gamma}$ arise from expanding the divergence of $G_\Lambda(\bar{g})$ in components, using the Christoffel symbols of $\bar{g}$, which involve metric components and their first partials.

From this exercise we see that the partials of λ^α satisfy a homogeneous linear hyperbolic system with vanishing initial data. Thus the λ^α vanish identically (see [190, Chapters 8 and 12]). Hence the gauge condition holds, and the solution to the reduced Einstein equation yields a solution to the Einstein equation, as desired.

5.3.2.1. *More on the evolution problem.* What we have discussed above is a local construction. One would like to piece local solutions together, by patching along a cover of Σ, to obtain an existence result for a vacuum spacetime containing (Σ, g, K); see [51; 190; 218] for details. In order to formulate this, we sketch a local uniqueness result for the evolution of initial data.

Suppose $(\mathcal{V}_1, \bar{g}_1)$ and $(\mathcal{V}_2, \bar{g}_2)$ are vacuum spacetimes containing respective coordinate neighborhoods U_1 and U_2 on Σ, with induced geometry (g, K). Set $U = U_1 \cap U_2 \subset \Sigma$. There is a coordinate system in a neighborhood of U_1 in $\mathcal{V}_1$ with respect to which the metric components of $\bar{g}_1$ along U_1 satisfy $(\bar{g}_1)_{ij} = g_{ij}$ and $(\bar{g}_1)_{0j} = 0$ for $i, j \geq 1$, and $(\bar{g}_1)_{00} = -1$; indeed, one can start from coordinates (x^i) for U_1 and build adapted coordinates by $(x^\mu) \mapsto \exp^{\bar{g}_1}_{p(x^i)}(x^0 n_1)$, where $p(x^i) \in U_1$ corresponds to coordinates (x^i) for U_1, and n_1 is a smooth timelike unit normal along U_1. We see that $x^0 = 0$ corresponds to U_1, and $n_1 = \partial/\partial x^0$ along U_1. We have then that for $i, j \geq 1$, $(\bar{g}_1)_{ij,0} = -2K_{ij}$ along U_1. Actually in these specific coordinates, the geodesic equations also yield $(\bar{g}_1)_{0\mu,0} = 0$ along U_1. The analogous construction can be made along U_2 with normal unit normal n_2 with respect to $\bar{g}_2$, and we can build such coordinate charts along U from a common coordinate domain to respective neighborhoods $\mathcal{W}_1 \subset \mathcal{V}_1$ and $\mathcal{W}_2 \subset \mathcal{V}_2$ of U. If we pullback $\bar{g}_1$ and $\bar{g}_2$ by these coordinate charts, respectively, then along $x^0 = 0$ we will have both $(\bar{g}_1)_{\mu\nu} = (\bar{g}_2)_{\mu\nu}$ and $(\bar{g}_1)_{\mu\nu,0} = (\bar{g}_2)_{\mu\nu,0}$ (again, this comes from the constraint equations, along with the geodesic equation). Said another way, if we define the diffeomorphism $\varphi : \mathcal{W}_1 \to \mathcal{W}_2$ by $\varphi(\exp^{\bar{g}_1}_{p(x^i)}(x^0 n_1)) = \exp^{\bar{g}_2}_{p(x^i)}(x^0 n_2)$ so that expressed in these coordinates, φ is the identity map, then $\varphi^*\bar{g}_2$ agrees with $\bar{g}_1$ along U, and $(\varphi^*\bar{g}_2)_{\mu\nu,0} = (\bar{g}_1)_{\mu\nu,0}$ on $x^0 = 0$ in these coordinates for $\mathcal{W}_1$.

We now move to wave coordinates, by modifying the given coordinates along U in $\mathcal{W}_1$ and $\mathcal{W}_2$, in exactly the way we did in Section 5.3.1: solve the wave equation using the original coordinate functions and their time derivatives at $x^0 = 0$ as the initial data; we can again restrict to a common coordinate domain neighborhood $\widetilde{\mathcal{U}}$ around the $x^0 = 0$ slice. It is not hard to see that changing to wave coordinates this way preserves the first-order partial derivative operators along U, so in particular $(\bar{g}_1)_{\mu\nu}$ and $(\bar{g}_1)_{ij,0}$ for $i, j \geq 1$ (which remains equal to $-2K_{ij}$) do not change at $x^0 = 0$. Note that in wave coordinates, we can then determine $(\bar{g}_1)_{0\mu,0}$ from $(\bar{g}_1)_{\mu\nu}$ and $(\bar{g}_1)_{ij,0}$, via (5.3.2)–(5.3.3). Similar comments apply to $\bar{g}_2$, and so in these wave coordinates, $(\bar{g}_1)_{\mu\nu} = (\bar{g}_2)_{\mu\nu}$ and $(\bar{g}_1)_{\mu\nu,0} = (\bar{g}_2)_{\mu\nu,0}$ along $x^0 = 0$. But as we have expressed vacuum Einstein metrics in wave coordinates, $(\bar{g}_1)_{\mu\nu}$ and $(\bar{g}_2)_{\mu\nu}$ are solutions of the reduced Einstein equation (5.3.1), with the same initial conditions. Uniqueness for the Cauchy problem for the reduced Einstein equation shows that the two solutions must in fact agree.

We formulate the above in terms of a local isometry. If the wave coordinate charts are given by $\widetilde{\varphi}_i : \widetilde{\mathcal{U}} \to \widetilde{\mathcal{W}}_i \subset \mathcal{W}_i$, let $\psi = \widetilde{\varphi}_2 \circ \widetilde{\varphi}_1^{-1} : \widetilde{\mathcal{W}}_1 \to \widetilde{\mathcal{W}}_2$ be the induced diffeomorphism. In $\widetilde{\mathcal{W}}_1$, $(\psi^* \bar{g}_2)$ and $\bar{g}_1$ both satisfy the vacuum Einstein equation. The components of the coordinate map $\widetilde{\varphi}_2^{-1}$ are harmonic for $\bar{g}_2$, so that the components of the pullback $\widetilde{\varphi}_1^{-1} = \widetilde{\varphi}_2^{-1} \circ \psi$, which are harmonic for $\bar{g}_1$, are also harmonic for $\psi^*(\bar{g}_2)$. Thus both $((\bar{g}_1)_{\mu\nu})$ and $(\psi^*(\bar{g}_2)_{\mu\nu})$ solve the reduced Einstein equation on $\widetilde{\mathcal{U}}$, and we can argue as above that $(\psi^* \bar{g}_2)_{\mu\nu} = (\bar{g}_1)_{\mu\nu}$ and $(\psi^* \bar{g}_2)_{\mu\nu,0} = (\bar{g}_1)_{\mu\nu,0}$ along $t = 0$. Of course in the $\widetilde{\varphi}_1$ coordinates, the components $\psi^*(\bar{g}_2)_{\mu\nu}$ are just the $\widetilde{\varphi}_2$-components $(\bar{g}_2)_{\mu\nu}$. Thus we see that $\psi^* \bar{g}_2 = \bar{g}_1$, so that $\bar{g}_1$ and $\bar{g}_2$ are isometric in neighborhoods around a common solution (U, g, K) of the constraint equations.

One can use local existence together with the above uniqueness result to build a solution of the vacuum Einstein equation containing (Σ, g, K). A natural question is whether there is a *maximal* (in a certain sense) such solution which is determined from the initial data by the Einstein equation. To formulate this one utilizes causality notions, some of which were introduced in Chapter 3, to establish the existence of a maximal globally hyperbolic spacetime development of the initial data (see [51; 190; 218], and [191, Chapter 23] for a detailed exposition). We will be content here with considering the following question: given a solution (Σ, g, K) of the vacuum Einstein constraint equations (with $\Lambda = 0$, say), does there exist a globally hyperbolic spacetime $(\mathcal{S}, \bar{g})$ (i.e., a connected, time-oriented Lorentzian manifold with a Cauchy hypersurface — see Corollary 3-38 and its converse), with the following conditions: (i) $\mathrm{Ric}(\bar{g}) = 0$;

(ii) there is an imbedding of Σ in $\mathcal{S}$, the image of which we identify with Σ, with induced fundamental forms (g, K); and (iii) $\Sigma \subset (\mathcal{S}, \bar{g})$ is a Cauchy hypersurface? By Remark 3-31, (iii) means Σ is achronal and $D(\Sigma) = \mathcal{S}$.

We build on the causality discussion in Chapter 3, and give a roadmap of the argument, citing facts from [174]. Let $(\bar{M}, \bar{g})$ be a connected time-oriented Lorentzian manifold. If a subset $A \subset \bar{M}$ is achronal, then int $D(A)$, if nonempty, is globally hyperbolic, cf. [174, Theorem 14.38]. Moreover, if $A \subset \bar{M}$ is a topological hypersurface which is *acausal* (i.e., no causal curve meets A more than once, a stronger condition than achronality), then $D(A)$ is open [174, Lemma 14.43]. As a consequence, we see that if $(\bar{M}, \bar{g})$ satisfies conditions (i) and (ii) above, such that $\Sigma \subset (\bar{M}, \bar{g})$ is acausal, then $(\mathcal{S}, \bar{g}) = (D(\Sigma), \bar{g})$ will satisfy (i), (ii) and (iii).

By [174, Lemma 14.42], an achronal spacelike hypersurface is in fact acausal. Moreover, [174, Lemma 14.45] shows that in case a closed, connected subset $\Sigma \subset \bar{M}$ is a spacelike hypersurface such that $\bar{M} \setminus \Sigma$ is disconnected, then Σ is in fact achronal, and hence acausal, and hence by the above, $D(\Sigma)$ is open.

As an upshot of these causality results, we see that given a connected solution (Σ, g, K) of the vacuum constraints, for an open set $\bar{M} \subset \mathbb{R} \times \Sigma$ with $\{0\} \times \Sigma \subset \bar{M}$, if we have (say, by solving Einstein's equation locally and patching solutions together as sketched above) a Lorentzian metric $\bar{g}$ on $\bar{M}$ with $\mathrm{Ric}(\bar{g}) = 0$, inducing on $\{0\} \times \Sigma$ the initial data (g, K), then $(D(\Sigma), \bar{g})$ will satisfy properties (i), (ii), (iii) as desired. Thus this spacetime $(\mathcal{S} = D(\Sigma), \bar{g})$ is determined by the evolution of the geometric data (Σ, g, K), a topic we discuss further in the next section.

5.3.3. *Evolution of the geometry.* Consider a Lorentzian spacetime $(\mathcal{S}, \bar{g}) = (I \times \Sigma^k, \bar{g})$, where $0 \in I \subset \mathbb{R}$ is an interval, and where the *slices* $\Sigma_t = \{t\} \times \Sigma$ are spacelike. We let $g = g(t)$ be the induced metric on Σ_t. Let the unit vector n be normal to the slices, parallel to the spacetime gradient of t, pointing in the same time direction as $\partial/\partial t = Nn + X$, where X is tangent to the slices and $N > 0$. We call N the *lapse* function and X the *shift* vector field. We can write the metric $\bar{g}$ in local coordinates x^i for Σ as follows (summation over $1, \ldots, k$):

$$\bar{g} = -N^2 \, dt^2 + g_{ij}(dx^i + X^i \, dt) \otimes (dx^j + X^j \, dt).$$

The first and second fundamental forms of the slices form a family of solutions to the Einstein constraint equations. In our discussion of solving the Einstein equations from initial data, we chose $N = 1$ and $X = 0$ on the initial slice. One can use the solution of the initial value problem to determine a lapse and shift for a spacetime splitting. You might also consider suitably prescribing N and X, possibly solving an auxiliary set of equations which might for instance impose gauge conditions, and solve for the induced metric and second fundamental form of the slices; cf. [7]. Given N and X, we indicate below the evolution equations of these geometric quantities on the slices.

It is useful to have the analogous formulas in the Riemannian setting as well. So, with $\frac{\partial}{\partial t} = Nn + X$, we let $\epsilon = \langle n, n \rangle = \pm 1$, and write the metric in the form

$$\bar{g} = \epsilon N^2 \, dt^2 + g_{ij}(dx^i + X^i \, dt) \otimes (dx^j + X^j \, dt)$$
$$= (\epsilon N^2 + |X|_g^2)dt^2 + X^\flat \otimes dt + dt \otimes X^\flat + g,$$

where $N^2 > |X|_g^2$ in the Lorentzian case. With our convention on K, we have $\nabla_X Y = \nabla_X^\Sigma Y + \epsilon K(X, Y)n$, where X and Y are tangent to a slice (and, as here, we may suppress the subscript on Σ_t).

5.3.3.1. *The ADM equations.* We compute the time derivative of the induced metric. Let $e_i = \frac{\partial}{\partial x^i}$, $1 \le i \le k$, be a coordinate frame for Σ, and let $e_0 = \frac{\partial}{\partial t}$. Using metric compatibility, the torsion-free property of the connection, and the fact that all the e_μ commute, we have

$$\frac{\partial g_{ij}}{\partial t} = \bar{g}(\nabla_{e_i} e_0, e_j) + \bar{g}(e_i, \nabla_{e_j} e_0)$$
$$= \bar{g}(\nabla_{e_i}(Nn + X), e_j) + \bar{g}(e_i, \nabla_{e_j}(Nn + X))$$
$$= N\bar{g}(\nabla_{e_i} n, e_j) + N\bar{g}(e_i, \nabla_{e_j} n) + \bar{g}(\nabla_{e_i} X, e_j) + \bar{g}(e_i, \nabla_{e_j} X)$$
$$= -2NK_{ij} + g(\nabla_{e_i}^\Sigma X, e_j) + g(e_i, \nabla_{e_j}^\Sigma X)$$
$$= -2NK_{ij} + (L_X g)_{ij} \tag{5.3.4}$$

where $L_X g$ is the Lie derivative, $(L_X g)_{ij} = X_{i;j} + X_{j;i}$, where the semicolon indicates covariant differentiation for the Levi-Civita connection of g. Note that we can solve for the second fundamental form:

$$K_{ij} = -\tfrac{1}{2} N^{-1} \left(\frac{\partial g_{ij}}{\partial t} - (L_X g)_{ij} \right).$$

A more laborious exercise determines the time evolution of K:

$$\frac{\partial K_{ij}}{\partial t} = \epsilon N_{;ij} + (L_X K)_{ij} + N\big(\epsilon(R_{ij} - R_{ij}^\Sigma) - 2K_i^\ell K_{j\ell} + K_\ell^\ell K_{ij}\big), \tag{5.3.5}$$

where $N_{;ij}$ are the components of $\mathrm{Hess}_g N$, R_{ij} are components of $\mathrm{Ric}(\bar{g})$, R_{ij}^Σ are components of $\mathrm{Ric}(g)$, and $L_X K$ is the Lie derivative of K, $(L_X K)(Y, Z) = X[K(Y, Z)] - K([X, Y], Z) - K(Y, [X, Z])$. Note that there are no time derivatives of N or X in (5.3.4)–(5.3.5). We outline the proof of (5.3.5) in the remainder of the section.

Exercise 5-17. a. Show that if T is a $(0, 2)$-tensor, and W, Y and Z are vector fields, then $(L_W T)(Y, Z) = (\nabla_W T)(Y, Z) + T(\nabla_Y W, Z) + T(Y, \nabla_Z W)$.

b. With e_i and $\frac{\partial}{\partial t} = Nn + X$ as above, if $Y = Y^i e_i$, then

$$[n, Y] = \left(n[Y^i] + N^{-1}Y[X^i]\right) e_i + N^{-1}Y[N]n.$$

Conclude that the tangential component of $[n, e_\ell]$ is $-N^{-1}[X, e_\ell]$.

We will require a couple of formulas relating g and K to the ambient geometry of $\bar{g} = \langle \cdot, \cdot \rangle$. Define the tensor K^2 by $K_{ij}^2 = K_{ik}K_{j\ell}g^{k\ell}$, a metric contraction of $K \otimes K$. Note that $\nabla_{e_i} n = -K_i^{\,\ell} e_\ell$, since $K_{ij} = \langle -\nabla_{e_i} n, e_j \rangle = K_i^{\,\ell} g_{\ell j}$, and thus $\langle \nabla_{e_i} n, \nabla_{e_j} n \rangle = \langle -\nabla_{e_i} n, K_j^{\,\ell} e_\ell \rangle = K_{i\ell}K_j^{\,\ell} = K_{ij}^2$. In other words, $K^2(Y, Z) = \langle \nabla_Y n, \nabla_Z n \rangle = -K(\nabla_Y n, Z)$.

We extend K, i.e., the family of tensors $K(t)$, as a tensor on S as follows: for $p \in \Sigma_t$ and $v, w \in T_p S$, we let $K(v, w) := K(t)(v^T, w^T)$, where $v^T = v - \epsilon \langle v, n \rangle n$ is the tangential projection of v to $T_p \Sigma_t$.

The following identity we leave as an exercise, whose derivation is similar to that of the Hamiltonian constraint equation (5.2.2).

Exercise 5-18. For Y and Z orthogonal to n, we have

$$\bar{g}(R(Y, n, n), Z) = \epsilon(\mathrm{Ric}_{\bar{g}}(Y, Z) - \mathrm{Ric}_g(Y, Z)) + (\mathrm{tr}_g K)K(Y, Z) - K^2(Y, Z).$$

(Hint: Start by writing out $\mathrm{Ric}_{\bar{g}}(Y, Z)$ in an adapted orthonormal frame.)

We will also make use of the following form of what is sometimes referred to as the *Mainardi equation*.

Lemma 5-19. *For Y and Z orthogonal to n,*

$$\bar{g}(R(Y, n, n), Z) = (\nabla_n K)(Y, Z) - K^2(Y, Z) - \epsilon N^{-1}\mathrm{Hess}_g N(Y, Z). \quad (5.3.6)$$

Proof. We first note the elementary identity

$$\begin{aligned}
(\nabla_n K)(Y, Z) &= \nabla_n(K(Y, Z)) - K(\nabla_n Y, Z) - K(Y, \nabla_n Z) \\
&= \nabla_n(K(Y, Z)) - K(\nabla_Y n, Z) - K([n, Y], Z) + \langle \nabla_Y n, \nabla_n Z \rangle \\
&= \nabla_n(K(Y, Z)) + K^2(Y, Z) - K([n, Y], Z) + \langle \nabla_Y n, \nabla_n Z \rangle.
\end{aligned}$$

Now use $[n, Y] = [n, Y]^T + \epsilon \langle [n, Y], n \rangle n = [n, Y]^T + N^{-1}Y[N]n$ (Exercise 5-17) and $K([n, Y], Z) = K([n, Y]^T, Z) = \langle -\nabla_{[n, Y]^T} n, Z \rangle$ (by definition) to obtain

$$\begin{aligned}
\langle R(Y, n, n), Z\rangle &= \langle \nabla_Y \nabla_n n - \nabla_n \nabla_Y n, Z\rangle - \langle \nabla_{[Y,n]} n, Z\rangle \\
&= \langle \nabla_Y \nabla_n n, Z\rangle - \nabla_n \langle \nabla_Y n, Z\rangle + \langle \nabla_Y n, \nabla_n Z\rangle + \langle \nabla_{[n,Y]} n, Z\rangle \\
&= \langle \nabla_Y \nabla_n n, Z\rangle + \nabla_n (K(Y, Z)) + \langle \nabla_Y n, \nabla_n Z\rangle \\
&\quad - K([n, Y], Z) + N^{-1} Y[N] \langle \nabla_n n, Z\rangle \\
&= (\nabla_n K)(Y, Z) - K^2(Y, Z) + \nabla_Y \langle \nabla_n n, Z\rangle \\
&\quad - \langle \nabla_n n, \nabla_Y Z\rangle + N^{-1} Y[N] \langle \nabla_n n, Z\rangle.
\end{aligned}$$

Now, for a vector field Z normal to n,

$$\begin{aligned}
\langle \nabla_n n, Z\rangle &= -\langle n, \nabla_n Z\rangle = -\langle n, \nabla_Z n + [n, Z]\rangle = -\langle n, [n, Z]\rangle \\
&= -\epsilon N^{-1} Z[N]. \tag{5.3.7}
\end{aligned}$$

Thus $\langle \nabla_n n, \nabla_Y Z\rangle = \langle \nabla_n n, \nabla_Y^\Sigma Z\rangle = -\epsilon N^{-1} \nabla_Y^\Sigma Z[N]$, and so

$$\begin{aligned}
\nabla_Y \langle \nabla_n n, Z\rangle - \langle \nabla_n n, \nabla_Y Z\rangle &+ N^{-1} Y[N] \langle \nabla_n n, Z\rangle \\
&= -\epsilon \big(Y[N^{-1} Z[N]] - N^{-1} \nabla_Y^\Sigma Z[N] + N^{-2} Y[N] Z[N] \big) \\
&= -\epsilon N^{-1} \mathrm{Hess}_g N(Z, Y) = -\epsilon N^{-1} \mathrm{Hess}_g N(Y, Z),
\end{aligned}$$

as desired (the Hessian is with respect to the metric g, with connection ∇^Σ). $\square$

To derive (5.3.5), we have with $\partial/\partial t = Nn + X$, and recalling the tangential component of $[n, e_\ell]$ is $-N^{-1}[X, e_\ell]$, as well as $K^2(Y, Z) = -K(\nabla_Y n, Z)$,

$$\begin{aligned}
\frac{\partial K_{ij}}{\partial t} &= \nabla_{Nn+X}[K(e_i, e_j)] \\
&= N(\nabla_n K)_{ij} + NK(\nabla_n e_i, e_j) + NK(e_i, \nabla_n e_j) + X[K_{ij}] \\
&= N(\nabla_n K)_{ij} + NK(\nabla_{e_i} n + [n, e_i], e_j) \\
&\quad + NK(e_i, \nabla_{e_j} n + [n, e_j]) + X[K_{ij}] \\
&= N(\nabla_n K)_{ij} - 2NK_{ij}^2 \\
&\quad + NK(-N^{-1}[X, e_i], e_j) + NK(e_i, -N^{-1}[X, e_j]) + X[K_{ij}] \\
&= N(\nabla_n K)_{ij} - 2NK_{ij}^2 + (L_X K)_{ij}.
\end{aligned}$$

Putting this together with Exercise 5-18 and Lemma 5-19, we obtain (5.3.5).

The system (5.3.4)–(5.3.5) is known as the *ADM equations*, after Arnowitt, Deser and Misner [10]. If the spacetime $(S, \bar{g})$ as above satisfies the vacuum Einstein equation, then $R_{ij} = 0$ in (5.3.5), with $\epsilon = -1$. Turning this around, suppose we have some lapse N and shift X, and we want to build a vacuum spacetime metric $\bar{g}$ by solving for $g = g(t)$. The spacetime scalar curvature is given by $R(\bar{g}) = -\mathrm{Ric}_{\bar{g}}(n, n) + g^{ij} R_{ij}$, and the Einstein tensor is given by $G(n, n) = \frac{1}{2}\mathrm{Ric}_{\bar{g}}(n, n) + \frac{1}{2}g^{ij} R_{ij}$, $G(n, e_i) = \mathrm{Ric}_{\bar{g}}(n, e_i)$, and $G_{ij} = R_{ij} - \frac{1}{2}R(\bar{g})g_{ij}$. We want to encode $G_{ij} = 0$ into the dynamical equations, since as we have seen earlier $G(n, \cdot)$ is encoded in the constraint equations. With $G(n, n) = \frac{1}{2}\big(R(g) - |K|_g^2 + (\mathrm{tr}_g K)^2\big)$, we have

$$\mathrm{Ric}_{\bar{g}}(n, n) = R(g) - |K|_g^2 + (\mathrm{tr}_g K)^2 - g^{ij} R_{ij},$$

and so $R(\bar{g}) = 2g^{ij} R_{ij} - (R(g) - |K|_g^2 + (\mathrm{tr}_g K)^2)$. Note that if $G_{ij} = 0$, then $(k \geq 2)$ $g^{ij} R_{ij} = \frac{k}{2}R(\bar{g})$, and so $R(\bar{g}) = \frac{1}{k-1}(R(g) - |K|_g^2 + (\mathrm{tr}_g K)^2)$. Thus $G_{ij} = 0$ can be written $R_{ij} = \frac{1}{2(k-1)}\big(R(g) - |K|_g^2 + (\mathrm{tr}_g K)^2\big)g_{ij}$.

If we thus insert R_{ij} as above into the ADM equations, and if we have a PDE framework that allows us to solve the system (cf., e.g., [7], and [17; 51] and references therein), then we can construct $\bar{g}$ from the solution. The first equation (5.3.4) guarantees that the tensor $K = K(t)$ is in fact the second fundamental form of $\{t\} \times \Sigma \subset (\mathcal{S}, \bar{g})$, and so (5.3.5) with R_{ij} as above really does imply that the spatial components G_{ij} (tangent to $\{t\} \times \Sigma$) vanish. As we recalled above, the other components will be given by the constraints. Indeed, we have not yet mentioned the initial data $(g(0), K(0))$; we of course will impose the vacuum constraints on this data, and in so doing, we will have $G_{\mu\nu} = 0$ on the $t = 0$ slice.

The Einstein equation $G_{\mu\nu} = 0$ will come from the dynamical equations (5.3.4)–(5.3.5) together with the constraint equations. The constraints are imposed at $t = 0$, and we need to verify (as we did in Section 5.3.2) that they are propagated. This is done as before, using the Bianchi identity $\bar{g}^{\alpha\beta} G_{\mu\alpha;\beta} = 0$, which, given $G_{ij} = 0$, can be written as a (symmetrizable hyperbolic: see [51, Chapter VI] or [214, Chapter 18]) system of linear homogeneous evolution equations for $G_{\mu 0}$, with vanishing initial conditions (imposed by the constraints). Thus $G_{\mu 0}$ must vanish identically.

Exercise 5-20. Suppose for simplicity that the lapse N is 1 and the shift X is 0, and that we have solved for the spacetime metric $\bar{g}$ from (5.3.4)–(5.3.5) imposing $G_{ij} = 0$ for $i, j \geq 1$ as above. Suppose furthermore that (g, K) solve the constraint equations at $t = 0$. Using $G_{ij} = 0$ and the Bianchi identity, show that $G_{00,0}$ is a linear combination (with coefficients depending on $\bar{g}$ and its Christoffel symbols) of $G_{0\mu}$ and $G_{i0,j}$. Show that $G_{i0,0}$ is such a linear combination, but only with terms G_{j0}; you might want to observe $\Gamma^0_{i0} = 0$. Because of this, you can derive a system of the form $\partial Z / \partial t = AZ$, where A is a matrix function depending smoothly (algebraically in fact) on $(\bar{g}, \partial\bar{g}, \partial^2\bar{g})$, and where the components of the vector Z are the components $G_{0\mu}$ and $G_{i0,j}$ for $\mu \geq 0$ and $i, j \geq 1$.

5.3.3.2. *Hamiltonian formulation.* The ADM equations are related to a *Hamiltonian formulation* of the Einstein equations, for which we have already seen a Lagrangian formulation. We will indicate some basic notions in this regard, referring the interested reader to references such as [10; 17; 90; 51; 218].

In classical mechanics, one can often recast an action generated by a Lagrangian $L = L(q, \dot{q}, t)$ in *Hamiltonian* form, using the *Legendre transform* $H(q, p, t) = p\dot{q} - L(q, \dot{q}, t)$ with $p = \partial L/\partial \dot{q}$ to define the associated Hamiltonian function [8; 220]; q is the vector of *state variables* and p is the vector of the *conjugate momenta* (we have abbreviated by not including subscripts to number the components, e.g., $p\dot{q} = \sum_{j=1}^{N} p_j \dot{q}_j$). Stationarity of the action $\int_{t_1}^{t_2} L(q, \dot{q}, t)\, dt = \int_{t_1}^{t_2} (p\dot{q} - H(q, p, t))\, dt$, with respect to variations of $q(t)$ fixing the endpoint values $q(t_1)$ and $q(t_2)$, leads naturally to the Euler–Lagrange equations

$$\frac{d}{dt}\left(\frac{\partial L}{\partial \dot{q}}\right) = \frac{\partial L}{\partial q};$$

using these along with $p = \partial L/\partial \dot{q}$ while examining the differential dH leads to *Hamilton's equations*

$$\dot{q} = \frac{\partial H}{\partial p}, \quad \dot{p} = -\frac{\partial H}{\partial q}.$$

We remark that precisely the same equations characterize critical points of the Hamiltonian form of the action over more general curves $(q(t), p(t))$ in momentum phase space fixing the endpoint values $q(t_1)$ and $q(t_2)$, not only over curves for which $p = \partial L/\partial \dot{q}$ [8].

We can recast the Einstein–Hilbert action $\int_{\mathcal{S}} R(\bar{g})\, dv_{\bar{g}}$ on $(\mathcal{S} = I \times \Sigma, \bar{g})$, where Σ is a k-manifold ($k \geq 2$), in an analogous fashion, connecting the ADM equations with Hamilton's equations. We assume we can integrate over the time interval $I = (t_1, t_2)$ (restricting from a larger interval if needed), with fields extending smoothly to the boundary $\partial\mathcal{S}$, which in case Σ has empty boundary, consists of two slices $\Sigma_{t_1} = \{t_1\} \times \Sigma$ and $\Sigma_{t_2} = \{t_2\} \times \Sigma$. If $\mathcal{S}$ is not compact, then as we have noted before, one can derive the field equations by considering stationary points $\bar{g}$ of a regularized action defined by integrating $(R(\bar{g}+\bar{h})-R(\bar{g}))$ for compactly supported variations $\bar{h}$, or one can work in spaces of metrics (with associated deformations) in which the scalar curvature is integrable. Though we generally take the field variations to vanish near the boundary (if nonempty) or near infinity (in the noncompact case) when deriving the Euler–Lagrange equations, we do remark that an analysis of the boundary terms (including the behavior at infinity) is of interest (see [229; 113; 218], for example), and in particular plays a role in the study of isolated gravitational systems, and in the notion of quasilocal mass [158]. In fact, for isolated systems, the natural decay rates are such that not all the boundary terms should be discarded in the Hamiltonian analysis; as we will see in Section 7.2.1, the ADM energy-momentum is defined in terms of flux integrals at infinity (i.e., as a limit of flux integrals over large spheres in an asymptotic end), and these terms must be included to get a well-defined Hamiltonian in the natural phase space [9; 16; 75; 187]. For simplicity on the first pass, we will have in mind the case when Σ is compact with no boundary, but see Remark 5-22 and Exercise 5-30 for a brief discussion of boundary terms, and see Section 7.2.1 for a discussion in the asymptotically flat setting. Finally, we note that to get the units of energy, we want to consider the Hamiltonian related to the Lagrangian $\int_S \frac{1}{2\kappa} R(\bar{g}) \, dv_{\bar{g}}$; see Remark 5-23. For simplicity, we work below in units where $2\kappa = 1$, though in later chapters we will use units where $G = 1$ and $c = 1$, so that the energy of the Schwarzschild spacetime is m; in spacetime dimension four, for instance, $2\kappa = 16\pi$.

We begin by finding expressions for the the scalar curvature $R(\bar{g})$ of $(S, \bar{g})$ with $\bar{g} = -N^2 \, dt^2 + g_{ij}(dx^i + X^i \, dt) \otimes (dx^j + X^j \, dt)$.

Exercise 5-21. Prove that $\nabla_n (\mathrm{tr}_g K) = \mathrm{tr}_g (\nabla_n K)$. Then use Lemma 5-19 to show

$$R(\bar{g}) = -2\nabla_n(\mathrm{tr}_g K) - 2N^{-1}\Delta_g N + 2|K|_g^2 + \left(R(g) - |K|_g^2 + (\mathrm{tr}_g K)^2\right) \quad (5\text{-}21a)$$

$$= -2\mathrm{div}_{\bar{g}}((\mathrm{tr}_g K)n) - 2(\mathrm{tr}_g K)^2 - 2\mathrm{div}_{\bar{g}}(\nabla_n n) + 2|K|_g^2$$
$$+ \left(R(g) - |K|_g^2 + (\mathrm{tr}_g K)^2\right). \quad (5\text{-}21b)$$

(Hint: You might use (5.3.7) to help show $N^{-1}\Delta_g N = \mathrm{div}_{\bar{g}}(\nabla_n n)$.)

To rewrite the Einstein–Hilbert action, we will use the ADM equations (5.3.4) and (5.3.5), and we also note the following fact: $\sqrt{|\det \bar{g}|} = N\sqrt{\det g}$. To derive this by direct expansion of the matrix for $\bar{g}$ in lapse-shift form, let $\xi^{(i)}$ be the matrix obtained by replacing column i of the matrix for g with the column whose entry in row j is $X_j = g_{j\ell}X^\ell$, so that

$$\det \bar{g} = (-N^2 + |X|_g^2) \det g - \sum_{i=1}^{k} X_i \det(\xi^{(i)})$$

$$= (-N^2 + |X|_g^2) \det g - \sum_{i,j=1}^{k} (-1)^{i+j} X_i X_j M_{ij},$$

where M_{ij} is the (i, j)-minor determinant of the matrix for g. Cramer's rule for the inverse (Section 2.3.4) now can be employed to get $g^{ij} \det g = (-1)^{i+j} M_{ji}$, from which the determinant formula follows.

We can thus write the Einstein–Hilbert action over the spacetime product as $\int_S R(\bar{g})\,dv_{\bar{g}} = \int_I \int_\Sigma R(\bar{g})N\,dv_g\,dt$. Before we insert the formula (5-21b) for $R(\bar{g})$ and integrate, we note that with Hamilton's equations, as well as the evolution equation (5.3.4) for $\partial g/\partial t$, in mind, we let $\pi^{ij} = K^{ij} - (\mathrm{tr}_g K)g^{ij}$, and $\hat{\pi}^{ij} = \hat{K}^{ij} - (\mathrm{tr}_g \hat{K})g^{ij} = -\pi^{ij}$, and we consider the following expression:

$$\hat{\pi}^{ij}\frac{\partial g_{ij}}{\partial t} = -(K^{ij} - (\mathrm{tr}_g K)g^{ij})(-2NK_{ij}) + \hat{\pi}^{ij}(L_X g)_{ij}$$
$$= -2N(\mathrm{tr}_g K)^2 + 2N|K|_g^2 + \hat{\pi}^{ij}(X_{i;j} + X_{j;i})$$
$$= -2N(\mathrm{tr}_g K)^2 + 2N|K|_g^2 + 2\hat{\pi}^{ij}X_{i;j}. \qquad (5.3.8)$$

We insert this into (5-21b) to obtain

$$R(\bar{g}) = N^{-1}\hat{\pi}^{ij}\frac{\partial g_{ij}}{\partial t} - 2N^{-1}\hat{\pi}^{ij}X_{i;j} + \big(R(g) - |K|_g^2 + (\mathrm{tr}_g K)^2\big)$$
$$- 2\mathrm{div}_{\bar{g}}((\mathrm{tr}_g K)n) - 2\mathrm{div}_{\bar{g}}(\nabla_n n).$$

Thus the action $\int_S R(\bar{g})\,dv_{\bar{g}}$ (modulo any boundary terms) is equal to

$$\int_I \int_\Sigma \left(\hat{\pi}^{ij}\frac{\partial g_{ij}}{\partial t} + 2(\mathrm{div}_g \hat{\pi})_i X^i + N\big(R(g) - |K|_g^2 + (\mathrm{tr}_g K)^2\big)\right) dv_g\,dt.$$

Remark 5-22. In case $\partial\Sigma$ is empty, the boundary terms are easily seen to be

$$-\left(\int_{\Sigma_{t_2}} 2\mathrm{tr}_g K\,dv_g - \int_{\Sigma_{t_1}} 2\mathrm{tr}_g K\,dv_g\right) =: -\left(\int_{\Sigma_{t_2}} - \int_{\Sigma_{t_1}}\right) 2\mathrm{tr}_g K\,dv_g.$$

For more on boundary terms, see Exercise 5-30.

We now assemble this into Hamiltonian form. We have $\mathrm{tr}_g \hat{\pi} = (k-1)\,\mathrm{tr}_g K$ and $|\hat{\pi}|_g^2 = |K|_g^2 + (k-2)(\mathrm{tr}_g K)^2$. Define $\mathcal{H}_{\mathrm{ADM}} = \int_\Sigma \widehat{\mathcal{H}}\,dv_g$, with

$$\widehat{\mathcal{H}} = \widehat{\mathcal{H}}(g, \hat{\pi}) = -N\big(R(g) - |K|_g^2 + (\mathrm{tr}_g K)^2\big) + 2\big(\mathrm{div}_g K - d(\mathrm{tr}_g K)\big)_i X^i$$
$$= -N\left(R(g) - |\hat{\pi}|_g^2 + \frac{1}{k-1}(\mathrm{tr}_g\hat{\pi})^2\right) - 2(\mathrm{div}_g \hat{\pi})_i X^i \qquad (5.3.9)$$
$$= (N, X)\cdot\widehat{\Phi}(g, \hat{\pi}), \qquad (5.3.10)$$

where

$$\widehat{\Phi}(g, \hat{\pi}) = -\left(R(g) - |\hat{\pi}|_g^2 + \frac{1}{(k-1)}(\mathrm{tr}_g\hat{\pi})^2,\ 2(\mathrm{div}_g \hat{\pi})^\flat\right),$$

with the second component a one-form field. We emphasize that the vacuum constraint equations (with $\Lambda = 0$) are just $\widehat{\Phi}(g, \hat{\pi}) = 0$.

Thus we can write $\int_{\mathcal{S}} R(\bar{g})\,dv_{\bar{g}}$ (once again, modulo boundary terms) as

$$\int_I \left(\int_\Sigma \hat{\pi}^{ij} \frac{\partial g_{ij}}{\partial t} dv_g - \mathcal{H}_{\mathrm{ADM}} \right) dt$$

$$= \int_I \int_\Sigma \left(\hat{\pi}^{ij} \frac{\partial g_{ij}}{\partial t} - (N, X) \cdot \widehat{\Phi}(g, \hat{\pi}) \right) dv_g\, dt. \quad (5.3.11)$$

N and X appear as Lagrange multipliers for a constrained optimization problem. Indeed, stationarity of the action with respect to variations of lapse and shift yield the vacuum constraint equations.

With an eye toward Hamilton's equations, we choose a background volume element $d\mathring{v}$ on Σ, induced from a background metric $\mathring{g}$ on Σ. The function $\theta_g = \sqrt{\det g}/\sqrt{\det \mathring{g}}$ encodes the ratio of the volume elements (and in particular is independent of coordinates). We let $\tilde{\pi}^{ij} = \theta_g \hat{\pi}^{ij}$, $\widetilde{\mathcal{H}}(g, \tilde{\pi}) = \theta_g \widehat{\mathcal{H}}(g, \hat{\pi})$, and $\widetilde{\Phi}(g, \tilde{\pi}) = \theta_g \widehat{\Phi}(g, \hat{\pi})$, so that the action can be written

$$\int_I \left(\int_\Sigma \hat{\pi}^{ij} \frac{\partial g_{ij}}{\partial t} dv_g - \mathcal{H}_{\mathrm{ADM}} \right) dt$$

$$= \int_I \int_\Sigma \left(\hat{\pi}^{ij} \frac{\partial g_{ij}}{\partial t} - \widehat{\mathcal{H}} \right) dv_g\, dt = \int_I \int_\Sigma \left(\tilde{\pi}^{ij} \frac{\partial g_{ij}}{\partial t} - \widetilde{\mathcal{H}} \right) d\mathring{v}\, dt$$

$$= \int_I \int_\Sigma \left(\tilde{\pi}^{ij} \frac{\partial g_{ij}}{\partial t} - (N, X) \cdot \widetilde{\Phi}(g, \tilde{\pi}) \right) d\mathring{v}\, dt. \quad (5.3.12)$$

Starting from the stationarity of the action in Hamiltonian form, we can obtain the ADM equations. Indeed consider a one-parameter variation (compactly supported away from the boundary) $\hat{\pi}^{ij} + \epsilon \hat{\sigma}^{ij}$, keeping g fixed, or equivalently $\tilde{\pi}^{ij} + \epsilon \tilde{\sigma}^{ij} = \theta_g(\hat{\pi}^{ij} + \epsilon \hat{\sigma}^{ij})$, and define

$$\hat{\sigma}^{ij} \frac{\delta \widehat{\mathcal{H}}}{\delta \hat{\pi}^{ij}} = \frac{d}{d\epsilon}\Big|_{\epsilon=0} \widehat{\mathcal{H}}(g, \hat{\pi} + \epsilon \hat{\sigma}), \quad \tilde{\sigma}^{ij} \frac{\delta \widetilde{\mathcal{H}}}{\delta \tilde{\pi}^{ij}} = \frac{d}{d\epsilon}\Big|_{\epsilon=0} \widetilde{\mathcal{H}}(g, \tilde{\pi} + \epsilon \tilde{\sigma}).$$

We then observe that $\delta \widehat{\mathcal{H}}/\delta \hat{\pi}^{ij} = \delta \widetilde{\mathcal{H}}/\delta \tilde{\pi}^{ij}$, and moreover with $\mathcal{H}_{\mathrm{ADM}}(g, \hat{\pi}) := \int_\Sigma \widehat{\mathcal{H}}\, dv_g := \int_\Sigma \widetilde{\mathcal{H}}\, d\mathring{v} =: \mathcal{H}_{\mathrm{ADM}}(g, \tilde{\pi})$,

$$\frac{d}{d\epsilon}\Big|_{\epsilon=0} \mathcal{H}_{\mathrm{ADM}}(g, \hat{\pi} + \epsilon \hat{\sigma}) = \int_\Sigma \hat{\sigma}^{ij} \frac{\delta \widehat{\mathcal{H}}}{\delta \hat{\pi}^{ij}} dv_g$$

$$= \int_\Sigma \tilde{\sigma}^{ij} \frac{\delta \widetilde{\mathcal{H}}}{\delta \tilde{\pi}^{ij}} d\mathring{v} = \frac{d}{d\epsilon}\Big|_{\epsilon=0} \mathcal{H}_{\mathrm{ADM}}(g, \tilde{\pi} + \epsilon \tilde{\sigma}).$$

From the first two lines of (5.3.12), the corresponding variation of the action is

$$\int_I \int_\Sigma \hat{\sigma}^{ij} \left(\frac{\partial g_{ij}}{\partial t} - \frac{\delta \widehat{\mathcal{H}}}{\delta \hat{\pi}^{ij}} \right) dv_g\, dt = \int_I \int_\Sigma \tilde{\sigma}^{ij} \left(\frac{\partial g_{ij}}{\partial t} - \frac{\delta \widetilde{\mathcal{H}}}{\delta \tilde{\pi}^{ij}} \right) d\mathring{v}\, dt. \quad (5.3.13)$$

For this to vanish for all $\hat{\sigma}$ (compactly supported away from the boundary), we conclude one of Hamilton's equations

$$\frac{\partial g_{ij}}{\partial t} = \frac{\delta \widehat{\mathcal{H}}}{\delta \hat{\pi}^{ij}} = \frac{\delta \widetilde{\mathcal{H}}}{\delta \tilde{\pi}^{ij}}. \tag{5.3.14}$$

If we now consider a (compactly supported) variation $g_{ij} + \epsilon h_{ij}$, with $\tilde{\pi}^{ij}$ *fixed* (so $\hat{\pi}$ may change with g), we define

$$h_{ij} \frac{\delta \widetilde{\mathcal{H}}}{\delta g_{ij}} = \frac{d}{d\epsilon}\bigg|_{\epsilon=0} \widetilde{\mathcal{H}}(g + \epsilon h, \tilde{\pi}),$$

so that

$$\frac{d}{d\epsilon}\bigg|_{\epsilon=0} \mathcal{H}_{\mathrm{ADM}}(g + \epsilon h, \tilde{\pi}) =: \int_{\Sigma} h_{ij} \frac{\delta \widetilde{\mathcal{H}}}{\delta g_{ij}} d\mathring{v}.$$

The variation of the action $\int_I \int_{\Sigma} \left(\tilde{\pi}^{ij} \frac{\partial g_{ij}}{\partial t} - \widetilde{\mathcal{H}} \right) d\mathring{v}\, dt$ is then

$$\int_I \int_{\Sigma} \left(\tilde{\pi}^{ij} \frac{\partial h_{ij}}{\partial t} - h_{ij} \frac{\delta \widetilde{\mathcal{H}}}{\delta g_{ij}} \right) d\mathring{v}\, dt.$$

We integrate by parts in t; since the expression above must vanish for all compactly supported h, we get another of Hamilton's equations

$$\frac{\partial \tilde{\pi}^{ij}}{\partial t} = -\frac{\delta \widetilde{\mathcal{H}}}{\delta g_{ij}}. \tag{5.3.15}$$

We now determine what (5.3.14) means. We compute from (5.3.9)

$$\hat{\sigma}^{ij} \frac{\delta \widehat{\mathcal{H}}}{\delta \hat{\pi}^{ij}} = 2N\hat{\sigma}^{ij}\left(\hat{\pi}_{ij} - \frac{1}{k-1}(\mathrm{tr}_g \hat{\pi})g_{ij} \right) - 2\hat{\sigma}^{ij}_{\ ;j} X_i = 2N\hat{\sigma}^{ij} \hat{K}_{ij} - 2\hat{\sigma}^{ij}_{\ ;j} X_i.$$

Thus with an integration by parts we find

$$\int_{\Sigma} \hat{\sigma}^{ij} \frac{\partial \widehat{\mathcal{H}}}{\partial \hat{\pi}^{ij}}\, dv_g - \int_{\Sigma} \hat{\sigma}^{ij} (2N\hat{K}_{ij} + X_{i;j} + X_{j;i})\, dv_g.$$

By (5.3.13), we thus recover the first of the ADM equations, $\partial g_{ij}/\partial t = -2NK_{ij} + (L_X g)_{ij}$. A similar (but more laborious) analysis produces the second ADM equation, which we leave as an exercise; cf. [51] or [218, Appendix E].

We can write the ADM equations in a symplectic form using the constraint map $\hat{\Phi}$, as follows. If we let

$$D\hat{\Phi}_{(g,\hat{\pi})}(h, \hat{\sigma}) = \frac{d}{d\epsilon}\bigg|_{\epsilon=0} \hat{\Phi}(g + \epsilon h, \hat{\pi} + \epsilon\hat{\sigma}),$$

we define a formal $L^2(dv_g)$-adjoint operator by the prescription

$$\int_\Sigma (N, X) \cdot D\widehat{\Phi}_{(g,\hat\pi)}(h, \hat\sigma)\, dv_g = \int_\Sigma (h, \hat\sigma) \cdot_g D\widehat{\Phi}^*_{(g,\hat\pi)}(N, X)\, dv_g$$

for all compactly supported (vanishing near the boundary) $(h, \hat\sigma)$. We similarly define $D\widetilde{\Phi}_{(g,\tilde\pi)}(h, \tilde\sigma) = \frac{d}{d\epsilon}\big|_{\epsilon=0}\widetilde{\Phi}(g + \epsilon h, \tilde\pi + \epsilon\tilde\sigma)$ and

$$\int_\Sigma (N, X) \cdot D\widetilde{\Phi}_{(g,\tilde\pi)}(h, \tilde\sigma)\, d\mathring{v} = \int_\Sigma (h, \tilde\sigma) \cdot_g D\widetilde{\Phi}^*_{(g,\tilde\pi)}(N, X)\, d\mathring{v}.$$

One needs to take care when computing this adjoint, since the volume element for the integral is induced from $\mathring{g}$, but the derivative operators would naturally be taken with respect to g (factors of θ_g are used to compensate; cf. Exercise 5-28).

Using (5.3.11) and (5.3.12), we obtain the first variation of the action fixing g and varying $\hat\pi^{ij} + \epsilon\hat\sigma^{ij}$ (equivalently $\tilde\pi^{ij} + \epsilon\tilde\sigma^{ij}$) to be

$$\int_I \int_\Sigma \left(\hat\sigma^{ij}\frac{\partial g_{ij}}{\partial t} - (N, X) \cdot D\widehat{\Phi}_{(g,\hat\pi)}(0, \hat\sigma) \right) dv_g\, dt$$

$$= \int_I \int_\Sigma \left(\hat\sigma^{ij}\frac{\partial g_{ij}}{\partial t} - (0, \hat\sigma) \cdot_g D\widehat{\Phi}^*_{(g,\hat\pi)}(N, X) \right) dv_g\, dt;$$

equivalently,

$$\int_I \int_\Sigma \left(\tilde\sigma^{ij}\frac{\partial g_{ij}}{\partial t} - (N, X) \cdot D\widetilde{\Phi}_{(g,\tilde\pi)}(0, \tilde\sigma) \right) d\mathring{v}\, dt$$

$$= \int_I \int_\Sigma \left(\tilde\sigma^{ij}\frac{\partial g_{ij}}{\partial t} - (0, \tilde\sigma) \cdot_g D\widetilde{\Phi}^*_{(g,\tilde\pi)}(N, X) \right) d\mathring{v}\, dt. \qquad (5.3.16)$$

Similarly, if we keep $\tilde\pi^{ij}$ fixed and vary g by $g_{ij} + \epsilon h_{ij}$ we get the first variation of the action to be

$$\int_I \int_\Sigma \left(\tilde\pi^{ij}\frac{\partial h_{ij}}{\partial t} - (N, X) \cdot D\widetilde{\Phi}_{(g,\tilde\pi)}(h, 0) \right) d\mathring{v}\, dt$$

$$= \int_I \int_\Sigma \left(-h_{ij}\frac{\partial \tilde\pi^{ij}}{\partial t} - (h, 0) \cdot_g D\widetilde{\Phi}^*_{(g,\tilde\pi)}(N, X) \right) d\mathring{v}\, dt. \qquad (5.3.17)$$

By (5.3.17) and (5.3.16), for the action to be stationary for variations of g_{ij}, $\tilde\pi^{ij}$, as well as N and X (which yield the constraints), we have, with the components of the adjoint written vertically and with the indices raised or lowered by g to match the tensor type,

$$\begin{bmatrix} \partial g/\partial t \\ \partial \tilde\pi/\partial t \end{bmatrix} = \begin{bmatrix} 0 & 1 \\ -1 & 0 \end{bmatrix} D\widetilde{\Phi}^*_{(g,\tilde\pi)}(N, X). \qquad (5.3.18)$$

Compare [51, p. 154–155] (slightly different notation), or [17] (check signs). This illustrates how the constraints operator generates the evolution; furthermore, an element in the kernel of $D\widetilde{\Phi}^*_{(g,\tilde{\pi})}$ (cf. Exercise 5-28) corresponds to a Killing field of the vacuum spacetime, which is then *stationary* (if the Killing field is timelike), see [163]. Such an element (N, X) in the kernel is called a *KID*, for *Killing initial data*, and the presence (or lack thereof) of KIDs has implications for constructions involving the constraints, such as linearization stability [88; 163], as well as for gluing constructions of solutions to the constraints [61; 69].

Remark 5-23. As noted, to get the units right, we consider the Hamiltonian related to the Lagrangian $\int_S \frac{1}{2\kappa} R(\bar{g}) dv_{\bar{g}}$. We can readily modify the above to incorporate κ. Indeed, the canonical momentum would be $\hat{\pi}^{ij}_\kappa = \frac{1}{2\kappa}\hat{\pi}^{ij}$, with $\tilde{\pi}^{ij}_\kappa = \frac{1}{2\kappa}\tilde{\pi}^{ij}$. The Hamiltonian would change to $\mathcal{H}^\kappa_{\text{ADM}}(g, \hat{\pi}_\kappa) = \frac{1}{2\kappa}\mathcal{H}_{\text{ADM}}(g, \hat{\pi})$ with the weighted Hamiltonian $\widetilde{\mathcal{H}}_\kappa(g, \tilde{\pi}_\kappa) = \frac{1}{2\kappa}\widetilde{\mathcal{H}}(g, \tilde{\pi}) = \frac{1}{2\kappa}\theta_g\widehat{\mathcal{H}}(g, \hat{\pi})$, so that $\widehat{\mathcal{H}}_\kappa(g, \hat{\pi}_\kappa) = \frac{1}{2\kappa}\widehat{\mathcal{H}}(g, \hat{\pi})$, and finally $\widehat{\Phi}_\kappa(g, \hat{\pi}_\kappa) = \frac{1}{2\kappa}\widehat{\Phi}(g, \hat{\pi})$, with the weighted form $\widetilde{\Phi}_\kappa(g, \tilde{\pi}_\kappa) = \frac{1}{2\kappa}\widetilde{\Phi}(g, \tilde{\pi})$. Hamilton's equations (5.3.14)–(5.3.15) retain the same form, since one can readily show that $\delta\widehat{\mathcal{H}}/\delta\hat{\pi}^{ij} = \delta\widehat{\mathcal{H}}_\kappa/\delta\hat{\pi}^{ij}_\kappa = \delta\widetilde{\mathcal{H}}_\kappa/\delta\tilde{\pi}^{ij}_\kappa$, while $\partial\tilde{\pi}^{ij}_\kappa/\partial t = -\delta\widetilde{\mathcal{H}}_\kappa/\delta g_{ij}$ is obtained simply by multiplying (5.3.15) by $\frac{1}{2\kappa}$. Finally, (5.3.18) would retain the same form, as it can be readily seen to be equivalent to the same equation with $\widetilde{\Phi}_\kappa$ and $\tilde{\pi}_\kappa$ in place of $\widetilde{\Phi}$ and $\tilde{\pi}$. See also Exercise 5-28.

Exercises

Exercise 5-24. Let $g_{\mathbb{E}^3}$ be the Euclidean metric. Let $I \subset \mathbb{R}$ be an interval, and let $a: I \to (0, +\infty)$ be a smooth function. Let $S = I \times \mathbb{R}^3$, with $\bar{g} = -dt^2 + (a(t))^2 g_{\mathbb{E}^3}$. On the slices $\{t\} \times \mathbb{R}^3$, $g(t) = (a(t))^2 g_{\mathbb{E}^3}$, and $n = \partial/\partial t$ is a global unit normal field along the slices. Let a zero index denote the $\partial/\partial t$-component direction.

a. Use the ADM equations to compute the second fundamental form K, as well as the components of $\text{Ric}(\bar{g})$ in directions tangent to the slices.

b. Use the relation between g and K and $G_{\mu 0}$ to compute $G_{\mu 0}$; then compute $G_{\mu\nu}$.

Exercise 5-25 (shape operator). Suppose $\Sigma \subset (M, \bar{g} = \langle \cdot, \cdot \rangle)$ is an embedded oriented hypersurface, with smooth unit normal vector field ν, and with induced metric (first fundamental form) g with Levi-Civita connection ∇^Σ. For $W \in T_p\Sigma$, let $S_p(W) = -\nabla_W \nu$.

a. Show that for all $p \in \Sigma$, $S = S_p$ gives a linear operator $S : T_p\Sigma \to T_p\Sigma$, *the shape operator*. Show that S is self-adjoint: $\langle S(V), W \rangle = \langle V, S(W) \rangle$ for $V, W \in T_p\Sigma$.

The corresponding bilinear form, given by $K(V, W) = \langle S(V), W \rangle$, is the *second fundamental form* of Σ, and is symmetric.

b. The covariant derivative of S as a $(1, 1)$-tensor field on Σ produces a $(1, 2)$-tensor field $\nabla^\Sigma S$. Show that for V and W in $T_p\Sigma$,

$$(\nabla^\Sigma_V S)(W) = \nabla^\Sigma_V (S(W)) - S(\nabla^\Sigma_V W).$$

c. Consider the case where $\Sigma \subset (M, \bar{g}) = (\mathbb{R}^n, g_{\mathbb{E}^n})$ is an oriented hypersurface in Euclidean space. Prove $(\nabla^\Sigma_V S)(W) = (\nabla^\Sigma_W S)(V)$ for V and W in $T_p\Sigma$, and conclude $\mathrm{div}_g S = d(\mathrm{tr}_g S)$ on Σ.

d. Σ is *totally umbilic* at $p \in \Sigma$ if there is a normal vector Z at p for which $\mathrm{III}(V, W) = \langle V, W \rangle Z$ for all $V, W \in T_p\Sigma$. Σ is *totally umbilic* if it is totally umbilic at each point. In this case, show that Z is a uniquely defined smooth vector field along Σ. Show that Σ is totally umbilic if and only if there is a (smooth) function f on Σ such that $S_p(V) = f(p)V$ for all $p \in \Sigma$ and all $V \in T_p\Sigma$.

e. If Σ is a connected, totally umbilic hypersurface in Euclidean space $(\mathbb{R}^n, g_{\mathbb{E}^n})$, $n \geq 3$, show that the function f from part d. must be constant. If Σ is also compact (and without boundary), one can show furthermore that Σ is a round sphere; see [174], for instance.

Exercise 5-26 (some classical surface geometry). Let $\mathbb{X} : U \to \mathbb{R}^3$ be an embedding of an open subset U of $\mathbb{R}^2$ onto a surface $\Sigma = \mathbb{X}(U)$ in Euclidean space; we let D be the Levi-Civita connection for $g_{\mathbb{E}^3} = \langle \cdot, \cdot \rangle$. The components of the induced metric (first fundamental form) on Σ are written $g_{11} = E = \langle \mathbb{X}_u, \mathbb{X}_u \rangle > 0$, $g_{12} = F = \langle \mathbb{X}_u, \mathbb{X}_v \rangle = g_{21}$, and $g_{22} = G = \langle \mathbb{X}_v, \mathbb{X}_v \rangle > 0$. Let

$$v = \frac{\mathbb{X}_u \times \mathbb{X}_v}{\|\mathbb{X}_u \times \mathbb{X}_v\|}$$

be a smooth unit normal field. For each $p \in \Sigma$, let $S = S_p$ be the *shape operator*, defined for $W \in T_p\Sigma$ by $S_p(W) = -D_W v$. By Exercise 5-25, S_p defines a self-adjoint linear operator on $T_p\Sigma$, so that the associated bilinear form given by $K(V, W) = \langle S(V), W \rangle$, for $V, W \in T_p\Sigma$, is symmetric. S is diagonalizable, with the *principal curvatures* defined to be the eigenvalues κ_1 and κ_2. The *Gauss curvature* is defined as $\kappa_1 \kappa_2$.

a. Suppose $S = S_p$ is represented by the matrix $\left(\begin{smallmatrix} a & c \\ b & d \end{smallmatrix}\right)$ in the basis $\{\mathbb{X}_u, \mathbb{X}_v\}$, and that K is represented in the same basis by $\left(\begin{smallmatrix} \ell & m \\ m & n \end{smallmatrix}\right)$. Relate these two matrices to the first fundamental form matrix $\left(\begin{smallmatrix} E & F \\ F & G \end{smallmatrix}\right)$. These are the *Weingarten equations*.

b. What do you get if you decompose the vector equations $\mathbb{X}_{uuv} - \mathbb{X}_{uvu} = \mathbf{0}$ and $\mathbb{X}_{vuv} - \mathbb{X}_{vvu} = \mathbf{0}$ into tangential and normal components? (Answer: The Gauss–Codazzi equations. From the tangential components comes the Gauss equation, which shows the Gauss curvature $\kappa_1\kappa_2$ (derived from the second fundamental form) is the sectional curvature of Σ, which can be computed from the first fundamental form. This is known as Gauss's *theorema egregium*, or "remarkable theorem".)

Exercise 5-27. a. Prove that Euclidean space $(\mathbb{R}^3, g_{\mathbb{E}^3})$ does not admit a closed immersed minimal surface (one for which $\boldsymbol{H} = \mathbf{0}$). To do this, show that there must be a point on the surface where the Gaussian curvature is strictly positive. Generalize this result to closed immersed hypersurfaces in $(\mathbb{R}^n, g_{\mathbb{E}^n})$ for $n \geq 3$.

b. Conclude that any embedding of a two-torus $\mathbb{T}^2$ into Euclidean $\mathbb{R}^3$ is not flat. Show that there is an embedding of a two-torus into Euclidean $\mathbb{R}^4$ for which the induced metric is flat. Can you likewise embed a flat torus isometrically in the product $\mathbb{S}^2 \times \mathbb{S}^2$ of round spheres?

c. Consider the surface of revolution in Euclidean space given by $\sqrt{x^2 + y^2} - \cosh z$. The surface is a *catenoid*. Show the catenoid is minimal, and compute its Gaussian curvature.

Exercise 5-28. Let (Σ, g) be Riemannian, and let $\hat{\pi}$ be a symmetric $(2, 0)$-tensor on Σ. Recalling notation from Section 5.3.3.2, we let $\tilde{\pi} = \theta_g \hat{\pi}$ and $\widetilde{\Phi}(g, \tilde{\pi}) = \theta_g \widehat{\Phi}(g, \theta_g^{-1}\tilde{\pi})$, from which we have

$$\mathcal{H}_{\mathrm{ADM}} = \int_\Sigma (N, X) \cdot \widetilde{\Phi}(g, \tilde{\pi})d\mathring{v} = \int_\Sigma (N, X) \cdot \widehat{\Phi}(g, \theta_g^{-1}\tilde{\pi})dv_g.$$

a. Prove that, with $\tilde{\sigma} = \theta_g \hat{\sigma}$,

$$D\widetilde{\Phi}_{(g,\tilde{\pi})}(h, \tilde{\sigma})$$
$$= \theta_g D\widehat{\Phi}_{(g,\hat{\pi})}(h, \hat{\sigma}) + \tfrac{1}{2}\theta_g\big((\mathrm{tr}_g h)\,\widehat{\Phi}(g, \hat{\pi}) - D\widehat{\Phi}_{(g,\hat{\pi})}(0, (\mathrm{tr}_g h)\hat{\pi})\big). \quad (5\text{-}28\mathrm{a})$$

From here you can conclude that if $(g, \hat{\pi})$ solves the vacuum constraint equations, then (N, X) is in the kernel of $D\widehat{\Phi}^*_{(g,\hat{\pi})}$ if and only if it is in the kernel of $D\widetilde{\Phi}^*_{(g,\tilde{\pi})}$.

b. Show that the kernel of $D\widehat{\Phi}^*_{(g,\hat{\pi})}$ agrees with the kernel of $(D\widehat{\Phi}_K)^*_{(g,\hat{\pi}_K)}$, and if (g, π) solves the vacuum constraint equations, find a simple bijection between the kernel of $D\widehat{\Phi}^*_{(g,\hat{\pi})}$ and the kernel of the map $D\Phi^*_{(g,\pi)}$ defined in (5-29a).

Exercise 5-29. Consider the vacuum constraints operator on (M^n, g, π),

$$\Phi(g, \pi) = \big(R(g) - |\pi|_g^2 + \tfrac{1}{n-1}(\mathrm{tr}_g \pi)^2, \mathrm{div}_g\pi\big),$$

where g is a Riemannian metric, and π^{ij} is a $(2, 0)$-tensor, whose divergence defines a *vector field*. We define the operator $D\Phi^*_{(g,\pi)}$, from the relation

$$\int_M (N, X) \cdot_g D\Phi_{(g,\pi)}(h, \sigma) \, dv_g = \int_M D\Phi^*_{(g,\pi)}(N, X) \cdot_g (h, \sigma) \, dv_g, \quad (5\text{-}29\text{a})$$

assuming the integrand has compact support (in the interior of M, should ∂M be nonempty).

a. Suppose $\Phi(g, 0) = 0$. Find $D\Phi_{(g,0)}(h, \sigma)$, where h is a symmetric $(0, 2)$-tensor and σ is a symmetric $(2, 0)$-tensor, and then find the operator $D\Phi^*_{(g,0)}$. Since $\pi = 0$, the linearization simplifies dramatically from the general case.

b. Find the kernel of $D\Phi^*_{(g_{\mathbb{E}^n},0)}$ at the Minkowski data on $M = \mathbb{R}^n$. (Hint: for $n = 3$, the kernel is ten-dimensional.)

c. Let $m \neq 0$. Find the kernel of $D\Phi^*_{(g_S,0)}$ at the data for a constant t-slice in Schwarzschild $g_S = \left(1 + \frac{m}{2|x|}\right)^4 g_{\mathbb{E}^3}$ on $M = \mathbb{R}^3 \setminus \left\{x : |x| \leq \max\left(0, -\frac{m}{2}\right)\right\}$. (Hint: the kernel is four-dimensional; cf. Exercise 2-53.)

d. Find $D\Phi_{(g,\pi)}(h, \sigma)$ and $D\Phi^*_{(g,\pi)}(N, X)$ in general, proving the following formulas:

$$D\Phi_{(g,\pi)}(h, \sigma)$$
$$= \Big(L_g h - 2h_{ij}\pi^i_{\ell}\pi^{j\ell} - 2\pi^j_{k}\sigma^k_{j} + \tfrac{2}{n-1}\mathrm{tr}_g\pi\,(h_{ij}\pi^{ij} + \mathrm{tr}_g\sigma),$$
$$(\mathrm{div}_g\,\sigma)^i - \tfrac{1}{2}\pi^{jk}h_{jk;\ell}g^{\ell i} + \pi^{jk}h^i_{j;k} + \tfrac{1}{2}\pi^{ij}(\mathrm{tr}_g h)_{,j} \Big), \quad (5\text{-}29\text{b})$$

$$D\Phi^*_{(g,\pi)}(N, X)$$
$$= \Big((L_g^* N)_{ij} + N\big(\tfrac{2}{n-1}(\mathrm{tr}_g\pi)\pi_{ij} - 2\pi_{ik}\pi^k_{j}\big) + \tfrac{1}{2}\big(g_{i\ell}g_{jm}(L_X\pi)^{\ell m} + (\mathrm{div}_g X)\pi_{ij}\big)$$
$$- \tfrac{1}{2}(X_i\pi^k_{j;k} + X_j\pi^k_{i;k} + X_{k;m}\pi^{km}g_{ij} + X_k\pi^{km}_{;m}g_{ij})$$
$$- \tfrac{1}{2}(X^i_{;\ell}g^{\ell j} + X^j_{;\ell}g^{\ell i}) + N\big(\tfrac{2}{n-1}(\mathrm{tr}_g\pi)g^{ij} - 2\pi^{ij}\big) \Big), \quad (5\text{-}29\text{c})$$

where L_g is the linearization of the scalar curvature operator (cf. (2.3.11)), and L_X is the Lie derivative (cf. Exercise 5-17).

Exercise 5-30 (boundary terms in the Hamiltonian). We briefly explore the boundary terms in the Hamiltonian formulation for the (vacuum) Einstein equation, cf. [113; 158; 218; 229]. The boundary integral $\left(\int_{\Sigma_{t_2}} - \int_{\Sigma_{t_1}}\right) 2\,\mathrm{tr}_g K\, dv_g$ is often added to the Einstein–Hilbert action, and from Remark 5-22, we see it cancels out the first boundary term we discarded. We now assume $\partial\Sigma$ is nonempty, so that ∂S has an additional piece given by the timelike hypersurface $I \times \partial\Sigma$.

a. In general the outward unit conormal field v to the slices Σ_t does not agree with the outward unit normal of $\partial \mathcal{S}$. To analyze the boundary terms, you might rewrite the scalar curvature from (5-21a), and combine with (5.3.8), to obtain

$$R(\bar{g}) = 2N^{-1}\left(-\frac{\partial}{\partial t}(\operatorname{tr}_g K) + \nabla_X(\operatorname{tr}_g K)\right) - 2N^{-1}\Delta_g N + N^{-1}\hat{\pi}^{ij}\frac{\partial g_{ij}}{\partial t}$$
$$- 2N^{-1}\hat{\pi}^{ij}X_{i;j} + 2(\operatorname{tr}_g K)^2 + \left(R(g) - |K|_g^2 + (\operatorname{tr}_g K)^2\right). \quad \text{(5-30a)}$$

(i) We can write the integral over $\mathcal{S}$ as a product over $I \times \Sigma$. This allows us to apply integration by parts to the first term in the expression (5-30a) for the scalar curvature. Using the equality (2.3.10) in the form

$$\frac{\partial}{\partial t}\sqrt{\det g} = \frac{1}{2}\operatorname{tr}_g\left(\frac{\partial g}{\partial t}\right)\sqrt{\det g}$$

and (5.3.4), show that

$$\int_I \int_\Sigma -2N^{-1}\frac{\partial}{\partial t}(\operatorname{tr}_g K)\, dv_{\bar{g}}$$
$$= \int_I \int_\Sigma \left(-2N(\operatorname{tr}_g K)^2 + 2(\operatorname{tr}_g K)\operatorname{div}_g X\right) dv_g\, dt - \left(\int_{\Sigma_{t_2}} - \int_{\Sigma_{t_1}}\right) 2\operatorname{tr}_g K\, dv_g.$$

(ii) Show that the modified action $\int_\mathcal{S} R(\bar{g})\, dv_{\bar{g}} + \left(\int_{\Sigma_{t_2}} - \int_{\Sigma_{t_1}}\right) 2\operatorname{tr}_g K\, dv_g$ can be written

$$\int_I \int_\Sigma \left(\hat{\pi}^{ij}\frac{\partial g_{ij}}{\partial t} + 2(\operatorname{div}_g \hat{\pi})_i\, X^i + N\left(R(g) - |K|_g^2 + (\operatorname{tr}_g K)^2\right)\right) dv_g\, dt$$
$$- 2\int_I \int_{\partial\Sigma}\left(\frac{\partial N}{\partial v} + \hat{\pi}^{ij}X_i v_j\right) d\sigma_g\, dt + \int_I \int_{\partial\Sigma} 2(\operatorname{tr}_g K)X_j v^j\, d\sigma_g\, dt.$$

Note that we could use (5.3.7) to replace $\partial N/\partial v$ by $N\langle \nabla_n n, v\rangle$ on the last line.

To interpret the boundary integrals from the divergence terms in (5-21b), it would be convenient for n to be tangent to this timelike boundary, i.e., for the outward unit conormal v to the slices Σ_t to be the outward unit normal to $\partial \mathcal{S}$; however, as $\partial/\partial t$ is tangent to $\partial \mathcal{S}$, this would restrict X to satisfy $\langle X, v\rangle = 0$ along $\partial \mathcal{S}$. So as in [113; 158], we restrict if necessary to a spacetime $(\mathcal{S}', \bar{g})$, with $\mathcal{S}' = \bigcup_{t \in I} \Sigma_t'$, with smooth hypersurfaces $\Sigma_t' \subset \Sigma_t$, where the boundary of each Σ_t' meets the timelike boundary $\partial \mathcal{S}'$ orthogonally, so that the outward pointing unit conormal v of each Σ_t' is normal to this timelike boundary. Then if we let γ be the induced metric on $\partial \mathcal{S}'$, the orthogonality gives us that, if we have a local form field ω_g which restricts to a local volume form along each $\partial \Sigma_t'$ (in a suitable neighborhood), then a local volume form for $(\partial \mathcal{S}', \gamma)$ is $n^\flat \wedge \omega_g = N\, dt \wedge \omega_g$.

If we extend v as $v = -n$ on Σ_{t_2}' and $v = n$ on Σ_{t_1}', the term that is added to the action is then $\int_{\partial \mathcal{S}'} 2\operatorname{tr}_\gamma A\, dv_\gamma$, where if $\mathrm{I\!I}$ is the second fundamental form of the boundary $\partial \mathcal{S}'$, $\mathrm{II}(X, Y) = (\nabla_X Y)^N$, then $A(X, Y) = \langle \mathrm{II}(X, Y), -\nu \rangle$. Thus $A = K$ on Σ'_{t_2}, but $A = -K$ on Σ'_{t_1}, so that $\left(\int_{\Sigma'_{t_2}} - \int_{\Sigma'_{t_1}} \right) 2 \operatorname{tr}_g K \, dv_g$ is indeed part of the boundary term $\int_{\partial \mathcal{S}'} 2 \operatorname{tr}_\gamma A \, dv_\gamma$. The modified action $\int_{\mathcal{S}'} R(\bar{g}) \, dv_{\bar{g}} + \int_{\partial \mathcal{S}'} 2 \operatorname{tr}_\gamma A \, dv_\gamma$ gives rise to a modified Hamiltonian with boundary terms, as follows.

b. In the setting above, show the modified action $\int_{\mathcal{S}'} R(\bar{g}) \, dv_{\bar{g}} + \int_{\partial \mathcal{S}'} 2 \operatorname{tr}_\gamma A \, dv_\gamma$ can be written

$$\int_I \left(\int_{\Sigma'_t} \hat{\pi}^{ij} \frac{\partial g_{ij}}{\partial t} \, dv_g - \mathcal{H}_{\mathrm{ADM}} - 2 \int_{\partial \Sigma'_t} (NH + \hat{\pi}^{ij} X_i \nu_j) \, d\sigma_g \right) dt,$$

where H is the mean curvature of the boundary $\partial \Sigma'_t$ inside Σ'_t, with respect to the normal ν (i.e., if Π is the vector-valued second fundamental form of $\partial \Sigma'_t$ inside Σ'_t, then H is the trace along $\partial \Sigma'_t$ of $\kappa(X, Y) := \langle \Pi(X, Y), \nu \rangle$). Thus the modified Hamiltonian is $\mathcal{H}_{\mathrm{ADM}} + 2 \int_{\partial \Sigma'_t} (NH + \hat{\pi}^{ij} X_i \nu_j) \, d\sigma_g$.

Remark. If you try to compare boundary terms from the analyses in parts a. and b., you have to take care; in part a. we obtained one such term from an integration by parts in the t-direction; if you tried to carry over the derivation from part (ii) for $\mathcal{S}'$, there may be an additional term in the t-derivative of the slice integrals, since the slices in $\mathcal{S}'$ are generally not just $\Sigma_t = \{t\} \times \Sigma$, which had allowed us to easily write the slice integrals in $\mathcal{S}$ over a fixed domain Σ.

CHAPTER 6

Scalar curvature deformation and the Einstein constraint equations

We have derived the Einstein constraint equations and pinpointed their place in terms of the initial value formulation for Einstein's equation. From one point of view, to construct interesting spacetimes, one could construct interesting solutions of the Einstein constraint equations, and then study their evolution. If one could then effectively parametrize the space of solutions to the Einstein constraint equations, one could identify the parameters as the true gravitational degrees of freedom; cf. [226; 227]. A classical approach to this is the *conformal method* to construct solutions of the constraint equations, and we introduce this in Section 6.2.

Once we set up the conformal method, we will focus on the constant mean curvature (CMC) case. In this case, the momentum constraint decouples, and the problem will be reduced to the analysis of a single nonlinear equation, the *Lichnerowicz equation* (6.2.8), to solve the Hamiltonian constraint, as we will see. The Hamiltonian constraint brings the analysis of the scalar curvature operator to the fore, and we will study the range of the scalar curvature operator, focusing on scalar curvature deformation, sometimes within a conformal class of metrics. As we have seen, in the time-symmetric case, the scalar curvature is proportional to energy density, and the dominant energy condition then focuses our attention to positive (nonnegative) scalar curvature metrics. There are in fact topological obstructions to admitting positive scalar curvature metrics, and as developed by R. M. Schoen and S.-T. Yau, this gives an approach to the positive mass theorem, as we will see in Chapter 7.

The remaining chapters will make heavier use of analysis and PDE (partial differential equations), such as properties of the Laplace operator on Euclidean space or more general Riemannian manifolds. We will discuss some aspects of elliptic PDE theory in various amounts of detail. Rather than relegating the PDE aspects to an appendix or simply listing bullet points of facts, we prefer to cover various aspects of the theory that we will use, giving many references to possible sources for further reading, and sketching a number of missing details as exercises. We suggest the reader at least peruse the PDE sections to set some notations and expectations. Hopefully someone who has had a course in elliptic PDE will get a better appreciation for some basic applications of the theory, while those without the background can get some feel for the place of elliptic PDE in the study of the Einstein constraint equations.

6.1. A primer on elliptic PDE

This section contains some of the required fundamentals of elliptic PDE theory. Given the fundamental role of the Laplace and Poisson equations in geometric analysis and mathematical physics, the Laplace operator will be our primary, though not exclusive, example of an elliptic operator. While we assume the reader is familiar with basic tools of analysis, including basic functional analysis, found in standard texts such as those mentioned in the preface, we start by collecting some of the facts and notation about function spaces that we will use.

6.1.1. *Sobolev and Hölder spaces.* Many results on elliptic PDE are cast in Sobolev and Hölder spaces. We define them briefly, and encourage the reader to review their basic properties from references such as [2; 86; 107; 144].

We let Ω be an open subset of $\mathbb{R}^n$, sometimes called a *domain in* $\mathbb{R}^n$. For a real number $p \geq 1$ and a nonnegative integer k, the space $W^{k,p}(\Omega)$ is the set of all Lebesgue measurable functions (up to an equivalence relation for functions which agree except for a set of measure zero) which are in $L^p(\Omega)$ along with weak derivatives up to order k. The space has a Banach space structure with norm given by

$$\|u\|_{W^{k,p}(\Omega)}^p = \sum_{|\beta| \leq k} \|\partial^\beta u\|_{L^p(\Omega)}^p,$$

where $\beta = (\beta_1, \ldots, \beta_n)$ is a *multi-index* (n-tuple of nonnegative integers), and we have set $|\beta| = \sum_{i=1}^n \beta_i$ and

$$\partial^\beta = \partial_x^\beta = \frac{\partial^\beta}{\partial x^\beta} = \left(\frac{\partial}{\partial x^1}\right)^{\beta_1} \cdots \left(\frac{\partial}{\partial x^n}\right)^{\beta_n}.$$

When $p = 2$, the norm is induced by an evident Hilbert space structure, and we let $H^k = W^{k,2}$. We let $W_{\text{loc}}^{k,p}(\Omega)$ (or just $W_{\text{loc}}^{k,p}$) be the space of functions in $W^{k,p}(\Omega')$ for all Ω' compactly contained in Ω. A straightforward cutoff and mollification argument proves that $C_c^\infty(\mathbb{R}^n)$ is dense in $W^{k,p}(\mathbb{R}^n)$, while the Meyers–Serrin theorem shows that $C^\infty(\Omega) \cap W^{k,p}(\Omega)$ is dense in $W^{k,p}(\Omega)$ for any open $\Omega \subset \mathbb{R}^n$. See, for example, [2, Theorem 3.16] or [107, Theorem 7.9].

There are ways to extend the definition to $s \in \mathbb{R}$, and we briefly discuss the case $p = 2$. The space $H^s(\mathbb{R}^n)$ is the completion of the set of Schwartz functions under the norm $\|u\|^2_{H^s(\mathbb{R}^n)} = \int_{\mathbb{R}^n} (1 + |\xi|^2)^s |\hat{u}(\xi)|^2 \, d\xi$, where $\hat{u}$ is the Fourier transform of u. By Plancherel's theorem, the norm on $H^0(\mathbb{R}^n)$ agrees with that of $L^2(\mathbb{R}^n)$ (up to a constant multiple), and in fact, for $k \in \mathbb{Z}_+$, $H^k(\mathbb{R}^n)$ agrees with the preceding definition (with equivalent norm), while for $s < 0$, $H^s(\mathbb{R}^n)$ is a subset of the set of *tempered* distributions (see [195], for example).

Let k be a nonnegative integer, and let $\alpha \in (0, 1]$. We recall that $C^k(\Omega)$ is the set of all functions u on Ω such that u and all its partials up through order k are continuous. For such u, let

$$\|u\|_{C^k(\Omega)} = \sum_{|\beta| \leq k} \sup_{x \in \Omega} |\partial^\beta u(x)|$$

(which could be infinite). There are different conventions for $C^k(\overline{\Omega})$. A natural one, adopted in [107], is that $C^k(\overline{\Omega})$ consists of all functions $u \in C^k(\Omega)$ such that u and all its partials through order k possess continuous extensions to $\overline{\Omega}$; for such functions, it is easy to see $\|u\|_{C^k(\Omega)} = \sum_{|\beta| \leq k} \sup_{x \in \overline{\Omega}} |\partial^\beta u(x)| =: \|u\|_{C^k(\overline{\Omega})}$, where we have not introduced new notation to denote the extension of $\partial^\beta u$. If $\overline{\Omega}$ is not compact, the functions $\partial^\beta u$, $|\beta| \leq k$, need not be bounded, however. To get a natural Banach space structure using $\|u\|_{C^k(\Omega)}$, one can consider the space $C^k_B(\Omega) \subset C^k(\Omega)$ of functions u for which $\partial^\beta u$ is bounded on Ω for $|\beta| \leq k$, and the corresponding space $C^k_B(\overline{\Omega})$ of functions $u \in C^k_B(\Omega)$ for which $\partial^\beta u$ also extends continuously to $\overline{\Omega}$, for $|\beta| \leq k$. (In a convex domain, for example, this extension condition is automatic for $|\beta| < k$ by applying the mean value theorem to infer uniform continuity.) Both $C^k_B(\Omega)$ and $C^k_B(\overline{\Omega})$ are Banach spaces under the norm $\|\cdot\|_{C^k(\Omega)}$, and $C^k_B(\overline{\Omega})$ agrees with $C^k(\overline{\Omega})$ when Ω is bounded. (We note that in some sources like [2], $C^k(\overline{\Omega})$ is defined to be the subspace of functions u in $C^k(\Omega)$ with $\partial^\beta u$ bounded and uniformly continuous on Ω, for $|\beta| \leq k$; such functions then possess continuous extensions to $\overline{\Omega}$. Of course, when Ω is bounded, these two notions of $C^k(\overline{\Omega})$ agree.)

For $X \subset \Omega$, we set

$$[u]_{k,\alpha;X} = \max_{|\beta|=k} \; \sup_{\substack{x,y \in X \\ x \neq y}} \frac{|\partial^\beta u(x) - \partial^\beta u(y)|}{|x-y|^\alpha},$$

and define $\|u\|_{C^{k,\alpha}(\Omega)} = \|u\|_{C^k(\Omega)} + [u]_{k,\alpha;\Omega}$, which may be infinite. We let $C^{k,\alpha}(\Omega)$ (or $C^{k,\alpha}_{\text{loc}}(\Omega)$ for emphasis) be the set of all functions u in $C^k(\Omega)$ such that for all compact $X \subset \Omega$, $[u]_{k,\alpha;X}$ is finite; this is equivalent to requiring $[u]_{k,\alpha;\overline{B}}$ be finite for all closed balls $\overline{B} \subset \Omega$. Note that if $[u]_{0,\alpha;\Omega}$ is finite, then u is uniformly continuous on Ω, and hence extends continuously to $\overline{\Omega}$, and if u denotes the extension, $[u]_{0,\alpha;\Omega} = [u]_{0,\alpha;\overline{\Omega}}$. In [107], $C^{k,\alpha}(\overline{\Omega})$ is then defined to be all $u \in C^k(\overline{\Omega})$ such that $[u]_{k,\alpha;\Omega}$ is finite. To have a Banach space with norm $\|u\|_{C^{k,\alpha}(\Omega)}$, one would naturally restrict to the subspace $C_B^{k,\alpha}(\overline{\Omega})$ of functions $u \in C_B^k(\overline{\Omega})$ with $[u]_{k,\alpha;\Omega}$ finite; when Ω is bounded $C_B^{k,\alpha}(\overline{\Omega}) = C^{k,\alpha}(\overline{\Omega})$, and we will generally work in bounded domains, or in unbounded domains with weighted norms. We note that in [2], the definition of $C^{k,\alpha}(\overline{\Omega})$ also includes the condition that $[u]_{\ell,\alpha;\Omega}$ is finite for all $0 \le \ell \le k$, and the norm adds the maximum of these seminorms to the C^k-norm. In the domains we will use, it is automatic that $u \in C_B^{k,\alpha}(\overline{\Omega})$ will satisfy this requirement, and in particular the seminorms $[u]_{\ell,\alpha;\Omega}$, $0 \le \ell \le k$, are bounded up a constant factor by

$$\|u\|_{C^{k,\alpha}(\overline{\Omega})} := \|u\|_{C^{k,\alpha}(\Omega)} = \|u\|_{C^k(\Omega)} + [u]_{k,\alpha;\Omega}.$$

For example, if $\overline{\Omega}$ is convex, this follows from the mean value theorem; for $\overline{\Omega}$ a smooth compact manifold-with-boundary, this again follows by the mean value theorem, using a finite covering by coordinate charts, each of which is a compactly contained restriction of a larger chart, and the image of which is a ball or half-ball.

Remark 6-1. Whereas smooth functions are dense in Sobolev spaces, this fails in Hölder spaces. For $0 < \alpha \le 1$, the function $f^\alpha(x) = |x|^\alpha$ is readily seen to be in $C^{0,\alpha}([-1, 1])$. For $0 < \alpha < 1$ and f Lipschitz, we have, for $0 < x \le 1$,

$$[f^\alpha - f]_{0,\alpha;[-1,1]} \ge \frac{|x^\alpha - (f(x) - f(0))|}{x^\alpha},$$

which approaches 1 as $x \searrow 0$. A similar two-sided argument shows the estimate $[f^1 - f]_{0,1,[-1,1]} \ge 1$ for $f \in C^1([-1, 1])$. Furthermore, while by Weierstrass approximation, $W^{k,p}((-1, 1))$ and $C^k([-1, 1])$ are separable, this fails for $C^{0,\alpha}([-1, 1])$ for $0 < \alpha \le 1$: the functions $f_p(x) = |x - p|^\alpha$ are in $C^{0,\alpha}([-1, 1])$, while for p and q distinct points in $[-1, 1]$, $[f_p - f_q]_{0,\alpha;[-1,1]} \ge 2$.

As we will see in this section, elliptic regularity theory is readily formulated in terms of Sobolev and Hölder spaces, rather than C^k spaces. (See also the discussion of the Newtonian potential in Section 7.1.1.2.)

The above definitions can be extended naturally to tensor fields (for example by taking the summation or maximum of the corresponding norms of Cartesian components of the tensor field). To extend to functions or tensors (or more generally to sections of vector bundles) on compact manifolds (smooth, and with or without boundary), one builds a norm using the Sobolev or Hölder norms coming from a covering by a finite number of appropriate coordinate charts (say, charts which are compactly contained restrictions of larger charts, so that overlap maps have uniformly bounded derivatives); any two such norms are easily seen to be equivalent. In fact one could define an equivalent norm by replacing the partial derivative with a smooth connection such as the Levi-Civita connection for a smooth Riemannian metric on M (or compatible connection for a Riemannian vector bundle [139]), using a Riemannian volume measure for integrals, and using parallel transport to construct the Hölder quotient (replacing distance with arclength). We remark that using the Levi-Civita connection to define the function spaces requires some regularity in the metric, which we assume (and sometimes but not always reiterate) is smooth (C^∞) unless noted otherwise. One is often led to consider metrics g that have finite regularity *a priori*, to consider an operator with coefficients depending on g (via curvature, or Christoffel symbols in local coordinates, say) or to define an operator acting on metrics g, so that it may in fact be more natural to define the function spaces using either a covering by charts or a smooth background metric $\mathring{g}$. In any case, one must pay attention to the regularity of g, which may naturally limit the regularity of operators under consideration.

We let $H^k(M)$ be the space of sections of class H^k (suppressing notation for bundle), with norm $\| \cdot \|_{H^k(M)}$ (suppressing notation for a metric if used for the norm), and similarly for other regularity classes of sections. The density results cited above can be readily used to show that the space of smooth sections is dense in the Sobolev space $W^{k,p}(M)$, for k a nonnegative integer and $1 \le p < \infty$.

One can also define suitable such function spaces in the noncompact setting, possibly with some control on the geometry, such as a positive injectivity radius, or asymptotic conditions. As we will see in Section 7.2.2.2, to define the relevant function spaces it is often natural to use an appropriately chosen family of coordinate charts, or to use a background metric adapted to the asymptotics.

6.1.2. *A few compactness results.* We will make use of a number of compact inclusion mappings of one space into another. We start by recalling some notions.

Definition 6-2. Let X and Y be normed vector spaces. A map $T : X \to Y$ is *compact* if the image of every bounded subset of X has compact closure in Y.

An easy exercise shows that (i) $T : X \to Y$ is compact if and only if the image of every bounded sequence in X has a convergent subsequence in Y, and (ii) if $T : X \to Y$ is a *linear* map, then T is compact if and only if the image of the unit ball in X has compact closure in Y, and such a compact linear map $T : X \to Y$ is necessarily continuous.

Definition 6-3. Let (X, d_X) and (Y, d_Y) be metric spaces. Let $\mathcal{F}$ be a set of functions from X to Y. For $x \in X$, $\mathcal{F}$ is *equicontinuous* at x if for all $\varepsilon > 0$, there is $\delta > 0$ such that for all $f \in \mathcal{F}$, and for all $x_1 \in X$ with $d_X(x_1, x) < \delta$, it follows that $d_Y(f(x_1), f(x)) < \varepsilon$. $\mathcal{F}$ is *pointwise equicontinuous* if it is equicontinuous at each $x \in X$. $\mathcal{F}$ is *(uniformly) equicontinuous* if for all $\varepsilon > 0$, there is $\delta > 0$ so that for all $f \in \mathcal{F}$, and for all $x_1, x_2 \in X$ with $d_X(x_1, x_2) < \delta$, it follows that $d_Y(f(x_1), f(x_2)) < \varepsilon$.

If (X, d_X) is a compact metric space, it is a simple exercise in compactness to show that (i) $\mathcal{F}$ is pointwise equicontinuous if and only if it is equicontinuous, and (ii) if $\mathcal{F}$ is equicontinuous, then $\{f(x) : x \in X, \ f \in \mathcal{F}\}$ is d_Y-bounded ($\mathcal{F}$ is *uniformly bounded*) if and only if for each $x \in X$, $\{f(x) : \ f \in \mathcal{F}\}$ is d_Y-bounded ($\mathcal{F}$ is *pointwise bounded* at each x).

A fundamental result which yields compact embeddings is the Ascoli–Arzelà theorem, a version of which is formulated as follows.

Theorem 6-4 (Ascoli–Arzelà). *A collection $\mathcal{F}$ of continuous real-valued functions on a compact metric space X has compact closure in the uniform norm $\|f\|_{C^0(X)} = \sup_{x \in X} |f(x)| = \max_{x \in X} |f(x)|$ if and only if $\mathcal{F}$ is bounded and equicontinuous.*

The Ascoli–Arzelà theorem yields a compactness result for Hölder spaces. For example, if (M, g) is a compact smooth Riemannian manifold (possibly with smooth boundary), then for $0 < \alpha < 1$, the inclusion of $C^1(M)$ into $C^{0,\alpha}(M)$ is a *compact* inclusion (i.e., any bounded sequence in $C^1(M)$ has a subsequence which converges in $C^{0,\alpha}(M)$), and similarly, $C^{0,\alpha}(M)$ is compactly included in $C^{0,\beta}(M)$ for $0 < \beta < \alpha \leq 1$, and in $C^0(M)$.

There are also compactness results in Sobolev spaces. We recall a version of the *Rellich–Kondrachov theorem* [86; 107]. A proof for $p = 2$ (*Rellich's lemma*) based on the Fourier transform and the Ascoli–Arzelà theorem can be found in [139], where the authors also discuss going from the setting of a domain in $\mathbb{R}^n$ to the manifold setting.

Lemma 6-5 (Rellich–Kondrachov). *Suppose (M, g) is a compact Riemannian manifold (possibly with smooth boundary), $1 \leq p < \infty$, and $k, j \geq 0$ are integers, with $k > j$. Then the inclusion $W^{k,p}(M) \hookrightarrow W^{j,p}(M)$ is compact.*

Other such embeddings are recalled in Section 7.2.2.1. Before we move on, we note that we can use compactness to interpolate between certain norms, a basic example of which is as follows.

Exercise 6-6. Suppose (M, g) is a compact Riemannian manifold (possibly with smooth boundary), $1 \leq p < \infty$ and $\ell > k > j \geq 0$ are integers. Show that for any $\varepsilon > 0$, there is $C > 0$ such that $\|u\|_{W^{k,p}(M)} \leq \varepsilon \|u\|_{W^{\ell,p}(M)} + C\|u\|_{W^{j,p}(M)}$ for all $u \in W^{\ell,p}(M)$. You can argue abstractly, by contradiction, using only that the inclusion $W^{\ell,p}(M) \hookrightarrow W^{k,p}(M)$ is compact and the inclusion $W^{k,p}(M) \hookrightarrow$

$W^{j,p}(M)$ is continuous. Apply the same method to get corresponding inequalities for Hölder and C^k-spaces.

6.1.3. *Elliptic operators.*

We will often apply elliptic theory to the Laplace operator Δ_g on a Riemannian manifold (M, g), given in coordinates by

$$\Delta_g = \frac{1}{\sqrt{\det g}} \frac{\partial}{\partial x^i} \left(g^{ij} \sqrt{\det g} \, \frac{\partial}{\partial x^j} \right) = g^{ij} \frac{\partial^2}{\partial x^i \partial x^j} - g^{ij} \Gamma_{ij}^k \frac{\partial}{\partial x^k}.$$

We first review a notion of ellipticity for second-order scalar operators on $\Omega \subset \mathbb{R}^n$, and then generalize from there.

An operator of the form

$$L = a^{ij}(x) \frac{\partial^2}{\partial x^i \partial x^j} + b^k(x) \frac{\partial}{\partial x^k} + c(x)$$

is *elliptic* at $x \in \Omega \subset \mathbb{R}^n$ if $a^{ij}(x)\xi^i\xi^j > 0$ for all $\xi \in \mathbb{R}^n \setminus \{0\}$. Observe that by symmetrization, we could arrange $a^{ij} = a^{ji}$ and define ellipticity to mean $(a^{ij}(x))$ is positive definite. In this case we let $\lambda(x) > 0$ be the greatest λ such that for all $\xi \in \mathbb{R}^n$, $a^{ij}(x)\xi_i\xi_j \geq \lambda|\xi|^2$; likewise we let $\mu(x) \geq \lambda(x) > 0$ be the least μ such that $a^{ij}(x)\xi_i\xi_j \leq \mu|\xi|^2$ for all $\xi \in \mathbb{R}^n$. We call L elliptic on $\Omega \subset \mathbb{R}^n$ if it is elliptic at each point in Ω, and call it *strictly* elliptic on Ω if there is a $\lambda > 0$ such that $a^{ij}(x)\xi_i\xi_j \geq \lambda|\xi|^2$ for all $x \in \Omega$ and all $\xi \in \mathbb{R}^n$. We could replace L by $(\lambda(x))^{-1}L$ to achieve strict ellipticity, but then we would likely want to impose appropriate bounds on the coefficients. For example, considering $\lambda(x)|\xi|^2 \leq a^{ij}(x)\xi_i\xi_j \leq \mu(x)|\xi|^2$, we say L is *uniformly* elliptic on Ω if μ/λ is bounded. If we have an elliptic operator with continuous coefficients, we get uniform ellipticity on any compactly contained subdomain. It is a simple exercise to check that ellipticity is preserved by a change of coordinates. With this in mind, note that we get an elliptic operator when we express the operator Δ_g for a metric g in a local chart; in fact in a suitable chart on a compactly contained domain, the operator written in local coordinates is uniformly elliptic. As you might imagine, we can generalize the notion of ellipticity to the manifold setting in a coordinate-invariant way as follows.

Suppose L is a linear differential operator of order $m \in \mathbb{Z}_+$, either operating on functions, or more generally between (sections of) two vector bundles over M, such as vector or tensor fields; for the moment we can consider smooth sections, but we will soon also want to allow sections in Sobolev or Hölder spaces. Motivated by the Fourier transform, we will define the *principal symbol* σ of L (at $p \in M$), which takes any $\xi \in T_p^*M$ to a linear transformation $\sigma(\xi)$ between the fibers at p of the relevant vector bundles; we illustrate with L taking vectors fields to vector fields, so that $\sigma(\xi)$ is a linear operator on T_pM, and the generalization will be immediate. To define $\sigma(\xi)$, first choose a smooth function φ with $\varphi(p) = 0$ and $d\varphi|_p = \xi$, and given $v \in T_pM$, choose a smooth vector field W with $W(p) = v$. Then

$$\sigma(\xi)(v) := \frac{i^m}{m!} L(\varphi^m W)(p)$$

is easily seen to depend only on ξ, v, and *the principal part* of L, i.e., the highest order (order m) derivative terms in L. (The factor i^m is not essential, but it becomes useful when considering the symbol of the adjoint operator; see p. 175.) To generalize to sections of other bundles, given v in the fiber at p of the relevant bundle, W would be chosen as a smooth section of the bundle for which $W(p) = v$.

If $\xi = \xi_i \, dx^i$, we define, in multi-index notation, $\xi^\alpha = \xi_1^{\alpha_1} \cdots \xi_n^{\alpha_n}$. For any multi-index α of order $|\alpha| = m$, the principal symbol $\sigma(\xi)$ of $\partial^\alpha/\partial x^\alpha$ is seen to be $i^m \xi^\alpha$, the same as the action of the Fourier transform; clearly, this correspondence between the symbol and the Fourier transform extends to all constant-coefficient linear differential operators. It is a simple exercise to check that if $L = \sum_{|\alpha| \leq m} A^\alpha(x) \partial^\alpha/\partial x^\alpha$, where each $A^\alpha(x)$ is a coefficient matrix of the same size for all α, then at x, $\sigma(\xi) = i^m \sum_{|\alpha|=m} A^\alpha(x)\xi^\alpha$, as a matrix linear transformation. For any such operator L between vector bundles on a manifold, in local coordinates the size of the matrices A^α depends on the fiber dimensions of the bundles, and the preceding formula yields the matrix representation $[\sigma(\xi)]$ for $\sigma(\xi)$ at the point p with coordinates x, in the corresponding local bases (for example, for tensor bundles, we would use bases built out of dx^i and $\partial/\partial x^j$). While the definition for the principal symbol is not coordinate-dependent, it is a good exercise to compute $[\sigma(\xi)]$ in two different coordinate systems and compare, taking care with the change of bases from dx^i and $\partial/\partial x^j$ to dy^k and $\partial/\partial y^\ell$, which affects the coefficient matrices and the components of ξ.

Definition 6-7. Let L be a linear differential operator as above, with principal symbol σ. L is an *elliptic operator* at p in case $\sigma(\xi)$ is an isomorphism for any $\xi \in T_p^*M \setminus \{0\}$. L is an elliptic operator if it is elliptic at each $p \in M$.

While this definition does generalize the one above for second-order linear scalar operators, one might consider the following generalization: suppose L is linear of order m and operates by mapping sections of a bundle E to sections of E, so that in a local representation, the symbol will be represented by a square matrix $[\sigma(\xi)]$. When working with operators with real coefficients, we could require the real matrix $i^{-m}[\sigma(\xi)]$ to be positive definite for all $\xi \neq 0$; this is called *strong ellipticity* in [22, Appendix I] and elsewhere. For such an operator, we see by changing ξ to $-\xi$ that m must be even.

There are related notions which include elliptic operators, but which can accommodate systems of differential equations involving bundles whose fibers may have different dimensions.

Definition 6-8. A linear differential operator L with principal symbol σ is *overdetermined-elliptic* if for each $\xi \in T_p^*M \setminus \{0\}$, $\sigma(\xi)$ is injective, while it is *underdetermined-elliptic* if for each $\xi \in T_p^*M \setminus \{0\}$, $\sigma(\xi)$ is surjective.

Suppose L is a linear differential operator between two bundles E and F each equipped with a smoothly varying fiber-wise inner product $\langle \cdot , \cdot \rangle$ (we use the same notation for both if it will not cause confusion). We will in particular focus on tensor bundles over a Riemannian manifold, and the inner product will be the natural inner product on tensors induced by the metric. The *formal adjoint* operator L^* between F and E is defined by integration by parts against compactly supported (away from the boundary if ∂M is nonempty) sections of the corresponding bundle: $\int_M \langle Lu, v \rangle \, dv_g = \int_M \langle u, L^*v \rangle \, dv_g$; note that $(L^*)^* = L$. We say that $Lu = f$ *weakly* if for all smooth sections v (compactly supported in the interior of M), $\int_M \langle f, v \rangle \, dv_g = \int_M \langle u, L^*v \rangle \, dv_g$; this can be suitably interpreted and defined for distributions u. It is not a hard exercise to show that the principal symbol of L^* is the conjugate transformation $\sigma(\xi)^*$: at each point $p \in M$ and $\xi \in T_p^*M$, we have $\langle \sigma(\xi)(v), w \rangle = \langle v, \sigma(\xi)^*(w) \rangle$. (Without the factor i^m in the definition, which gets conjugated on one side of the Hermitian product, this relation would hold only up to sign.)

Exercise 6-9. a. Prove the preceding claim, and conclude that L is underdetermined-elliptic if and only if L^* is overdetermined-elliptic.

b. Suppose Q is a linear differential operator between (sections of vector bundles) E and E_1, and likewise P is a linear differential operator between E_1 and F, so that $P \circ Q$ is a linear differential operator between E and F. Show that the principal symbols satisfy $\sigma_{P \circ Q}(\xi) = \sigma_P(\xi) \circ \sigma_Q(\xi)$.

c. Conclude that L^* is overdetermined-elliptic if and only if LL^* is elliptic.

We will now turn to two important examples; another will be met in the discussion of the TT decomposition (Section 6.2.1).

Example 6-10. Recall the linearization DR_g of the scalar curvature operator (2.3.11), and its formal adjoint $(DR_g)^*$, given by

$$(DR_g)^*(f) = -(\Delta_g f)g + \mathrm{Hess}_g f - f \, \mathrm{Ric}(g).$$

The principal symbol of $(DR_g)^*$ is then given by $\sigma(\xi)(1) = |\xi|_g^2 g - \xi \otimes \xi$, mapping a one-dimensional fiber to the space of symmetric $(0, 2)$-tensors on T_pM. If $\sigma(\xi)(1) = 0$, then upon taking a trace we get $(n-1)|\xi|_g^2 = 0$, which cannot hold for $\xi \neq 0$. In this case, the symbol of $(DR_g)^*$ is injective i.e., $(DR_g)^*$ is *overdetermined-elliptic*. Note that $DR_g(DR_g)^*$ is a fourth-order operator with principal part $(n-1)\Delta_g^2$, which is indeed elliptic, in agreement with the preceding exercise.

For nonlinear operators, one can define ellipticity (at any $p \in M$) *at a given section of the bundle* on which the operator acts, using the linearization of the operator at the given section. For example, the graphical mean curvature operator, cf. Exercise 6-58, is elliptic. As another example, we now consider the Ricci curvature operator, which does not have an elliptic linearization (see the next example). We saw in Section 5.3.1 that in the Lorentzian setting, the Ricci curvature operator can be put into hyperbolic form in wave coordinates. An analogue holds in the Riemannian setting, using a derivation following that of equation (5.3.1), with *harmonic coordinates* playing the role of wave coordinates.

Example 6-11. The operator $g \mapsto \mathrm{Ric}(g)$ is nonlinear, and we let $P = D\,\mathrm{Ric}_g$ be its linearization at a metric g. This operator acts between symmetric $(0, 2)$-tensor fields on M, and it fails to be elliptic *precisely* because of diffeomorphism invariance. We first note that if φ is a diffeomorphism on M, then $\varphi^*(\mathrm{Ric}(g)) = \mathrm{Ric}(\varphi^* g)$ (this is easy to compute in local coordinates $\psi_1 : U \subset \mathbb{R}^n \to V \subset M$ and $\psi_2 = \varphi \circ \psi_1 : U \to \varphi(V)$). If X is a smooth vector field, generating a flow φ_t, then differentiating $\varphi_t^*(\mathrm{Ric}(g)) = \mathrm{Ric}(\varphi_t^* g)$ with respect to t at $t = 0$ gives $L_X(\mathrm{Ric}(g)) = P(L_X g)$, where L_X is the Lie derivative. As differential operators on X, $P(L_X g)$ is of third order and $L_X(\mathrm{Ric}(g))$ is of first order. Taking the principal symbol yields $0 = \sigma_P(\xi) \circ \sigma_Q(\xi)$, where we let $Q(X) = L_X g$. Since $X \mapsto Q(X)$ maps vector fields to symmetric $(0, 2)$-tensor fields, this symbol equation means that $0 = \sigma_P(\xi)(\sigma_Q(\xi)(v))$ for any $v \in T_p M$ and $\xi \in T_p^* M$. From $(L_X g)_{ij} = X_{i;j} + X_{j;i}$, we see that $\sigma_Q(\xi)(v) = i\,(\xi \otimes v^\flat + v^\flat \otimes \xi)$. Thus for $\xi \neq 0$, we have a nontrivial kernel for $\sigma_P(\xi)$. In Exercise 6-59 we will see that the kernel elements of $\sigma_P(\xi)$ for $\xi \neq 0$ are precisely those symmetric $(0, 2)$-tensors of the form $\xi \otimes v^\flat + v^\flat \otimes \xi$.

The constraints operator $\Phi(g, K) = \big(R(g) - |K|_g^2 + (\mathrm{tr}_g K)^2,\ \mathrm{div}_g K - d(\mathrm{tr}_g K)\big)$ can be construed as a nonlinear, underdetermined-elliptic operator of *mixed order*, following [76], cf. [69; 61].

6.1.4. *Elliptic estimates.* We consider linear operators L on a manifold M, and while it will be important to understand the role of the regularity of the coefficients of L in the analysis, we assume the coefficients of L are smooth unless so noted. There are fundamental estimates for elliptic operators that will play an important role in what follows. We will see these *elliptic estimates* (sometimes paired with compactness results in appropriate function spaces) can yield convergence results, which might be used for instance in an iteration scheme to produce a solution of a PDE under consideration. We will begin by illustrating several such estimates.

Recall that a closed manifold is one which is compact and without boundary.

Proposition 6-12 (elliptic estimates). *Let L be a linear elliptic operator of order m, on a closed manifold M. Let $1 < p < \infty$ and $0 < \alpha < 1$. There is a constant $C > 0$ such that the following estimates hold for any smooth u:*

$$\|u\|_{H^m(M)} \leq C(\|Lu\|_{L^2(M)} + \|u\|_{L^2(M)}), \qquad (6.1.1)$$

$$\|u\|_{W^{m,p}(M)} \leq C(\|Lu\|_{L^p(M)} + \|u\|_{L^p(M)}), \qquad (6.1.2)$$

$$\|u\|_{C^{m,\alpha}(M)} \leq C(\|Lu\|_{C^{0,\alpha}(M)} + \|u\|_{C^0(M)}). \qquad (6.1.3)$$

Remark 6-13. We could have labeled the constant differently for each of the estimates above, since in addition to depending on L and M (including the background metric g, if used to define the norms), the constant in (6.1.2) depends on p, and in (6.1.3) depends on α.

Inequalities (6.1.1)–(6.1.3) are basic versions of, respectively, the L^2-estimates, L^p-estimates, and *Schauder estimates*. One often breaks out the case $p = 2$ separately, as it can be obtained by a more elementary treatment than the general L^p case. These are *a priori* estimates: we are assuming smoothness on u to begin with, and then the estimates show that u and all its derivatives up through order m can be estimated in the indicated norms in terms of the associated norm of Lu, and a lower-order norm on u; for instance the first and second derivatives of a function u can be controlled in terms of $\Delta_g u$, along with a lower-order norm.

In fact, the estimates do not need u to be C^∞; it is sufficient to assume that u lies in $\in H^m(M)$, $W^{m,p}(M)$ or $C^{m,\alpha}(M)$ as the case may be. For (6.1.1)–(6.1.2) this can be easily derived from the smooth case by taking a sequence of smooth u_i converging to u in the relevant Sobolev norm.

We next indicate a few basic facts one can get by combining the estimates with compactness results. These observations are used heavily; for instance they can be employed in the proof of the Fredholm property (Proposition 6-25). For the next proposition, the domain of L can be any of $C^{m,\alpha}(M)$, $0 < \alpha < 1$, or $W^{m,p}(M)$, $1 < p < \infty$, for which the above elliptic estimates hold. We will see shortly that the kernel is independent of which of these function spaces we take for the domain.

Proposition 6-14. *Let M be a closed manifold, on which L is a linear elliptic operator of order m. The kernel of L is finite-dimensional.*

Proof. The kernel of L is a closed subspace, and hence a Banach space. The claim will be established by showing that the closed unit ball in $\ker L$ is compact. Given any sequence in this unit ball, we can extract a subsequence u_i that converges appropriately, either in $L^p(M)$ (using Rellich–Kondrachov) or in $C^0(M)$ (using Ascoli–Arzelà). Since $Lu_i = 0$, the respective elliptic estimate may be applied to the differences $u_i - u_j$ to show that the sequence u_i converges in $\ker L$, from which the claim follows. $\qquad\square$

A second fundamental fact is an injectivity estimate transverse to the kernel; thus should L have no kernel, it gives a sharper overall estimate.

Proposition 6-15. *Let M be a closed manifold, on which L is a linear elliptic operator of order m. Let $0 < \alpha < 1$, respectively $1 < p < \infty$, and suppose $S \subset C^{m,\alpha}(M)$, respectively $S \subset W^{m,p}(M)$, is a closed subspace with $S \cap \ker L = \{0\}$. There is a $C > 0$ such that for all $u \in S$, $\|u\|_{C^{m,\alpha}(M)} \le C \|Lu\|_{C^{0,\alpha}(M)}$, respectively $\|u\|_{W^{m,p}(M)} \le C\|Lu\|_{L^p(M)}$.*

Exercise 6-16. Prove this proposition. Proceed by contradiction, starting with a sequence $u_i \in S$ with $\|u_i\|_{C^{m,\alpha}(M)} = 1$ but $\|Lu_i\|_{C^{0,\alpha}(M)} \to 0$. Use the basic elliptic estimate, together with Ascoli–Arzelà, to get convergence of a subsequence in $C^{m,\alpha}(M)$. Find a contradiction. The respective statement in Sobolev spaces follows the same way, via Rellich–Kondrachov.

We can get higher-order estimates by differentiation.

Proposition 6-17. *Let M be a closed manifold, on which L is a linear elliptic operator of order m. Let $0 < \alpha < 1$, $1 < p < \infty$, and $k \in \mathbb{Z}_+$. There exists a constant C such that $\|u\|_{W^{m+k,p}(M)} \le C(\|Lu\|_{W^{k,p}(M)} + \|u\|_{L^p(M)})$ for all $u \in W^{m+k,p}(M)$. Likewise there is a constant C such that $\|u\|_{C^{m+k,\alpha}(M)} \le C(\|Lu\|_{C^{k,\alpha}(M)} + \|u\|_{C^0(M)})$ for all $u \in C^{m+k,\alpha}(M)$.*

Proof. Observe that $L(\partial^\beta u) - \partial^\beta(Lu)$ can be expressed as a linear combination of $\partial^\gamma u$ for $|\gamma| < m + |\beta|$ (assuming the coefficients of L are smooth, or smooth enough, else we have to restrict $|\beta|$). Applying (6.1.2), for example, to $\partial^\beta u$ for $|\beta| \le k$, we get $\|u\|_{W^{m+k,p}(M)} \le C(\|Lu\|_{W^{k,p}(M)} + \|u\|_{W^{m+k-1,p}(M)})$ for all $u \in W^{m+k,p}(M)$ (adjusting C as necessary). The lower-order term $\|u\|_{W^{m+k-1,p}(M)}$ can be replaced by $\|u\|_{L^p(M)}$ via interpolation (see Exercise 6-6). The higher-order analogue of (6.1.3) follows likewise. $\qquad\square$

6.1.4.1. *On the proof of* (6.1.1)–(6.1.3) *and interior estimates.* We make only brief comments on the proofs of (6.1.1)–(6.1.3). One natural approach to proving (6.1.1) is to use Fourier analysis; see, e.g., the proof from [139, p. 193], which also uses a *parametrix* (for more on this notion see [114; 115]). For a beautiful approach to Schauder estimates by scaling, see [210]. A standard approach that works for all three estimates is to use corresponding *interior estimates* as in [86; 107; 3], along with a suitable finite cover of M. To state and motivate these, we proceed in the converse direction, getting interior estimates from (6.1.1)–(6.1.3) as follows. Let Ω' be open with compact closure contained in the open set $\Omega \subset M$. Consider a smooth cutoff function ζ that is identically 1 on Ω' and vanishes outside Ω. By applying any of (6.1.1)–(6.1.3) to ζu, and noting $L(\zeta u) - \zeta L u$ can be expressed as a linear combination of u and its derivatives up to order $m - 1$, with coefficients depending on ζ and its derivatives, one can get *interior* estimates. We illustrate with the L^2-estimate, with analogous comments applying to the others. Applying the cutoff and (6.1.1), we get, adjusting C as needed, $\|u\|_{H^m(\Omega')} \le C(\|Lu\|_{L^2(\Omega)} + \|u\|_{H^{m-1}(\Omega)})$. As you can imagine, C will depend on the two domains; in particular, the derivatives of the cutoff function will depend on the distance between the domains. As noted, one often derives such interior estimates first, and then from these, one can get an estimate $\|u\|_{H^m(M)} \le C(\|Lu\|_{L^2(M)} + \|u\|_{H^{m-1}(M)})$ with a suitable covering. From here, (6.1.1) follows using interpolation as in Exercise 6-6 to replace the norm $\|u\|_{H^{m-1}(M)}$ with $\|u\|_{L^2(M)}$. We remark that to replace the norm $\|u\|_{H^{m-1}(\Omega)}$ with $\|u\|_{L^2(\Omega)}$ in the *interior* estimate (and with the analogous replacement for the other estimates, respectively), one has to work a bit harder, because the domains are different on each side of the inequality in the interior estimate. See, e.g., [86, Section 6.3], or for more general scaling arguments, [107, Sections 6.1 and 9.5] and [209, Chapter 2, Lemma 2].

6.1.5. *Elliptic regularity and bootstrapping.* Suppose for the moment that the coefficients of the linear elliptic operator L are smooth, and u is a locally integrable function (or more generally a distribution) so that $Lu = f$ weakly. We would like to infer how smooth u is based on the regularity of f. A seminal result along these lines is *Weyl's lemma*, which we recall now.

Lemma 6-18 (Weyl's Lemma). *Suppose $\Omega \subset \mathbb{R}^n$ is open, and that u is a distributional solution of $\Delta u = 0$ in Ω, i.e., for all $\psi \in C_c^\infty(\Omega)$, $\int_\Omega u \Delta \psi \, dx = 0$. Then $u \in C^\infty(\Omega)$, i.e., the distribution u can be represented by integration against a smooth function.*

This immediately implies that if $L = a^{ij} \, \partial^2/\partial x^i \partial x^j$, for (a^{ij}) a positive definite $n \times n$ constant matrix, then distributions u with $Lu = 0$ are actually smooth. A classical proof of Weyl's Lemma that does not use elliptic estimates is a mollification argument based on the mean value property (Proposition 7-8), as in [86] in case u is continuous; this can be extended to harmonic distributions

u by showing that the mollification of u yields a smooth harmonic function, cf. [195, Theorem 6.30].

The interior estimates can be used to obtain more general *elliptic regularity* results for solutions of elliptic equations. For illustration, let us suppose L is linear elliptic of order m, with smooth coefficients, and M is a closed manifold (although as we will see presently, most of the action will be happening in a localized setting). Take $u \in H^m(M)$, so that $Lu \in L^2(M)$; if we suppose furthermore that $f := Lu \in H^1(M)$, we would like to conclude that $u \in H^{m+1}(M)$. We can turn this into a local question using a suitable covering by coordinate charts, and work on a bounded domain in $\mathbb{R}^n$.

Proposition 6-19. *Suppose L is a linear elliptic operator of order m with smooth coefficients on an open set $\Omega \subset \mathbb{R}^n$. Let $1 < p < \infty$, and $0 < \alpha < 1$, and $k \in \mathbb{Z}_+$. If $u \in W^{m,p}_{\mathrm{loc}}(\Omega)$ with $Lu = f \in W^{k,p}_{\mathrm{loc}}(\Omega)$, then $u \in W^{m+k,p}_{\mathrm{loc}}(\Omega)$. Likewise if $u \in C^{m,\alpha}_{\mathrm{loc}}(\Omega)$ with $Lu = f \in C^{k,\alpha}_{\mathrm{loc}}(\Omega)$, then $u \in C^{m+k,\alpha}_{\mathrm{loc}}(\Omega)$.*

Proof. Suppose $u \in H^m_{\mathrm{loc}}(\Omega)$ with $Lu = f \in H^1_{\mathrm{loc}}(\Omega)$, and take compactly contained domains $\Omega'' \subset \Omega' \subset \Omega$. While we might not be able to differentiate the equation, we can *difference* it to obtain an elliptic equation satisfied by any *difference quotient* $\Delta^h_i u(x) = h^{-1}(u(x+he_i) - u(x))$ $(h \neq 0)$, where $\{e_1, \ldots, e_n\}$ is the standard basis for $\mathbb{R}^n$. Indeed, we let $u^h(x) = u(x + he_i)$, and if $L = \sum_{|\alpha| \leq m} a_\alpha \, \partial^\alpha/\partial x^\alpha$, then we let $\Delta^h_i L = \sum_{|\alpha| \leq m} (\Delta^h_i a_\alpha) \, \partial^\alpha/\partial x^\alpha$. It is easy to show the product rule for differences: $\Delta^h_i(Lu) = L(\Delta^h_i u) + (\Delta^h_i L)(u^h)$. For $0 < |h| < \mathrm{dist}(\Omega', \partial\Omega)$, the interior estimate for L gives, since $\Delta^h_i u \in H^m(\Omega')$,

$$\|\Delta^h_i u\|_{H^m(\Omega'')} \leq C\big(\|L(\Delta^h_i u)\|_{L^2(\Omega')} + \|\Delta^h_i u\|_{H^{m-1}(\Omega')}\big)$$

$$\leq C'\big(\|\Delta^h_i(Lu)\|_{L^2(\Omega')} + \|u\|_{H^m(\Omega)} + \|\Delta^h_i u\|_{H^{m-1}(\Omega')}\big) \quad (6.1.4)$$

where we used a bound on the difference quotients $\Delta^h_i a_\alpha$, by the mean value theorem, say. To complete the proof, we invoke a *difference quotient lemma*, which roughly says that for $u \in H^m$, difference quotients of u are uniformly bounded if and only if $u \in H^{m+1}$. More precisely, for $v \in H^1(\Omega)$, and for any $\Omega' \subset \Omega$ with $0 < |h| < d(\Omega', \partial\Omega)$, $\|\Delta^h_i v\|_{L^2(\Omega')} \leq \|\partial v/\partial x^i\|_{L^2(\Omega)}$; conversely for $v \in L^2(\Omega)$, if there is a $K > 0$ such that for any $\Omega' \subset \Omega$ with $0 < h < d(\Omega', \partial\Omega)$, $\|\Delta^h_i v\|_{L^2(\Omega')} \leq K$, then $\partial v/\partial x^i \in L^2(\Omega)$ and $\|\partial v/\partial x^i\|_{L^2(\Omega)} \leq K$. For a proof of the analogous statement for all $1 < p < \infty$, see [107, Chapter 7]. Applying this to (6.1.4) yields a constant C'' such that (note where we use $Lu \in H^1(\Omega)$)

$$\|\Delta^h_i u\|_{H^m(\Omega'')} \leq C''(\|Lu\|_{H^1(\Omega)} + \|u\|_{H^m(\Omega)}).$$

By applying the difference quotient lemma again, we get $u \in H^{m+1}_{\mathrm{loc}}(\Omega)$ as desired.

The proof for $1 < p < \infty$ follows in the same way, and we can apply induction to *bootstrap regularity* for higher k. A similar approach can be used for Hölder regularity; see Exercise 6-60. $\qquad\square$

From here we see that for L linear elliptic of order m, with smooth coefficients, if $u \in H^m(M)$ and $Lu = f \in C^\infty(M)$, then by induction we can prove $u \in H^k(M)$ for all $k \in \mathbb{Z}_+$, and analogously if instead $u \in W^{m,p}(M)$ and $1 < p < \infty$, or $u \in C^{m,\alpha}(M)$ with $0 < \alpha < 1$. In the Hölder setting, then, we see directly that u is C^∞. In the case of Sobolev spaces, one invokes a *Sobolev embedding*: if $kp > n$, any $u \in W^{k,p}(M)$ can be represented by a continuous function $u \in C^0(M)$; in fact if ℓ is a nonnegative integer, $(k-1)p \le n < kp$, and $0 < \alpha < 1$ with $\alpha \le k - \frac{n}{p}$, then if $u \in W^{k+\ell,p}(M)$, it follows that $u \in C^{\ell,\alpha}(M)$, and there is a constant $C > 0$ so that for all $u \in W^{k,p}(M)$, $\|u\|_{C^{\ell,\alpha}(M)} \le C\|u\|_{W^{k+\ell,p}(M)}$. Section 7.2.2.1 discusses more such embeddings; see also [2; 86; 107], for example. We likewise conclude that if L is linear elliptic of order m on Ω, with smooth coefficients, and $u \in W^{m,p}_{\mathrm{loc}}(\Omega)$ for some $1 < p < \infty$ with $Lu = f \in C^\infty(\Omega)$, then $u \in C^\infty(\Omega)$.

Remark 6-20. We have been working with *a priori* regularity for u, say $u \in W^{m,p}_{\mathrm{loc}}$, for instance. We often naturally run into weak solutions that do not have this regularity *a priori*, and in fact the following elliptic regularity result holds: for a linear elliptic operator L of order m on $\Omega \subset \mathbb{R}^n$ with smooth coefficients, if u is a distribution with $Lu = f \in C^\infty(\Omega)$, then $u \in C^\infty(\Omega)$. Since smoothness is a local question, we can localize support near any point, and so consider a smooth bump function $\zeta \in C^\infty_c(\Omega)$ that is identically 1 on $B_\varepsilon(p)$ compactly contained in Ω, and let ψ_1 be a smooth bump function identically 1 on $B_{\varepsilon/2}(p)$ supported in $B_\varepsilon(p)$. By [195, Theorem 8.9], $\zeta u \in H^s(\mathbb{R}^n)$ for some s, so that since $\zeta u = u$ on the support of ψ_1, $L(\psi_1 u) - \psi_1 Lu \in H^{s-m+1}(\mathbb{R}^n)$. We can apply elliptic regularity in these Sobolev spaces as in [93, Lemma 6.32] (cf. [139, Theorem III.4.3] for a related approach) to conclude $\psi_1 u \in H^{s+1}(\mathbb{R}^n)$. The claim follows by further localization and bootstrapping regularity.

We could compactify the localized problem to a closed manifold if we wish: take an elliptic operator $\widetilde{L}$ for which $\widetilde{L} = L$ on $B_\varepsilon(p)$, while $\widetilde{L}$ has constant coefficients outside the support of ζ; this can be done with a smooth convex combination of L with the constant-coefficient operator whose coefficients are those of L at p, by taking sufficiently small ε if necessary. We can construe $\widetilde{L}(\psi_1 u) =: \tilde{f}$ on a domain compactified to an n-torus — at which point Fourier series methods could be employed, as in [219, Chapter 6].

As we may want to apply linear theory in situations where the coefficients are not necessarily smooth, it is very important to note the regularity required on the coefficients of L to obtain the elliptic estimates and regularity at a given level. For (6.1.1)–(6.1.2), one can impose a continuity requirement, at least for the top-order coefficients, to be able to locally approximate L with a constant-coefficient operator to which the estimates will apply. For the Schauder estimate (6.1.3), Hölder continuity of the coefficients is sufficient; cf. Chapters 6, 8, and 9 of [107] for such results in the second-order case. Given these results, for higher-order estimates and regularity, we observe directly from the difference quotient method above, that it is sufficient for L to have, for $k \in \mathbb{Z}_+$, coefficients in $C_{\mathrm{loc}}^{k-1,1}$ in order to conclude that $u \in H_{\mathrm{loc}}^{m+k}$ follows from $u \in H_{\mathrm{loc}}^m$ and $Lu = f \in H_{\mathrm{loc}}^k$, and analogously for $1 < p < \infty$. For Hölder regularity, given the basic estimate (6.1.3) holds for $C^{0,\alpha}$ coefficients, we can then prove that for $u \in C_{\mathrm{loc}}^{m,\alpha}$, for L with coefficients in $C_{\mathrm{loc}}^{k,\alpha}$, then from $Lu = f \in C_{\mathrm{loc}}^{k,\alpha}$ we can conclude $u \in C_{\mathrm{loc}}^{m+k,\alpha}$, cf. Exercise 6-60.

Some of the technicalities of the preceding discussion can be appreciated with a couple basic *nonlinear* examples. For a very simple example, suppose we have a solution to the equation $Lu = h(u)$ for some smooth function h, where L is linear elliptic of order m with smooth coefficients. If u is $C_{\mathrm{loc}}^{m,\alpha}$, say, then so is $h(u)$. By elliptic regularity, we have u, and hence then $h(u)$, in $C_{\mathrm{loc}}^{2m,\alpha}$. We can continue from here to get that $u \in C^\infty$. The equation $Lu = h(u)$ is *semilinear*: the highest order terms appear linearly. Certain operators of interest are *quasilinear*: the operator is linear in the highest-order derivatives, but the coefficients can depend on lower-order derivatives. A classical example of this is the minimal surface equation (see Exercise 6-58). Suppose u satisfies an equation of the form $a^{ij}(x, u, du)\, \partial^2 u/\partial x^i \partial x^j = f(x)$, where f is smooth, and a^{ij} is smooth in its arguments. Given the solution u, we can construe the coefficients as given and consider the equation in terms of linear theory; suppose $(a^{ij}(x, u, du))$ is positive definite, so that the operator is elliptic. If u is $C_{\mathrm{loc}}^{2,\alpha}$, say, then the coefficients $a^{ij}(x, u, du)$ are $C_{\mathrm{loc}}^{1,\alpha}$; by the above discussions, we see we can conclude $u \in C_{\mathrm{loc}}^{3,\alpha}$; this in turn allows us to bootstrap the regularity of the coefficients $a^{ij}(x, u, du)$ to $C_{\mathrm{loc}}^{2,\alpha}$, and continue.

6.1.6. *The Fredholm alternative/Hodge decomposition.*

In this section we discuss a result on solvability of elliptic equations. Suppose that (M, g) is a closed Riemannian manifold, and P is a linear elliptic operator of order m with smooth coefficients. From Proposition 6-14, P has a finite-dimensional kernel, which by elliptic regularity is comprised of smooth sections, and thus we can write the domain of P (i.e., $W^{m,p}(M)$ or $C^{m,\alpha}(M)$) as $S \oplus \ker P$, where S is a closed subspace. Applying Proposition 6-15, we can easily conclude P has closed range. Thus in the target of P (i.e., $L^p(M)$ or $C^{0,\alpha}(M)$) we have a direct sum of closed subspaces ran $P \oplus \ker P^*$, since by integration by parts any $Pw = f \in L^2(M)$ annihilates the kernel of P^*, which is itself a finite-dimensional space of smooth sections. In fact, the content of the Fredholm alternative/Hodge decomposition is that this direct sum is the entire target space. The proof uses tools from functional analysis to get a weak solution of $Ph = f$, for any f in the target space that annihilates $\ker P^*$ by integration; and then proving that the solution h can be represented by an element in the respective space $W^{m,p}(M)$ or $C^{m,\alpha}(M)$, by elliptic regularity. In fact, then, one has the following decompositions: for $1 < p < \infty$, $0 < \alpha < 1$, and k a nonnegative integer (again, we suppress the metric from the notation):

$$W^{k,p}(M) = P(W^{m+k,p}(M)) \oplus \ker P^* \tag{6.1.5}$$

$$C^{k,\alpha}(M) = P(C^{m+k,\alpha}(M)) \oplus \ker P^* \tag{6.1.6}$$

$$C^\infty(M) = P(C^\infty(M)) \oplus \ker P^*. \tag{6.1.7}$$

These decompositions are valid as well in case the linear operator P is underdetermined-elliptic, i.e., P^* is overdetermined-elliptic. To see this we apply the decomposition to the *elliptic* operator PP^*: for $p = 2$, say, we get $L^2(M) = PP^*(H^{2m}(M)) \oplus \ker(PP^*)$. Integrating $0 = \int_M \langle u, PP^*u \rangle \, dv_g$ by parts shows that $\ker(PP^*) = \ker P^*$. Also, $PP^*(H^{2m}(M)) \subset P(H^m(M)) \perp \ker P^*$, from which we conclude $L^2(M) = P(H^m(M)) \oplus \ker P^*$. We highlight that a solution h of $Ph = f$ can be taken to be of the form $h = P^*u$, which is $L^2(dv_g)$-orthogonal to $\ker P$.

As (6.1.5)–(6.1.7) are fundamental, we sketch a proof below, along lines that will be useful for us later. Before doing so, we ask you to recall the following basic fact that will be used in the proof.

Exercise 6-21. Suppose X is a Banach space, and x_i is a sequence converging weakly to $x \in X$. Use the Hahn–Banach theorem to prove that $\|x\|_X \leq \liminf_{i \to \infty} \|x_i\|_X$. When X is a Hilbert space with inner product $\langle \cdot, \cdot \rangle$, give a more elementary proof by considering the sequence $\langle x_i, x \rangle$.

Proof of the Fredholm alternative/Hodge decomposition. We now give a proof of (6.1.5) for the case $p = 2$. As we have seen, $\mathcal{H} = \ker P^*$ is a finite-dimensional space of smooth sections. Given an $f \in L^2(dv_g)$ which is L^2-orthogonal to $\mathcal{H}$, we seek to solve the equation $Ph = f$ for $h \in H^m(M)$. To do this, we define the functional $\mathcal{G}(u) = \int_M \left(\frac{1}{2}|P^*u|_g^2 - uf \right) dv_g$. Let $S = \{u \in H^m(M) : \int_M uw \, dv_g = 0 \text{ for all } w \in \mathcal{H}\}$, i.e., $S = H^m(M) \cap \mathcal{H}^\perp$, where the orthogonal complement is taken in $L^2(dv_g)$. It is easy to see S is closed in $H^m(M)$ and that $H^m(M) = S \oplus \mathcal{H}$.

By Proposition 6-15 and Cauchy–Schwarz, there is a constant C such that for all $u \in S$, $\mathcal{G}(u) \geq C\|u\|_{H^m}^2 - \|f\|_{L^2}\|u\|_{L^2}$. Thus we conclude that $\mathcal{G}$ is bounded from below on S, and that an infimizing sequence $u_i \in S$ must be H^m-bounded. By Riesz Representation and compactness (Banach–Alaoglu, Rellich), we can assume by re-indexing a suitable subsequence that there is a $u \in S$ (which we recall is closed) so that u_i converges to u weakly in $H^m(M)$, and in the L^2-norm. Thus P^*u_i converges to P^*u weakly in $L^2(dv_g)$, and then by Exercise 6-21 we have that $\|P^*u\|_{L^2} \leq \liminf_{i \to \infty} \|P^*u_i\|_{L^2}$. Thus we see $\mathcal{G}(u) \leq \liminf_{i \to \infty} \mathcal{G}(u_i)$, and hence $\mathcal{G}(u)$ is the minimum value of $\mathcal{G}$ on S.

We can thus compute the Euler–Lagrange equation: for $v \in S$, we have

$$0 = \frac{d}{dt}\bigg|_{t=0} \mathcal{G}(u + tv) = \int_M (\langle P^*u, P^*v \rangle - vf)\, dv_g.$$

At the same time we have $0 = \int_M (\langle P^*u, P^*v \rangle - vf)\, dv_g$ for $v \in \mathcal{K} = \ker P^*$, by the assumption on f. Thus this integral identity holds for all $v \in H^m(M)$, and hence $PP^*u = f$ weakly. Since P is assumed to have smooth coefficients and g is smooth, elliptic regularity will allow us to conclude the appropriate regularity for any of the splittings above. For example, for $f \in L^2(M)$, we conclude $u \in H^{2m}(M)$, so $h := P^*u \in H^m(M)$ solves $Ph = f$ as desired. See Exercise 6-60 for a proof of the splitting for $p > 1$, and for the Hölder setting. $\square$

Remark 6-22. We have observed that certain facts extend to overdetermined-elliptic operators L on a closed manifold M, such as finite-dimensionality of the kernel, which we gleaned by using a metric g to define $P = L^*$, and considering the elliptic operator $PP^* = L^*L$. Of course, if we could establish the elliptic estimates in Proposition 6-12 for overdetermined-elliptic operators, then the proofs of various facts would follow from the estimates, such as Proposition 6-15, which could then be used to prove the Fredholm/Hodge splitting directly for $L = P^*$ overdetermined-elliptic. In fact, the Fourier transform proof for the L^2-estimates works for injective symbol (cf. [114]); the scaling argument in [210] for the Schauder estimates uses *hypoellipticity*, that the elements in the kernel are smooth, which holds for overdetermined-elliptic operators. One could also establish estimates for L overdetermined-elliptic via the corresponding estimates for $L^*L = PP^*$, say in case $p = 2$ as a mapping $L^*L : H^m(M) \to H^{-m}(M)$, for which suitable mapping properties and estimates are available via Fourier analysis. For other spaces, see, e.g., [210] and [214, Chapters 13–14].

6.1.7. *Eigenvalue decomposition for self-adjoint elliptic operators.* Suppose P is a linear *self-adjoint* ($P = P^*$) elliptic operator of order m with smooth coefficients on a closed Riemannian manifold (M, g). Then we have $L^2(M) =$

$\ker P \oplus P(H^m(M))$. Furthermore for any $\lambda \in \mathbb{R}$, $P - \lambda I$ is also elliptic and self-adjoint, and so it either has nontrivial kernel E_λ (which then must also be finite-dimensional space of smooth sections), so that λ is an eigenvalue, or $P - \lambda I : H^m(M) \to L^2(M)$ is invertible. Although real matrices can have complex eigenvalues, you will recall that real symmetric matrices have real eigenvalues. Similarly, even if we consider the self-adjoint operator P operating on a Hermitian bundle, the same proof as in the matrix case will show that the eigenvalues can only be real. Let $\Sigma \subset \mathbb{R}$ be the set of eigenvalues. One can show, see, e.g., [139, Theorem III.5.8], that Σ is an infinite discrete set, and $L^2(M) = \bigoplus_{\lambda \in \Sigma} E_\lambda$ as a Hilbert space direct sum of orthogonal subspaces.

6.1.8. Fredholm operators. We now recall the notion of a *Fredholm operator*, which will be used later. Let $T : X \to Y$ be a bounded linear operator between Banach spaces. The term *cokernel* refers to a quotient space, which in this context can either mean the *algebraic* cokernel $Y/T(X)$, or the Banach space cokernel $Y/\overline{T(X)}$. Of course these notions agree for operators with closed range. Recall that by the Hahn–Banach theorem, if $T(X)$ is a closed, proper subspace of Y, there must be a nontrivial element θ in the dual space Y^* such that θ vanishes on $T(X)$.

Exercise 6-23. Let $T : X \to Y$ be a bounded linear operator between Banach spaces. Show that if there is a closed subspace $W \subset Y$ so that $Y = T(X) \oplus W$, then $T(X)$ is closed in Y. In particular, conclude that if $Y/T(X)$ is finite-dimensional, then $T(X)$ is closed.

Definition 6-24. A bounded linear operator $T : X \to Y$ between Banach spaces is *Fredholm* if T has finite-dimensional kernel, closed range, and finite-dimensional cokernel. We call $\operatorname{ind}(T) = \dim(\ker T) - \dim(\operatorname{cok} T)$ the *Fredholm index* of T.

It follows that a bounded linear operator $T : X \to Y$ between Banach spaces is Fredholm if and only if T has finite-dimensional kernel and finite-dimensional algebraic cokernel $Y/T(X)$, but for clarity regarding the cokernel, the definition is usually stated as above.

Elliptic operators between appropriate function spaces are often Fredholm, as in the following fundamental proposition, a corollary of the above discussion.

Proposition 6-25. *Suppose M is a closed manifold, and L is a linear elliptic operator of order m with smooth coefficients. For an integer $k > m$ and $1 < p < \infty$, $L : W^{k,p}(M) \to W^{k-m,p}(M)$ is Fredholm; the analogous statement holds in Hölder spaces $(0 < \alpha < 1)$. Moreover, by elliptic regularity, the index of L is independent of $k \geq m$ (and of $1 < p < \infty$, respectively $0 < \alpha < 1$).*

6.1.9. *The maximum principle and the Harnack inequality.* The last stop on this brief tour of elliptic PDE brings us to several important results for second-order scalar operators. We will review in the next chapter how these properties can be proved for the Euclidean Laplace operator using the mean value property. For now, consider an operator

$$L = a^{ij}(x)\frac{\partial^2}{\partial x^i \partial x^j} + b^i(x)\frac{\partial}{\partial x^i} + c(x).$$

We should impose some ellipticity and boundedness conditions on the coefficients a^{ij}, b^i and c. Assume that the coefficients are continuous on the closure $\overline{\Omega}$ of a bounded domain $\Omega \subset \mathbb{R}^n$, and that (a^{ij}) satisfies an ellipticity condition: there is $\lambda > 0$ so that for all $x \in \overline{\Omega}$ and all $\xi \in \mathbb{R}^n$, $a^{ij}(x)\xi_i\xi_j \geq \lambda|\xi|^2$. (The continuity assumption, plus ellipticity, is sufficient to give the required bounds, but not necessary; see [107], for example. As noted earlier, we may arrange by symmetrization that $a^{ij} = a^{ij}$.)

Weak maximum principle. We state first a *weak maximum principle* for L, which allows one to estimate certain u on a bounded open set in terms of its boundary values. Suppose $u \in C^2(\Omega) \cap C^0(\overline{\Omega})$. We write $u = u^+ - u^-$, $|u| = u^+ + u^-$, where $u^+ = \max(u, 0) \geq 0$, $u^- = -\min(u, 0) \geq 0$ (note that in [107], $u^- = \min(u, 0)$ is the opposite of what we have taken here).

(i) Suppose $c = 0$ on Ω. If $Lu \geq 0$ on Ω, then $\max_{\overline{\Omega}} u = \max_{\partial\Omega} u$, while if $Lu \leq 0$, then $\min_{\overline{\Omega}} u = \min_{\partial\Omega} u$.

(ii) Suppose $c \leq 0$ in Ω. If $Lu \geq 0$ on Ω, then $\max_{\overline{\Omega}} u \leq \max_{\partial\Omega} u^+$, while if $Lu \leq 0$ in Ω, then $\min_{\overline{\Omega}} u \geq -\max_{\partial\Omega} u^-$. Thus $\max_{\overline{\Omega}} |u| = \max_{\partial\Omega} |u|$ if $Lu = 0$.

The idea behind these is simple. To illustrate, if $Lu > 0$, we observe that u cannot have an interior maximum. If it does, say at $x_0 \in \Omega$, then each $\partial u/\partial x^i$ vanishes at x_0, and the Hessian matrix of u is nonnegative definite at x_0. If $c = 0$, we see, working in Cartesian coordinates diagonalizing $(a^{ij}(x_0))$ (if symmetric; else diagonalize the Hessian at x_0), that $Lu(x_0) \leq 0$. This observation also works when $c \leq 0$ and $u(x_0) \geq 0$, or in any case when $u(x_0) = 0$. This gives a contradiction. The general case where the inequality satisfied by Lu is not necessarily strict uses a perturbation to achieve the result; see [86, Chapter 6] or [107, Chapter 3].

The strong maximum principle and the Harnack inequality. We now state a *strong maximum principle* for operators L as above, which is more subtle (see the references just cited for proofs). It strengthens the weak maximum principle by ruling out certain interior extrema for nonconstant super- or subsolutions. Assume the bounded open set Ω is connected, and that $u \in C^2(\Omega) \cap C^0(\overline{\Omega})$.

(i) Suppose $c = 0$ on Ω. If $Lu \geq 0$ on Ω and if for some $x_0 \in \Omega$, $u(x_0) = \max_{\overline{\Omega}} u$ (or if $Lu \leq 0$ in Ω and for some $x_0 \in \Omega$, $u(x_0) = \min_{\overline{\Omega}} u$), then u is constant.

(ii) Suppose $c \leq 0$ on Ω. If $Lu \geq 0$ on Ω and u attains a *nonnegative* global maximum at a point $x_0 \in \Omega$, i.e., $u(x_0) = \max_{\overline{\Omega}} u \geq 0$ (or, if $Lu \leq 0$ on Ω and u attains a nonpositive global minimum $u(x_0) = \min_{\overline{\Omega}} u \leq 0$ for some $x_0 \in \Omega$), then u is constant. In fact, if the maximum (or minimum) satisfies $u(x_0) = 0$, then u is identically zero, regardless of the sign of c.

One last fundamental result is the *Harnack inequality* [86, Theorem 6.5] (see also [107, Corollary 9.25]): for any connected, compactly contained open subset W of Ω, there is a $C > 0$ such that if $u \geq 0$ is a *nonnegative* C^2 solution of $Lu = 0$ in Ω, then $\sup_W u \leq C \inf_W u$. In particular, if the supremum is positive on W, the function u is *strictly* positive in the connected component of Ω containing W.

The Harnack inequality can be established with weaker *a priori* assumptions on u, and for operators L with coefficients satisfying modest assumptions on ellipticity and boundedness; see [107, Chapter 9], for instance. For nonnegative weak solutions of analogous *divergence form* equations (where the leading order part of the operator has the form $\partial_i(a^{ij}\partial_j u)$), and under very general assumptions of ellipticity, boundedness and measurability on the coefficients, Moser established a Harnack inequality, from which he recovered oscillation and Hölder continuity estimates first established in the breakthrough discovery made independently by DeGiorgi and Nash, (see, for example, [107, Chapter 8]) or [214, Chapter 14]). We will not go into the details here, but in light of the discussion above (p. 182) about applications to nonlinear elliptic equations, the reader should still be able to appreciate that this result is fundamental in going from weak solutions to higher regularity. For applications to the mean curvature operator one can consult [107; 108], for example.

6.1.10. *The method of super- and subsolutions.* As a final topic in this section, we discuss a version of the method of super- and subsolutions that will suffice for our purposes. We will want to solve semilinear elliptic PDE of the form $\Delta_g \phi = f(x, \phi)$, for $f : M \times I \to \mathbb{R}$, where $I \subset \mathbb{R}$ is an open interval. A *subsolution* ϕ_- satisfies $\Delta_g \phi_- \geq f(x, \phi_-)$, and a *supersolution* ϕ_+ satisfies $\Delta_g \phi_+ \leq f(x, \phi_+)$. We reiterate that we take g to be smooth.

Theorem 6-26. *Suppose (M, g) is a closed Riemannian manifold, and suppose $f : M \times I \to \mathbb{R}$ is smooth. Suppose $\phi_- \leq \phi_+$ are smooth sub- and supersolutions* *for $\Delta_g \phi = f(x, \phi)$, such that $[\inf_M \phi_-, \sup_M \phi_+] \subset I$. Then there is a smooth function ϕ, with $\phi_- \leq \phi \leq \phi_+$, such that $\Delta_g \phi = f(x, \phi)$.*

For more general formulations of this method, see, e.g., [86; 127; 151]. We use the proof as a means to illustrate the utility of some of the tools for elliptic PDE that we have introduced above.

Proof. We first use continuity and compactness to choose a constant $\rho > 0$ so that, if we write the function f as $f(x, s)$, then $\rho - \partial f / \partial s$ is positive on $M \times [\inf_M \phi_-, \sup_M \phi_+] = M \times [\min_M \phi_-, \max_M \phi_+]$. We let

$$L = -\Delta_g + \rho \quad \text{and} \quad F(x, s) = -f(x, s) + \rho s,$$

so that $\Delta_g \phi = f(x, \phi)$ is equivalent to $L(\phi) = F(x, \phi)$. Note that $\partial F / \partial s > 0$ on $M \times [\min_M \phi_-, \max_M \phi_+]$.

Since we take g to be smooth, the Fredholm alternative/Hodge decomposition of the space of smooth functions for the linear, self-adjoint, elliptic operator L has the form $C^\infty(M) = \ker L \oplus \operatorname{ran} L$. Now, by considering $\int_M u L u \, dv_g$ and integrating by parts, we see the operator L has trivial kernel. Thus L is surjective as well. Since $\phi_\pm$ is smooth, we can thus define the following sequence of smooth functions recursively: ϕ_1 is the unique solution to $L\phi_1 = F(x, \phi_+)$, and for any $k \in \mathbb{Z}_+$, ϕ_{k+1} is the unique solution to $L\phi_{k+1} = F(x, \phi_k)$. We remark that for this to be well-defined, we have to make sure the range of each ϕ_k is inside the interval I. In fact, the sequence ϕ_k is not only well-defined, but satisfies

$$\phi_+ \geq \phi_1 \geq \phi_2 \geq \cdots \geq \phi_k \geq \phi_{k+1} \geq \cdots \geq \phi_- \, .$$

Indeed, since ϕ_+ is a supersolution, $L(\phi_+ - \phi_1) = L(\phi_+) - F(x, \phi_+)$ is nonnegative, i.e., $\Delta_g(\phi_+ - \phi_1) \leq \rho(\phi_+ - \phi_1)$. Clearly this implies $(\phi_+ - \phi_1)$ cannot have a negative minimum, so $\phi_+ \geq \phi_1$. Similarly, $(\phi_1 - \phi_-)$ cannot have a negative minimum, given that $L(\phi_1 - \phi_-) \geq F(x, \phi_+) - F(x, \phi_-) \geq 0$ (since $\partial F / \partial s > 0$). Thus the range of ϕ_1 is inside the interval I, and ϕ_2 can be defined as indicated above. To establish the required inequalities for ϕ_2, note $L(\phi_1 - \phi_2) = F(x, \phi_+) - F(x, \phi_1) \geq 0$, and $L(\phi_2 - \phi_-) \geq F(x, \phi_1) - F(x, \phi_-) \geq 0$, so the same considerations apply. We can proceed in the recursion, inductively establishing the required estimates.

We have a bounded monotonic sequence ϕ_k, which thus converges pointwise to a limit function ϕ. We want to show this limit function is smooth and satisfies the desired PDE. To do this, we apply elliptic estimates for the operator L. Indeed if we use the Schauder estimate (6.1.3) for Hölder norms, say, then we have a $C > 0$ such that $\|u\|_{C^{2,\alpha}(M)} \leq C(\|Lu\|_{C^{0,\alpha}(M)} + \|u\|_{C^0(M)})$ for all $u \in C^{2,\alpha}(M)$. In our setting, L has trivial kernel, in which case there is a constant $C > 0$ such that for all $u \in C^{2,\alpha}(M)$, $\|u\|_{C^{2,\alpha}(M)} \leq C\|Lu\|_{C^{0,\alpha}(M)}$, by Proposition 6-15. We have the analogous Sobolev space estimates (6.1.2), but again since L has trivial kernel, there is a $C > 0$ such that for all $u \in W^{2,p}(M)$, $\|u\|_{W^{2,p}(M)} \leq C\|Lu\|_{L^p(M)}$.

With this in hand, we note that

$$\|\phi_1\|_{W^{2,p}(M)} \leq C\|F(x, \phi_+)\|_{L^p(M)},$$

$$\|\phi_{k+1}\|_{W^{2,p}(M)} \leq C\|F(x, \phi_k)\|_{L^p(M)}.$$

Applying continuity, compactness, and boundedness of the sequence ϕ_k, we see there is a $K > 0$ such that $\|\phi_k\|_{W^{2,p}(M)} \leq K$ for all k. To get pointwise bounds, we cite a form of the *Sobolev embedding* recalled earlier: for $p > \frac{n}{2}$, any $u \in W^{2,p}(M)$ can be represented by a unique continuous function, and there is a constant $C > 0$ such that for all $u \in W^{2,p}(M)$, $\|u\|_{C^0(M)} \leq C\|u\|_{W^{2,p}(M)}$. Actually, for $0 < \gamma < \min\left(1, 2 - \frac{n}{p}\right)$, we can take $u \in C^{0,\gamma}(M)$, and there is a $C > 0$ for which $\|u\|_{C^{0,\gamma}(M)} \leq C\|u\|_{W^{2,p}(M)}$. Thus from $\|\phi_k\|_{W^{2,p}(M)} \leq K$, we see that ϕ_k is bounded in $C^{0,\gamma}(M)$, so that Ascoli–Arzelà yields a C^0-convergent subsequence, and so the limit ϕ is continuous; moreover, by monotonicity, the full sequence converges uniformly.

By applying the elliptic estimate judiciously, we can obtain more regularity on ϕ. For instance, for $k, \ell > 1$, we have

$$\|\phi_k - \phi_\ell\|_{W^{2,p}(M)} \leq C\|F(x, \phi_{k-1}) - F(x, \phi_{\ell-1})\|_{L^p(M)}.$$

By continuity of F, compactness, and uniform convergence, we have that $F(x, \phi_k)$ is Cauchy in $L^p(M)$, and so ϕ_k is $W^{2,p}(M)$-Cauchy; the limit must be ϕ, and thus $\phi \in W^{2,p}(M)$. Using the Sobolev embedding, we then have not only that $\phi \in C^{0,\gamma}(M)$, but we also see that ϕ_k is $C^{0,\gamma}(M)$-Cauchy, and so converges to ϕ in $C^{0,\gamma}(M)$. From this and the smoothness of F, we can infer that $F(x, \phi_k)$ is also $C^{0,\gamma}(M)$-Cauchy. Therefore, by the Schauder estimates, we have for $k, \ell > 1$, $\|\phi_k - \phi_\ell\|_{C^{2,\gamma}(M)} \leq C\|F(x, \phi_{k-1}) - F(x, \phi_{\ell-1})\|_{C^{0,\gamma}(M)}$. Thus ϕ_k is $C^{2,\gamma}(M)$-Cauchy, and hence converges in $C^{2,\gamma}(M)$, and the limit must be ϕ.

That ϕ solves the desired equation now follows from $L\phi = \lim_{k \to \infty} L\phi_k = \lim_{k \to \infty} F(x, \phi_{k-1}) = F(x, \phi)$. By elliptic regularity, we conclude ϕ is smooth. $\square$

6.1.11. *Application to conformal deformation of scalar curvature: Yamabe classes.* Before we return to the constraint equations, we give a geometric application of the PDE theory just discussed. We show that the space of (smooth) Riemannian metrics on a closed manifold M^n, $n \geq 3$, can be broken up into a union of three mutually disjoint sets, $\mathcal{Y}^+$, $\mathcal{Y}^0$ and $\mathcal{Y}^-$, each of which is a union of conformal classes. To do this, we begin by recalling the formula for the scalar curvature under a conformal change of metric.

Proposition 6-27. *If $\hat{g}$ and g are conformally related metrics on an n-manifold M ($n \geq 3$), say $\hat{g} = u^{\frac{4}{n-2}} g$ with $u > 0$, then*

$$R(\hat{g}) = -\tfrac{4(n-1)}{n-2} u^{-\frac{n+2}{n-2}} \left(\Delta_g u - \tfrac{n-2}{4(n-1)} R(g) u \right). \tag{6.1.8}$$

The proof is a straightforward but somewhat laborious calculation, which we highly recommend to the reader (Exercise 2-51). From it one will observe that with $\hat{g} = u^q g$, the choice of exponent $q = \frac{4}{n-2}$ is made to avoid $|du|_g^2$-terms in (6.1.8).

The linear operator in parentheses in (6.1.8) is the *conformal Laplacian* $\mathscr{L}_g$, given by $\mathscr{L}_g u = \Delta_g u - \tfrac{n-2}{4(n-1)} R(g) u$ (cf. p. 103). For (M, g) a closed Riemannian manifold, we define an energy associated to $\mathscr{L}_g$ by

$$\mathcal{E}_g(u) = \int_M -u \mathscr{L}_g u \, dv_g = \int_M \left(|\nabla_g u|_g^2 + \tfrac{n-2}{4(n-1)} R(g) u^2 \right) dv_g.$$

If $\hat{g} = u^{\frac{4}{n-2}} g$ with $u > 0$, then $dv_{\hat{g}} = u^{\frac{2n}{n-2}} dv_g$, and we have an identity for the total scalar curvature of $\hat{g}$ using (6.1.8):

$$\mathcal{R}(\hat{g}) = \int_M R(\hat{g}) \, dv_{\hat{g}} = \int_M -\tfrac{4(n-1)}{n-2} u^{-\frac{n+2}{n-2}} (\mathscr{L}_g u) \, u^{\frac{2n}{n-2}} \, dv_g = \tfrac{4(n-1)}{n-2} \mathcal{E}_g(u).$$

For (M, g) Riemannian, $\mathscr{L}_g$ is a self-adjoint elliptic operator, and so as in Section 6.1.7, we have that for M closed, $\mathscr{L}_g$ possesses a discrete set of eigenvalues, each of finite multiplicity, and the eigenfunctions can be chosen to form a complete orthonormal basis for $L^2(dv_g)$. As a consequence of the following lemma, we will see that, upon writing the eigenvalue equation as $\mathscr{L}_g u + \lambda u = 0$, the eigenvalues can be arranged in a monotonic sequence $\lambda_1 < \lambda_2 \leq \lambda_3 \leq \cdots$, with $\lim_{k \to \infty} \lambda_k = +\infty$, and for each $k \in \mathbb{Z}_+$, there is an eigenfunction u_k with $\mathscr{L}_g u_k + \lambda_k u_k = 0$. Of particular note, the first eigenvalue is simple, and a first eigenfunction u_1 can be chosen so that $u_1 > 0$. Recall that for u locally integrable (e.g., $u \in L^2(dv_g)$), du can be defined weakly, by requiring $\int_M du(X) \, dv_g = -\int_M u \operatorname{div}_g X \, dv_g$ for all smooth vector fields X. The Sobolev space $H^1(M, g)$ is comprised of those functions $u \in L^2(dv_g)$ with $|du|_g \in L^2(dv_g)$, forming a Hilbert space with inner product $\langle u, w \rangle_{H^1} = \int_M uw \, dv_g + \int_M \langle du, dw \rangle_g \, dv_g$.

Lemma 6-28. *For (M, g) a closed connected Riemannian manifold, $\mathscr{L}_g$ has a smallest eigenvalue λ_1, which is a simple eigenvalue, and there is an eigenfunction u with $u > 0$.*

Proof. There is a constant $C > 0$ (depending on n and g) so that $\mathcal{E}_g(u) \geq -C \int_M u^2 \, dv_g$. For an eigenfunction u_k for λ_k, $\mathcal{E}_g(u_k) = \lambda_k \int_M u_k^2 \, dv_g$, from which we then conclude the eigenvalues are bounded from below. Moreover, we see there is a variational characterization of λ_1, as the infimum of

$$\mathcal{G}_g(u) := \frac{\mathcal{E}_g(u)}{\int_M u^2 \, dv_g}$$

over all nontrivial $u \in H^1(M, g)$; the lower bound on $\mathcal{E}_g(u)$ shows that this infimum is finite. As such, we will argue that $\inf_{u \in H^1 \setminus \{0\}} \mathcal{G}_g(u)$ is achieved by a minimizer $u \in H^1(M, g)$, and then that this minimizer is a smooth eigenfunction which does not change sign. Given this, the minimizer has eigenvalue $\lambda_1 = \inf_{u \in H^1 \setminus \{0\}} \mathcal{G}_g(u)$, and furthermore this eigenvalue is *simple*: if there were two linearly independent eigenfunctions for an eigenvalue, we could arrange them to be orthogonal in $L^2(dv_g)$, but we will have shown that minimizers (first eigenfunctions) cannot change sign, so that two such functions cannot be orthogonal.

That a minimum is achieved can be proven using similar tools as in the proof above of the Hodge decomposition, based on elementary functional analysis, together with the fundamental compactness result from the *Rellich lemma*: the inclusion of $H^1(M, g) \hookrightarrow L^2(dv_g)$ is *compact*, which we recall means that given an H^1-bounded sequence, then there is a subsequence which converges in L^2. In addition, by Riesz representation for Hilbert spaces, together with the Banach–Alaoglu theorem, we can choose the subsequence to converge weakly in $H^1(M, g)$ as well (to the same limit).

With this in mind, consider an infimizing sequence $v_i \in H^1(M, g) \setminus \{0\}$ for $\mathcal{G}_g$. Since $\mathcal{G}_g(u) = \mathcal{G}_g(cu)$ for a constant $c \neq 0$, we may take $\int_M v_i^2 \, dv_g = 1$. The fact that $\mathcal{G}_g(v_i)$ approaches the finite infimum then implies that $\|v_i\|_{H^1}$ is a bounded sequence. Thus we may assume, reindexing the subsequence, that v_i converges strongly in L^2 and weakly in H^1 to a function $u \in H^1(M, g)$. u is nontrivial because $\int_M u^2 \, dv_g = 1$. That $\mathcal{G}_g(u)$ is the minimum of $\mathcal{G}_g$ comes from the fact that the norm is weakly lower semicontinuous, i.e., $\|u\|_{H^1} \leq \liminf_{i \to \infty} \|v_i\|_{H^1}$ (cf. Exercise 6-21). Together with the L^2-convergence of v_i to u, we can now conclude that $\mathcal{G}_g(u) \leq \liminf_{i \to \infty} \mathcal{G}_g(v_i)$, so that $\mathcal{G}_g(u)$ realizes the minimum. As with our variational formulation of the Einstein equation, for example, we can compute the Euler–Lagrange equation by setting

$$\frac{d}{dt}\Big|_{t=0} \mathcal{G}_g(u + tv) = 0,$$

for any $v \in H^1(M, g)$ (for small enough $|t|$, $\|u + tv\|_{L^2} \neq 0$ since $\|u\|_{L^2} = 1$).

Exercise 6-29. Compute the Euler–Lagrange equation, deriving the following identity: with u as above and $\lambda = \mathcal{G}_g(u)$, we have for all $v \in H^1(M, g)$,

$$\int_M \lambda u v \, dv_g = \int_M \left(\langle \nabla_g u, \nabla_g v \rangle + \tfrac{n-2}{4(n-1)} R(g) u v \right) dv_g.$$

The right-hand side is equal to $\int_M (-\mathscr{L}_g u) v \, dv_g$ for smooth enough u, and thus the Euler–Lagrange equation is the weak formulation of $\mathscr{L}_g u + \lambda u = 0$. As Δ_g, hence $\mathscr{L}_g$, is elliptic, and g is smooth, *elliptic regularity* gives that u is given by a smooth function. Thus u is an eigenfunction, and $\lambda = \lambda_1$, as desired.

We have only to show that u has a definite sign. Note that since $|\nabla_g u|_g = \left| \nabla_g |u| \right|_g$ almost everywhere, $|u| \in H^1(M, g)$ [86; 107], and so $\mathcal{G}_g(u) = \mathcal{G}_g(|u|)$. Thus $|u|$ is a minimizer, and the above argument shows that $|u| \geq 0$ is an eigenfunction. That $|u|$ cannot have a zero minimum value comes from the strong maximum principle. $\qquad\square$

In the next proposition, we show the sign of λ_1 is invariant across a conformal class. Suppose g is conformal to $\mathring{g}$, say $g = \theta^{\frac{4}{n-2}} \mathring{g}$ for $\theta > 0$. Then, for $u > 0$,

$$-\tfrac{4(n-1)}{n-2} u^{-\frac{n+2}{n-2}} \mathscr{L}_g u = R(u^{\frac{4}{n-2}} g) = R((u\theta)^{\frac{4}{n-2}} \mathring{g}) = -\tfrac{4(n-1)}{n-2} (u\theta)^{-\frac{n+2}{n-2}} \mathscr{L}_{\mathring{g}}(u\theta),$$

i.e.,

$$\mathscr{L}_g u = \theta^{-\frac{n+2}{n-2}} \mathscr{L}_{\mathring{g}}(u\theta). \tag{6.1.9}$$

This identity extends to all smooth u by a simple continuity argument about points where $u = 0$. Next we see that if $u > 0$ and $\hat{g} = u^{\frac{4}{n-2}} g = (u\theta)^{\frac{4}{n-2}} \mathring{g}$, then

$$\tfrac{4(n-1)}{n-2} \mathcal{E}_g(u) = \mathcal{R}(\hat{g}) = \tfrac{4(n-1)}{n-2} \mathcal{E}_{\mathring{g}}(u\theta). \tag{6.1.10}$$

We are now ready to summarize the above into a key proposition.

Proposition 6-30. *Suppose M^n ($n \geq 3$) is a closed connected manifold. Let $\mathcal{C}$ be a conformal class of Riemannian metrics. For $g \in \mathcal{C}$, let $\lambda_1(g)$ be the lowest eigenvalue of the conformal Laplacian $\mathscr{L}_g$. Then the sign of $\lambda_1(g)$ is the same for all $g \in \mathcal{C}$: it is positive (zero, negative) if and only if there is a metric $\hat{g} \in \mathcal{C}$ for which $R(\hat{g})$ is positive (zero, negative).*

Proof. We let $u > 0$ be a first eigenfunction of $\mathscr{L}_g$ for a metric $g \in \mathcal{C}$. If $\lambda_1(g) = 0$, and if $\mathring{g} \in \mathcal{C}$, then by (6.1.9) or (6.1.10), we see $\lambda_1(\mathring{g}) \leq 0$; similarly, if $\lambda_1(g) < 0$, then $\lambda_1(\mathring{g}) < 0$. Hence, if $\lambda_1(g) = 0$, then $\lambda_1(\mathring{g}) = 0$ too. Thus we can also conclude that if $\lambda_1(g) > 0$, then $\lambda_1(\mathring{g}) > 0$ as well.

We again let $u > 0$ be a first eigenfunction of $\mathscr{L}_g$. It follows from $R(u^{\frac{4}{n-2}} g) = -\tfrac{4(n-1)}{n-2} u^{-\frac{n+2}{n-2}} \mathscr{L}_g u = \tfrac{4(n-1)}{n-2} \lambda_1(g) u^{-\frac{4}{n-2}}$ that for any metric g, there is a conformally related metric $\hat{g} = u^{\frac{4}{n-2}} g$ with scalar curvature whose sign agrees with that of $\lambda_1(g)$. We have only left to show that if $\mathring{g}$ is conformal to g, so that $R(\mathring{g})$ has a definite sign (or is identically zero), then the sign is the same as that of $\lambda_1(g)$. To see this, if $g = \theta^{\frac{4}{n-2}} \mathring{g}$, and if $u > 0$ is a first eigenfunction of $\mathscr{L}_g$, then we have

$$\Delta_{\mathring{g}}(u\theta) - \tfrac{n-2}{4(n-1)} u\theta R(\mathring{g}) = \mathscr{L}_{\mathring{g}}(u\theta) = \theta^{\frac{n+2}{n-2}} \mathscr{L}_g u = -\lambda_1(g)\theta^{\frac{n+2}{n-2}} u.$$

If $\lambda_1(g) > 0$, then by considering a point p where $u\theta > 0$ has a minimum value on M (and hence $\Delta_{\mathring{g}}(u\theta)|_p \geq 0$), we see $R(\mathring{g})|_p > 0$. Similar consideration applies if $\lambda_1(g) < 0$. If $\lambda_1(g) = 0$, then $\Delta_{\mathring{g}}(u\theta) = \tfrac{n-2}{4(n-1)} u\theta R(\mathring{g})$. By considering points where $u\theta$ obtains a maximum and minimum on M, respectively, we see that since $R(\mathring{g})$ has a definite sign, then $R(\mathring{g})$ must be identically zero. $\qquad\square$

As a direct corollary, the set of Riemannian metrics on a closed and connected manifold can be written as a disjoint union $\mathcal{Y}^+ \cup \mathcal{Y}^0 \cup \mathcal{Y}^-$ of *Yamabe classes*, where $\mathcal{Y}^+$ is the set of all metrics which are conformally related to a metric with positive scalar curvature, and analogously for $\mathcal{Y}^0$ and $\mathcal{Y}^-$. Each of these three sets is a union of conformal classes. We note that $\mathcal{Y}^+$ and $\mathcal{Y}^0$ might in fact be empty, whereas $\mathcal{Y}^-$ is always nonempty (see [22, p. 123-4], for example). For example, if $M = \mathbb{T}^3$ is the three-torus, $\mathcal{Y}^+$ is empty [198] (Theorem 6-56 below), and if M is a compact hyperbolic manifold, then $\mathcal{Y}^+ \cup \mathcal{Y}^0$ is empty (see [139, Corollary IV.5.6], for instance), and similarly $\mathcal{Y}^+ \cup \mathcal{Y}^0$ is empty for $\mathbb{T}^3 \# \mathbb{T}^3$ (see comment in Section 6.3.1).

Before we move back to the constraint equations, we note a simple application of the method of super- and subsolutions.

Theorem 6-31. *Suppose $(M, \mathring{g})$ is a closed Riemannian n-manifold ($n \geq 3$). Let C be the conformal class of $\mathring{g}$, and suppose that $\mathring{g} \in \mathcal{Y}^-$. For any constant $\xi < 0$, there is a metric $g \in C$ such that $R(g) = \xi$.*

Proof. By the preceding proposition, we may assume without loss of generality that $R(\mathring{g}) < 0$. By (6.1.8), we see we want to show the following PDE has a positive solution:

$$\Delta_{\mathring{g}}\phi = \tfrac{n-2}{4(n-1)}\phi\big(R(\mathring{g}) - \phi^{\frac{4}{n-2}}\xi\big) =: F(x, \phi).$$

For $\delta > 0$ sufficiently small, $F(x, \delta) < 0$, and so $\phi_- = \delta$ is a subsolution. For $\Theta > 0$ sufficiently large, $F(x, \Theta) > 0$, so that we can choose $\phi_+ = \Theta \geq \delta = \phi_- > 0$ to be a supersolution. $\qquad\square$

From this and Proposition 6-30, we conclude that if a metric is in $\mathcal{Y}^0 \cup \mathcal{Y}^-$, there is a conformally related metric with constant scalar curvature. The $\mathcal{Y}^+$ case is *much* harder, but by the resolution of the *Yamabe problem* by R. M. Schoen (following work of Yamabe, Aubin and Trudinger, cf. [143]) each conformal class does contain a metric of constant scalar curvature.

6.2. Solving the constraint equations: the conformal method

In this section we discuss one method for producing solutions to the vacuum Einstein constraint equations, *the conformal method*, which dates back to Lichnerowicz, Choquet-Bruhat, York and Ó Murchadha [50; 52; 145; 172; 228]. The method has proved very useful in both theory and applications, and remains an active area of research. We will not try to give an up-to-date account of the latest developments or even an exhaustive list of references, but rather we will develop some of the basic formulation and results of the method, and the interested reader can find further results and references, such as [17; 64; 153; 154].

In terms of parametrizing the space of solutions (g, K) to the Einstein constraint equations, the conformal method has proved successful in the *constant mean curvature* (CMC) case, i.e., in case $\text{tr}_g K$ is constant.[5] The non-CMC case is not completely understood, with some near-CMC results, and a number of recent results which cast doubts on how successful the conformal method can be for parametrizing the moduli space of solutions to the constraints. We will set up the conformal method, and then study the CMC regime, following Isenberg's treatment [127], which unified a number of earlier results and completed the analysis in the CMC case.

The basic idea is to prescribe *part* of the initial data (g, K) as *free data*, and solve for the other components. For instance, the Riemannian metric g is not prescribed, but rather its *conformal class* is, and the method will involve solving for the metric in the class. If $\mathring{g}$ is a metric in the conformal class, the method will determine a function $\phi > 0$ such that $g = \phi^{\frac{4}{n-2}} \mathring{g}$ will be the desired metric. While this part was simple enough to describe, it takes more finesse to understand how to assemble the symmetric tensor K, part of which will be freely prescribed, with the remainder to be determined by the constraints. The construction of K will be motivated below with a discussion of the transverse-traceless (TT) decomposition of symmetric tensors. To establish rigorously the TT decomposition and to solve the equations that arise from the conformal formulation of the Einstein constraints, we will draw on our above tour of elliptic PDE.

6.2.1. *Transverse-traceless (TT) decomposition.* We now show how to write a symmetric $(0, 2)$-tensor Ψ on a closed Riemannian manifold (M^n, g), $n \geq 2$,

---

[5]See [153; 154] for a discussion of the conformal method outside the CMC regime.

as a sum of three terms, $\Psi = \Psi^{TT} + \Psi^{L} + \Psi^{Tr}$. Here $\Psi^{Tr} = \frac{1}{n}(\mathrm{tr}_g \Psi) g$ is a pure-trace term, while Ψ^{TT} is *transverse* (divergence-free) and *traceless* (trace-free). Hence we see that the *longitudinal part* Ψ^{L} must be trace-free. We give an ansatz for Ψ^{L} as $\Psi^{L} = L_g W$, where L_g is the *conformal Killing operator* $(L_g W)_{ab} = W_{a;b} + W_{b;a} - \frac{2}{n}(\mathrm{div}_g W) g_{ab}$ (the indices are raised and lowered using g, $W_a = g_{ab} W^b$). You will note the conformal Killing operator is the trace-free part of the *Lie derivative* $(L_W g)_{ab} = W_{a;b} + W_{b;a}$. Sometimes $L_g W$ is written as a $(2, 0)$-tensor, so that

$$(L_g W)^{ab} = g^{ac} g^{bd} (L_g W)_{cd} = g^{bd} W^a_{\;;d} + g^{ac} W^b_{\;;c} - \frac{2}{n}(\mathrm{div}_g W) g^{ab}.$$

Recall that a vector field W generates a flow $\varphi_t : M \to M$ by $\frac{d}{dt}\big|_{t=0}\varphi_t(p) = W(p)$; a vector field W is a *Killing field* for g if the flow it generates preserves g, i.e., if each φ_t is an isometry, whereas W is a *conformal Killing field* (CKV) if φ_t preserves the conformal class of g. It is well-known (see for instance [174; 218, Appendix C]) that W is a Killing field if and only if $L_W g = 0$, whereas W is a CKV if and only if $L_g W = 0$.

As $L_g W$ is already trace-free, we need only impose the divergence-free condition on Ψ^{TT}, $\mathrm{div}_g (\Psi - \Psi^{L} - \Psi^{Tr}) = 0$, to obtain an equation for W:

$$\mathrm{div}_g(L_g W) = \mathrm{div}_g (\Psi - \Psi^{Tr}). \tag{6.2.1}$$

We let $\mathcal{P}_g = \mathrm{div}_g \circ L_g$, and we write $\mathcal{P}_g(W) = \mathrm{div}_g (L_g W)$ in components as

$$(\mathcal{P}_g W)_a = (L_g W)_{ab;c} g^{bc} = W_{a;bc} g^{bc} + W_{b;ac} g^{bc} - \frac{2}{n} W^\ell_{\;;\ell c} g_{ab} g^{bc}. \tag{6.2.2}$$

We wish to commute the covariant derivatives on the second term, for which we use the *Ricci formula*.

Lemma 6-32 (Ricci formula). *If X is a vector field on a semi-Riemannian manifold (M, g), then*

$$X^i_{\;;jk} - X^i_{\;;kj} = X^\ell R^i_{kj\ell}, \tag{6.2.3}$$

so that for a one-form α,

$$\alpha_{i;jk} - \alpha_{i;kj} = \alpha_\ell R^\ell_{jki}. \tag{6.2.4}$$

Proof. For vector fields X, Y and Z and one-form field θ,

$$\nabla(\nabla X)(\theta, Y, Z) - \nabla_Z(\nabla X)(\theta, Y)$$
$$= \nabla_Z(\nabla_Y X(\theta)) - \nabla X(\nabla_Z \theta, Y) - \nabla X(\theta, \nabla_Z Y)$$
$$= (\nabla_Z(\nabla_Y X))(\theta) + \nabla_Y X(\nabla_Z \theta) - \nabla X(\nabla_Z \theta, Y) - \nabla X(\theta, \nabla_Z Y)$$
$$= (\nabla_Z(\nabla_Y X))(\theta) - \nabla_{\nabla_Z Y} X(\theta).$$

Therefore we have

$$\nabla(\nabla X)(\theta, Y, Z) - \nabla(\nabla X)(\theta, Z, Y)$$
$$= (\nabla_Z(\nabla_Y X))(\theta) - \nabla_{\nabla_Z Y} X(\theta) - (\nabla_Y(\nabla_Z X))(\theta) + \nabla_{\nabla_Y Z} X(\theta)$$
$$= R(Z, Y, X)(\theta)$$

where we used the fact that $\nabla_Z Y - \nabla_Y Z = [Z, Y]$.

For the one-form version, we note that because $\nabla g = 0$, raising and lowering commute with covariant differentiation, and we use symmetries of the curvature tensor to deduce $R_{kj\ell i} = -R_{jk\ell i} = -R_{\ell ijk} = R_{i\ell jk} = R_{jki\ell}$. $\qquad\square$

We apply this in (6.2.2) to obtain

$$(\mathscr{P}_g W)_a = W_{a;bc} g^{bc} + (W_{b;ca} + W_\ell R^\ell_{acb}) g^{bc} - \tfrac{2}{n} W^\ell_{;\ell c} g_{ab} g^{bc}$$
$$= W_{a;bc} g^{bc} + \tfrac{n-2}{n}(\operatorname{div}_g W)_{;a} + W_\ell R^\ell_{acb} g^{bc}.$$

The operator $\mathscr{P}_g$ enjoys several integral properties, which we collect in the next proposition. For simplicity of exposition, we assume the vector and tensor fields are smooth, though clearly less regularity is needed.

Proposition 6-33. *Let Z and W be vector fields on a closed Riemannian manifold (M, g). For any symmetric and trace-free $(0, 2)$-tensor field S,*

$$\langle Z, \operatorname{div}_g S\rangle_{L^2(dv_g)} = -\tfrac{1}{2}\langle L_g Z, S\rangle_{L^2(dv_g)},$$

from which follows that

$$\langle W, \mathscr{P}_g W\rangle_{L^2(dv_g)} = -\tfrac{1}{2}\|L_g W\|^2_{L^2(dv_g)},$$
$$\langle Z, \mathscr{P}_g W\rangle_{L^2(dv_g)} = \langle \mathscr{P}_g Z, W\rangle_{L^2(dv_g)}.$$

Thus $\mathscr{P}_g$ is formally self-adjoint.

Proof. We apply integration by parts (divergence theorem) to obtain

$$\langle Z, \operatorname{div}_g S\rangle_{L^2(dv_g)} = \int_M Z^a S_{ab;c} g^{bc}\, dv_g = -\int_M Z^a_{;c} S_{ab} g^{bc}\, dv_g$$
$$= -\tfrac{1}{2}\int_M (Z_{a;b} + Z_{b;a}) S^{ab}\, dv_g$$
$$= -\tfrac{1}{2}\int_M (L_g Z)_{ab} S^{ab}\, dv_g = -\tfrac{1}{2}\langle L_g Z, S\rangle_{L^2(dv_g)},$$

where in the last line we used that S is trace-free, and in the previous line we used symmetry. Applying this with $S = L_g W$ we obtain $\langle Z, \mathscr{P}_g W\rangle_{L^2(dv_g)} = -\tfrac{1}{2}\langle L_g Z, L_g W\rangle_{L^2(dv_g)}$, which shows we can switch the roles of Z and W, to yield the self-adjoint property. If $Z = W$, we obtain $\langle W, \mathcal{P}_g W \rangle_{L^2(dv_g)} = -\frac{1}{2}\|L_g W\|^2_{L^2(dv_g)}$. $\qquad\qquad\square$

We note an immediate corollary, recalling from above that a conformal Killing field (CKV) W solves $L_g W = 0$.

Corollary 6-34. *On a closed Riemannian manifold (M, g), the image of $\mathcal{P}_g$ is $L^2(dv_g)$-orthogonal to its kernel. The kernel of $\mathcal{P}_g$ is precisely the space of CKV fields, each of which is $L^2(dv_g)$-orthogonal to the divergence of any symmetric trace-free tensor field S.*

We want to apply the Fredholm alternative/Hodge decomposition, for which we show the operator $\mathcal{P}_g$ is *elliptic*. Indeed, the principal symbol σ of $\mathcal{P}_g$ at $\xi \in T_p^* M$ is a linear transformation $\sigma(\xi) : T_p M \to T_p M$ satisfying (recall the factor $i^2 = -1$)

$$-\sigma(\xi) : V \mapsto V^a \xi_b \xi_c g^{bc} + \tfrac{n-2}{n} V^\ell \xi_\ell \xi_c g^{ac} = |\xi|_g^2 V + \tfrac{n-2}{n}\xi(V)\xi^\sharp. \qquad (6.2.5)$$

If $\sigma(\xi)(V) = 0$, we apply ξ to the equation to get $|\xi|_g^2 \xi(V) + \tfrac{n-2}{n}\xi(V)|\xi|_g^2 = 0$. For $\xi \neq 0$ (and $n \geq 2$), we obtain $\xi(V) = 0$, which together with (6.2.5) implies $V = 0$. Thus for each $\xi \in T_p^* M \setminus \{0\}$, $\sigma(\xi)$ is an isomorphism, which is the definition of ellipticity.

We saw above that the image of $\mathcal{P}_g$ is orthogonal to the kernel; in fact the image is complementary to the kernel. Indeed, from (6.1.5)–(6.1.7), the Fredholm alternative for an elliptic operator like $\mathcal{P}_g$ gives a Hodge decomposition of a suitable function space, either a Banach space of vector fields with Sobolev or Hölder regularity, or by elliptic regularity, a splitting of the space of smooth vector fields, as an L^2-orthogonal direct sum $\ker \mathcal{P}_g^* \oplus \operatorname{ran} \mathcal{P}_g$. Now, because $\mathcal{P}_g = \mathcal{P}_g^*$ (self-adjointness) this becomes $\ker \mathcal{P}_g \oplus \operatorname{ran} \mathcal{P}_g$.

Corollary 6-35. *For any symmetric $(0, 2)$-tensor Ψ on a closed Riemannian manifold (M, g), there is a vector field W such that $\mathcal{P}_g W = \operatorname{div}_g(\Psi - \Psi^{Tr})$, and W is determined up to the addition of a CKV.*

Proof. By Corollary 6-34, if $S = \Psi - \Psi^{Tr}$, then $\operatorname{div}_g S$ is $L^2(dv_g)$-orthogonal to $\ker \mathcal{P}_g$. By the Fredholm alternative, then, $\operatorname{div}_g S$ must be in the image of $\mathcal{P}_g$. $\qquad\square$

Recalling (6.2.1), we summarize as follows.

Proposition 6-36. *Any symmetric $(0, 2)$-tensor Ψ on a closed Riemannian manifold (M, g) can be uniquely decomposed into the $L^2(dv_g)$-orthogonal sum $\Psi = \Psi^{TT} + \Psi^L + \Psi^{Tr}$.*

Proof. Given that we solved (6.2.1), this gives the existence of the decomposition. By the preceding corollary, the vector field W giving $\Psi^L = L_g W$ is determined up to addition of a CKV, which plainly does not affect the term Ψ^L.   □

We note that we could have done the analysis for a symmetric $(2, 0)$-tensor, and the resulting TT decomposition is obtained from that of the corresponding metrically equivalent $(0, 2)$-tensor. Furthermore, while we assumed M to be closed, there are analogous results for certain noncompact manifolds, such as in suitable weighted spaces on (M, g) asymptotically flat (see [39], for example).

6.2.2. *The conformal data.* We now define the *conformal data* on an n-manifold ($n \geq 3$). The essential idea is to prescribe certain parts of g and K, using the idea of the decomposition we have just seen above, and to solve for the remaining parts to constitute g and K satisfying the constraint equations. In fact we specify the metric up to a conformal factor, prescribing a Riemannian metric $\mathring{g}$ such that g will be in the conformal class of metrics $\mathcal{C}$ containing $\mathring{g}$. In light of the decomposition of symmetric tensors we have discussed, we also prescribe a symmetric TT (with respect to $\mathring{g}$) tensor σ along with a scalar function τ, as giving part of the tensor K (up to conformal rescaling). We will specify K in terms of this data and derive the corresponding form of the constraint equations, but first we note some useful conformal identities.

6.2.2.1. *Some conformal identities.* We collect here some identities enjoyed by metrics g and $\mathring{g}$ conformally related by $g = \phi^q \mathring{g}$, where $\phi > 0$, with Levi-Civita connections ∇ and $\mathring{\nabla}$. The computations collected in this section are relatively straightforward, with some care. We start with a simple but useful formula.

Exercise 6-37. Recall that the difference $T := \nabla - \mathring{\nabla}$ in two connections is tensorial, and show that $T_{ij}^k = \frac{q}{2}\big(\delta_j^k (d\log\phi)_i + \delta_i^k (d\log\phi)_j - \mathring{g}_{ij}(\mathrm{grad}_{\mathring{g}}\log\phi)^k\big)$. Use this to show that if X is a vector field, then $\mathrm{div}_g(\phi^{-qn/2}X) = \phi^{-qn/2}\mathrm{div}_{\mathring{g}}X$.

Recall that conformal Killing vector (CKV) fields are those fields whose flow preserves the conformal class of the given metric. Thus a vector field W is a CKV for g if and only if it is a CKV for $\mathring{g}$. This can also be seen through the identity

$$(L_g W)^{ab} = \phi^{-q}(L_{\mathring{g}} W)^{ab}. \tag{6.2.6}$$

Exercise 6-38. Prove the preceding identity. Leave indices on W up.

Remark 6-39. One must take care when proving identities for conformally related metrics, in terms of raising and lowering indices. For example, if W is a given vector field, we can lower the index to get a one-form using either metric, which generally would be called "W" with indices down when only one metric is under consideration. For instance, if we write

$$(L_g W)_{ab} = W_{a;b} + W_{b;a} - \tfrac{2}{n} W^c_{\;;c}\, g_{ab} = g_{ac} g_{bd} (L_g W)^{cd},$$

we have $W_a = g_{ab} W^b$, whereas if we write the same formula with each g replaced by $\mathring{g}$, then $W_a = \mathring{g}_{ab} W^b$ (and of course the semicolon would now mean covariant differentiation using $\mathring{\nabla}$). Suppose to be clear we let $\mathring{\omega}_u = \mathring{g}_{ab} W^b$ and $\omega_a = g_{ab} W^b = \phi^q \mathring{g}_{ab} W^b = \phi^q \mathring{\omega}_a$, and for a one-form θ we let $(L_g \theta)_{ab} = \theta_{a;b} + \theta_{b;a} - \tfrac{2}{n} \theta_{c;d} g^{cd} g_{ab}$. Then using the identity (6.2.6), we get $(L_g W)_{ab} = \phi^q (L_{\mathring{g}} W)_{cd}$, i.e.,

$$(L_g \omega)_{ab} = g_{ac} g_{bd} (L_g W)^{cd} = \phi^q \mathring{g}_{ac} \mathring{g}_{bd} (L_{\mathring{g}} W)^{cd} = \phi^q (L_{\mathring{g}} \mathring{\omega})_{cd}.$$

We now turn to TT tensors in conformally related metrics. If σ is a symmetric $(0, 2)$-tensor on M which is trace-free with respect to $\mathring{g}$, then it is trace-free with respect to any $g \in \mathcal{C}$. The corollary following the next lemma says that an analogue holds if σ is TT with respect to $\mathring{g}$; note that the equality below is a one-form identity (indices down).

Lemma 6-40. *Let $g = \phi^q \mathring{g}$. For any symmetric $(0, 2)$-tensor S on $(M, \mathring{g})$, and for any $\xi \in \mathbb{R}$,*

$$\phi^{q-\xi} (\mathrm{div}_g (\phi^\xi S))_a$$
$$= (\mathrm{div}_{\mathring{g}} S)_a - \tfrac{q}{2} (\mathrm{tr}_{\mathring{g}} S)(d \log \phi)_a + \left(\xi - \tfrac{q}{2}(2 - n)\right) S_{ab} (d \log \phi)_c \mathring{g}^{bc}.$$

As we have seen (Proposition 6-27), we find it convenient to use $q = \frac{4}{n-2}$. With this choice of q, the following corollary is immediate.

Corollary 6-41. *Let $g = \phi^q \mathring{g}$ with $q = \frac{4}{n-2}$. For any trace-free symmetric $(0, 2)$-tensor σ on $(M, \mathring{g})$, we have the one-form identity $\mathrm{div}_g (\phi^{-2} \sigma) = \phi^{-2-q} \mathrm{div}_{\mathring{g}} \sigma$. Thus if σ is TT for $\mathring{g}$, then $\phi^{-2} \sigma$ is TT for g.*

Exercise 6-42. Prove Lemma 6-40. You might find Exercise 6-37 helpful.

There is an analogous formula for trace-free symmetric $(2, 0)$-tensors: if Ξ is a trace-free symmetric $(2, 0)$-tensor, then the following *vector* equation (indices up) holds, again with $q = \frac{4}{n-2}$: $\mathrm{div}_g (\phi^{-2-2q} \Xi) = \phi^{-2-2q} \mathrm{div}_{\mathring{g}} \Xi$.

Exercise 6-43. Prove the preceding formula, based on the corresponding $(0, 2)$-formula. You might let $S_{cd} = \Xi^{ab} g_{ac} g_{bd}$ and $\mathring{S}_{cd} = \Xi^{ab} \mathring{g}_{ac} \mathring{g}_{bd}$ be the corresponding trace-free $(0, 2)$-tensors for Ξ. Note that $S = \phi^{2q} \mathring{S}$, and (as either form or vector identities) $\mathrm{div}_g \Xi = \mathrm{div}_g S$ and $\mathrm{div}_{\mathring{g}} \Xi = \mathrm{div}_{\mathring{g}} \mathring{S}$. Use the preceding lemma to compute $\mathrm{div}_g (\phi^{-2-2q} \Xi)$ (take care when raising and lowering indices).

6.2.2.2. *The constraint equations.* We will now determine the data g and K in terms of the conformal data on M. Fix $q = \frac{4}{n-2}$ for the rest of the chapter. We specify a metric $\overset{\circ}{g}$ to be in the conformal class of g, a symmetric TT (for $\overset{\circ}{g}$) tensor σ, along with a scalar function τ. We can construe σ to be $(0,2)$ or $(2,0)$, depending on how we want to present K (indices up or down). Given this data, construct the *physical data* (g, K) as follows: for $\phi > 0$ and W a vector field, let

$$g = \phi^q \overset{\circ}{g} \quad \text{and} \quad K = \phi^{-2}(\sigma + L_{\overset{\circ}{g}} W) + \tfrac{\tau}{n}\phi^q \overset{\circ}{g},$$

where g and K are $(0,2)$-tensors (indices down). Note that $\operatorname{tr}_g K = \tau$. If we let the indices on K and σ be up, the corresponding formulation would be $K^{cd} = \phi^{-2-2q}(\sigma^{cd} + (L_{\overset{\circ}{g}} W)^{cd}) + \tfrac{\tau}{n}\phi^{-q}\overset{\circ}{g}{}^{cd}$.

We now want to evaluate the constraint map (with $\Lambda = 0$)

$$\Phi(g, K) = \left(R(g) - |K|_g^2 + (\operatorname{tr}_g K)^2, \operatorname{div}_g K - d(\operatorname{tr}_g K)\right)$$

for this conformally constituted data. Note that $|K|_g^2 = \phi^{-4-2q}|\sigma + L_{\overset{\circ}{g}} W|_{\overset{\circ}{g}}^2 + \frac{\tau^2}{n}$ (the trace-free part is pointwise orthogonal to the pure trace part), so that using (6.1.8) yields

$$R(g) - |K|_g^2 + (\operatorname{tr}_g K)^2$$
$$= -\tfrac{4(n-1)}{n-2}\phi^{-\frac{n+2}{n-2}}\left(\Delta_{\overset{\circ}{g}}\phi - \tfrac{n-2}{4(n-1)} R(\overset{\circ}{g})\phi\right) - \phi^{-4-2q}|\sigma + L_{\overset{\circ}{g}} W|_{\overset{\circ}{g}}^2 + \tfrac{n-1}{n}\tau^2.$$

Using Corollary 6-41 with $S = (\sigma + L_{\overset{\circ}{g}} W)$, we have

$$\operatorname{div}_g (\phi^{-2}(\sigma + L_{\overset{\circ}{g}} W)) = \phi^{-2-q}\operatorname{div}_{\overset{\circ}{g}}(\sigma + L_{\overset{\circ}{g}} W) = \phi^{-2-q}\operatorname{div}_{\overset{\circ}{g}}(L_{\overset{\circ}{g}} W). \quad (6.2.7)$$

Since $\operatorname{div}_g (\tfrac{\tau}{n} g) = \tfrac{1}{n} d\tau$, we see that

$$\operatorname{div}_g K - d(\operatorname{tr}_g K) = \phi^{-2-q}\operatorname{div}_{\overset{\circ}{g}}(L_{\overset{\circ}{g}} W) - \tfrac{n-1}{n} d\tau.$$

Remark 6-44. We see from (6.2.7) that the power of ϕ scaling $L_{\overset{\circ}{g}} W$ in K is chosen so that there are no $d\phi$-terms in $\operatorname{div}_g K$; cf. Lemma 6-40. This gives the momentum constraint a simple form, which will be exploited below. Strictly speaking, we are employing *Method A*: indeed, there is another natural conformal rescaling (*Method B*), which gives the resulting formulation a natural conformal invariance, cf. Exercise 6-61.

Thus the vacuum constraint equations $\Phi(g, K) = 0$ can be written as follows, using $q + 1 = \frac{n+2}{n-2}$, i.e., $q + 2 = \frac{2n}{n-2}$ and $q + 3 = \frac{3n-2}{n-2}$:

$$\Delta_{\overset{\circ}{g}}\phi - \tfrac{1}{q(n-1)} R(\overset{\circ}{g})\phi + \tfrac{1}{q(n-1)}|\sigma + L_{\overset{\circ}{g}} W|_{\overset{\circ}{g}}^2 \phi^{-q-3} - \tfrac{1}{qn}\tau^2\phi^{q+1} = 0, \quad (6.2.8)$$

$$\operatorname{div}_{\overset{\circ}{g}}(L_{\overset{\circ}{g}} W) = \tfrac{n-1}{n}\phi^{q+2} d\tau. \quad (6.2.9)$$

The scalar equation (6.2.8) is known as the *Lichnerowicz equation*.

Whereas the constraint equations $\Phi(g, K) = 0$ can be construed as a nonlinear, underdetermined-elliptic system, the equations above constitute a second-order, semilinear, elliptic system for (ϕ, W). The equations are coupled in general. There are a number of known results about solving the system under certain smallness conditions on $d\tau$, which essentially make the coupling weak enough for iteration schemes to converge to a solution; there are too many results to cite here, but see [128] for a seminal such result. In fact, we will in the next section proceed to discuss the case when $d\tau = 0$, i.e., τ is constant (for M connected). In this case, the system decouples: we can take $W = 0$ and proceed to study the Lichnerowicz equation.

6.2.3. *The CMC case.*

We consider a closed manifold M of dimension $n \geq 3$. As noted above, we will discuss the case of conformal data with τ constant; as $\tau = \mathrm{tr}_g K$, this is the *constant mean curvature* (CMC) case of the conformal method. We will follow the paper of Isenberg [127], which completed the CMC case, building on works of Lichnerowicz, Choquet-Bruhat, Ó Murchadha and York. Isenberg's paper deals with the case $n = 3$, although the analysis easily extends to $n \geq 3$.

In case τ is constant, (6.2.9) becomes the linear equation $\mathrm{div}_{\mathring{g}}(L_{\mathring{g}} W) = 0$. We saw earlier that solutions to this equation are conformal Killing fields W, so that $L_{\mathring{g}} W = 0$. Thus we see that W will not in any way affect the solution K, or the Lichnerowicz equation. Thus, we might as well just choose $W = 0$ (which might be the only solution anyway). Thus in the CMC case, we are left with finding a *positive* solution $\phi > 0$ to the Lichnerowicz equation (again, $q = \frac{4}{n-2}$):

$$\Delta_{\mathring{g}}\phi = \tfrac{1}{q(n-1)} R(\mathring{g})\phi - \tfrac{1}{q(n-1)} |\sigma|_{\mathring{g}}^2 \phi^{-q-3} + \tfrac{1}{qn}\tau^2 \phi^{q+1}. \tag{6.2.10}$$

Remark 6-45. For the analysis it can be useful for $R(\mathring{g})$ to have a definite sign, and to arrange this we will change to another metric in the conformal class. Bearing this in mind, we point out that while Method B (Remark 6-44 and Exercise 6-61) enjoys a natural conformal invariance, the Lichnerowicz equation for the CMC case of Method A also enjoys a conformal invariance property: namely, given $(\mathring{g}, \sigma, \tau)$ with τ constant and σ a $(0, 2)$-tensor, for any function $\theta > 0$, the Lichnerowicz equation for $(\mathring{g}, \sigma, \tau)$ admits a solution $\phi > 0$ if and only if the Lichnerowicz equation for $(\theta^q \mathring{g}, \theta^{-2}\sigma, \tau)$ admits a solution $\phi\theta^{-1} > 0$, and these solutions lead to the same data (g, K). Indeed, by Corollary 6-41, $\theta^{-2}\sigma$ is TT for $\theta^q \mathring{g}$, so that from (6.2.8) and (6.2.9), we see $g = \phi^q \mathring{g} = (\phi\theta^{-1})^q (\theta^q \mathring{g})$ and $K = (\phi\theta^{-1})^{-2}(\theta^{-2}\sigma) + \frac{\tau}{n}(\phi\theta^{-1})^q (\theta^q \mathring{g})$.

So in discussing the solvability of (6.2.10), we are free to change the conformal data to equivalent data, which will produce the same physical data (g, K), but may have some useful property, such as constant scalar curvature, or scalar curvature with a definite sign: see Proposition 6-30 and Theorem 6-31.

6.2.3.1. *Main theorem on the CMC conformal method on closed manifolds.* We are now at long last ready to state the main result on the CMC conformal method on closed manifolds. We will follow [127], using the method of super- and subsolutions to solve the Lichnerowicz equation (6.2.10). Prior to Isenberg's work, fixed-point theorems were generally used to solve this semilinear elliptic equation, covering all but one case of the CMC conformal method (see below). Isenberg employed the barrier method (super- and subsolutions), and Maxwell [151] later gave a streamlined approach with the barrier method (and with lower regularity of the conformal data, which we do not discuss here).

Theorem 6-46 (Choquet-Bruhat, O'Murchadha, York, Isenberg). *On a closed connected manifold M^n, $n \geq 3$, let $(\mathring{g}, \sigma, \tau)$ be CMC conformal data (τ constant). The Lichnerowicz equation (6.2.10) admits a positive solution ϕ as indicated in the table below, where the indication is whether the tensor σ is identically zero. The solution is unique except in the case $(\mathring{g} \in \mathcal{Y}^0, \sigma = 0, \tau = 0)$, in which case any positive constant is a solution.*

	$\sigma = 0$ $\tau = 0$	$\sigma = 0$ $\tau \neq 0$	$\sigma \neq 0$ $\tau = 0$	$\sigma \neq 0$ $\tau \neq 0$
$\mathring{g} \in \mathcal{Y}^+$	no	no	yes	yes
$\mathring{g} \in \mathcal{Y}^0$	yes	no	no	yes
$\mathring{g} \in \mathcal{Y}^-$	no	yes	no	yes

Proof. We present a fairly complete proof, except for the uniqueness statement, for which we refer you to [127; 151]. By Remark 6-45, we can move $\mathring{g}$ within a conformal class without changing the required solvability. In particular, we may assume without further comment that a representative $\mathring{g}$ of the conformal class has been chosen so that $R(\mathring{g})$ has a definite sign (or vanishes identically). We then seek a *positive* solution to the Lichnerowicz equation (6.2.10).

The nonexistence cases are readily shown. Consider the case $\sigma = 0$, so that the equation reduces to $\Delta_{\mathring{g}} \phi = \frac{1}{q(n-1)} R(\mathring{g})\phi + \frac{1}{qn}\tau^2 \phi^{q+1}$. If $\tau \neq 0$, then since by Stokes' theorem $\int_M \Delta_{\mathring{g}} \phi \, dv_{\mathring{g}} = 0$ for M closed, we see there can be no positive solution ϕ if $R(\mathring{g}) \geq 0$. If $\tau = 0$ as well, we get $\Delta_{\mathring{g}} \phi = \frac{1}{q(n-1)} R(\mathring{g})\phi$, from which we again see there can be no positive solution $\phi > 0$ when $R(\mathring{g}) > 0$ or

$R(\mathring g) < 0$. Similar reasoning yields the other two nonexistence cases for $\sigma \neq 0$, $\tau = 0$ and $\mathring g \notin \mathcal{Y}^+$.

There is one very easy existence case, when $\sigma = 0$, $\tau = 0$ and $\mathring g \in \mathcal{Y}^0$, for which we can assume $R(\mathring g) = 0$. Then the equation reduces to $\Delta_{\mathring g}\phi = 0$, whose solutions are constants.

Consider the case $\mathring g \in \mathcal{Y}^-$, $\sigma = 0$, $\tau \neq 0$. By (6.1.8) and Theorem 6-31, we can solve

$$-\frac{4(n-1)}{n-2}\phi^{-\frac{n+2}{n-2}}\mathcal{L}_g\phi = R(\phi^q\mathring g) = -\frac{n-1}{n}\tau^2$$

for $\phi > 0$. This equation can easily be rewritten $\Delta_{\mathring g}\phi - \frac{1}{q(n-1)}R(\mathring g)\phi = \frac{1}{qn}\tau^2\phi^{q+1}$, which is the Lichnerowicz equation in this case.

For the case $\mathring g \in \mathcal{Y}^-$, $\sigma \neq 0$, $\tau \neq 0$, we follow [151]. We assume we have applied Theorem 6-31 to choose a conformal representative $\mathring g$ so that $R(\mathring g) = -\frac{n-1}{n}\tau^2$. Hence the Lichnerowicz equation becomes

$$\Delta_{\mathring g}\phi = -\frac{\tau^2}{qn}\phi - \frac{1}{q(n-1)}|\sigma|_{\mathring g}^2\phi^{-q-3} + \frac{1}{qn}\tau^2\phi^{q+1}. \tag{6.2.11}$$

We make another preliminary conformal change: consider the linear operator $S_{\mathring g}\theta := \Delta_{\mathring g}\theta - \frac{n-2}{4n}\tau^2\theta$. $S_{\mathring g}$ is self-adjoint and elliptic, and from the equation $0 = \int_M w\, S_{\mathring g}w\, dv_{\mathring g}$, we can conclude $w = 0$, so that $S_{\mathring g}$ has trivial kernel. We can thus consider the solution θ of $S_{\mathring g}\theta := \Delta_{\mathring g}\theta - \frac{n-2}{4n}\tau^2\theta = -\frac{n-2}{4(n-1)}|\sigma|_{\mathring g}^2$, which exists by the Fredholm alternative. By elliptic regularity, the solution θ is smooth, for smooth $\mathring g$ and σ. Furthermore, we observe that θ cannot have a negative minimum, so then by the strong maximum principle, $\theta > 0$. Recalling that $q+1 = \frac{n+2}{n-2}$ and $R(\mathring g) = -\frac{n-1}{n}\tau^2$, note that

$$R(\theta^q\mathring g) = -\frac{4(n-1)}{n-2}\theta^{-\frac{n+2}{n-2}}\left(\Delta_{\mathring g}\theta - \frac{n-2}{4(n-1)}R(\mathring g)\theta\right) = -\frac{2(n-1)}{n}\tau^2\theta^{-q} + \theta^{-q-1}|\sigma|_{\mathring g}^2.$$

We replace $\mathring g$ and σ in the Lichnerowicz equation (6.2.10) with $\tilde g = \theta^q\mathring g$ and $\tilde\sigma = \theta^{-2}\sigma$ (as $(0, 2)$-tensors), respectively, to obtain

$$\Delta_{\tilde g}\phi - \frac{1}{q(n-1)}\left(-\frac{2(n-1)}{n}\tau^2\theta^{-q} + \theta^{-q-1}|\sigma|_{\mathring g}^2\right)\phi - \frac{1}{q(n-1)}|\tilde\sigma|_{\tilde g}^2\phi^{-q-3} + \frac{1}{qn}\tau^2\phi^{q+1}.$$

Since $|\sigma|_{\mathring g}^2 = \theta^{2q+4}|\tilde\sigma|_{\tilde g}^2$, this becomes

$$\Delta_{\tilde g}\phi = -\frac{2}{qn}\tau^2\theta^{-q}\phi + \frac{1}{q(n-1)}\theta^{q+3}|\tilde\sigma|_{\tilde g}^2\phi - \frac{1}{q(n-1)}|\tilde\sigma|_{\tilde g}^2\phi^{-q-3} + \frac{1}{qn}\tau^2\phi^{q+1},$$

which we write as $\Delta_{\tilde g}\phi = F(x, \phi)$. We will find constants $0 < \phi_- \leq \phi_+$ to serve as sub- and supersolutions. Choose $\phi_- > 0$ so that $\phi_-^{q+4} \leq \min_M \theta^{-q-3}$ and $\phi_-^q \leq \min_M 2\theta^{-q}$. Then a little algebra shows $F(x, \phi_-) \leq 0$ as desired. Likewise, choosing ϕ_+ so that $\phi_+^{q+4} \geq \max_M \theta^{-q-3}$ and $\phi_+^q \geq \max_M 2\theta^{-q}$, we have $\phi_+ \geq \phi_-$, and $F(x, \phi_+) \geq 0$. Thus we can solve the Lichnerowicz equation;

compare the equation we solved to (6.2.11) to see how the preliminary step added a term that allowed a simple choice of super- and subsolutions.

For the three cases with $\lambda_1(\mathring{g}) \geq 0$ and $\sigma \neq 0$, we again proceed following Maxwell [151]. We arrange by conformal invariance to work with $\mathring{g}$ in the conformal class with $R(\mathring{g}) > 0$ or $R(\mathring{g}) = 0$, and note that $R(\mathring{g}) + \frac{n-1}{n}\tau^2 > 0$ for these three cases. We now define an operator $S_{\mathring{g}}$ analogous to that above:

$$S_{\mathring{g}} := \Delta_{\mathring{g}} - \frac{n-2}{4(n-1)}\left(R(\mathring{g}) + \frac{n-1}{n}\tau^2\right) = \mathscr{L}_{\mathring{g}} - \frac{n-2}{4n}\tau^2.$$

Again $S_{\mathring{g}}$ is linear, elliptic, self-adjoint, and has trivial kernel. Thus, by the Fredholm alternative and elliptic regularity, we can solve the following with u smooth:

$$S_{\mathring{g}}u = -\frac{n-2}{4(n-1)}|\sigma|_{\mathring{g}}^2.$$

At a minimum point p for u, $\Delta_{\mathring{g}}u|_p \geq 0$, so that by the strict positivity of $(R(\mathring{g}) + \frac{n-1}{n}\tau^2)$, $u \geq 0$. As we did earlier, we can apply the strong maximum principle to conclude $u > 0$. We use conformal invariance again, and transform the conformal data to $\tilde{g} = u^q \mathring{g}$, $\tilde{\sigma} = u^{-2}\sigma$ (as $(0,2)$-tensors), and τ. Note that

$$R(\tilde{g}) = -\frac{4(n-1)}{n-2}u^{-\frac{n+2}{n-2}}\mathscr{L}_{\mathring{g}}u = -\frac{4(n-1)}{n-2}u^{-\frac{n+2}{n-2}}\left(\frac{n-2}{4n}\tau^2 u - \frac{n-2}{4(n-1)}|\sigma|_{\mathring{g}}^2\right)$$
$$= -u^{-(q+1)}\left(\frac{n-1}{n}\tau^2 u - |\sigma|_{\mathring{g}}^2\right)$$
$$= |\tilde{\sigma}|_{\mathring{g}}^2 u^{q+3} - \frac{n-1}{n}\tau^2 u^{-q}.$$

We write the Lichnerowicz equation for $(\tilde{g}, \tilde{\sigma}, \tau)$ in the form

$$\Delta_{\tilde{g}}\phi = \frac{n-2}{4(n-1)}\left(R(\tilde{g})\phi - |\tilde{\sigma}|_{\mathring{g}}^2\phi^{-q-3} + \frac{n-1}{n}\tau^2\phi^{q+1}\right)$$
$$= \frac{1}{q(n-1)}\left(|\tilde{\sigma}|_{\mathring{g}}^2(u^{q+3}\phi - \phi^{-q-3}) + \frac{n-1}{n}\tau^2\phi(\phi^q - u^{-q})\right) =: F(x, \phi).$$

Now we will again find constants $0 < \phi_- \leq \phi_+$ to be sub- and supersolutions. Choose $\phi_- > 0$ so that $\phi_-^{q+4} \leq \min_M u^{-q-3}$ and $\phi_-^q \leq \min_M u^{-q}$. Likewise, choosing ϕ_+ so that $\phi_+^{q+4} \geq \max_M u^{-q-3}$ and $\phi_+^q \geq \max_M u^{-q}$, we have $\phi_+ \geq \phi_-$, and a little algebra shows $F(x, \phi_-) \leq 0$ and $F(x, \phi_+) \geq 0$ as before. Thus we can solve the Lichnerowicz equation. $\square$

Remark 6-47. Isenberg [127] used the resolution of the Yamabe problem to handle some cases in $\mathcal{Y}^+$, in particular the case when $\sigma \neq 0$ but $\tau = 0$, which was the one case not handled by previous works using fixed-point methods. We briefly note Isenberg's approach to this case. For simplicity, we work (as he did) in case $n = 3$, so that $q = 4$. We use the resolution of the Yamabe problem to pick a conformal representative $\mathring{g}$ with $R(\mathring{g}) = 8$. Then the Lichnerowicz equation can be written $\Delta_{\mathring{g}}\phi = \phi - \frac{1}{8}|\sigma|_{\mathring{g}}^2\phi^{-7} =: F(x, \phi)$. If σ *never* vanishes, then appropriate sub- and supersolutions are given by

$$\phi_- := \sqrt[8]{\tfrac{1}{8} \min_M |\sigma|_{\mathring{g}}^2}, \qquad \phi_+ := \sqrt[8]{\tfrac{1}{8} \max_M |\sigma|_{\mathring{g}}^2}.$$

For the general case, however, one has to be more clever to pick the subsolution. Following [127], define $\phi_+ = \max\left(1, \tfrac{1}{8}\max_M |\sigma|_{\mathring{g}}^2\right) \geq 1$. It is easy to check ϕ_+ is a supersolution. Just as in cases above, the operator $S_{\mathring{g}} = \Delta_{\mathring{g}} - 1$ is self-adjoint elliptic with trivial kernel, so we can let ϕ_- be the solution to $S_{\mathring{g}}(\phi_-) = -\tfrac{1}{8}\phi_+^{-7}|\sigma|_{\mathring{g}}^2$. Again, we see ϕ_- cannot have a negative minimum, so $\phi_- > 0$ by the strong maximum principle. Furthermore, since $\tfrac{1}{8}\phi_+^{-7}|\sigma|_{\mathring{g}}^2 \leq \phi_+$, we have $(\Delta_{\mathring{g}} - 1)(\phi_-) = S_{\mathring{g}}(\phi_-) \geq -\phi_+$, from which we see that the maximum of ϕ_- cannot be greater than ϕ_+, i.e., $\phi_- \leq \phi_+$. Finally, $\Delta_{\mathring{g}}\phi_- = \phi_- - \tfrac{1}{8}\phi_+^{-7}|\sigma|_{\mathring{g}}^2 \geq \phi_- - \tfrac{1}{8}\phi_-^{-7}|\sigma|_{\mathring{g}}^2 = F(x, \phi_-)$, as desired.

6.3. Scalar curvature deformation on closed manifolds

In this section, we discuss two results on the image of the scalar curvature map on closed manifolds. We will develop in detail how to prescribe perturbations of the scalar curvature function on a closed manifold M using inverse function theorem methods, following Fischer and Marsden [89]. First we note the following theorem of Kazdan and Warner [135; 134] on the range of the scalar curvature map.

6.3.1. *The Kazdan–Warner classification.*

Theorem 6-48 (Kazdan–Warner). *Suppose M^n, $n \geq 3$ is a closed connected manifold. There are three possibilities for the set $\mathcal{S}$ of scalar curvatures of smooth Riemannian metrics on M. Indeed, a function $f \in C^\infty(M)$ is a scalar curvature of a smooth Riemannian metric on M according to one of the following three cases:*

(i) $\mathcal{S} = C^\infty(M)$.

(ii) $f \in \mathcal{S}$ *if and only if* $\{f < 0\} \neq \varnothing$.

(iii) $f \in \mathcal{S}$ *if and only if either* $\{f < 0\} \neq \varnothing$ *or f is identically zero.*

From this theorem, we immediately see that M admits a metric with positive scalar curvature (PSC) if and only if it can carry any smooth function as a scalar curvature for some metric. For an example of case (iii) above, the torus $\mathbb{T}^n$ admits a metric with zero scalar curvature (in fact a flat metric), but no metrics with positive scalar curvature: see Theorem 6-56 (cf. [198]) for the three-dimensional case, and for general dimension see [200; 205] for the Schoen–Yau minimal hypersurface approach, or see [139, Theorem IV.5.5] for the Gromov–Lawson approach via the Dirac operator. For case (ii), an example is provided by $\mathbb{T}^n \# \mathbb{T}^n$, on which, as with the torus, there is no metric of positive scalar curvature, and so by Corollary 6-53, any scalar-flat metric would be Ricci flat; however, by Bochner's theorem (see [182], for instance), a closed n-manifold with nonnegative Ricci curvature has first Betti number at most n.

The analogous result in dimension two also holds [135; 134]; in this case, the three possibilities are governed by the Euler characteristic $\chi(M)$, consistent with the Gauss–Bonnet formula.

Theorem 6-49 (Kazdan–Warner). *Suppose M is a closed connected surface. There are three possibilities for the set S of scalar curvatures of smooth Riemannian metrics on M. Indeed, a function $f \in C^\infty(M)$ is a scalar curvature of a smooth Riemannian metric on M according to one of the following three cases:*

(i) *$f \in S$ if and only if $\{f > 0\} \neq \varnothing$, in case $\chi(M) > 0$.*

(ii) *$f \in S$ if and only if either f changes sign or f is identically zero, in case $\chi(M) = 0$.*

(iii) *$f \in S$ if and only if $\{f < 0\} \neq \varnothing$, in case $\chi(M) < 0$.*

6.3.2. *The Fischer–Marsden theorem.* A fruitful technique for studying the scalar curvature on a manifold is to fix a metric g, and then consider metrics conformal to g, i.e., to consider the scalar curvature functions across the conformal class of g, as we have seen earlier. For our purposes here, we will consider the full range of metric deformations, and consider the scalar curvature map on the space of all metrics. For this we restate Lemma 2-7, recalling that a dot between tensor fields denotes the metric contraction of their tensor product, e.g., $h \cdot k = g^{i\ell} g^{jm} h_{ij} k_{\ell m}$.

Lemma 6-50 (linearization of the scalar curvature). *Let h be a symmetric $(0, 2)$-tensor such that, for some $\varepsilon > 0$, $g + th$ is a metric on M for $|t| < \varepsilon$. Then*

$$L_g h := \frac{d}{dt}\bigg|_{t=0} R(g + th) = -\Delta_g(\operatorname{tr}_g h) + \operatorname{div}_g \operatorname{div}_g h - h \cdot \operatorname{Ric}(g).$$

In case M is compact, then any sufficiently smooth h can be used in the above formula; if M were noncompact, the formula could be interpreted locally, or additional conditions might be imposed on h (such as compact support) so that $g + th$ is a metric for small $|t|$.

We also recall equation (2.4.5): we can use integration by parts to define the formal L^2-adjoint $L_g^* f$, which satisfies $\int_M h \cdot L_g^* f \, dv_g = \int_M f L_g h \, dv_g$ for all h compactly supported in the interior of M, so that

$$L_g^* f = -(\Delta_g f)g + \operatorname{Hess}_g f - f \operatorname{Ric}(g).$$

Note that $L_g L_g^*$ is a fourth-order *elliptic* operator, with principal part $(n-1)\Delta_g^2 f$, as we will show later in this section (you might try it now as an easy exercise).

We have seen the operator L_g^* before, when we introduced static vacuum metrics in Section 2.4.5. Such metrics are precisely those for which the kernel of L_g^* is nontrivial. We know from Corollary 2-45 (or by Exercise 2-52a), if (M, g) is connected and static vacuum, then M has constant scalar curvature. On a closed manifold we can say more, as in the following result from [89].

Proposition 6-51. *Suppose (M^n, g) $(n \geq 2)$ is a closed, connected Riemannian manifold, for which there is a nontrivial function $f : M \to \mathbb{R}$ with $L_g^* f = 0$. Then the scalar curvature is a nonnegative constant, the function f is smooth and is an eigenfunction of the Laplacian Δ_g. In case $R(g) = 0$, we have that $\mathrm{Ric}(g) = 0$.*

Proof. We have already recalled that the scalar curvature is constant. Tracing the equation $L_g^* f = 0$ we obtain $-(n-1)\Delta_g f - R(g) f = 0$, which shows f is an eigenfunction with eigenvalue $\frac{R(g)}{n-1}$, which must be nonnegative by Exercise 1-11e. In case $R(g) = 0$, then, $\Delta_g f = 0$, so that f is a constant, which by scaling we can take to be $f = 1$. Since $L_g^*(1) = -\mathrm{Ric}(g)$, the claim follows. $\square$

We stress that being static vacuum is a very special condition on a metric g, and that most metrics are not static. Careful formulations of this and related results about the existence of *Killing initial data* (analogous kernel elements in the case of the Einstein constraints, cf. Section 5.3.3.2) are established in [19].

We now state a scalar curvature deformation result for metrics which are not static vacuum [89]. For a nonnegative integer k, we let $\|\cdot\|_k$ be the norm for either a Sobolev space $W^{k,p}(M)$ for some $p \in (1, \infty)$, or for a Hölder space $C^{k,\alpha}(M)$ for some $\alpha \in (0, 1)$. For the Sobolev case of the next result, we use $k + 2 > \frac{n}{p}$ for the existence, and while we take $k > \frac{n}{p}$ for higher regularity for convenience, the interested reader can check that $k + 1 > \frac{n}{p}$ suffices. We recall that we take g to be smooth.

Theorem 6-52 (Fischer–Marsden). *Suppose (M^n, g), for $n \geq 2$, is a closed Riemannian manifold for which L_g^* has trivial kernel. There is an $\varepsilon > 0$ and $C > 0$ such that for all $S \in C^\infty(M)$ with $\|S\|_k < \varepsilon$, there a smooth symmetric $(0, 2)$-tensor field h such that $g + h$ is a smooth Riemannian metric with $R(g + h) - R(g) + S$, and $\|h\|_{k+2} \leq C\|S\|_k$. In fact, there is such an h of the form $h = L_g^* f$ for some $f \in C^\infty(M)$ with $\|f\|_{k+4} \leq C\|S\|_k$.*

The proof uses the implicit function theorem along with some elliptic PDE theory. We sketch the main ideas of the proof below. We first observe the following direct corollary of Theorem 6-52 and Proposition 6-51, attributed to

J.-P. Bourguignon [136], which we will utilize in the discussion of positive scalar curvature and the positive energy theorem.

Corollary 6-53. *Suppose (M, g) is a closed connected Riemannian manifold with nonnegative scalar curvature. Either* $\mathrm{Ric}(g) = 0$ *or M admits a metric with positive scalar curvature.*

As we noted above, the torus $\mathbb{T}^n$ does not admit a metric with positive scalar curvature, so that any zero scalar curvature metric on $\mathbb{T}^n$ must be Ricci-flat. It is not hard to show from here that the metric must actually be *flat*; see, e.g., [182, Chapters 7 and 9]. Cf. [91] and [139, Chapter IV, Section 5] for related results.

6.3.3. *Outline of the proof of Theorem 6-52.*

6.3.3.1. *Elliptic splitting/Hodge decompositions.* Let X^k denote either function space $W^{k,p}(M)$ or $C^{k,\alpha}(M)$ (for k a nonnegative integer, $1 < p < \infty$, and $\alpha \in (0, 1)$). As we saw in Section 6.1, we can use the Fredholm alternative/Hodge decomposition for the elliptic operator $L_g L_g^* : X^{k+4} \to X^k$ to split the relevant function space X^k as

$$X^k = \mathrm{ran}(L_g L_g^*) \oplus \ker(L_g L_g^*),$$

where $\mathrm{ran}(L_g L_g^*) = (L_g L_g^*)(X^{k+4})$. Since g is smooth, elliptic regularity implies that this decomposition holds in the smooth setting. We observe that $\ker(L_g L_g^*) = \ker L_g^*$, because multiplying the equation $L_g L_g^* u = 0$ by u and integrating by parts, we see $\int_M |L_g^* u|^2 dv_g = 0$. Then, since $\mathrm{ran}\, L_g$ is $L^2(dv_g)$-orthogonal to $\ker L_g^*$, the splitting becomes

$$X^k = \mathrm{ran}(L_g L_g^*) \oplus \ker L_g^* = \mathrm{ran}\, L_g \oplus \ker L_g^*. \tag{6.3.1}$$

We note that the kernel is finite-dimensional, and the range is closed.

We can also split the corresponding function spaces Y^k, $k \geq 2$, of symmetric $(0, 2)$-tensors (of class $W^{k,p}(M)$ or $C^{k,\alpha}(M)$) as (cf. [20], [22, Appendix I]) $Y^k = \mathrm{ran}\, L_g^* \oplus \ker L_g$, or more precisely,

$$Y^k = L_g^*(X^{k+2}) \oplus \{h_0 \in Y^k : L_g h_0 = 0\}. \tag{6.3.2}$$

We note that whereas for L_g^*, for which $\ker L_g^* = \ker(L_g L_g^*)$ is the kernel of an elliptic operator and thus is comprised of smooth sections, and so is k-independent, the same may not be true of the distributional kernel of L_g.

To obtain the splitting of Y^k, for $h \in Y^k$, we have $L_g h \in X^{k-2}$. As we saw with the earlier splitting of X^{k-2}, then, there is $f \in X^{k+2}$ such that $L_g h = L_g L_g^* f$. Thus $h_0 := h - L_g^* f \in Y^k \cap \ker L_g$. Thus we have an algebraic sum $Y^k = L_g^*(X^{k+2}) \oplus \{h_0 \in Y^k : L_g h_0 = 0\}$, which is direct because the summands are $L^2(dv_g)$-orthogonal. Since $L^2(dv_g)$-orthogonality is preserved under convergence in Y^k, we see that $L_g^*(X^{k+2})$ is closed; as the kernel of L_g in Y^k is closed, the direct sum is topological as well.

6.3.3.2. *Local surjectivity: the implicit function theorem.* From the splitting (6.3.1), we see that to prove Theorem 6-52, we can basically invoke the implicit function theorem [1], once we observe that the scalar curvature map is smooth in the appropriate spaces. Let Y^k and X^k be as above; as noted earlier, the norms can be defined with respect to a smooth background metric $\mathring{g}$, or via a suitable covering. Let $\mathcal{M}^k$ denote either the set $\mathcal{M}^{k,p}$ of Riemannian metrics in $W^{k,p}(M)$ or the set $\mathcal{M}^{k,\alpha}$ of Riemannian metrics in $C^{k,\alpha}(M)$. By Sobolev embedding, for $k > \frac{n}{p}$ and $0 < \alpha < \min(1, k - \frac{n}{p})$, the map $W^{k,p}(M) \hookrightarrow C^{0,\alpha}(M)$ is a continuous inclusion, and hence $\mathcal{M}^{k,p}$ is open in $W^{k,p}(M)$ and included in $\mathcal{M}^{0,\alpha}$.

In the Hölder setting with $k \geq 2$, one readily checks that the scalar curvature map $R : \mathcal{M}^{k,\alpha} \to C^{k-2,\alpha}(M)$ is a smooth map of Banach manifolds. We have the following proposition for the Sobolev setting.

Proposition 6-54. *Assume $s \geq 2$ and $s > \frac{n}{p}$. Then the scalar curvature map $R : \mathcal{M}^{s,p} \to W^{s-2,p}(M)$ is a smooth map of Banach manifolds.*

Sketch of proof. In local coordinates, the scalar curvature is given by

$$R(g) = g^{jk}\left(\Gamma_{jk,\ell}^{\ell} - \Gamma_{\ell k,j}^{\ell} + \Gamma_{jk}^{m}\Gamma_{\ell m}^{\ell} - \Gamma_{\ell k}^{m}\Gamma_{jm}^{\ell}\right).$$

The expression is polynomial in the derivatives of the metric components, with coefficients that are rational functions in g_{ij}; in fact, the second derivatives of metric components appear linearly, and the first derivatives appear quadratically; the coefficients are rational, due to the presence of $(\det g)^{-1}$ in the inverse metric and its derivative. In the Hölder regularity setting, smoothness of $g \mapsto R(g)$ is immediate. In the Sobolev setting with $s > \frac{n}{p}$, by judicious use of Sobolev embedding, one proves that $W^{s,p}(M)$ is a ring under function multiplication. The proof of this fact in [2, Theorem 5.23] can be readily adapted to show that $(\det g)^{-1} \subset W^{s,p}(M)$ and that the following multiplication maps are continuous:

$$W^{s-1,p}(M) \times W^{s-1,p}(M) \to W^{s-2,p}(M), \quad W^{s,p}(M) \times W^{s-2,p}(M) \to W^{s-2,p}(M).$$

This allows us to estimate the second derivative and the quadratic first derivative terms in the local expression for the scalar curvature map, and smoothness follows.

To illustrate, we consider the case $s = 2$. If $p > n$, then $W^{1,p}(M) \hookrightarrow C^0(M)$ by Sobolev embedding. If $1 \leq p < n$, the Sobolev embedding takes the form of a continuous inclusion $W^{1,p}(M) \hookrightarrow L^r(M)$ for any $1 \leq r \leq p^* = \frac{np}{n-p}$, while in the borderline case $p = n$, $W^{1,n}(M) \hookrightarrow L^r(M)$ for $1 \leq r < \infty$ (see Section 7.2.2.1;

cf. [2; 86; 107; 144], for more on Sobolev embedding). Now if $n > p > \frac{n}{2}$, then $\frac{n}{n-p} > 2$. Thus we can choose $2 \le s_1 < \frac{n}{n-p}$, so that $1 < s_2 < \frac{n}{n-p}$, with $s_2 = \frac{s_1}{s_1 - 1}$. Using Hölder's inequality, we obtain, for $u, v \in W^{1,p}(M)$,

$$\int_M |u|^p |v|^p \, dv_{\mathring{g}} \le \left(\int_M |u|^{s_1 p} \, dv_{\mathring{g}} \right)^{1/s_1} \left(\int_M |v|^{s_2 p} \, dv_{\mathring{g}} \right)^{1/s_2}.$$

The Sobolev embedding $W^{1,p}(M) \hookrightarrow L^{s_i p}(M)$ then allows us to estimate the product $\|uv\|_{L^p} \le \|u\|_{L^{s_1 p}} \|v\|_{L^{s_2 p}} \le C \|u\|_{W^{1,p}} \|v\|_{W^{1,p}}$. Note that we could have taken $s_1 = 2 = s_2$ above. The same argument works in the borderline case, or, using the product rule, we could show that $\|uv\|_{W^{1,\frac{n}{2}}} \le C \|u\|_{W^{1,n}} \|v\|_{W^{1,n}}$, and use the continuous embedding $W^{1,\frac{n}{2}}(M) \hookrightarrow L^{(\frac{n}{2})^*}(M) = L^n(M)$.

Thus in all cases for $2 > \frac{n}{p}$, multiplication is a continuous bilinear mapping $W^{1,p}(M) \times W^{1,p}(M) \to L^p(M)$, and likewise for $W^{2,p}(M) \times L^p(M) \to L^p(M)$. That $W^{2,p}(M)$ is a ring, with continuous multiplication, follows by the applying the product rule. $\square$

Exercise 6-55. Consider the case $s = 3 > \frac{n}{p}$ of the above argument. Prove that multiplication gives continuous maps $W^{3,p}(M) \times W^{3,p}(M) \to W^{3,p}(M)$, $W^{2,p}(M) \times W^{2,p}(M) \to W^{1,p}(M)$ and $W^{3,p}(M) \times W^{1,p}(M) \to W^{1,p}(M)$. In fact, observe that this can quickly be reduced to showing multiplication gives a continuous mapping $W^{2,p}(M) \times W^{1,p}(M) \to W^{1,p}(M)$, and prove this using Sobolev embedding in a similar manner as above.

To summarize, if L_g^* has trivial kernel, as in the setting of Theorem 6-52, then $L_g : Y^{k+2} \to X^k$ is surjective, by (6.3.1). The implicit function theorem [1, Section 2.5] now implies that the map $R : \mathcal{M}^{k+2} \to X^k$ is a surjection from a neighborhood of g in $\mathcal{M}^{k+2}$ to a neighborhood of $R(g)$ in X^k. That is, there is $\varepsilon > 0$ and $C > 0$ such that, given a function $S \in X^k$ with $\|S\|_k < \varepsilon$, there is an $h \in Y^{k+2}$ with $\|h\|_{k+2} \le C\|S\|_k$ and such that $g + h \in \mathcal{M}^{k+2}$ with $R(g + h) = R(g) + S$. In fact, by the splitting (6.3.2), h can be chosen in the form $h = L_g^* f$ for $f \in X^{k+4}$, and the estimate $\|f\|_{k+4} \le C\|S\|_k$ (adjusting C if necessary) then follows from Proposition 6-15 with $L = L_g L_g^*$ (or with $L = L_g^*$ using also Remark 6-22). This gives a localized deformation result in the finite regularity spaces. To complete the proof, we next consider higher regularity.

6.3.3.3. *Regularity.* In the setting of the preceding paragraph, we consider a smooth metric g with $\ker L_g^* = \{0\}$, a smooth S with $\|S\|_k < \varepsilon$, and $f \in X^{k+4}$ with $R(g + L_g^* f) - R(g) = S$ and $\|f\|_{k+4} \le C\|S\|_k$. To motivate why f should be smooth, for small S (and hence for small f), the left-hand side is roughly $L_g L_g^* f$, which is an elliptic operator of order four on f, with smooth coefficients.

We note that if $\sim$ denotes equality up to terms in f and its derivatives up to order three, we have

$$
\begin{aligned}
L_g L_g^* f &\sim -\Delta_g(-(n-1)\Delta_g f) + \operatorname{div}_g \operatorname{div}_g (-(\Delta_g f)g + \operatorname{Hess}_g f) \\
&\sim (n-1)\Delta_g^2 f - g^{ij} f_{;ijk\ell} g_{rs} g^{kr} g^{\ell s} + f_{;ijk\ell} g^{ik} g^{j\ell} \\
&= (n-1)\Delta_g^2 f - g^{ij} f_{;ijk\ell} g^{\ell k} + f_{;ijk\ell} g^{ik} g^{j\ell} \\
&\sim (n-1)\Delta_g^2 f,
\end{aligned}
$$

where we commuted derivatives to cancel out the last two terms, either replacing the derivatives with partial derivatives (up to $\sim$), or keeping the covariant derivatives and using the Ricci identity $\alpha_{i;jk} - \alpha_{i;kj} = \alpha_\ell R^\ell_{jki}$ applied to $\alpha = df$; in either case, the principal order term in $L_g L_g^*$ is $(n-1)\Delta_g^2$. We note that the symbol of this operator is $(n-1)|\xi|_g^4$.

With this in mind, to prove that f is smooth, we write the left-hand side of $R(g + L_g^* f) - R(g) = S$ as a quasilinear operator on f. Indeed, if we let γ be the metric $\gamma = g + L_g^* f$, then with a short computation we leave as an exercise,

$$
R(g + L_g^* f) - R(g) = \gamma^{i\ell} \gamma^{jk} g^{ub} (-g_{j\ell} f_{,abik} + g_{jk} f_{,abi\ell}) + \mathcal{S}(g, f)
$$

where $\mathcal{S}(g, f)$ depends smoothly on (g, f) and involves derivatives of f up to third order. You can easily observe that if $\gamma = g$, and the derivatives are covariant, the leading order term is of course $(n-1)\Delta_g^2 f$. For $\varepsilon > 0$ small, $\|L_g^* f\|_{k+2} \le C\varepsilon$, so that (by Sobolev embedding, if needed), $\|L_g^* f\|_{C^0}$ is small; thus γ and g are sufficiently close so that the leading order term of $R(g + L_g^* f) - R(g)$ is elliptic as a fourth-order operator, where the coefficients depend on lower-order derivatives of f. We can treat the operator as linear in f, with coefficients depending on f. It follows that f is smooth: for some $0 < \alpha < 1$, $f \in C^{4,\alpha}(M)$ (by Sobolev embedding, if needed). Then $L_g^* f \in C^{2,\alpha}(M)$, and $\mathcal{S}(g, f) \in C^{1,\alpha}(M)$. Elliptic regularity on the fourth-order equation implies $f \in C^{5,\alpha}(M)$; hence the coefficients of the principal term as well as $\mathcal{S}(g, f)$ now gain a degree of regularity. We can thus bootstrap our way up to $f \in C^\infty(M)$; of course, if S were less smooth, this process would terminate at a finite regularity stage. (If allowing g to be less smooth, one must be especially careful, since derivatives of g appear in the operators L_g and L_g^*.)

6.3.4. *A topological obstruction to positive scalar curvature.* The theorems of Bonnet–Myers and of Bochner give classical topological obstructions to manifolds admitting Riemannian metrics with positive Ricci curvature; information on the sectional or Ricci curvature can give some control on the behavior of geodesics. While from Exercise 2-56 we see how the scalar curvature governs the leading-order deviation of the volumes of small geodesic balls from their Euclidean counterparts (see also [142, Theorem 1.16] for a comparison result on the conjugate radius), its influence on the geometry, and certainly the topology, seems less direct. Thus it was a breakthrough when Schoen and Yau established a celebrated fundamental group obstruction to positive scalar curvature [198], which we will turn to presently. At the time of their work, the only known topological restrictions on positive scalar curvature involved harmonic spinors, from the Dirac operator on spin manifolds, dating from works of Lichnerowicz and Hitchin (as discussed in [139], where one can also explore Gromov and Lawson's development of this line of research). Of interest for us here will be the connection made by Schoen and Yau in relating the positivity of the mass of an asymptotically flat metric of nonnegative scalar curvature to topological obstructions to positive scalar curvature (PSC), a topic we will explore in the next chapter. For now we state one of the main theorems from [198].

Theorem 6-56 (Schoen–Yau). *Suppose M is a closed, connected, orientable three-manifold. Suppose furthermore that either $\pi_1(M)$ contains a finitely-generated non-cyclic abelian subgroup, or that $\pi_1(M)$ contains a subgroup abstractly isomorphic to the fundamental group of a closed orientable surface of positive genus. Then M admits no metric with positive scalar curvature, and any metric on M having nonnegative scalar curvature is* flat.

We discuss some elements of the proof. If Σ is a smooth, closed two-sided minimally immersed hypersurface in (M, g) with unit normal ν along the immersion, and second fundamental form $A = A^\nu$, then by (X.14) on p. 224, the second variation of area for variation field $V = \varphi\nu$ is

$$-\int_\Sigma \varphi \mathcal{L}_\Sigma \varphi \, d\sigma = \int_\Sigma \left(|\nabla^\Sigma \varphi|^2 - (|A|^2 + \mathrm{Ric}_g(\nu, \nu))\varphi^2\right) d\sigma \,, \qquad (6.3.3)$$

where $\mathcal{L}_\Sigma \varphi = \Delta_\Sigma \varphi + (|A|^2 + \mathrm{Ric}_g(\nu, \nu))\varphi$ is the *Jacobi operator* on Σ, with Δ_Σ, ∇^Σ, $|\cdot|$ and $d\sigma$ induced on Σ from (M, g). A minimal surface Σ is called *stable* if the second variation of area is nonnegative for all variations, which translates into the *stability inequality* $-\int_\Sigma \varphi \mathcal{L}_\Sigma \varphi \, d\sigma \geq 0$, for all sufficiently smooth φ on Σ. One of the key components of the proof of Theorem 6-56 is the following beautiful observation [198], cf. [106], using the stability inequality.

Proposition 6-57. *Let (M, g) be a closed, oriented Riemannian three-manifold with positive scalar curvature. Then (M, g) admits no stable minimal immersion from a closed orientable surface Σ with positive genus.*

Proof. Suppose Σ is a closed oriented surface, with a stable minimal immersion to M. Choose a local orthonormal frame $\{E_1, E_2, E_3\}$ adapted to Σ, with E_1

and E_2 tangential, and $E_3 = \nu$, the oriented unit normal to Σ. For $i, j = 1, 2$, let $A_{ij} = g(\nabla_{E_i} E_j, E_3)$ be the components of the second fundamental form in the frame. Let K_{ij} denote the sectional curvature in (M, g) of the E_i-E_j two-plane, so that $\mathrm{Ric}_g(\nu, \nu) = \mathrm{Ric}_g(E_3, E_3) = K_{13} + K_{23}$, and the scalar curvature is given by $R(g) = 2(K_{12} + K_{13} + K_{23})$.

Let K_Σ denote the Gauss curvature of Σ (the sectional curvature of its tangent plane), which by the Gauss equation satisfies $K_\Sigma = K_{12} + A_{11}A_{22} - A_{12}^2$. From minimality we have $A_{11} + A_{22} = 0$, and using this along with symmetry $A_{12} = A_{21}$, we get $K_\Sigma = K_{12} - \frac{1}{2}\sum_{i,j} A_{ij}^2 = K_{12} - \frac{1}{2}|A|^2$. Therefore, the coefficient in the Jacobi operator can be written

$$|A|^2 + \mathrm{Ric}_g(\nu, \nu) = \tfrac{1}{2}|A|^2 + K_{12} - K_\Sigma + K_{13} + K_{23} = \tfrac{1}{2}|A|^2 + \tfrac{1}{2}R(g) - K_\Sigma.$$

Using (6.3.3), the stability inequality for the variation field $V = \varphi\nu$ takes the form

$$\int_\Sigma \left(\tfrac{1}{2}|A|^2 + \tfrac{1}{2}R(g) - K_\Sigma\right)\varphi^2\, d\sigma \leq \int_\Sigma |\nabla^\Sigma \varphi|^2\, d\sigma.$$

We let $\varphi = 1$, and apply the Gauss–Bonnet Theorem $\int_\Sigma K_\Sigma\, d\sigma = 2\pi \chi(\Sigma)$, where $\chi(\Sigma)$ is the Euler characteristic of Σ, to obtain

$$\frac{1}{2} \int_\Sigma \left(R(g) + |A|^2\right) d\sigma \leq 2\pi \chi(\Sigma). \tag{6.3.4}$$

Thus $\chi(\Sigma)$ is positive, and so must equal 2; that is, Σ has genus zero. $\qquad\square$

By (6.3.4), we see that in case $R(g) \geq 0$, we still get $\chi(\Sigma) \geq 0$, and in case Σ is a torus, we see $|A|^2 = 0$, i.e., the immersion is actually totally geodesic (see Exercise 6-62 for more on this, as well as [92; 99; 37; 100]).

To prove Theorem 6-56, then, Schoen and Yau establish the existence of a stable minimally immersed surface of positive genus, given the condition on the fundamental group. For example, given an abelian subgroup of rank two in $\pi_1(M)$, one gets a continuous map of a torus $\mathbb{T}^2$ into M which maps $\pi_1(\mathbb{T}^2)$ onto this subgroup as follows: consider a torus as a rectangle with opposite sides suitably identified, and map these opposite sides to generators of the rank-two subgroup; the boundary of this rectangle maps to a null-homotopic curve in M, and so the continuous map extends to the interior of the rectangle, and hence to the torus. (A similar procedure works for higher genus g by representing the surface as a suitable quotient of a $4g$-gon.) Amongst all maps inducing the same action on $\pi_1(\mathbb{T}^2)$ (a conjugation may be invoked to keep track of the base point), one finds an energy-minimizing, hence harmonic, map. The energy is defined with respect to a surface metric, and is conformally invariant. By varying across conformal classes of $\mathbb{T}^2$ (or the higher genus surface), one finds a map of least energy whose action on $\pi_1(\mathbb{T}^2)$ is the same as the original (and in particular, then, the map is nontrivial). The energy-minimizer can be shown to be a stable minimal immersion (without branch points). We can now apply the preceding proposition. See [198] for details. By (6.3.4), one may allow $R(g) = 0$, for example in case $(\mathbb{T}^3, g)$ were the flat torus. By scalar curvature deformation (Theorem 6-52 and Corollary 6-53, cf. [198]), g must be Ricci-flat (hence flat in dimension three), else the scalar curvature may be made positive.

Exercises

Exercise 6-58 (graphical minimal surface equation). Suppose $u : \Omega \subset \mathbb{R}^n \to \mathbb{R}$ is a smooth function on an open set Ω, and let $\operatorname{grad} u$ be its Euclidean gradient vector in $\mathbb{R}^n$. The *graph* of u is a hypersurface $\Sigma \subset \mathbb{R}^{n+1}$ with unit normal field v given by $v\sqrt{1 + |\operatorname{grad} u|^2} = -\operatorname{grad} u + \partial/\partial x^{n+1}$, and a basis coordinate frame for $T\Sigma$ given by

$$X_i = \frac{\partial}{\partial x^i} + \frac{\partial u}{\partial x^i}\frac{\partial}{\partial x^{n+1}},$$

for $i \in \{1, \ldots, n\}$. In this frame, the first fundamental form g has components

$$g_{ij} = \langle X_i, X_j \rangle = \delta_{ij} + \frac{\partial u}{\partial x^i}\frac{\partial u}{\partial x^j},$$

while the second fundamental form K has components (where ∇ is the ambient Euclidean connection)

$$K_{ij} = \langle X_j, -\nabla_{X_i} v \rangle = \langle \nabla_{X_i} X_j, v \rangle = \left\langle X_i\left[\frac{\partial u}{\partial x^j}\right]\frac{\partial}{\partial x^{n+1}}, v \right\rangle = \frac{\dfrac{\partial^2 u}{\partial x^i \partial x^j}}{\sqrt{1 + |\operatorname{grad} u|^2}}.$$

a. Find the components g^{ij} explicitly, and use this to show that the mean curvature $H = H(u) = g^{ij} K_{ij}$ satisfies

$$(1 + |\operatorname{grad} u|^2)^{3/2} H(u) = (1 + |\operatorname{grad} u|^2)\Delta u - \sum_{i,j=1}^{n} \frac{\partial u}{\partial x^i}\frac{\partial u}{\partial x^j}\frac{\partial^2 u}{\partial x^i \partial x^j}.$$

(Hint: Let $G = (g_{ij})$ and $B = (\operatorname{grad} u)(\operatorname{grad} u)^T$, where $\operatorname{grad} u$ is a column vector (which may be zero). Find GB, and noting $G = I + B$, infer G^{-1} from here.)

b. Show that

$$H = \operatorname{div} \frac{\operatorname{grad} u}{\sqrt{1 + |\operatorname{grad} u|^2}},$$

where div is the Euclidean divergence in $\mathbb{R}^n$, by comparing to the result of the calculation in part a.

c. Show how the form of the minimal surface equation $H = 0$ using part b. follows naturally from the variational characterization of minimal surfaces as critical points of area functional, using the fact that the area $\mathcal{A}(u)$ of the graph of u is $\mathcal{A}(u) = \int_\Omega \sqrt{1 + |\operatorname{grad} u|^2}\, dx$. (If the area is infinite, you can still consider the variation, via the regularization expressed schematically as $\mathcal{A}(u + tv) - \mathcal{A}(u)$ for some v compactly supported in Ω.)

d. Show that the mean curvature operator $u \mapsto H = H(u)$ is elliptic, or equivalently, $u \mapsto \mathcal{M}(u) = (1 + |\operatorname{grad} u|^2)^{3/2} H(u)$ is elliptic.

Exercise 6-59. Suppose (M^n, g) is Riemannian.

a. Using the form (2-8a)–(2-8b) for the curvature tensor in coordinates, show the symbol of the linearization $P = D\mathrm{Ric}_g$ is given by (note that the symbol as given in some references differs from this up to sign)

$$\sigma(\xi)(h) = \tfrac{1}{2}\big(h|\xi|_g^2 + \operatorname{tr}_g h\, \xi \otimes \xi - h(\,\cdot\,, \xi^\sharp) \otimes \xi - \xi \otimes h(\xi^\sharp, \cdot\,)\big),$$

or equivalently, in orthonormal coordinates at $p \in M$, this is just

$$(\sigma(\xi)(h))_{ij} = \tfrac{1}{2}\Big(h_{ij}|\xi|_g^2 + \sum_{k=1}^n h_{kk}\xi_i\xi_j - \sum_{k=1}^n (h_{ik}\xi_j\xi_k + h_{kj}\xi_i\xi_k)\Big).$$

b. Suppose $\xi \in T_p^*M \setminus \{0\}$, with $\sigma(\xi)(h) = 0$; this still holds if we rescale so that $|\xi|_g = 1$. By using an orthonormal frame for T_pM with the first vector $\xi^\sharp$, show that h vanishes on $(\xi^\sharp)^\perp \times (\xi^\sharp)^\perp$. Then show that there is a form $\eta \in T_p^*M$ such that $h = \xi \otimes \eta + \eta \otimes \xi$. We know from Example 6-11 that indeed such h are in the kernel of $\sigma(\xi)$.

c. Adjust the preceding argument for the situation where g is a Lorentzian metric on M; you might break up into cases depending on the causal character of ξ.

Exercise 6-60 (elliptic estimates and regularity). Suppose $\Omega \subset \mathbb{R}^n$ is open. Let $0 \le \phi \le 1$ be a smooth function supported on the unit ball in $\mathbb{R}^n$, satisfying $\int_{\mathbb{R}^n} \phi(x)\, dx = 1$, and for any $\sigma > 0$, let $\phi_\sigma(x) = \sigma^{-n}\phi(\tfrac{x}{\sigma})$. If $u \in L^p_{\mathrm{loc}}$, and if we define the convolution $u^\sigma = \phi_\sigma * u$, then u^σ is smooth, and u^σ converges to u as $\sigma \searrow 0$ almost everywhere and in L^p_{loc}, while for $u \in C^k(\Omega)$ with k a nonnegative integer, u^σ converges to u in $C^k_{\mathrm{loc}}(\Omega)$ [86]. Suppose L is linear elliptic of order m, and the domain is Ω in the first two parts of the exercise.

a. Suppose the coefficients of L are constant, and that for some $p > 1$, $u \in L^p_{\mathrm{loc}}$ weakly solves $Lu = f \in L^p_{\mathrm{loc}}$. Consider a ball B_1 compactly contained in a ball B_2, compactly contained in Ω. Employ the interior elliptic estimate between B_1 and B_2 to show that the functions u^σ form a bounded set in $W^{m,p}(B_1)$, and then conclude that $u \in W^{m,p}_{\mathrm{loc}}$. If instead $f \in L^q_{\mathrm{loc}}$ for some $q > 1$, prove that $u \in W^{m,q}_{\mathrm{loc}}$

(using the above argument and Sobolev embedding to handle the case $q > p$). Finally, if moreover $f \in C^{0,\alpha}_{\text{loc}}$ for some $0 < \alpha < 1$, argue that $u \in C^{m,\alpha}_{\text{loc}}$. To do this, you might use the preceding to first show u is continuous (in fact, you can get $u \in C^{m-1,\alpha}_{\text{loc}}$); argue that f^σ is uniformly bounded in $C^{0,\alpha}$ on any compact subset, and apply the interior Schauder estimate.

b. You may assume the Schauder estimate (6.1.3) holds for L with $C^{0,\alpha}$ coefficients. Suppose for some $0 < \alpha < 1$, $u \in C^{m,\alpha}_{\text{loc}}$ solves $Lu = f$. Suppose that for some $k \in \mathbb{Z}_+$, the coefficients of L as well as the function f are in $C^{k,\alpha}_{\text{loc}}$. Show that $u \in C^{m+k,\alpha}_{\text{loc}}$. To do this for $k = 1$, bound $L(\Delta^h_i u)$ in $C^{0,\alpha}(B)$ for a ball B compactly contained in Ω. Use interior estimates to show that for any B' compactly contained in B, and for some $\delta > 0$, the set $\left\{\partial^\beta_x \Delta^h_i u : 0 < |h| < \delta, \; |\beta| \leq m\right\}$ is bounded and equicontinuous on B'. From here, conclude $u \in C^{m+1,\alpha}_{\text{loc}}$.

c. Suppose L is linear elliptic of order m with smooth coefficients on a closed Riemannian manifold (M, g). The Fredholm alternative/Hodge decomposition was established earlier in the L^2 and smooth cases. Prove the decomposition $W^{k,p}(M) = L(W^{m+k,p}(M)) \oplus \ker L^*$ for $1 < p < \infty$ and k a nonnegative integer, as follows. Let $^{\perp}\mathcal{K} := \{f \in L^p(M) : \int_M f w \, dv_g = 0 \text{ for all } w \in \mathcal{K}\}$, where $\mathcal{K} = \ker L^*$. Prove that the subspace $(^{\perp}\mathcal{K}) \cap C^\infty(M)$ is dense in the closed subspace $(^{\perp}\mathcal{K}) \cap W^{k,p}(M)$ of $W^{k,p}(M)$. Apply an approximation argument, the $C^\infty(M)$-splitting, and elliptic estimates to show that for $f \in (^{\perp}\mathcal{K}) \cap W^{k,p}(M)$, there is a solution $h \in W^{m+k,p}(M)$ of $Lh = f$; conclude that any distributional solution w of $Lw = f$ must lie in $W^{m+k,p}(M)$. By a similar method, show that for $f \in (^{\perp}\mathcal{K}) \cap C^{k,\alpha}(M)$, with k a nonnegative integer and $0 < \alpha < 1$, there is a solution $h \in C^{m+k,\alpha}(M)$ of $Lh = f$; conclude that any distributional solution w of $Lw = f$ must lie in $C^{m+k,\alpha}(M)$. (Hint for the last part: show that there is a sequence $f_j \in (^{\perp}\mathcal{K}) \cap C^\infty(M)$ uniformly bounded in $C^{k,\alpha}(M)$ and converging to f in $C^k(M)$.)

(Note: Part a. holds more generally. Suppose the coefficients of L were merely (sufficiently) smooth. Then while $Lu^\sigma - f^\sigma$ goes to zero distributionally, one needs an estimate in L^p_{loc} to apply the above argument, and in fact, K. O. Friedrichs [98, (3.8)] showed that the limit is zero in L^p_{loc}. For the Hölder case, to go from $u \in C^{m-1,\alpha}_{\text{loc}}$ to $u \in C^{m,\alpha}_{\text{loc}}$, one could localize and compactify as in Remark 6-20, and then apply part c.)

Exercise 6-61 (conformal method B: conformally covariant split). The conformal method introduced in Section 6.2 used a splitting of the tensor K which has been called the *semidecoupling split*, and indeed, the Hamiltonian and momentum constraints completely decouple in the CMC case. This method (*Method A*) also enjoys a natural conformal invariance in the CMC case, cf. Remark 6-45. There is another splitting for the conformal method which enjoys a natural conformal invariance more generally, and we introduce this method, *Method B*, now, cf. [17; 65; 172]. In this method, the freely prescribed TT tensor gives, up to conformal factor, the TT part of K, and the vector field W is used to generate the longitudinal part of K itself.

We start with a Riemannian metric $\mathring{g}$ on M^n, a symmetric TT tensor σ, and a function τ. We let $q = \frac{4}{n-2}$.

a. For $\phi > 0$, we let $K_{ab} = \phi^{-2}\sigma_{ab} + \phi^q (L_{\mathring{g}} W)_{ab} + \frac{\tau}{n}\phi^q \mathring{g}_{ab}$. This last term is pure trace. Observe that the first two summands are TT and longitudinal for $g = \phi^q \mathring{g}$, respectively. Show that the vacuum constraints $\Phi(g, K) = 0$ can be written

$$\Delta_{\mathring{g}}\phi = \frac{1}{q(n-1)} R(\mathring{g})\phi + \frac{1}{qn}\tau^2 \phi^{1+q} - \frac{1}{q(n-1)}|\sigma|_{\mathring{g}}^2 \phi^{-3-q}$$
$$- \frac{1}{q(n-1)}|L_{\mathring{g}} W|_{\mathring{g}}^2 \phi^{1+q} - \frac{2}{q(n-1)}\langle L_{\mathring{g}} W, \sigma \rangle_{\mathring{g}} \phi^{-1},$$
$$(\operatorname{div}_{\mathring{g}}(L_{\mathring{g}} W))_a = \frac{n-1}{n} d\tau_a - (q+2)(L_{\mathring{g}} W)^{bc}(d \log \phi)_c \mathring{g}_{ab}.$$

b. Show that this method enjoys a natural conformal invariance: for $\theta > 0$, $(\mathring{g}, \sigma, \tau)$ admits a solution $\phi > 0$ and W to the above system if and only if $(\theta^q \mathring{g}, \theta^{-2}\sigma, \tau)$ admits a solution $\phi_\theta = \phi\theta^{-1} > 0$ and $W_\theta = W$ to the corresponding system.

Exercise 6-62. Suppose that, in the setting of Proposition 6-57, we instead have $R(g) \geq 0$. As remarked after the proof, in case there is a stable minimal immersion from a torus Σ, then Σ is also totally geodesic, and $R(g) = 0$ along Σ as well, by (6.3.4). Show that more is true: prove that $K_\Sigma = 0$, and conclude $\operatorname{Ric}_g(\nu, \nu) = 0$ along Σ. (Hint: Following [92], consider the functional $\int_\Sigma (|\nabla^\Sigma \varphi|^2 + K_\Sigma \varphi^2)\, d\sigma$. Argue that $\varphi = 1$ is a minimizer, and consider the Euler–Lagrange equation.)

Excursus:
First and second variation of area

The theory of minimal surfaces is fundamental in geometry, and plays a significant role in aspects of mathematical relativity, in particular, in the study of the geometry of initial data sets. The minimal surface equation arises as the Euler–Lagrange equation for the area functional, as we now recall. The second variation at a minimal surface is important for geometric analysis, as we have seen in Section 6.3.4. The formulas we present here are standard, but they are so important that we derive them in detail.

Variation of a submanifold. Let Σ be a smooth manifold (with or without boundary). Consider an immersion $f : \Sigma \to M$, where $\dim \Sigma = k < n = \dim M$. We consider a *variation* of Σ (or more precisely, of f) to be a smooth map $F : I \times \Sigma \to M$, where $I \subset \mathbb{R}$ is an open interval around 0, so that $F(0, \cdot) = f$, and so that for each $t \in I$, if we let $f^t = F(t, \cdot) : \Sigma \to M$, then f^t is an immersion. We note that if we only assume the immersion condition at $t = 0$, there would be an open set $\mathcal{U} \supset \{0\} \times \Sigma$ such that for each $(t, p) \in \mathcal{U}$, the pushforward map f^t_* would be injective on $T_p\Sigma$. In case Σ were compact, or in case $f^t = f$ outside a compact set, which are two cases of interest for studying the area functional, then the set $\mathcal{U}$ could be chosen to be a product $J \times \Sigma$, where $0 \in J \subset I$ is an open interval. We assume going forward that we are in one of these cases, and thus we have just arranged each f^t to be an immersion.

If $\bar{g} = \langle \cdot, \cdot \rangle$ is a Riemannian metric on M, with Levi-Civita connection ∇, we let $g = f^*\bar{g}$ be the pullback metric on Σ with Levi-Civita connection ∇^Σ, and we let $g(t) := (f^t)^*\bar{g}$, a smooth curve of metrics on Σ with $g(0) = g$. If $\bar{g}$ is semi-Riemannian, we make the hypothesis that for each t, $g(t)$ is semi-Riemannian too. (We remark that even when $(M, \bar{g})$ is a spacetime, the variation variable t is just an independent parameter.) We let $d\sigma_{g(t)}$ be the area measure on $(\Sigma, g(t))$, and if Σ is compact, we let $A(t)$ be the area of $(\Sigma, g(t))$. We want to study the *variation* of area $A'(t)$, which makes sense even when Σ is noncompact but $f^t = f$ outside a compact subset, by replacing $A(t)$ by $\int_\Sigma (d\sigma_{g(t)} - d\sigma_g)$.

We recall that a vector field V *along the map* F is a map $V : I \times \Sigma \to TM$ such that $V|_{(t,p)} \in T_{F(t,p)}M$ for all $(t, p) \in I \times \Sigma$. A vector field along f^t is defined analogously.

We let $x = (x^i)$ be local coordinates on Σ, so that (t, x) give local coordinates on $I \times \Sigma$. We let $\partial F / \partial t = F_*(\partial / \partial t)$, and $\partial F / \partial x^i = F_*(\partial / \partial x^i)$ be the pushforward vectors, which define vector fields along the map F. Since $f = F(0, \cdot)$ is an immersion, the vectors $E_i := \partial F / \partial x^i$ form a local frame for $T\Sigma$. In particular, E_i is nonzero, so for a vector field W along Σ, we have $DW / \partial x^i = \nabla_{E_i} W$.

We start with a simple lemma. Since f is locally an embedding, we can use it to identify $T_p \Sigma$ inside $T_{f(p)}M$, as we do without additional notation in the next lemma.

Lemma X-1. *Let $f : \Sigma \to M$ be an immersion, and let $p \in \Sigma$. Suppose L is a scalar-valued or vector-valued bilinear form on $T_p \Sigma$ (i.e., a $(0, 2)$-tensor with scalar values or values in $T_{f(p)}M$). If $\{E_1, \ldots, E_k\}$ is a basis for $T_p \Sigma$, then the quantity $g^{ij} L(E_i, E_j)$ is independent of basis chosen.*

Proof. Let $\{E_1, \ldots, E_k\}$ and $\{\hat{E}_1, \ldots, \hat{E}_k\}$ be bases for $T_p \Sigma$, so that the metric g is represented by $g_{ij} = \langle E_i, E_j \rangle$ and $\hat{g}_{ij} = \langle \hat{E}_i, \hat{E}_j \rangle$. We have the change of basis given by $E_i = M_i^j \hat{E}_j$, and $\hat{E}_j = \hat{M}_j^\ell E_\ell$, so that $M_i^j \hat{M}_j^\ell = \delta_i^\ell$, and $\hat{g}_{ij} = \hat{M}_i^m g_{m\ell} \hat{M}_j^\ell$. Thus we see $\hat{g}^{jc} = M_a^j g^{ab} M_b^c$, and hence

$$g^{ij} L(E_i, E_j) = g^{ij} M_i^m M_j^\ell L(\hat{E}_m, \hat{E}_\ell) = \hat{g}^{m\ell} L(\hat{E}_m, \hat{E}_\ell). \qquad \square$$

As a corollary, for any vector field W defined along Σ, we can define $\mathrm{div}_\Sigma W :=$ $g^{ij} \langle \nabla_{E_i} W, E_j \rangle$, corresponding to the $(0, 2)$-tensor $L(X, Y) = \langle \nabla_X W, Y \rangle$ defined along Σ. We note that if W is tangent to Σ, then $\mathrm{div}_\Sigma W = g^{ij} \langle \nabla_{E_i}^\Sigma W, E_j \rangle = \mathrm{div}_g W$ is the usual divergence of a tangential vector field. For another example, if Σ is a hypersurface with unit normal field v and respective mean curvature H, then

$$\mathrm{div}_\Sigma v = g^{ij} \langle \nabla_{E_i} v, E_j \rangle = -g^{ij} \langle \mathrm{I\!I}(E_i, E_j), v \rangle = -H,$$

where as before $\mathrm{I\!I}(X, Y) = (\nabla_X Y)^N$ is the second fundamental form of Σ.

First variation. We compute the variation of the area element, written in local coordinates on Σ as $\sqrt{|\det(g_{ij}(t))|}\, dx = \sqrt{|\det g(t)|}\, dx$, as we have done in (2.3.10):

$$\frac{d}{dt} \sqrt{|\det g(t)|} = \tfrac{1}{2} g^{ij} \frac{d}{dt} \left\langle \frac{\partial F}{\partial x^i}, \frac{\partial F}{\partial x^j} \right\rangle \sqrt{|\det g(t)|},$$

with

$$\frac{d}{dt}\left\langle \frac{\partial F}{\partial x^i}, \frac{\partial F}{\partial x^j}\right\rangle = \left\langle \frac{D}{\partial t}\frac{\partial F}{\partial x^i}, \frac{\partial F}{\partial x^j}\right\rangle + \left\langle \frac{\partial F}{\partial x^i}, \frac{D}{\partial t}\frac{\partial F}{\partial x^j}\right\rangle$$
$$= \left\langle \frac{D}{\partial x^i}\frac{\partial F}{\partial t}, \frac{\partial F}{\partial x^j}\right\rangle + \left\langle \frac{\partial F}{\partial x^i}, \frac{D}{\partial x^j}\frac{\partial F}{\partial t}\right\rangle,$$

where we used (2.3.8). By symmetry we conclude that

$$\frac{d}{dt}\sqrt{|\det g(t)|} = g^{ij}\left\langle \frac{D}{\partial x^i}\frac{\partial F}{\partial t}, \frac{\partial F}{\partial x^j}\right\rangle \sqrt{|\det g(t)|}. \tag{X.1}$$

We evaluate this at $t = 0$, where $V := \frac{\partial F}{\partial t}\big|_{t=0}$ is the variation field. Since $F(0, \cdot) = f$ is an immersion, we can write at $t = 0$

$$\frac{d}{dt}\Big|_{t=0}\sqrt{|\det g(t)|} = g^{ij}\left\langle \nabla_{E_i} V, E_j\right\rangle \sqrt{|\det g(0)|}.$$

Along Σ we divide up the variation field V into tangential and normal components, $V = V^T + V^N$. Then we have

$$g^{ij}\left\langle \nabla_{E_i} V, E_j\right\rangle = g^{ij}\left\langle \nabla_{E_i}(V^T), E_j\right\rangle + g^{ij}\left\langle \nabla_{E_i}(V^N), E_j\right\rangle$$
$$= g^{ij}\left\langle \nabla^\Sigma_{E_i}(V^T), E_j\right\rangle - g^{ij}\left\langle V^N, \nabla_{E_i} E_j\right\rangle$$
$$= \mathrm{div}_\Sigma(V^T) - g^{ij}\left\langle V^N, \mathrm{I\!I}(E_i, E_j)\right\rangle$$
$$= \mathrm{div}_\Sigma(V^T) - \langle V, \boldsymbol{H}\rangle, \tag{X.2}$$

where we recall the vector-valued mean curvature $\boldsymbol{H} = g^{ij}\mathrm{I\!I}(E_i, E_j)$.

Thus we arrive at the *first variation of area formula*

$$A'(0) = \int_\Sigma \left(\mathrm{div}_\Sigma(V^T) - \langle V, \boldsymbol{H}\rangle\right) d\sigma_g$$
$$= -\int_\Sigma \langle V, \boldsymbol{H}\rangle d\sigma_g + \int_{\partial\Sigma} \langle V^T, \eta\rangle d\xi_g, \tag{X.3}$$

where $d\xi_g$ is the induced area element of $\partial\Sigma$ and η is the outward-pointing unit conormal vector along $\partial\Sigma$, the tangent vector to Σ which is a unit outward-pointing normal to $\partial\Sigma$. Note that the boundary term in (X.3) will pick up changes to the area obtained by pushing the boundary along Σ. Since each f^t is an immersion, the derivation above can be applied at any t-value, and yields

$$A'(t) = \int_\Sigma \left(\mathrm{div}_\Sigma\left(\frac{\partial F}{\partial t}\right)^T - \left\langle \frac{\partial F}{\partial t}, \boldsymbol{H}_{\Sigma_t}\right\rangle\right) d\sigma_{g(t)}$$
$$= -\int_\Sigma \left\langle \frac{\partial F}{\partial t}, \boldsymbol{H}_{\Sigma_t}\right\rangle d\sigma_{g(t)} + \int_{\partial\Sigma} \left\langle \left(\frac{\partial F}{\partial t}\right)^T, \eta^t\right\rangle d\xi_{g(t)}, \tag{X.4}$$

where $\boldsymbol{H}_{\Sigma_t}$ is the mean curvature of the immersion f^t, and η^t is the corresponding conormal.

By (X.3), we see that the first variation of area vanishes for tangential variations that vanish along the boundary. This also follows from the following exercise, which shows why one sometimes restricts to *normal* variations.

Exercise X-2. a. Suppose W is a compactly supported tangent vector field to Σ, which vanishes along $\partial \Sigma$ (if nonempty). Show that W is the variation field for a variation $\Phi : I \times \Sigma \to \Sigma$, where each $\phi^t := \Phi(t, \cdot) : \Sigma \to \Sigma$ is a diffeomorphism.

b. Consider a variation of Σ with area function $A(t)$ and variation field V which is compactly supported on Σ with $V^T = 0$ along $\partial \Sigma$ (if nonempty). Show that there is a variation of Σ with the same area function $A(t)$ and such that the variation field is V^N.

Second variation. For the second variation we take the derivative of (X.1), obtaining

$$
\frac{d^2}{dt^2} \sqrt{|\det g(t)|}
$$
$$
= \sqrt{|\det g(t)|} \left[-g^{im} \frac{\partial g_{m\ell}}{\partial t} g^{\ell j} \left\langle \frac{D}{\partial x^i} \frac{\partial F}{\partial t}, \frac{\partial F}{\partial x^j} \right\rangle + \left(g^{ij} \left\langle \frac{D}{\partial x^i} \frac{\partial F}{\partial t}, \frac{\partial F}{\partial x^j} \right\rangle \right)^2 \right.
$$
$$
\left. + g^{ij} \left(\left\langle \frac{D}{\partial t} \frac{D}{\partial x^i} \frac{\partial F}{\partial t}, \frac{\partial F}{\partial x^j} \right\rangle + \left\langle \frac{D}{\partial x^i} \frac{\partial F}{\partial t}, \frac{D}{\partial t} \frac{\partial F}{\partial x^j} \right\rangle \right) \right]. \quad \text{(X.5)}
$$

We evaluate the three terms inside the bracket at $t = 0$. The second is just

$$
\left(g^{ij} \left\langle \frac{D}{\partial x^i} \frac{\partial F}{\partial t}, \frac{\partial F}{\partial x^j} \right\rangle \right)^2 \bigg|_{t=0} = \left(\operatorname{div}_\Sigma (V^T) - \langle V, \boldsymbol{H} \rangle \right)^2. \quad \text{(X.6)}
$$

For the term on the second line of (X.5) we apply (2.3.8) and (2.3.9):

$$
g^{ij} \left(\left\langle \frac{D}{\partial t} \frac{D}{\partial x^i} \frac{\partial F}{\partial t}, \frac{\partial F}{\partial x^j} \right\rangle + \left\langle \frac{D}{\partial x^i} \frac{\partial F}{\partial t}, \frac{D}{\partial t} \frac{\partial F}{\partial x^j} \right\rangle \right)
$$
$$
= g^{ij} \left(\left\langle \frac{D}{\partial x^i} \frac{D}{\partial t} \frac{\partial F}{\partial t}, \frac{\partial F}{\partial x^j} \right\rangle + \left\langle R\left(\frac{\partial F}{\partial t}, \frac{\partial F}{\partial x^i}, \frac{\partial F}{\partial t} \right), \frac{\partial F}{\partial x^j} \right\rangle + \left\langle \frac{D}{\partial x^i} \frac{\partial F}{\partial t}, \frac{D}{\partial x^j} \frac{\partial F}{\partial t} \right\rangle \right).
$$

Evaluation at $t = 0$ yields

$$
\operatorname{div}_\Sigma \frac{DV}{\partial t} - g^{ij} \langle R(E_i, V, V), E_j \rangle + g^{ij} \left\langle \frac{DV}{\partial x^i}, \frac{DV}{\partial x^j} \right\rangle. \quad \text{(X.7)}
$$

Just as in (X.2) above, we can write

$$
\operatorname{div}_\Sigma \frac{DV}{\partial t} = \operatorname{div}_\Sigma \left(\frac{DV}{\partial t} \right)^T + \operatorname{div}_\Sigma \left(\frac{DV}{\partial t} \right)^N = \operatorname{div}_\Sigma \left(\frac{DV}{\partial t} \right)^T - \left\langle \frac{DV}{dt}, \boldsymbol{H} \right\rangle. \quad \text{(X.8)}
$$

As for the first term in brackets in (X.5), we get that it equals the negative of

$$
g^{im} \left(\left\langle \frac{D}{\partial x^m} \frac{\partial F}{\partial t}, \frac{\partial F}{\partial x^\ell} \right\rangle + \left\langle \frac{\partial F}{\partial x^m}, \frac{D}{\partial x^\ell} \frac{\partial F}{\partial t} \right\rangle \right) g^{\ell j} \left\langle \frac{D}{\partial x^i} \frac{\partial F}{\partial t}, \frac{\partial F}{\partial x^j} \right\rangle,
$$

which at $t = 0$ is

$$g^{im}\left(\left\langle\frac{DV}{\partial x^m}, E_\ell\right\rangle + \left\langle E_m, \frac{DV}{\partial x^\ell}\right\rangle\right)g^{\ell j}\left\langle\frac{DV}{\partial x^i}, E_j\right\rangle. \qquad \text{(X.9)}$$

We can simplify the expressions in (X.7) and (X.9) when the variation field is *normal* to Σ. To facilitate this, we introduce some terminology. The *normal Ricci curvature* $\mathcal{R}$ is a linear operator on each normal space $N_p\Sigma$, defined for vectors W normal to Σ by $\mathcal{R}(W) = g^{ij}R(W, E_i, E_j)^N$, where $\{E_1, \ldots, E_k\}$ is a basis for $T_p\Sigma$. This is well-defined by Lemma X-1. Note that by symmetry-by-pairs, if W is normal to Σ, $\langle\mathcal{R}(W), W\rangle = g^{ij}\langle R(E_i, W, W), E_j\rangle$.

For each vector W normal to Σ, we define a scalar-valued $(0, 2)$-tensor A^W on Σ by

$$A^W(X, Y) = \langle-\nabla_X W, Y\rangle = \langle W, \nabla_X Y\rangle = \langle W, \text{I\!I}(X, Y)\rangle.$$

If $\{E_1, \ldots, E_k\}$ is a basis for $T_p\Sigma$ as above, then $A^W(E_i, E_j) = \langle-\nabla_{E_i} W, E_j\rangle = \langle-(\nabla_{E_i} W)^T, E_j\rangle$, so that $g^{jm}A^W(E_i, E_j)E_m = -(\nabla_{E_i} W)^T$. In case $W = \nu$ is a unit vector, A^ν is the scalar valued second fundamental form K with respect to ν.

Finally, we recall the *normal connection* ∇^N on the normal bundle of Σ as follows: for X tangent to Σ, and W normal along Σ, we let $\nabla_X^N W = (\nabla_X W)^N$. If we apply Lemma X-1 to $L(X, Y) = \langle\nabla_X^N W, \nabla_Y^N W\rangle$, we let the result be $\langle\nabla^N W, \nabla^N W\rangle_g = g^{ij}\langle\nabla_{E_i}^N W, \nabla_{E_j}^N W\rangle$, or $|\nabla^N W|_g^2$ in the Riemannian case.

So, if $V = \frac{\partial F}{\partial t}\big|_{t=0}$ is normal to Σ, then (X.7) can be written, using $\frac{DV}{\partial x^\ell} = \nabla_{E_\ell} V$ and (X.8), as

$$\text{div}_\Sigma\frac{DV}{\partial t} - \langle\mathcal{R}(V), V\rangle + g^{ij}\left\langle(\nabla_{E_i} V)^T, (\nabla_{E_j} V)^T\right\rangle + g^{ij}\langle\nabla_{E_i}^N V, \nabla_{E_j}^N V\rangle$$

$$= \text{div}_\Sigma\frac{DV}{\partial t} - \langle\mathcal{R}(V), V\rangle + g^{ij}g^{a\ell}A^V(E_i, E_a)g^{sm}A^V(E_j, E_s)g_{\ell m}$$

$$+ g^{ij}\langle\nabla_{E_i}^N V, \nabla_{E_j}^N V\rangle$$

$$= \text{div}_\Sigma\left(\frac{DV}{\partial t}\right)^T - \left\langle\frac{DV}{dt}, H\right\rangle - \langle\mathcal{R}(V), V\rangle + \langle A^V, A^V\rangle_g + \langle\nabla^N V, \nabla^N V\rangle_g. \quad \text{(X.10)}$$

At last we write (X.9) in the case that V is normal to Σ as

$$g^{im}\left(\langle\nabla_{E_m} V, E_\ell\rangle + \langle E_m, \nabla_{E_\ell} V\rangle\right)g^{\ell j}\langle\nabla_{E_i} V, E_j\rangle$$

$$= g^{im}\left(\langle V, \nabla_{E_m} E_\ell\rangle + \langle V, \nabla_{E_\ell} E_m\rangle\right)g^{\ell j}\langle V, \nabla_{E_i} E_j\rangle$$

$$= g^{im}\left(\langle V, \text{I\!I}(E_m, E_\ell)\rangle + \langle V, \text{I\!I}(E_\ell, E_m)\rangle\right)g^{\ell j}\langle V, \text{I\!I}(E_i, E_j)\rangle$$

$$= 2\langle A^V, A^V\rangle_g. \qquad \text{(X.11)}$$

Applying (X.11), (X.10) and (X.6) (with $V^T = 0$) to (X.5), we obtain for normal variations V

$$A''(0) = \int_\Sigma \left(-2\langle A^V, A^V \rangle_g + \mathrm{div}_\Sigma \left(\frac{DV}{\partial t} \right)^T - \left\langle \frac{DV}{dt}, \boldsymbol{H} \right\rangle \right.$$
$$\left. - \langle \mathcal{R}(V), V \rangle + \langle A^V, A^V \rangle_g + \langle \nabla^N V, \nabla^N V \rangle_g + \langle V, \boldsymbol{H} \rangle^2 \right) d\sigma_g$$
$$= \int_\Sigma \left(\langle \nabla^N V, \nabla^N V \rangle_g - \langle A^V, A^V \rangle_g - \langle \mathcal{R}(V), V \rangle \right) d\sigma_g$$
$$- \int_\Sigma \left(\left\langle \frac{DV}{dt}, \boldsymbol{H} \right\rangle - \langle V, \boldsymbol{H} \rangle^2 \right) d\sigma_g + \int_{\partial\Sigma} \left\langle \frac{DV}{\partial t}, \eta \right\rangle d\xi_g. \quad \text{(X.12)}$$

This is the *second variation formula for normal variations*.

A number of comments are in order. By the first variation formula (X.3), since any V can be the variation field for some variation, the area is critical for all compactly supported normal variations if and only if $\boldsymbol{H} = \boldsymbol{0}$. If we consider the second variation of such an immersion, and if either $\partial\Sigma$ is empty, or if the variation $F = f$ on the boundary, i.e., $F(t, p) = f(p)$ for $(t, p) \in I \times \partial\Sigma$ (so that the variation field $\frac{\partial F}{\partial t}$ vanishes there), then we just get

$$A''(0) = \int_\Sigma \left(\langle \nabla^N V, \nabla^N V \rangle_g - \langle A^V, A^V \rangle_g - \langle \mathcal{R}(V), V \rangle \right) d\sigma_g. \quad \text{(X.13)}$$

We have not yet restricted the signatures of $\bar{g}$ and g, and thus the first two terms above do not have a (semi)-definite sign. In case g is Riemannian, then $|A^V|_g^2 := \langle A^V, A^V \rangle_g \geq 0$. If $\bar{g}$ is Lorentzian and Σ is a spacelike hypersurface, then $\langle \nabla^N V, \nabla^N V \rangle_g = g^{ij} \langle \nabla_{E_i}^N V, \nabla_{E_j}^N V \rangle \leq 0$.

In the case (Σ, g) is a Riemannian hypersurface, we consider a smooth unit normal field ν to Σ; either we assume $\Sigma \subset M$ is two-sided and ν is defined globally on Σ, or ν is a local unit normal field. We let $A = A^\nu$, which is just the scalar-valued second fundamental form K with respect to ν. Consider a normal variation $V = \varphi\nu$, where φ is a smooth function of compact support on Σ, vanishing on the boundary. In this case, $\langle \mathcal{R}(V), V \rangle = \varphi^2 \mathrm{Ric}_{\bar{g}}(\nu, \nu)$, $|A^V|_g^2 = \varphi^2 |A|_g^2$, and $\langle \nabla^N V, \nabla^N V \rangle_g = \langle \nu, \nu \rangle |\nabla^\Sigma \varphi|_g^2$, so that we have

$$A''(0) = \int_\Sigma \left(\langle \nu, \nu \rangle |\nabla^\Sigma \varphi|_g^2 - \varphi^2 |A|_g^2 - \varphi^2 \mathrm{Ric}_{\bar{g}}(\nu, \nu) \right) d\sigma_g. \quad \text{(X.14)}$$

If $(M, \bar{g})$ is a spacetime satisfying the Einstein vacuum equation $\mathrm{Ric}(\bar{g}) = 0$, or more generally $\mathrm{Ric}_{\bar{g}}(\nu, \nu) \geq 0$ (which follows from the timelike convergence condition, cf. Section 2.3.2), then we see from (X.14) why a spacelike hypersurface Σ with $\boldsymbol{H} = \boldsymbol{0}$ is sometimes called a *maximal* hypersurface (as opposed to a *minimal* hypersurface in the Riemannian context). In the Riemannian setting, we integrate (X.14) by parts and use the boundary condition on φ to obtain $A''(0) = -\int_\Sigma \varphi \mathcal{L}_\Sigma \varphi \, d\sigma_g$, where $\mathcal{L}_\Sigma = \Delta_\Sigma + |A|_g^2 + \mathrm{Ric}_{\bar{g}}(\nu, \nu)$ is the *Jacobi operator*, whose properties play an important role in applying minimal surface theory to geometry.

One can express the Ricci term in the second variation in terms of scalar curvature, using the Gauss equation, as in the proof of Proposition 6-57 and as in the derivation of the Hamiltonian constraint equation. Indeed, using (5.2.4)–(5.2.5) and accounting for signature, we have

$$|A|_g^2 + \mathrm{Ric}_{\bar{g}}(\nu, \nu) = \tfrac{1}{2}\langle \nu, \nu \rangle \big(R(\bar{g}) - R(g) + \langle \nu, \nu \rangle (|A|_g^2 + H^2)\big),$$

which simplifies at a minimal ($H = 0$) hypersurface.

Exercises

Exercise X-3. Consider the three-dimensional Riemannian Schwarzschild manifold (M, g_S), where

$$g_S = \left(1 + \frac{m}{2|x|}\right)^4 g_{\mathbb{E}^3} \quad \text{and} \quad M = \begin{cases} \{x \in \mathbb{R}^3 : x \neq 0\} & \text{if } m > 0, \\ \mathbb{R}^3 & \text{if } m = 0, \\ \{x \in \mathbb{R}^3 : |x| > -\tfrac{m}{2}\} & \text{if } m < 0. \end{cases}$$

For any $r > \max(0, -m/2)$, let $S_r = \{x \in \mathbb{R}^3 : |x| = r\} \subset M$.

a. Find the second fundamental form and the mean curvature vector H of $S_r = \{x : |x| = r\}$ in the metric g_S. Conclude that S_r is minimal only for $r = m/2 > 0$. (It is straightforward to do the computation using the relevant Christoffel symbols for g_S in spherical coordinates; compare Exercise X-5, and cf. Exercise 2-49.)

b. Let $A(r)$ be the g_S-area of S_r. Show directly that $A'(r) = -\int_{S_r} \langle H, X \rangle_{g_S} \, d\sigma$, where $X = \partial/\partial r$ and $d\sigma$ is the area measure induced by g_S, thus verifying the first variation of area formula (X.3).

c. The *Hawking mass* of a surface Σ with area $|\Sigma|$ is given by

$$m_H(\Sigma) = \sqrt{\frac{|\Sigma|}{16\pi}} \left(1 - \frac{1}{16\pi} \int_\Sigma H^2 \, d\sigma\right).$$

Show that $m_H(S_r) = m$.

d. Generalize the argument in part a. of Exercise 5-27 to show that if there is a closed minimal surface Σ in (M, g_S), then $m > 0$ and $\Sigma = S_{m/2}$.

Exercise X-4. Suppose (M, g) is Riemannian. Let $I \subset \mathbb{R}$ be an open interval containing 0. Suppose $F : I \times M \to M$ is smooth, with $F(0, \cdot) : M \to M$ the identity map. Suppose $\Omega \subset M$ is an open set with compact closure and smooth hypersurface boundary Σ, and outward unit normal field ν along Σ. Let $V(t)$ be the volume of $\Omega_t := F(\{t\} \times \Omega)$, let $A(t)$ be the area of $F(\{t\} \times \Sigma)$, and let $d\sigma$ be the induced surface measure.

a. Suppose that along Σ we have

$$\left.\frac{\partial F}{\partial t}\right|_{t=0} := F_*\left(\left.\frac{\partial}{\partial t}\right|_{t=0}\right) = X^T + \varphi\nu,$$

where X^T is tangent to Σ. Show that $V'(0) = \int_\Sigma \varphi \, d\sigma$. (Hint: Compute the derivative of the pullback of the volume measure $d(F(t, \cdot)^*(dv_g))/dt$.)

b. Show that $A'(0) = 0$ for all F as above for which $V'(0) = 0$ means that Σ has constant mean curvature (CMC).

c. Show that the CMC condition from part b. can be slightly rephrased to $A'(0) = 0$ for all F for which $V(t) = V(0)$ is constant (i.e., the area is stationary for all volume-preserving deformations.) (Hint: Given φ on Σ with $\int_\Sigma \varphi \, d\sigma = 0$, construct a volume-preserving deformation as follows: start with an F with variation field $\varphi\nu$ along $\{0\} \times \Sigma$; you can even arrange F to be the identity off of a neighborhood of $\{0\} \times \Sigma$. Extend ν off of $\{0\} \times \Sigma$ and construct a suitable $O(t^2)$ deformation of F; one suggestion is to employ the exponential map and the implicit function theorem.)

Exercise X-5. Suppose M is a n-dimensional manifold, $n \geq 3$, with $\Sigma \subset M$ a smooth embedded hypersurface. Let $u > 0$ be a smooth positive function on M. Suppose $\mathring{g}$ is a metric on M with Levi-Civita connection $\mathring{\nabla}$, and $g = u^q \mathring{g}$ is a conformal metric, with Levi-Civita connection ∇; in case $\mathring{g}$ is semi-Riemannian, we assume that g induces a semi-Riemannian metric on Σ. Note that for $p \in \Sigma$, the splitting $T_p M = T_p \Sigma \oplus (T_p \Sigma)^\perp$ is the same in both metrics. Let $\mathring{II}$ and II be the second-fundamental forms of Σ with respect to $\mathring{g}$ and g, respectively, i.e., for X, Y tangent to Σ, $\mathring{II}(X, Y) = (\mathring{\nabla}_X Y)^\perp$ and $II(X, Y) = (\nabla_X Y)^\perp$, with respective mean curvature vectors $\mathring{H} = \mathrm{tr}_{\mathring{g}} \mathring{II}$ and $H = \mathrm{tr}_g II$.

a. Verify that the tensor $\nabla - \mathring{\nabla}$ (Exercise 1-6) is given by (cf. Exercise 6-37)

$$\nabla_X Y - \mathring{\nabla}_X Y = \tfrac{1}{2} q u^{-1}\left(du(X)Y + du(Y)X - \mathring{g}(X, Y)\,\mathrm{grad}_{\mathring{g}} u\right).$$

From here it is easy to express II in terms of $\mathring{II}$, $\mathring{g}$ and u.

If $\mathring{\nu}$ is a smooth local unit normal field to Σ with respect to $\mathring{g}$, then $\nu = u^{-\frac{q}{2}} \mathring{\nu}$ is a smooth local unit normal field to Σ with respect to g. Recall our convention on the mean curvature is $H = g(H, \nu)$, $\mathring{H} = \mathring{g}(\mathring{H}, \mathring{\nu})$; the opposite convention changes the sign in the next formula.

b. Prove that $H = u^{-q}\mathring{H} - \frac{1}{2}q(n-1)u^{-1-q}(\mathrm{grad}_{\mathring{g}}\,u)^{\perp}$, and conclude that $H = u^{-q/2}\mathring{H} - \frac{1}{2}q(n-1)u^{-1-q/2}\partial u/\partial\mathring{v}$.

c. Use the above formula to find the mean curvatures of the spheres $S_r = \{|x| = r\}$ in the Riemannian Schwarzschild manifold (M, g_S), with

$$g_S = \left(1 + \frac{m}{2|x|^{n-2}}\right)^{\frac{4}{n-2}} g_{\mathbb{E}^n},$$

and for $m > 0$, find r_m for which S_{r_m} is minimal in (M, g_S), cf. Exercise 2-49.

d. In the Riemannian Schwarzschild metric from part c., find the mean curvature vector of the coordinate hyperplanes $x^n = \pm\xi$.

Exercise X-6. Suppose $(M, \bar{g} = \langle\,\cdot\,,\,\cdot\,\rangle)$ is a Riemannian or Lorentzian manifold with Levi-Civita connection ∇, and $f : \Sigma \to M$ is an immersed hypersurface. Suppose $F : I \times \Sigma \to M$ is a variation of f, and let Σ_t be the immersed surface given by $f^t := F(t, \cdot) : \Sigma \to M$, with induced Riemannian metric $g(t)$, and with a smooth section v of the pullback bundle $F^*(TM)$ whose values $v(t, p) \in T_{F(t,p)}M$ give a unit normal to Σ_t along f^t. We now define a *shape operator* along f^t. For any $(t, p) \in I \times \Sigma$, there is a neighborhood V_t of p in Σ on which f^t is an embedding, and we let $\Sigma_t' = f^t(V_t)$. For $q = F(t, p) \in \Sigma_t'$, we have $f^t_*(T_p\Sigma) = T_q\Sigma_t'$, and we let $S_q : T_q\Sigma_t' \to T_q\Sigma_t'$ be the shape operator of Σ_t' at q, given by $S_q(X) = (-\nabla_X v)_q \in T_q\Sigma_t'$. As in Exercise 5-25, the shape operator S_q is a $(1, 1)$-tensor with associated bilinear form K on $T_q\Sigma_t'$. We can extend S_q to T_qM by letting $S(W) = -\nabla_X v = S_q(X)$, for $W = X + \langle v, v\rangle\langle W, v\rangle v$, i.e., $X = W^T$, which is compatible with the way we extended K in Section 5.3.3.1.

Define $\nabla_v S$ by $(\nabla_v S)(W) = \nabla_v(S(W)) - S(\nabla_v W)$, which is readily seen to be tensorial in W.

a. Suppose the variation field is given by $\frac{\partial F}{\partial t}\big|_{t=0} = v(0, \cdot)$. Show that at $t = 0$, $(\nabla_v S)(W)$ is orthogonal to v, and verify the *Riccati equation* along Σ

$$\nabla_v S - S^2 = R(\,\cdot\,, v, v), \tag{X-6a}$$

where $S^2 = S \circ S$. You could do this by interpreting in terms of more general equations we derived in Chapter 5.

b. Recall that the mean curvature vector field H along Σ is independent of (local) orientation, and is given as the normal vector $H = \mathrm{tr}_\Sigma(\mathrm{I\!I})$. Note that $\langle H, v\rangle = \mathrm{tr}_\Sigma S = \mathrm{tr}_g K = H$. Consider a variation $F : I \times \Sigma \to M$ as above, for which $\frac{\partial F}{\partial t}\big|_{t=0} = v(0, \cdot)$. Let $H(t, p)$ be the mean curvature of the immersion $F(t, \cdot)$ at p. Show that the variation in the mean curvature is $\frac{\partial H}{\partial t}\big|_{t=0} = |K|_g^2 + \mathrm{Ric}_{\bar{g}}(v, v)$, where the curvature term is evaluated at $F(0, p) = f(p)$.

c. For an example of such a variation, suppose again that $(M, \bar{g} = \langle \cdot, \cdot \rangle)$ is a Riemannian or Lorentzian manifold, and $\Sigma \subset M$ is an embedded hypersurface with induced Riemannian metric g, and with smooth unit normal field v. For each p in Σ, let γ_p be the $\bar{g}$-geodesic in M with $\gamma_p'(0) = v|_p$. We note that given coordinates $x = (x^1, \ldots, x^k)$ for the hypersurface Σ, we let $(x^1, \ldots, x^k, t) \mapsto \gamma_{p(x)}(t) = \exp_{p(x)}(tv|_{p(x)}) \in M$. This map gives local coordinates for M (Fermi coordinates), and along Σ (i.e., $t = 0$), $\partial/\partial t = v$. By the geodesic equations, $\partial/\partial t$ is unit length, and by the Gauss Lemma (cf. [109], e.g.), $\partial/\partial t$ is orthogonal to the level sets Σ_t of t. We can thus extend v locally as $v = \partial/\partial t$, and the shape operator S is defined for level sets of t in a neighborhood of Σ.

Prove that for $q \in \Sigma_t$ and for all $W \in T_q M$, $S(W) = -\nabla_W v$. Then verify the Riccati equation $\nabla_v S - S^2 = R(\cdot, v, v)$ directly, without appealing to results from Chapter 5. Conclude that $\frac{\partial H}{\partial t} = |K|^2_g + \mathrm{Ric}_{\bar{g}}(v, v)$ holds for all t.

d. Suppose Σ is a closed manifold (compact without boundary), and consider a variation F with $\frac{\partial F}{\partial t}\big|_{t=0} = v(0, \cdot)$. Let $A(t)$ denote the area of Σ_t. Use the first variation formula derived earlier to write $A'(t)$ in terms of $H(t, p)$, and then, assuming $A'(0) = 0$, derive $A''(0)$ using the formula for $\frac{\partial H}{\partial t}\big|_{t=0}$ to see it agrees with the second variation of area formula, (X.14).

e. Now suppose instead that the variation field is given by $\frac{\partial F}{\partial t}\big|_{t=0} = \varphi v$. Show that $\frac{\partial H}{\partial t}\big|_{t=0} = \langle v, v \rangle \Delta_\Sigma \varphi + (|K|^2_g + \mathrm{Ric}_{\bar{g}}(v, v))\varphi$, which in the Riemannian case ($\langle v, v \rangle = 1$) is $\mathcal{L}_\Sigma \varphi$, where we recall $\mathcal{L}_\Sigma \varphi$ is the Jacobi operator. If Σ is closed, if $A(t)$ is the area function for the variation, and if $A'(0) = 0$, note that the above identity is consistent with second variation of area formula, (X.14). (Hint: To do this, you might prove the variation of mean curvature formula on the open set $U = \{\varphi \neq 0\} \subset \Sigma$; it is clearly true on the open set $\Sigma \setminus \overline{U}$ where φ is identically zero, and the identity extends to $\overline{U}$ by continuity. It might be instructive to describe how φ can vanish at a point $p \in \overline{U} \setminus U$ where $\frac{\partial H}{\partial t}\big|_{t=0}$ does not vanish.)

Exercise X-7. We have from (X.14) that a hyperplane in Euclidean space is stable. Let Σ be the totally geodesic plane $x^3 = 0$ in the Riemannian Schwarzschild manifold

$$g_S = \left(1 + \frac{m}{2|x|}\right)^4 g_{\mathbb{E}^3},$$

with smooth unit normal field v_{g_S}. In case $m > 0$, show that Σ is unstable. In fact, show this plane is unstable for volume-preserving deformations (cf. Exercise X-4) by finding a compactly supported function φ with $\int_\Sigma \varphi \, d\sigma_{g_S} = 0$ and such that a variation with $V = \varphi v_{g_S}$ decreases area. (Hint: Construct φ by modifying a suitable cutoff of the constant function 1, and then reflecting over the minimal sphere. One needs to control the integral of $|\nabla^\Sigma \varphi|^2_{g_S}$; to do this one can employ a *logarithmic cutoff* $\log(\theta^2|x|^{-1})/\log\theta$, which interpolates between 1 and 0 on an annulus $\theta \le |x| \le \theta^2$, for $\theta > 1$ sufficiently large, as in [199, p. 54]. One can then modify φ by adding a suitable term supported on a large annulus to arrange that $\int_\Sigma \varphi d\sigma_{g_S} = 0$, cf. [79, Proposition 3.3] (thanks to Otis Chodosh for the reference). We remark that the logarithmic cutoff is Lipschitz, which suffices for the argument. If one feels the need to smooth it out, one can replace $|x|$ with $\rho(x)$, where $\rho(x) = \theta$ for $|x| \le \theta + \delta_0$, $\rho(x) = |x|$ for $\theta + \delta_1 \le |x| \le \theta^2 - \delta_1$, and $\rho(x) = \theta^2$ for $|x| \ge \theta^2 - \delta_0$, where $0 < \delta_0 < \delta_1$, and δ_0 and δ_1 are fixed, for all $\theta > 1$ sufficiently large. For $\theta + \delta_0 \le |x| \le \theta + \delta_1$, we can let ρ interpolate smoothly between θ and $|x|$ by letting $\rho'(r) = \psi(r - \theta)$, where $\psi \ge 0$ is smooth, $\psi = 0$ for $t \le \delta_0$, $\psi(t) = 1$ for $t \ge \delta_1$, and $\int_{\delta_0}^{\delta_1} \psi(t)\,dt = \delta_1$. For $\theta^2 - \delta_1 \le |x| \le \theta^2 - \delta_0$, we can interpolate analogously, with $\rho'(r) = \psi(\theta^2 - r)$ for $\theta^2 - \delta_1 \le r \le \theta^2 - \delta_0$.)

CHAPTER 7

Asymptotically flat solutions of the Einstein constraint equations

This chapter develops the geometry of and analysis on initial data sets that arise in models of isolated gravitational systems. While such models should behave in the far field like Minkowski spacetime, there are important physical and mathematical issues to consider when specifying precisely what the spacetime asymptotic behavior should be, and the devil is in the details. A desirable, but rather strong, requirement would be to have the spacetime admit a conformal compactification in the spirit of that of Minkowski spacetime from Section 1.4.2. (See, for instance, [218, Section 11.1], [112], and [176] for the notion of *asymptotically simple* spacetimes, ideas from the analysis of which are often used to model gravitational radiation.) One often considers weaker notions that still capture in some sense how the spacetime approaches the Minkowski spacetime.

A natural question to ask is what kind of spacetime behavior is inherited from the evolution of initial data which approaches, say, Euclidean geometry, or, initial data which approaches that of a hyperboloid in Minkowski spacetime, in the far field. With the conformal compactification of Minkowski spacetime in mind, hypersurfaces in the former class might be thought of as tending to spacelike infinity, and those in the latter class as tending to null infinity, though such notions would come from the spacetime evolution of the data. There are results, cf. [56; 55; 97], which establish properties about the evolution of such initial data. We will not discuss such results here, but rather we will focus on the analysis and geometry of the initial data, with particular emphasis on features of the asymptotic structure imposed by the constraint equations (vacuum, or under the dominant energy condition), leaving the interested reader to pursue the fascinating questions and results for the evolution problem, which has seen spectacular progress in recent years.

Throughout this chapter, $n \geq 3$, $g_{\mathbb{E}^n}$ is the Euclidean metric on $\mathbb{R}^n$, $\overset{\circ}{g}_{\mathbb{S}^{n-1}}$ is the unit round metric on the sphere, and we take the cosmological constant Λ to be 0. We let dx be Euclidean volume measure, and let $|\cdot|$ be the Euclidean norm, while for a submanifold of Euclidean space, $d\sigma$ is Euclidean surface measure, and ν is a Euclidean normal vector, whereas dv_g, $|\cdot|_g$, $d\sigma_g$ and ν_g will be the analogous quantities with respect to a metric g.

The first two sections of the chapter contain some detailed discussion and analysis involving the Laplace operator on Euclidean space and on asymptotically flat manifolds. While it is not necessary to understand and fill in every detail on a first pass, we hope that the reader can follow some of the discussion to get a sense of what is important, and then be able to apply the results in later sections. Indeed, rather than just state results and point to a panoply of references, we have attempted to motivate the results and sketch proofs, hopefully providing a reader's guide to some of the references.

7.1. Harmonically flat solutions of the constraint equations

In Chapter 2 we obtained the Schwarzschild spacetime by imposing the vacuum Einstein equation along with rotational symmetry. The resulting spacetime is static and asymptotic to Minkowski spacetime in a far field regime. While static metrics are rather special, there are many spacetimes metrics which asymptote to Minkowski spacetime, and are often used to model isolated gravitational systems. In terms of initial data, certain spacelike slices of the Schwarzschild spacetime have induced metric g_S and vanishing second fundamental form, (cf. Exercise 2-34) and hence give *time-symmetric* solutions to the vacuum constraint equations. In the time-symmetric case, the vacuum constraints reduce to the vanishing of the scalar curvature. For $n \geq 3$, the metric g_S with mass m can be written in isotropic coordinates as

$$g_S = \left(1 + \frac{m}{2|x|^{n-2}}\right)^{\frac{4}{n-2}} g_{\mathbb{E}^n}. \tag{7.1.1}$$

This metric is conformally flat, and *asymptotically flat* (or *asymptotically Euclidean*), in that it approaches $g_{\mathbb{E}^n}$ as $|x| \to \infty$. (Though it is not immediately obvious, for $m > 0$ it is also asymptotically flat as $|x| \searrow 0$; see Exercise 2-49.)

If we consider $g = u^{\frac{4}{n-2}} g_{\mathbb{E}^n}$, then the conformal change of scalar curvature (6.1.8) simplifies, since $R(g_{\mathbb{E}^n}) = 0$, to

$$R(g) = -\frac{4(n-1)}{n-2} u^{-\frac{n+2}{n-2}} \Delta u,$$

where Δ is the Euclidean Laplacian. Thus for the scalar curvature to vanish, we require $\Delta u = 0$. We remark that if we only require the dominant energy condition (Section 2.3.2, see also Section 5.2), then in the time-symmetric case we get $R(g) \geq 0$, which for $g = u^{\frac{4}{n-2}} g_{\mathbb{E}^n}$ translates into $\Delta u \leq 0$, i.e., u is *superharmonic*.

The conformal factor u for the Riemannian Schwarzschild metric is, up to constants, the *fundamental solution* for the Laplace operator on $\mathbb{R}^n$. If more generally we consider positive harmonic functions on domains $\Omega \subset \mathbb{R}^n$, we obtain solutions $(g, K) = (u^{\frac{4}{n-2}} g_{\mathbb{E}^n}, 0)$ to the time-symmetric vacuum constraints on Ω. Such metrics might be called *harmonically flat*, though the term is sometimes reserved for the case when Ω is a neighborhood of infinity (i.e., contains the complement of a compact subset), and when u limits to a positive constant (which can be rescaled to unity) at infinity, so that g is also asymptotically flat; in this case we say that g is *harmonically flat at infinity* [25].

Such metrics play an important role in the study of isolated gravitational systems, as we will see. For now, we will review some properties of the Laplace operator and harmonic functions on $\mathbb{R}^n$, and apply them to harmonically flat metrics.

7.1.1. *On the Laplacian in $\mathbb{R}^n$.*

As we have seen, the gravitational potential Φ in Newtonian gravity satisfies Poisson's equation $\Delta \Phi = 4\pi G\sigma$ on $\mathbb{R}^3$, where σ is the matter density. For an isolated system, we can consider σ to be compactly supported, or more generally to decay suitably at infinity. We want to know how the potential Φ behaves at infinity, and so we will begin by studying the Laplace operator on $\mathbb{R}^n$. In the case of a point mass $\sigma = M\delta_c$ (a multiple of the Dirac distribution at c) if we impose that Φ decays at infinity, we have $\Phi(x) = -GM/|x - c|$.

In this section, *measurable* functions are defined with respect to Lebesgue measure on $\mathbb{R}^n$, and for an integrable function, $\int_{\mathbb{R}^n} f(x)\, dx$ is the integral with respect to Lebesgue measure.

7.1.1.1. *The fundamental solution.*

On a Riemannian manifold (M, g), the trace of the covariant Hessian $\Delta_g u = \mathrm{div}_g (\nabla_g u) = g^{ij} u_{;ij}$ can be expressed as (recall the summation convention is in force) (cf. Exercise 1-10)

$$\Delta_g u = \frac{1}{\sqrt{\det g}} \frac{\partial}{\partial x^i} \left(g^{ij} \sqrt{\det g}\, \frac{\partial u}{\partial x^j} \right). \tag{7.1.2}$$

We recall the rotationally symmetric harmonic functions on Euclidean space. If we consider (7.1.2) for the Euclidean metric $(\mathbb{R}^n, g_{\mathbb{E}^n} = dr^2 + r^2 \mathring{g}_{\mathbb{S}^{n-1}})$, applied to a function u which only depends on the radial distance $r = |x|$ from the origin, we obtain (writing $u = u(r)$)

$$\Delta u = \frac{1}{r^{n-1}} \frac{d}{dr} \left(r^{n-1} \frac{du}{dr} \right) = u''(r) + \tfrac{n-1}{r} u'(r).$$

For such u, to be harmonic is thus equivalent to $\frac{d}{dr}\left(r^{n-1}\frac{du}{dr}\right)=0$, that is,

$$u(r) = \begin{cases} Ar + B & \text{for } n = 1, \\ A\log r + B & \text{for } n = 2, \\ Ar^{2-n} + B & \text{for } n > 2, \end{cases} \tag{7.1.3}$$

for some constants A and B. In dimension one the function u extends continuously to the whole line; for $n = 2$, $\log|x|$ is unbounded both as $r = |x|$ tends to 0 or ∞, whereas for $n > 2$, $u(x)$ decays to 0 as $|x|$ tends to infinity. We will generally be interested in the case $n \geq 3$, but will also include some discussion of the $n = 2$ case for comparison; recall that in the plane, harmonic function theory is intimately tied to the theory of holomorphic functions.

These special harmonic functions are essentially the *fundamental solutions* for the Laplace operator, as we now recall. First, note that these functions are locally integrable, even around the origin. Therefore they define distributions, and thus have distributional derivatives, and hence the Laplace operator can be applied distributionally to these functions. To be precise, if f is locally integrable, Δf is the distribution T defined as follows: for any $\psi \in C_c^\infty(\mathbb{R}^n)$, $T(\psi) = \int_{\mathbb{R}^n} f \Delta \psi \, dx$.

Let Ω_n be the volume of the unit ball B in $\mathbb{R}^n$, and ω_{n-1} the surface area of the unit round sphere $\mathbb{S}^{n-1} = \partial B$. Note that $2\Omega_2 = 2\pi = \omega_1$, $3\Omega_3 = 4\pi = \omega_2$, and in general $n\Omega_n = \omega_{n-1}$. The proof of the following is a generalization of that of (2.1.2), and is left as an exercise.

Proposition 7-1. *Let*

$$\Gamma(x-y) = \begin{cases} \dfrac{1}{(2-n)n\Omega_n}|x-y|^{2-n} & \text{if } n > 2, \\ \dfrac{1}{2\pi}\log|x-y| & \text{if } n = 2. \end{cases}$$

Then, for all $u \in C_c^2(\mathbb{R}^n)$,

$$u(x) = \int_{\mathbb{R}^n} \Gamma(x-y)\Delta u(y)\,dy. \tag{7.1.4}$$

In other words, the following distributional identities hold: in dimension two,

$$\Delta\left(\tfrac{1}{2\pi}\log|x|\right) = \delta_0 ,$$

while for $n > 2$,

$$\Delta\left(\frac{1}{(2-n)n\Omega_n}|x|^{2-n}\right) = \delta_0.$$

Furthermore, $\Delta\Gamma(x-y) = \delta_0(x-y)$, consistent with (7.1.4).

We remark that $\dfrac{\partial \Gamma(x)}{\partial x^i} = \dfrac{1}{\omega_{n-1}} \dfrac{x^i}{|x|^n}$ for all $n \geq 2$.

7.1.1.2. *The Poisson equation and the Newtonian potential.* For functions $f \in L^\infty_{\mathrm{loc}}(\mathbb{R}^n) \cap L^1(\mathbb{R}^n)$, for instance f locally bounded, measurable and decaying suitably at infinity, the integral $N_f(x) = \int_{\mathbb{R}^n} \Gamma(x - y) f(y)\,dy$ converges, and constitutes the *Newtonian potential* of f.

Exercise 7-2. Let $n \geq 3$. Prove that the integral defining N_f converges for functions $f \in L^\infty_{\mathrm{loc}}(\mathbb{R}^n) \cap L^1(\mathbb{R}^n)$. (Hint: Break up the integral into two regions and exploit the fact that Γ is both locally integrable, and bounded near infinity.) Using Young's inequality for convolution, adapt the proof to show that if $f \in L^1(\mathbb{R}^n)$, the integral for N_f converges a.e. to a locally integrable function.

For f integrable with compact support, a simple Fubini-type argument, along with Proposition 7-1, yields that $\Delta N_f = f$ distributionally: for all $\psi \in C_c^\infty(\mathbb{R}^n)$,

$$\int_{\mathbb{R}^n} \Delta\psi(x) N_f(x)\,dx = \int_{\mathbb{R}^n} \psi(y) f(y)\,dy.$$

Proposition 7-3. *For $n \geq 3$ and $f \in L^1(\mathbb{R}^n)$, $\Delta N_f = f$ distributionally.*

Sketch of proof. For $f \in L^1(\mathbb{R}^n)$ and $R > 0$, we let $f_R = f \chi_{B_R(0)}$, where χ_A is the characteristic function of the set A. It is not hard to see that as $R \nearrow \infty$, f_R and $N_{f_R} = f_R * \Gamma$ converge as distributions to f and N_f; the latter fact can be proven along the lines of Exercise 7-2. That $\Delta N_f = f$ distributionally then follows from the analogous equation $\Delta(f_R * \Gamma) = f_R$, which is valid since f_R has compact support. $\qquad\square$

One can more generally define N_f when f is, say, a compactly supported distribution, and the distribution so defined also solves the Poisson equation $\Delta N_f = f$. See [93] or [195, Theorem 6.37].

A fundamental issue is the regularity of the Newtonian potential N_f. For instance, Proposition 7-1 shows us that $N_f \in C^2(\mathbb{R}^n)$ *for certain* continuous functions f, namely those $f \in \Delta(C_c^2(\mathbb{R}^n))$. We stress, however, that for general $f \in C_c(\mathbb{R}^n)$, while the following lemma indicates that the solution $N_f \in C_{\mathrm{loc}}^{1,\alpha}(\mathbb{R}^n)$ for $0 < \alpha < 1$, N_f might fail to be in $C^2(\mathbb{R}^n)$ (see [111, Example 4.4.4], for instance; for a more general result see [115, Theorem 7.9.8], and compare [107, Exercise 4.9]). Given $f \in C_c(\mathbb{R}^n)$ and $\Omega \subset \mathbb{R}^n$ open such that N_f fails to be in $C^2(\Omega)$, there is in fact *no* solution $v \in C^2(\Omega)$ of $\Delta v = f$: if there were such a v, then the continuous function $w = N_f - v$ would be harmonic, and hence a smooth function by Weyl's lemma (Lemma 6-18), which yields a contradiction.

We have the following lemma on the regularity of the Newtonian potential. The interested reader can formulate a modification for the case $n = 2$.

Lemma 7-4. *Let $n \geq 3$. Suppose $f \in L^{\infty}_{\text{loc}}(\mathbb{R}^n) \cap L^1(\mathbb{R}^n)$. For any $0 < \alpha < 1$, N_f lies in $C^{1,\alpha}_{\text{loc}}(\mathbb{R}^n)$, and for each $j \in \{1, 2, \ldots, n\}$,*

$$\partial_j N_f(x) = \int_{\mathbb{R}^n} \partial_j \Gamma(x - y) f(y) \, dy.$$

If in fact $f \in C^{0,\alpha}_{\text{loc}}(\mathbb{R}^n) \cap L^1(\mathbb{R}^n)$, then $N_f \in C^{2,\alpha}_{\text{loc}}(\mathbb{R}^n)$, and $\Delta N_f = f$ holds classically.

Sketch of proof. We will briefly indicate the issues involved in proving this lemma. First, a straightforward dominated convergence argument shows that for $f \in C^k_c(\mathbb{R}^n)$ and for any multi-index α with $|\alpha| \leq k$,

$$\partial^\alpha N_f(x) = \int_{\mathbb{R}^n} \Gamma(y) \partial^\alpha f(x - y) \, dy = \int_{\mathbb{R}^n} \Gamma(x - y) \partial^\alpha f(y) \, dy = N_{\partial^\alpha f}(x).$$

One can then use uniform continuity of $\partial^\alpha f$ to conclude that $N_f \in C^k(\mathbb{R}^n)$, and of course, if $f \in C^\infty_c(\mathbb{R}^n)$, then $N_f \in C^\infty(\mathbb{R}^n)$. Suppose we dial the regularity down to $f \in C^1_c(\mathbb{R}^n)$, so that $\partial_j N_f(x) = \int_{\mathbb{R}^n} \Gamma(x - y) \partial_j f(y) \, dy$, as above. Γ and $\partial_j \Gamma$ are locally integrable, and a very simple limiting argument around the singularity as in the derivation of (2.1.2) justifies integration by parts to yield

$$\partial_j N_f(x) = \int_{\mathbb{R}^n} \partial_j \Gamma(x - y) f(y) \, dy = \int_{\mathbb{R}^n} \partial_j \Gamma(y) f(x - y) \, dy.$$

Since $f \in C^1_c(\mathbb{R}^n)$ and $\partial_j \Gamma$ is locally integrable, dominated convergence can be applied again to give $\partial^2_{ij} N_f(x) = \int_{\mathbb{R}^n} \partial_j \Gamma(y) \partial_i f(x - y) \, dy$, which is likewise seen to be continuous. Thus $N_f \in C^2(\mathbb{R}^n)$.

While the C^1-regularity of f made this argument fairly simple, if we are willing to do a bit more work, we actually can get even more with less. We indicate how this works. Since $\partial_j \Gamma$ is locally integrable, one can use this along with a regularization argument (cf. [107, Lemma 4.1]) to show that $f \in L^{\infty}_{\text{loc}}(\mathbb{R}^n) \cap L^1(\mathbb{R}^n)$ implies that $\partial_j N_f(x) = \int_{\mathbb{R}^n} \partial_j \Gamma(x - y) f(y) \, dy$, and indeed that $N_f \in C^1(\mathbb{R}^n)$. Since $\partial^2_{ij} \Gamma$ is not locally integrable, we cannot run the analogous argument again; however, if we have suitable control on the difference $|f(x) - f(y)|$, such as from local Hölder continuity, say $f \in C^{0,\alpha}_{\text{loc}}(\mathbb{R}^n) \cap L^1(\mathbb{R}^n)$, then we can use the fact that $\partial^2_{ij} \Gamma(x - y)(f(y) - f(x))$ *is* locally integrable to establish $N_f \in C^2(\mathbb{R}^n)$; see [107, Lemma 4.2]. As for Hölder continuity of N_f, we sketch in Exercise 7-101 how to show $N_f \in C^{1,\alpha}_{\text{loc}}(\mathbb{R}^n)$ for $f \in L^{\infty}_{\text{loc}}(\mathbb{R}^n) \cap L^1(\mathbb{R}^n)$, and leave the $C^{2,\alpha}_{\text{loc}}$-proof to [107, Lemma 4.4]. $\square$

With Proposition 7-1 in mind, we point out the obvious fact that even if Δu is compactly supported, u generally will not be, even modulo an additive constant; as an example, consider u constructed by smoothing out $\Gamma(x)$ near $x = 0$. That said, for $n \geq 3$, if f is bounded, measurable and compactly supported, there is a constant C such that $|N_f(x)| \leq C|x|^{2-n}$; in particular, then, $N_f(x) \to 0$ as $|x| \to \infty$. Thus, if $f \in C_c^{0,\alpha}(\mathbb{R}^n)$, and if $u \in C^2(\mathbb{R}^n)$ with $\Delta u = f$ *and* such that u tends to 0 at infinity, an application of the maximum principle (which we recall in Proposition 7-10 below), along with the regularity and decay of u and N_f and the fact that $\Delta(u - N_f) = 0$, shows that (7.1.4) holds, i.e., $u = N_f = N_{\Delta u}$.

We have the following important corollary, for which we set up some notation.

Definition 7-5. For $s \in \mathbb{R}$, let $h(x) = O_\infty(|x|^s)$ indicate that h is a function defined on a domain $\Omega \subset \mathbb{R}^n$ with compact complement, and that for all $k \in \mathbb{Z}_+ \cup \{0\}$, there is a constant C_k such that for all $x \in \Omega \cap \{|x| > 1\}$ and all multi-indices β with $|\beta| = k$, we have $|\partial_x^\beta h(x)| \leq C_k |x|^{s-k}$. For $\ell \in \mathbb{Z}_+ \cup \{0\}$, we let $O_\ell(|x|^s)$ be defined analogously, just for $0 \leq k = |\beta| \leq \ell$. We write $O_\ell(|x|^s; \Omega)$ if we want to specify the domain.

Recall that $d\sigma$ is Euclidean surface measure. As usual, $r = |x|$, and $\partial u/\partial r = (x^i/r)\,\partial u/\partial x^i$, and for $B \in \mathbb{R}^n$, $x \cdot B = \sum_{i=1}^n x^i B^i$.

Corollary 7-6. *Let $n \geq 3$. Suppose $v \in C^2(\mathbb{R}^n)$ is such that Δv has compact support, and $\lim_{|x| \to \infty} v(x) = 0$. Then v admits an expansion*

$$v(x) = \frac{A}{|x|^{n-2}} + \frac{x \cdot B}{|x|^n} + O_\infty(|x|^{-n}),$$

for constants $A \in \mathbb{R}$ and $B \in \mathbb{R}^n$, which are given by

$$A = \frac{1}{(2-n)\omega_{n-1}} \lim_{r \to \infty} \int_{\{|x|=r\}} \frac{\partial v}{\partial r}\, d\sigma = \frac{1}{(2-n)\omega_{n-1}} \int_{\mathbb{R}^n} \Delta v \, dx \qquad (7.1.5)$$

$$B^i = -\frac{1}{\omega_{n-1}} \lim_{r \to \infty} \int_{\{|x|=r\}} \left(x^i \frac{\partial v}{\partial r} - \frac{x^i}{r} v \right) d\sigma = -\frac{1}{\omega_{n-1}} \int_{\mathbb{R}^n} x^i \Delta v \, dx. \qquad (7.1.6)$$

Exercise 7-7. Prove Corollary 7-6. (Hint: Use $v = N_{\Delta v}$ and the expansion

$$|x - y|^{2-n} = (|x|^2 - 2x \cdot y + |y|^2)^{\frac{2-n}{2}} = |x|^{2-n}\left(1 - 2\frac{x}{|x|} \cdot \frac{y}{|x|} + \frac{|y|^2}{|x|^2}\right)^{\frac{2-n}{2}}$$

$$= |x|^{2-n}\left(1 + (n-2)\frac{x}{|x|} \cdot \frac{y}{|x|} + O_\infty(|x|^{-2})\right)$$

for $|x|$ large relative to $|y|$.)

For such a function v as in the corollary, there is a neighborhood Ω of infinity for which $u = 1 + v > 0$ and such that $(\Omega, u^{\frac{4}{n-2}} g_{\mathbb{E}^n})$ is harmonically flat at infinity. In analogy with the Schwarzschild metric, we will identify $m = 2A$ as a mass. The vector B will be related to a *center of mass* in Exercise 7-28 below. At this point it might prove instructive to compare the Newtonian analogue: recall from (2.1.3) that the gravitational potential Φ satisfies the Poisson equation $\Delta\Phi = 4\pi\sigma = 4\pi\rho/c^2$, where ρ is the energy density, and $\sigma = \rho/c^2$ is the mass density, and we have taken $G = 1$ for convenience. If we assume ρ is smooth and compactly supported, and that $\lim_{|x|\to\infty} \Phi(x) = 0$, then Φ admits an expansion of the above form, which we write $\Phi(x) = -m/|x| - \beta \cdot x/|x|^3 + O_\infty(|x|^{-3})$. In fact, from (7.1.5) we see that

$$m = \frac{1}{4\pi} \lim_{r\to\infty} \int_{\{|x|=r\}} \frac{\partial\Phi}{\partial r}\, d\sigma = \frac{1}{4\pi} \int_{\mathbb{R}^3} \Delta\Phi\, dx = \int_{\mathbb{R}^3} \sigma\, dx \qquad (7.1.7)$$

and from (7.1.6)

$$\beta^i = \frac{1}{4\pi} \lim_{r\to\infty} \int_{\{|x|=r\}} \left(x^i \frac{\partial\Phi}{\partial r} - \frac{x^i}{r}\Phi \right) d\sigma = \frac{1}{4\pi} \int_{\mathbb{R}^3} x^i \Delta\Phi\, dx$$

$$= \int_{\mathbb{R}^3} x^i \sigma\, dx. \qquad (7.1.8)$$

Thus m is in fact the total mass of the matter distribution, and $\beta^i = mc^i$ is the first moment of the distribution, giving (for $m \neq 0$) the center of mass c^i.

7.1.2. *Harmonic functions.* We collect here some well-known facts about Euclidean harmonic functions on domains in $\mathbb{R}^n$, $n \geq 2$, a reference for which is [12]; see also [86; 107; 111]. We give some of the proofs, and refer others to exercises or these references.

For $a \in \mathbb{R}^n$ and $R > 0$, let $B_R(a) = \{x \in \mathbb{R}^n : |x - a| < R\}$, so that $\bar{B}_R(a) := \overline{B_R(a)} = \{x \in \mathbb{R}^n : |x - a| \leq R\}$, the boundary $\partial B_R(a)$ of which is a sphere. When a is the origin, we may abbreviate by omitting the center point.

If Ω is a bounded open set with smooth hypersurface boundary $\partial\Omega$, ν will be the outward unit normal, and for a differentiable function u, $\frac{\partial u}{\partial\nu}$ is the directional derivative in this normal direction. In this section, we may use a subscript on the surface measure to denote the variable of integration.

Classically, one defines u to be harmonic in an open set Ω if $u \in C^2(\Omega)$ and $\Delta u = 0$ in Ω. As we mentioned for the more general Poisson equation above, we could interpret $\Delta u = 0$ distributionally: if u is locally integrable, or more generally a distribution, then $\Delta u = 0$ on Ω distributionally means for all $\psi \in C_c^\infty(\Omega)$, $\int_\Omega u\Delta\psi\, dx = 0$. An important classical result is *Weyl's Lemma*: if such a u satisfies Laplace's equation distributionally, then u can be represented by a function in $C^\infty(\Omega)$. In fact, more is true: harmonic functions are real-analytic, and hence obey the unique continuation property [12]. In what follows, then, functions harmonic in an open set can be taken to be smooth there, without further mention; we may, however, note any assumptions about behavior of the function up to the boundary, as in the following proposition.

Proposition 7-8 (mean value property, MVP). *If $u \in C^0(\bar{B}_R(a))$ is harmonic on $B_R(a)$, then*

$$u(a) = \frac{1}{n\Omega_n R^{n-1}} \int_{\{|x-a|=R\}} u(x)\,d\sigma = \frac{1}{\Omega_n R^n} \int_{B_R(a)} u(x)\,dx.$$

Exercise 7-9. Prove the MVP, as follows. Apply Green's identity

$$\int_\Omega (u\,\Delta v - v\,\Delta u)\,dx = \int_{\partial\Omega} \left(u\frac{\partial v}{\partial v} - v\frac{\partial u}{\partial v} \right) d\sigma$$

with $v = |x - a|^{2-n}$ $(n > 2)$ and $\Omega = \{x : \varepsilon < |x - a| < R - \varepsilon\}$. Let $\varepsilon \searrow 0$ and use continuity. The case $n = 2$ is handled analogously.

As a corollary of the MVP, we immediately obtain a *maximum principle*, in both a weak and a strong form.

Proposition 7-10 (maximum principle). *If u is harmonic on a connected open set Ω, and if u attains an absolute maximum or minimum inside Ω, then u is constant. If Ω is a bounded open set for which u is harmonic on Ω and continuous on $\bar{\Omega}$, then the maximum and minimum of u are attained on the boundary of Ω.*

7.1.2.1. *Dirichlet problem.* We now recall how to solve the *Dirichlet problem* on a ball, to determine a harmonic function from its boundary values.

By the maximum principle (Proposition 7-10), there is at most one harmonic function in $C^0(\bar{B}_R)$ with given boundary values, and in fact there is precisely one. For this we use the *Poisson kernel*

$$P_R(x, y) = \frac{R^2 - |x|^2}{n\Omega_n R|x - y|^n}.$$

This is the generalization to all dimensions of the classical Poisson kernel in two dimensions, in which case the kernel can be derived using the series expansion for a holomorphic function on a disk [12].

Proposition 7-11 (Dirichlet problem). *If $\varphi \in C^0(\partial B_R)$, then*

$$u(x) = \int_{\partial B_R} \varphi(y) P_R(x, y)\,d\sigma_y$$

is harmonic on B_R, continuous on $\bar{B}_R$, and $u = \varphi$ on ∂B_R.

For the proof, see [12]. It is not hard to show that $\int_{\partial B_R} P_R(x, y)\,d\sigma_y = 1$ for $x \in B_R$; i.e., the formula holds for $\varphi = 1$ and $u = 1$. Moreover, as x approaches a point ξ on the boundary, $P_R(x, y)$ becomes sharply peaked about ξ, so that its unit integral is concentrated around $y = \xi$, and thus $\int_{\partial B_R} \varphi(y) P_R(x, y) \, d\sigma_y \approx \varphi(\xi)$.

Exercise 7-12 (Green's function). For any $x \in B_R$, let $h_x(y)$ solve $\Delta h_x(y) = 0$ on B_R with $h_x(y) = -\Gamma(x - y)$ for $y \in \partial B_R$. Let $G(x, y) = \Gamma(x - y) + h_x(y)$ be the *Green's function* for B_R. Show that $P_R(x, y) = \partial G / \partial \nu_y$ for $y \in \partial B_R$ by using Green's identity, applying what we know about the distributional Laplacian of Γ (Proposition 7-1), along with the fact that $G = 0$ on ∂B_R.

One can compute a formula for G, using inversion in the sphere, and then derive the above relation to the Poisson kernel [86; 107; 111]; we do not do this here, but we recall inversion in the sphere in Section 7.1.2.3 below.

We now arrive at a key fact about the behavior of harmonic functions at an isolated singularity.

Proposition 7-13 (removable singularities theorem). *An isolated singularity of a bounded harmonic function is removable.*

Proof. By translation and scaling we can, without loss of generality, assume that u is bounded and harmonic on $\bar{B} \setminus \{0\}$, where $B = B_1(0)$. Let v be harmonic on B, continuous on $\bar{B}$, with the same boundary values as u. For $n > 2$ and $\varepsilon > 0$, let $v_\varepsilon(x) = u(x) - v(x) + \varepsilon(|x|^{2-n} - 1)$; for $n = 2$, let $v_\varepsilon(x) = u(x) - v(x) - \varepsilon \log |x|$. Then v_ε is harmonic on the punctured ball, with $v_\varepsilon = 0$ on ∂B, while (since u and v are bounded) $v_\varepsilon(x) \to +\infty$ as $x \to 0$. By the maximum principle, $v_\varepsilon(x) \geq 0$ on $\bar{B} \setminus \{0\}$. Letting $\varepsilon \searrow 0$, we conclude $u - v \geq 0$ on $\bar{B} \setminus \{0\}$, and by replacing u with $(-u)$ in the argument, we can also conclude $u - v \leq 0$ on this domain. Hence $u = v$ on $\bar{B} \setminus \{0\}$, and v is harmonic in a neighborhood of the origin. $\square$

7.1.2.2. *Positive harmonic functions.* We will often be solving for a conformal factor for a metric, and we want such a factor to be positive. So we gather a few basic facts about positive harmonic functions in this section.

Proposition 7-14 (Liouville's theorem). *Any bounded harmonic function on $\mathbb{R}^n$ is constant. In fact, any positive harmonic function on $\mathbb{R}^n$ is a constant. Thus any harmonic function on $\mathbb{R}^n$ bounded either above or below must be constant.*

Proof. Let $u > 0$ be harmonic on $\mathbb{R}^n$. For any $x \in \mathbb{R}^n$, let $R > |x|$ and use the MVP to write

$$u(x) - u(0) = \frac{1}{\Omega_n R^n} \left(\int_{B_R(x)} u(y) \, dy - \int_{B_R(0)} u(y) \, dy \right).$$

Since $u > 0$ and $\big(B_R(x) \setminus B_R(0)\big) \cup \big(B_R(0) \setminus B_R(x)\big) \subset B_{R+|x|}(0) \setminus B_{R-|x|}(0)$, there is a constant $C = C(n, |x|) > 0$ such that

$$|u(x) - u(0)| \leq \frac{1}{\Omega_n R^n} \int_{B_{R+|x|}(0) \setminus B_{R-|x|}(0)} u(y)\, dy$$

$$= \frac{1}{R^n}\big((R + |x|)^n - (R - |x|)^n\big)u(0) \leq C R^{-1}.$$

Let $R \to \infty$ to conclude. $\qquad\square$

Exercise 7-15. Liouville's theorem implies that a bounded harmonic function on $\mathbb{R}^n$ must be constant. What can you say about a harmonic function $u \in L^p(\mathbb{R}^n)$, $1 \leq p < \infty$?

The following corollary is a key property of positive harmonic functions, and its extension to other second-order elliptic operators is a deep result (see [107, Theorem 8.20], for example). The simple proof here just relies on the MVP (cf. [111, p. 109–110]), so that a different proof must be used for other operators.

Proposition 7-16 (Harnack inequality). *Suppose $\Omega \subset \mathbb{R}^n$ is a connected open set, and that $K \subset \Omega$ is compact. There exists $C > 0$ such that for all nonnegative harmonic functions u on Ω, and for all $x, y \in K$,*

$$C^{-1}u(y) \leq u(x) \leq Cu(y).$$

Sketch of proof. We indicate two ways to prove a local inequality; the proof can be completed by a covering argument [12; 107].

Suppose $x \in B_R(y)$ and $\bar{B}_{3R}(y) \subset \Omega$. Then $B_R(y) \subset B_{2R}(x) \subset B_{3R}(y)$, so that since $u \geq 0$, we obtain

$$\frac{1}{\Omega_n R^n} \int_{B_R(y)} u\, dx \leq \frac{2^n}{\Omega_n (2R)^n} \int_{B_{2R}(x)} u\, dx \leq \frac{3^n}{\Omega_n (3R)^n} \int_{B_{3R}(y)} u\, dx.$$

Applying the MVP we obtain $u(y) \leq 2^n u(x) \leq 3^n u(y)$.

Once could also use the Poisson kernel to get a local estimate. To illustrate, suppose $u \geq 0$ is harmonic on $\Omega \supset \bar{B}$, where $B = B_1(0)$. From the Poisson kernel, the MVP, and the sign of u, we obtain for $x \in B$ the inequality

$$\frac{1 - |x|^2}{(1 + |x|)^n}u(0) \leq \underbrace{\frac{1}{\omega_{n-1}} \int_{\partial B} \frac{1 - |x|^2}{|x - \xi|^n} u(\xi)\, d\sigma_\xi}_{u(x)} \leq \frac{1 - |x|^2}{(1 - |x|)^n}u(0).$$

$\qquad\square$

For application to the asymptotics of harmonic functions, we consider positive harmonic functions on $\mathbb{R}^n \setminus \{0\}$. We start with a fundamental result, for proof of which we refer to [12].

Proposition 7-17 (Bôcher's theorem). *Let $\Omega \subset \mathbb{R}^n$ be open and $a \in \Omega$. Suppose u is harmonic on $\Omega \setminus \{a\}$ and positive in a punctured neighborhood of a. Then there is a unique constant $b \geq 0$ and a function v harmonic on Ω, such that on $\Omega \setminus \{a\}$,*

$$u(x) = \begin{cases} b\log(|x-a|^{-1}) + v(x) & \text{for } n = 2, \\ b|x-a|^{2-n} + v(x) & \text{for } n > 2. \end{cases}$$

Liouville's theorem tells us that positive harmonic functions on $\mathbb{R}^n$ must be constant. We can apply Bôcher's theorem to understand positive harmonic functions on punctured Euclidean space.

Proposition 7-18. *If $u > 0$ is harmonic on $\mathbb{R}^n \setminus \{0\}$, where $n \geq 3$, there exist nonnegative numbers a and b such that $u(x) = b|x|^{2-n} + a$.*

A positive harmonic function on $\mathbb{R}^2 \setminus \{0\}$ must be constant.

Proof. If $n \geq 3$, by Bôcher's theorem, we have a function v harmonic in $\mathbb{R}^n$ such that on $\mathbb{R}^n \setminus \{0\}$, $u(x) = b|x|^{2-n} + v(x)$. Thus $\liminf_{|x| \to \infty} v(x) \geq 0$, from which we conclude that v must be constant.

The case $n = 2$ follows directly from Liouville's theorem applied to the harmonic function $z \mapsto u(e^z)$ on $\mathbb{R}^2 \approx \mathbb{C}$, but we suggest as an exercise to prove it using Bôcher's theorem. $\qquad\square$

Example 7-19. Let $n \geq 3$. For $u > 0$, we consider a metric $g = u^{\frac{4}{n-2}} g_{\mathbb{E}^n}$ with zero scalar curvature on $\Omega \subset \mathbb{R}^n$. As we have seen, this translates into u being a positive harmonic function on Ω. Thus with the Liouville and Bôcher theorems in mind, we see that for such u nonconstant, the largest set on which we can define u is $\mathbb{R}^n$ minus a point. By translation we move the puncture to the origin, in which case $u(x) = b|x|^{2-n} + a$ for $a \geq 0$. If $a = 0$ then $b > 0$, and as one should readily check, the metric g is in then (up to constant scaling) the pullback of $g_{\mathbb{E}^n}$ under the inversion in the unit sphere, which we recall in the next section. On the other hand, if $a > 0$, then up to a constant rescaling and a coordinate shift, the metric is a Riemannian Schwarzschild metric (7.1.1).

7.1.2.3. *Harmonic functions at ∞.* In light of the preceding example, it is natural to consider functions which are harmonic outside a compact set, as in Corollary 7-6. As it turns out, there is a device which transforms such functions into functions harmonic in a punctured neighborhood of the origin. To develop this, we recall the formula for *inversion in the unit sphere*, as a mapping on $\mathbb{R}^n \cup \{\infty\}$:

$$x \mapsto x^* = \frac{x}{|x|^2}, \qquad 0 \mapsto \infty, \qquad \infty \mapsto 0.$$

In the $n = 2$ case, with $\mathbb{R}^2 \approx \mathbb{C}$, this map can be expressed as $z \mapsto 1/\bar{z}$. We leave it as an exercise to show that inversion in the sphere is a homeomorphism on the one-point compactification of $\mathbb{R}^n$ (which by stereographic projection is homeomorphic to $\mathbb{S}^n$), and is a *conformal* transformation, so pulling back the Euclidean metric under inversion produces a conformally Euclidean metric.

We now use inversion in the sphere to transform harmonic functions.

Definition 7-20 (Kelvin transform). For functions u defined on an open set $\Omega \subset \mathbb{R}^n \setminus \{0\}$, define $K[u]$ on $\Omega^* = \{x : x^* \in \Omega\}$ by

$$K[u](x) = |x|^{2-n} u(x^*).$$

Exercise 7-21 (a key exercise). Prove that u harmonic on Ω if and only if $K[u]$ is harmonic on Ω^*. One way to do this it to show $\Delta(K[u])(x) = |x|^{-n-2}\Delta u(x^*)$ by direct computation; also see [12, Theorem 4.4].

The Kelvin transform preserves harmonicity, transforming a harmonic function in a neighborhood of infinity to one defined in a neighborhood of the origin. We analyze conditions on the harmonic function near infinity that are encoded in the transform near the origin.

Definition 7-22. We say a function u is *harmonic near infinity* if there is a compact set K such that u is harmonic on $\mathbb{R}^n \setminus K$. If u is harmonic near infinity, we say u is *harmonic at infinity* provided $K[u]$ has a removable singularity at the origin.

The function in Corollary 7-6 is harmonic at infinity:

Proposition 7-23. *Let $n \geq 3$ and suppose u is harmonic near infinity. Then u is harmonic at infinity if and only if* $\lim_{|x|\to\infty} u(x) = 0$.

Proof. From $u(x) = |x|^{2-n} K[u](x^*)$, one direction above is trivial. For the converse, assuming $\lim_{|x|\to\infty} u(x) = 0$, we can adapt the proof of the removable singularities theorem above to conclude. $\qquad\square$

We remark that the Kelvin transform can be used to solve the exterior Dirichlet problem, to find a harmonic function on the exterior of a ball with given values on the boundary sphere, by inverting the solution of the interior Dirichlet problem on the ball. The result is a function which is, by definition, harmonic at infinity; by the preceding proposition, it decays to zero at infinity. Thus, given a continuous function on the boundary of a ball, there is a unique function that goes to zero at infinity with given boundary values. Compare the following simple example.

Example 7-24. Let $n \geq 3$ and $B = B_1(0)$. The function $u(x) = 1 - |x|^{2-n}$ is harmonic on $|x| \neq 0$, and $u = 0$ on ∂B. The unique function harmonic at infinity which solves the exterior Dirichlet problem with zero boundary data on ∂B is the zero function. The harmonic function u is not harmonic *at* infinity, but $u - 1$ is.

Proposition 7-25. *Let $n \geq 3$, and let u be harmonic near infinity. The following statements are equivalent:*

- *u is bounded near infinity.*

- *u is bounded above or below near infinity.*

- *There is a constant a such that $u - a$ is harmonic at infinity.*

- *u has a finite limit as $|x| \to \infty$.*

Proof. Most of the steps are straighforward. One step involves using Bôcher's theorem: if $u > 0$ near infinity, $K[u] > 0$ near the origin, so by Bôcher, there is an a such that $K[u](x) - a|x|^{2-n} = v(x)$ is harmonic near the origin. So $K[v](x) = u(x) - a$ is harmonic near infinity. $\qquad\square$

7.1.2.4. *Spherical harmonic expansion.* Let $n \geq 3$. We note that for a function harmonic near the origin, if you group the terms in the Taylor expansion into homogeneous polynomials, each such is harmonic [12]. If v is harmonic at infinity, by applying the Taylor expansion of $K[v]$ about $x^* = 0$ we have (recall that $B \cdot x = \sum_{i=1}^{n} B^i x^i$),

$$|x^*|^{2-n} v(x) = K[v](x^*) = A + B \cdot x^* + \cdots .$$

Thus

$$v(x) = A|x^*|^{n-2} + B \cdot x^* |x^*|^{n-2} + \cdots = \frac{A}{|x|^{n-2}} + \frac{B \cdot x}{|x|^n} + O_\infty(|x|^{-n});$$

cf. Corollary 7-6. One can compute higher-order terms in the expansion, which come from the homogeneous harmonic polynomials of degree higher than one.

Exercise 7-26. In the case $n = 3$, compute the next term in the expansion of v. What is the dimension of the space of homogeneous harmonic polynomials of degree two in dimension three? (The answer is five; see [12].)

We next explain why this is called the *spherical harmonic expansion*.

Remark 7-27. The eigenfunctions of $\Delta_{\mathring{g}}$ on the round unit sphere $(\mathbb{S}^n, \mathring{g})$ (i.e., nontrivial solutions of $\Delta_{\mathring{g}} u + \lambda u = 0$ for some constant λ) are given by restrictions of the homogeneous harmonic polynomials on $\mathbb{R}^{n+1}$. The lowest eigenvalue is $\lambda_0 = 0$ (constant eigenfunctions), and the next eigenvalue is $\lambda_1 = n$, with eigenfunctions x^i, $i = 1, \ldots, n+1$, restricted to the sphere [12], cf. Exercise 2-43.

Suppose $g = u^{\frac{4}{n-2}} g_{\mathbb{E}^n}$ is harmonically flat at infinity. We may assume, rescaling if necessary, that $\lim_{|x| \to \infty} u(x) = 1$. Then $u(x)$ admits a spherical harmonic expansion near infinity, as $u(x) = 1 + A/|x|^{n-2} + B \cdot x/|x|^n + O_\infty(|x|^{-n})$, cf. Corollary 7-6. As we remarked earlier, we let $m = 2A$ in analogy with the Schwarzschild metric, and the vector B is related to the center of mass, as we point out in the next exercise. Note that metrics which are harmonically flat at infinity asymptote to the Euclidean metric, with the difference on the order $O(|x|^{-(n-2)})$. The leading order deviation in the metric g from the flat metric comes from the mass term in the expansion. Harmonically flat metrics might be thought of as *asymptotic to Schwarzschild*, since $g - g_S$, for g_S given in (7.1.1) with mass $m = 2A$, decays as $O(|x|^{-(n-1)})$, one order better than the deviation from Euclidean. Actually, if we recenter coordinates, we can make the decay another order better, as in the next exercise.

Exercise 7-28. Suppose $g = u^{\frac{4}{n-2}} g_{\mathbb{E}^n}$ is harmonically flat at infinity, with expansion $u(x) = 1 + A/|x|^{n-2} + B \cdot x/|x|^n + O_\infty(|x|^{-n})$ for $|x| > r_0$. Translate coordinates to $y = x - a$, for $a \in \mathbb{R}^n$. For $|y + a| > r_0$, find the asymptotic expansion of $u(x) = u(y + a)$ in terms of y. Show that in case $A \neq 0$, there is a unique $c \in \mathbb{R}^n$ for which $u(y + c) = 1 + A/|y|^{n-2} + O(|y|^{-n})$.

With this important class of initial data sets in mind, we move to more general asymptotically flat initial data in the next section.

7.2. Asymptotically flat initial data

We will give a starting definition of asymptotically flat initial data (g, K), and remark later on natural generalizations of the definition in terms of how the regularity and decay assumptions are imposed. For simplicity, we will assume g and K are smooth; the interested reader can discern when a lower regularity level is sufficient. Recall the notation $O_\ell(|x|^s; \Omega)$ from Definition 7-5.

Definition 7-29. Let $n \geq 3$, $\ell \in \mathbb{Z}_+$, and $q > 0$. An n-dimensional Riemannian manifold $(\mathcal{E}, g)$ is called an *asymptotically flat* (or *asymptotically Euclidean*) *end*, with rate q and order $\ell + 1$, if $\mathcal{E}$ admits *asymptotically flat coordinates* x giving a diffeomorphism of $\mathcal{E}$ and $\Omega := \mathbb{R}^n \setminus \{|x| \leq 1\}$, in which, for $i, j \in \{1, \ldots, n\}$,

$$g_{ij}(x) - \delta_{ij} = O_{\ell+1}(|x|^{-q}; \Omega).$$

An *asymptotically flat initial data set* (g, K) on $\mathcal{E}$ will consist of an asymptotically flat metric g on $\mathcal{E}$ as above, along with an associated symmetric $(0, 2)$-tensor K (equivalently, $\pi = K - (\operatorname{tr}_g K)g$) satisfying for $i, j \in \{1, \ldots, n\}$

$$K_{ij}(x) = O_\ell(|x|^{-q-1}; \Omega).$$

We remark that sometimes in the definition one also imposes some extra decay requirement on $\Phi(g, K)$, i.e., on the energy-momentum density (ρ, J) of (5.2.2)–(5.2.3), beyond that which comes from the conditions on g and K above.

We may refer to either $(\mathcal{E}, g)$, $(\mathcal{E}, g, K)$ or $(\mathcal{E}, g, \pi)$ as an *asymptotically flat end*. (We emphasize that q here is different than in the conformal method from last chapter.) Furthermore, while the definition makes sense for $q > 0$, we will (unless otherwise noted) restrict to $q > \frac{n-2}{2}$, so as to have a well-defined energy-momentum vector of an asymptotically flat end. We will often remind the reader where computations in an end are done with respect to asymptotically flat coordinates x, though we may carry on without explicit remark, and we note that in the definition it is equivalent to employ asymptotic coordinates x for which the end $\mathcal{E}$ corresponds to $|x| > r_0 \geq 1$.

Definition 7-30. A connected complete n-dimensional Riemannian manifold (M, g) is called *asymptotically flat* (or *asymptotically Euclidean*) if there is a compact set $C \subset M$ and a $k \in \mathbb{Z}_+$ such that the components of $M \setminus C$ are given by the (open) sets $\mathcal{E}_1, \ldots, \mathcal{E}_k$, each of which is an asymptotically flat end $(\mathcal{E}_j, g)$. An asymptotically flat initial data set (M, g, K) (or (M, g, π)) will consist of an asymptotically flat (M, g) along with an associated symmetric $(0, 2)$-tensor K (or $\pi = K - (\operatorname{tr}_g K)g$) so that g and K satisfy that asymptotic conditions in Definition 7-29 on each end.

Metrics on $\mathbb{R}^n$ that are harmonically flat at infinity are also asymptotically flat with $q = n - 2$. For example, the Riemannian Schwarzschild metric, given on $\mathbb{R}^n \setminus \{0\}$ by (7.1.1) with $m > 0$, is asymptotically flat with two ends (Exercise 2-37). For $m < 0$, the Riemannian Schwarzschild metric g_S on $\mathbb{R}^n \setminus \left\{ |x| \leq -\frac{m}{2} \right\}$ includes one asymptotically flat end, but as we discussed earlier, this space is not complete. If we excise the singularity in g_S for $m < 0$ by restricting to a manifold $N = \mathbb{R}^n \setminus \left\{ |x| \leq \delta - \frac{m}{2} \right\}$ for any $\delta > 0$, we see the mean curvature vector H of ∂N points out of N (cf. Exercise X-3). On the other hand, in case $m > 0$, if we let $N = \mathbb{R}^n \setminus \left\{ |x| \leq \frac{m}{2} - \delta \right\}$ for $0 < \delta < \frac{m}{2}$, then the mean curvature vector of ∂N points *into* N, in contrast with the $m < 0$ case. Thus should one seek to formulate natural conditions to guarantee a nonnegative mass, this example clearly shows if we allow boundary components in initial data sets, we would need to impose some such condition on the boundary. For simplicity, we will consider the case without boundary in this chapter (whereas asymptotically flat metrics with minimal surface boundary components will play a role in the formulation of the *Riemannian Penrose inequality* in Chapter 9). Even without boundary, we also need to impose some condition on the constraints operator, i.e., on the energy-momentum density (ρ, J), else we could simply patch together any initial data (g, K) to a Schwarzschild end in an elementary manner, producing an asymptotically flat initial data set, with whatever mass m we choose. We will discuss this further in the section on the *positive mass theorem* below, where we will also see that under a natural energy condition, requiring a decay rate $q > n - 2$ above that of harmonically flat metrics will yield only trivial solutions to the constraints, i.e., data from a slice in Minkowski spacetime.

For certain results it will be equivalent to compute quantities in an asymptotic end (possibly sufficiently far out in an end) with respect to g or with respect to the background Euclidean metric induced from the asymptotically flat coordinates. Sometimes we will just refer to such coordinates, but we could also refer to a chosen background metric $\mathring{g}$ on M for which $\mathring{g}_{ij} = \delta_{ij}$ in the asymptotic coordinates near infinity in each end. For example, norms $|\cdot|_g$ and $|\cdot|_{\mathring{g}}$ are uniformly equivalent, and are asymptotic to each other near infinity in any end. The difference tensor $(\nabla - \mathring{\nabla})$ between the two Levi-Civita connections is easily seen to be $O(|x|^{-q-1})$, with norm measured equivalently with respect to either metric, and so in some asymptotic calculations, the background Euclidean connection can be used in place of the metric connection from g.

Likewise, for an asymptotically flat end $(\mathcal{E}, g)$, since g and $g_{\mathbb{E}^n}$ are uniformly equivalent, so are their corresponding volume measures. Thus, for a (Borel measurable) function f, $f \in L^1(\mathcal{E}, dv_g)$ if and only if $f \in L^1(\mathcal{E}, dx)$, and so we say f is *integrable* on $\mathcal{E}$ ($f \in L^1(\mathcal{E})$) under these equivalent conditions. For (M, g) an asymptotically flat manifold, a locally integrable function f is in $L^1(M, dv_g)$ precisely when it belongs to $L^1(\mathcal{E})$ for each end $\mathcal{E}$ of M. In terms of a chosen background metric $\mathring{g}$ on M for which $\mathring{g} = g_{\mathbb{E}^n}$ near infinity in each end as above, $f \in L^1(M, dv_g)$ if and only if $f \in L^1(M, dv_{\mathring{g}})$, which we define to be simply $f \in L^1(M)$. We note that with $2q + 2 > n$, $|x|^{-2q-2} \in L^1(\mathcal{E})$, so that $|K|^2_g \in L^1(\mathcal{E})$ for such asymptotically flat $(\mathcal{E}, g, K)$. Thus from the Einstein constraint equation (5.2.2) (with $\Lambda = 0$), we see that $R(g) \in L^1(\mathcal{E})$ if and only if $\rho \in L^1(\mathcal{E})$. In terms of modeling an isolated system, it seems to be a reasonable assumption to pose that the total energy density of the matter fields is integrable across an asymptotically flat spatial slice, and similarly, for momentum density J, which is proportional to $\mathrm{div}_g \pi$ by the constraint (5.2.3).

Before moving on, we note that while the Einstein summation convention remains in effect, when we have certain expressions that require a summation notation symbol for at least one index, we may, for the sake of clarity, extend the summation notation to indices that are already covered by the Einstein convention.

7.2.1. *The ADM energy and momenta.* We now want to define an energy-momentum vector for an isolated gravitational system, to capture the energy-momentum of the matter fields along with the gravitational field. This will not amount, then, to simply integrating the energy density ρ, since, for example, the Schwarzschild spacetime is vacuum, $\rho = 0$, while we have seen that far field observers will detect what they might describe as a gravitational field outside a massive object. As we will see, the energy-momentum vector will be defined as a limit of flux integrals, for which (7.1.7) gives a Newtonian analogue. We require $q > \frac{n-2}{2}$ in various places in this section, with or without explicit mention.

That said, one might look to a Hamiltonian to define an appropriate notion of energy of a system. We saw the ADM Hamiltonian in Section 5.3.3.2, cf. (5.3.10), and how it generated the equations of motion (5.3.18); we also noted (cf. Remark 5-22 and Exercise 5-30) that the Hamiltonian is modified when boundary terms need to be taken into account. In the asymptotically flat case, metrics we want to consider with finite nonzero energy have decay rates which necessitate modifying the Hamiltonian by such boundary terms (at infinity) (see [16; 187], and earlier works [9; 10; 75]).

Indeed, we consider g and $\pi = -\hat{\pi}$ with decay rates in asymptotic coordinates of the orders $g_{ij} - \delta_{ij} = O_{\ell+1}(|x|^{-q})$ and $\pi^{ij} = O_\ell(|x|^{-q-1})$. Given a timelike vector field V along M, we can extend these coordinates $(x \mapsto \varphi(x) \in M)$ to spacetime coordinates (t, x) such that $\partial/\partial t = V$ along M, for example by using the exponential map $(t, x) \mapsto \exp_{\varphi(x)}(tV|_{\varphi(x)})$. Doing this with a timelike *unit normal* field along M, we get Fermi (Gaussian normal) coordinates $(\mathring{t}, x)$, in which the metric has the form $\bar{g} = -d\mathring{t}^2 + \gamma_{ij} dx^i dx^j$, with $\gamma_{ij}|_{\mathring{t}=0} = g_{ij}$. Given such Fermi coordinates we let $V = N\,\partial/\partial\mathring{t} + X^i\,\partial/\partial x^i$ be a timelike vector field along M, and let (t, x) be adapted coordinates with $\partial/\partial t = V$. We extend N, X^i and g_{ij} off of M ($t = 0$) so that $N\,\partial/\partial\mathring{t} + X^i\,\partial/\partial x^i = \partial/\partial t$ and the metric $\bar{g}$ assumes the *lapse-shift* form $\bar{g} = -N^2\,dt^2 + g_{ij}(dx^i + X^i dt) \otimes (dx^j + X^j dt)$; Fermi coordinates correspond to $N = 1$, $X = 0$. More generally, we will take $N - X^0_\infty = O_{\ell+1}(|x|^{-q})$, $X^i - X^i_\infty = O_{\ell+1}(|x|^{-q})$, for some constants X^μ_∞; cf. [16]. We let $\mathring{g}$ be a background Riemannian metric as above, with $\mathring{g}_{ij} = \delta_{ij}$ in asymptotically flat coordinates near infinity, and let $d\mathring{v} = dv_{\mathring{g}}$. As in Section 5.3.3.2, we let $\theta_g = \sqrt{\det g}/\sqrt{\det \mathring{g}}$, and $\tilde{\pi} = \theta_g\hat{\pi}$.

From the derivation that led to equation (5.3.18), we see that what is needed to generate the correct equations of motion in the asymptotically flat context is a Hamiltonian whose first variation in the direction $(h, \tilde{\sigma})$ is equal to $\int_M (h, \tilde{\sigma}) \cdot_g D\widetilde{\Phi}^*_{(g,\tilde{\pi})}(N, X)\, d\mathring{v}$. The variation $\delta\mathcal{H}_{\mathrm{ADM}}$ of $\mathcal{H}_{\mathrm{ADM}}$ is given by

$\delta \mathcal{H}_{\text{ADM}} = (D\mathcal{H}_{\text{ADM}})_{(g,\tilde{\pi})}(h, \tilde{\sigma}) = \int_M (N, X) \cdot D\widetilde{\Phi}_{(g,\tilde{\pi})}(h, \tilde{\sigma}) \, d\mathring{v}$. (There is another complication, since for such data as above the integral for $\mathcal{H}_{\text{ADM}}$ might not converge; we can either restrict to data for which it does converge, or regularize by subtracting off some background terms, as in [16].) To try to put $\delta \mathcal{H}_{\text{ADM}}$ into the desired form, we integrate by parts, but for bounded (N, X), and for h decaying at the same rate as $(g - \mathring{g})$ and $\tilde{\sigma}$ decaying at the same rate as $\tilde{\pi}$, this does not necessarily give us the required identity in general: if we use a spherical exhaustion of the asymptotic ends, say, then integration by parts leaves terms on the boundary that may not limit to zero as the spheres tend to infinity.

We claim that the terms from $\delta \mathcal{H}_{\text{ADM}} = (D\mathcal{H}_{\text{ADM}})_{(g,\tilde{\pi})}(h, \tilde{\sigma})$ that require analysis are $\int_M -\big(N L_g h + 2(\text{div}_g \hat{\sigma})_i X^i\big) \, dv_g$, with the other terms either not involving derivatives of $(h, \tilde{\sigma})$ (and so not contributing a boundary term) or decaying fast enough so that the contributed boundary integral upon integration by parts tends to zero at infinity. For instance, we infer from equation (5-28a) (Exercise 5-28) that the boundary terms to focus on will come from integration by parts on $\int_M (N, X) \cdot D\widehat{\Phi}_{(g,\hat{\pi})}(h, \hat{\sigma}) \, dv_g$. It is easy to see the only boundary terms from the N-component of the integrand will come from $\,-N L_g h$, cf. equation (5-29b). As for the X-component, the relevant terms arise from the variation of $-2(\text{div}_g \hat{\pi})_i X^i = -2g_{ij} X^i (\hat{\pi}^{jk}_{\ ,k} + \hat{\pi}^{mk}\Gamma^j_{mk} + \hat{\pi}^{jm}\Gamma^k_{km})$, given by (see (5.3.9)–(5.3.10))

$$-2h_{ij} X^i (\text{div}_g \hat{\pi})^j - 2X^i (\text{div}_g \hat{\sigma})_i - 2g_{ij} X^i (\hat{\pi}^{mk}\delta\Gamma^j_{mk} + \hat{\pi}^{jm}\delta\Gamma^k_{km}).$$

The first of these terms will not contribute to the boundary integral, while for the last term we recall that $\delta\Gamma^j_{mk} = \frac{1}{2}g^{js}(h_{sk;m} + h_{ms;k} - h_{mk;s})$. Integrating a term like $2g_{ij} X^i \hat{\pi}^{mk}\delta\Gamma^j_{mk}$ by parts results in several boundary integral terms, the first of which is $X^i \hat{\pi}^{mk} h_{ik}(v_g)_m = O(|x|^{-2q-1})$, which tends to zero in a limit of integrals over large coordinate spheres $\{|x| = r\}$ as $r \to \infty$ $\big(\text{with } q > \frac{n-2}{2}\big)$; the other terms behave similarly.

As for the integral $\int_M -(N L_g h + 2(\text{div}_g \hat{\sigma})_i X^i) \, dv_g$, the term $-N L_g h = -N(-\Delta_g \text{tr}_g h + g^{ik}g^{jm}h_{ij;km} - h \cdot_g \text{Ric}(g))$ contributes a boundary integrand

$$(N(\text{tr}_g h)_{,j} - N_{,j}\text{tr}_g h - Ng^{ik}h_{ij;k} + N_m g^{sm}h_{js})v_g^j$$

$$= N \sum_{i,j=1}^{n} (h_{ii,j} - h_{ij,i})v^j + O(|x|^{-2q-1}),$$

where we swapped out the normal vector for the Euclidean normal (see Exercise 7-35 below). For the divergence term, the boundary integrand is simply

$$-2\hat{\sigma}_{ij} X^i v_g^j = -2\hat{\sigma}_{ij} X^i v^j + O(|x|^{-2q-1}).$$

In order to get the desired linearization for the Hamiltonian, and subsequently get the correct equations of motion, we need to add terms to $\mathcal{H}_{\mathrm{ADM}}$ to cancel out these boundary contributions, resulting in what we will call $\mathcal{H}_{\mathrm{RT}}$, so that the variation $\delta\mathcal{H}_{\mathrm{RT}}$ in the direction $(h, \tilde{\sigma}) = (h, \theta_g \hat{\sigma})$ is given by, using Exercise 7-35 to exchange the measure $d\sigma_g$ with the Euclidean measure $d\sigma$ without affecting the existence and value of the limit,

$$\delta\mathcal{H}_{\mathrm{RT}} = \lim_{r\to\infty} \int_{\{|x|=r\}} \left(N \sum_{i,j=1}^{n} (h_{ij,i} - h_{ii,j})v^j + 2\hat{\sigma}_{ij}X^i v^j \right) d\sigma + \delta\mathcal{H}_{\mathrm{ADM}}.$$

Thus we define, when the limit exists,

$$\mathcal{H}_{\mathrm{RT}} = \lim_{r\to\infty} \int_{\{|x|=r\}} \left(N \sum_{i,j=1}^{n} (g_{ij,i} - g_{ii,j})v^j + 2\hat{\pi}_{ij}X^i v^j \right) d\sigma + \mathcal{H}_{\mathrm{ADM}}.$$

We write

$$\mathcal{H}_{\mathrm{RT}} = 2\kappa\, \mathcal{H}^\kappa_{\mathrm{RT}},$$

so that $\mathcal{H}^\kappa_{\mathrm{RT}}$ has units of energy (see Remark 5-23). For solutions of the vacuum constraint equations, we have $\mathcal{H}_{\mathrm{ADM}} = 0$, and so should the boundary integrals exist, the value of the Hamiltonian $\mathcal{H}^\kappa_{\mathrm{RT}}$ would be

$$X^0_\infty \cdot \frac{1}{2\kappa} \lim_{r\to\infty} \int_{\{|x|=r\}} \sum_{i,j=1}^{n} (g_{ij,i} - g_{ii,j})v^j d\sigma$$
$$-X^i_\infty \cdot \frac{1}{\kappa} \lim_{r\to\infty} \int_{\{|x|=r\}} \pi_{ij}X^i v^j d\sigma. \qquad (7.2.1)$$

We will henceforth use units for which $c = 1$, $G = 1$ and $\kappa = (n-1)\omega_{n-1}$. If $X^\mu_\infty \partial/\partial x^\mu$ is a timelike unit vector (relative to the asymptotic background Minkowski spacetime) tangent to the spacetime path associated to an observer "at infinity" (far out in the asymptotic end), then (7.2.1) should represent the energy the observer would ascribe to the isolated gravitational system. Thus we can isolate boundary integrals which yield the ADM energy-momentum vector associated to the isolated gravitational system, which we do below.

Before defining the ADM energy-momentum vector, we analyze the convergence of the asymptotic boundary integrals more carefully, starting with an expansion of the constraint operator in an asymptotic end.

Proposition 7-31. *Suppose $(\mathcal{E}, g)$ is an asymptotically flat end, with asymptotically flat coordinates x with respect to which we take a background Euclidean metric $g_{\mathbb{E}^n}$ with components δ_{ij}, and we write $g_{ij} = \delta_{ij} + h_{ij}$. Then the scalar* *curvature $R(g)$ of g satisfies the expansion*

$$R(g) = \sum_{i,j=1}^{n} (g_{ij,ij} - g_{ii,jj}) + O(|x|^{-2q-2}) = L_{g_{\mathbb{E}^n}} h + O(|x|^{-2q-2}),$$

where $L_{g_{\mathbb{E}^n}}$ is the linearization of the scalar curvature operator at the Euclidean metric, $L_{g_{\mathbb{E}^n}} h = -\Delta_{g_{\mathbb{E}^n}} (\mathrm{tr}_{g_{\mathbb{E}^n}} h) + \mathrm{div}_{g_{\mathbb{E}^n}} h$ (cf. Lemma 2-7).

Proof. The second formula follows directly from the first. In the asymptotic coordinates, $\Gamma_{ij}^k = O(|x|^{-q-1})$. Thus for the scalar curvature we find

$$R(g) = g^{ij}\left(\Gamma_{ij,k}^k - \Gamma_{ik,j}^k + \Gamma_{km}^k \Gamma_{ij}^m - \Gamma_{jk}^m \Gamma_{im}^k\right) = g^{ij}(\Gamma_{ij,k}^k - \Gamma_{ik,j}^k) + O(|x|^{-2q-2}).$$

From $g_{,\ell}^{km} = -g^{ki} g_{ij,\ell} g^{jm} = O(|x|^{-q-1})$, we find

$$\Gamma_{ij,\ell}^k = \tfrac{1}{2} g^{km}(g_{mj,i\ell} + g_{im,j\ell} - g_{ij,m\ell}) + O(|x|^{-2q-2}).$$

Since $g_{ij} - \delta_{ij} = O(|x|^{-q})$ we have $g^{ij} - \delta^{ij} = O(|x|^{-q})$ as well, so that

$$R(g)$$
$$= g^{ij}(\Gamma_{ij,k}^k - \Gamma_{ik,j}^k) + O(|x|^{-2q-2}) = \delta^{ij}(\Gamma_{ij,k}^k - \Gamma_{ik,j}^k) + O(|x|^{-2q-2})$$
$$= \tfrac{1}{2}\delta^{ij}\delta^{km}\left((g_{mj,ik} + g_{im,jk} - g_{ij,mk}) - (g_{mk,ij} + g_{im,kj} - g_{ik,mj})\right) + O(|x|^{-2q-2}),$$

from which the desired result follows. $\square$

Exercise 7-32. Show that if $(\mathcal{E}, g, \pi)$ is an asymptotically flat end, then in asymptotically flat coordinates x,

$$\mathrm{div}_g \pi = \mathrm{div}_{g_{\mathbb{E}^n}} \pi + O(|x|^{-2q-2}).$$

Thus we see that, with $q > \frac{n-2}{2}$, if ρ and J from the constraint equations are integrable on $\mathcal{E}$, then so are $\sum_{i,j=1}^{n} (g_{ij,ij} - g_{ii,jj}) = L_{g_{\mathbb{E}^n}} h$ and $\mathrm{div}_{g_{\mathbb{E}^n}} \pi$, which immediately gives us the following proposition.

Proposition 7-33. *Suppose $(\mathcal{E}, g, K)$ is an asymptotically flat initial data set on an end $\mathcal{E}$, with $q > \frac{n-2}{2}$ and $R(g) - |K|_g^2 + (\mathrm{tr}_g K)^2 = 2\kappa\rho$ and $\mathrm{div}_g \pi = \kappa J$ integrable on $\mathcal{E}$. Then the following limits, computed in asymptotically flat coordinates x, exist, where $v^j = x^j/|x|$:*

$$\lim_{r \to \infty} \int_{\{|x|=r\}} \sum_{i,j=1}^{n} (g_{ij,i} - g_{ii,j}) \, v^j \, d\sigma, \qquad \lim_{r \to \infty} \int_{\{|x|=r\}} \sum_{j=1}^{n} \pi_{ij} \, v^j \, d\sigma.$$

Proof. The proof is by application of the divergence theorem: for $r > r_0 > 1$,

$$\int_{\{|x|=r\}} \sum_{i,j=1}^{n} (g_{ij,i} - g_{ii,j})\, v^j\, d\sigma - \int_{\{|x|=r_0\}} \sum_{i,j=1}^{n} (g_{ij,i} - g_{ii,j})\, v^j\, d\sigma$$

$$= \int_{\{r_0 \le |x| \le r\}} \sum_{i,j=1}^{n} (g_{ij,ij} - g_{ii,jj})\, dx.$$

Since, as remarked just above, the integrand is integrable, we see that the difference in flux integrals can be made small by taking r_0 sufficiently large. Similarly, the second limit follows from

$$\int_{\{|x|=r\}} \sum_{j=1}^{n} \pi_{ij} v^j\, d\sigma - \int_{\{|x|=r_0\}} \sum_{j=1}^{n} \pi_{ij}\, v^j\, d\sigma = \int_{\{r_0 \le |x| \le r\}} (\operatorname{div}_{g_{\mathbb{E}^n}} \pi)_i\, dx. \qquad \square$$

We now define the ADM energy-momentum vector.

Definition 7-34. Let $(\mathcal{E}, g, \pi)$ be an n-dimensional asymptotically flat end, with asymptotically flat coordinates x, and with ρ and J integrable. We define the ADM energy E and linear momentum P as follows:

$$E = \frac{1}{2(n-1)\omega_{n-1}} \lim_{r \to \infty} \int_{\{|x|=r\}} \sum_{i,j=1}^{n} (g_{ij,i} - g_{ii,j})\, v^j\, d\sigma, \qquad (7.2.2)$$

$$P_i = \frac{1}{(n-1)\omega_{n-1}} \lim_{r \to \infty} \int_{\{|x|=r\}} \sum_{j=1}^{n} \pi_{ij}\, v^j\, d\sigma. \qquad (7.2.3)$$

Here ω_{n-1} is the area of the unit $(n-1)$-sphere, so that in case $n = 3$, say, the constant in front of the E-integral is $\frac{1}{16\pi}$.

If the ADM energy-momentum vector is future-pointing causal, i.e. $E \ge |P|$ $(c = 1)$, we let $m = \sqrt{E^2 - |P|^2}$ define the *ADM mass* m. In the time-symmetric case $(\mathcal{E}, g)$, it is customary to call $m = E$ the ADM mass (regardless of its sign).

The integrand in (7.2.2) equals $\sum_{i,j=1}^{n} (h_{ij,i} - h_{ii,j})v^j$, for $h_{ij} = g_{ij} - \delta_{ij}$. As a corollary of the next simple exercise, the normal vector and the measure in the above integrals (7.2.2)–(7.2.3) can be replaced by v_g and $d\sigma_g$, respectively, without altering the limit, for $q > \frac{n-2}{2}$.

Exercise 7-35. a. Show that (as measures on $\{|x| = r\}$) $|d\sigma_g - d\sigma| \le O(r^{-q})d\sigma$.

b. Suppose that $v = x^j |x|^{-1} \partial/\partial x^j$, while v_g is the g-normal unit vector field to the coordinate spheres of constant $|x|$ in an asymptotic end $\mathcal{E}$, with $\langle v_g, v \rangle_g > 0$ (equivalently $\langle v_g, v \rangle_{g_{\mathbb{E}^n}} > 0$). Show that $|v_g - v| = O(|x|^{-q})$.

c. Conclude as noted above that ν and $d\sigma$ can be replaced by ν_g and $d\sigma_g$, respectively, in (7.2.2)–(7.2.3) without changing the values of the limits.

As you might expect from the way in which the existence of the limits is established, one can show that they can be computed on more general surfaces tending to infinity, and moreover the values are independent of the asymptotically flat coordinate chart in which they are computed; see, e.g., [15; 16].

For coordinates x^μ in which a spacetime metric asymptotes appropriately to a Minkowski metric with these as inertial coordinates, we let $E = P^0 = -P_0$ and $P^i = P_i$ be defined using these asymptotic coordinates on a constant time slice, and we define the ADM energy-momentum four-vector $\mathbb{P} = P^\mu \partial/\partial x^\mu$. Using (7.2.1), we see that if $\mathbb{U}_{\mathrm{obs}} = X_\infty^\mu \partial/\partial x^\mu$ is a timelike unit vector tangent to the spacetime path of an observer "at infinity", then the energy of the isolated gravitational system as measured by the observer should be $E_{\mathrm{obs}} = -\langle \mathbb{U}_{\mathrm{obs}}, \mathbb{P}\rangle = -X_\infty^\mu P_\mu = X_\infty^0 E - X_\infty^i P_i$.

In the next exercise we see the energy integral in fact picks up the mass m in the Schwarzschild metric, and more generally it picks up the relevant coefficient in the spherical harmonic expansion of the conformal factor for harmonically flat metrics. Note that pairing $K = 0$ with each of these metrics gives solutions to the *vacuum* constraint equations, and thus E can be nonzero even in vacuum.

Exercise 7-36. a. Show that $E = m$ for an asymptotically flat end in the Riemannian Schwarzschild metric g_S (7.1.1).

b. Show that $E = 2A$ for any harmonically flat end $(\mathcal{E}, u^{\frac{4}{n-2}} g_{\mathbb{E}^n})$ with $\Delta u = 0$ and $u(x) = 1 + A/|x|^{n-2} + O_\infty(|x|^{-(n-1)})$.

c. Show that $E = 2ab$ for $(\mathcal{E}, u^{\frac{4}{n-2}} g_{\mathbb{E}^n})$, with $u > 0$, $\Delta u = 0$, and $u(x) = a + b/|x|^{n-2} + O_\infty(|x|^{-(n-1)})$. (Hint: Rescale to asymptotically flat coordinates.)

Part b. of the preceding exercise is a special case of a result about the effect of a conformal change on the ADM energy that will be important later in the chapter.

Lemma 7-37. *Suppose $(\mathcal{E}, g)$ is an asymptotically flat end with ADM energy $E(g)$. Suppose $u > 0$ on $\mathcal{E}$, admitting an expansion in asymptotically flat coordinates x of the form $u(x) = 1 + A/|x|^{n-2} + O_1(|x|^{-(n-2)-\gamma})$, for some $\gamma > 0$. If $\tilde{g} = u^{\frac{4}{n-2}} g$, then $E(\tilde{g}) = E(g) + 2A$.*

Proof. This follows immediately from the following identity, which you should readily verify, with $\xi := \min(q, \gamma) > 0$:

$$\sum_{i=1}^{n} (\tilde{g}_{ij,i} - \tilde{g}_{ii,j}) = \sum_{i=1}^{n} (g_{ij,i} - g_{ii,j}) - \frac{4(n-1)}{n-2} \frac{\partial u}{\partial x^j} + O(|x|^{-(\xi+n-1)}). \quad \square$$

The ADM energy-momentum has been defined in terms of the initial data, whereas one might also like to understand these quantities relative to the spacetime picture. For instance, one could ask how these quantities transform under a *boost* $y^\mu = \Lambda^\mu{}_\nu x^\nu$, or slightly more generally $y^\mu = \Lambda^\mu{}_\nu x^\nu + a^\mu$, where $\Lambda^\mu{}_\nu$ is a proper (time-orientation preserving) Lorentz transformation, and a^μ is a constant spacetime vector representing a shift in coordinates (e.g., a time translation). Consider a suitably asymptotically flat (Minkowskian) spacetime with asymptotic coordinates x^μ, in which the initial data set corresponds to $x^0 = 0$ (at least near infinity, say), and contains for some $\theta_0 > 0$ the spacelike hypersurfaces defined near infinity by $y^0 = 0$, for $y^\mu = \Lambda^\mu{}_\nu x^\nu$, for proper Lorentz transformations with boost angle $|\theta| < \theta_0$; cf. [56], e.g., for the existence of such boosted slices in the spacetime evolution from asymptotically Euclidean initial data. One can then show as in Exercise 7-98, following [211, Chapter 7], [161, Chapter 20], cf. [61, Appendix E], that the energy-momentum vector transforms as a Minkowskian vector under a Lorentz transformation: if P^μ are the components of the energy-momentum computed in the x-coordinate chart, and $\widetilde{P}^\mu$ are those computed in the y-coordinate chart, then $\widetilde{P}^\mu = \Lambda^\mu{}_\nu P^\nu$, i.e.,

$$\tilde{P}^\mu \frac{\partial}{\partial y^\mu} = P^\nu \Lambda^\mu{}_\nu \frac{\partial}{\partial y^\mu} = P^\nu \frac{\partial y^\mu}{\partial x^\nu} \frac{\partial}{\partial y^\mu} = P^\nu \frac{\partial}{\partial x^\nu}.$$

See Exercise 7-97 for explicit confirmation of this for a family of boosted slices in the Schwarzschild spacetime. In terms of thinking of the spacetime geometry and boosted slices, it might be a good time to recall that in the conformal compactification of Minkowski spacetime (see Section 1.4.2), spacelike curves with $r \to +\infty$ all "end" at a single point i^0, spacelike infinity. Finally, note that if you simply switch the time orientation in the asymptotic regime, E does not change, but since the second fundamental form changes sign, the linear momentum changes sign.

As we see from the form of the expansion of $R(g)$ above, and from the Hamiltonian $\mathcal{H}_{RT}$, the energy and momentum integrals are directly related to the linearized constraint operator $D\Phi_{(g_{\mathbb{E}^n},0)}(h,\sigma) = (L_{g_{\mathbb{E}^n}}h, \mathrm{div}_{g_{\mathbb{E}^n}}\sigma)$. By integration of the linearized constraints operator against elements of the kernel of the adjoint operator $D\Phi^*_{(g_{\mathbb{E}^n},0)}$ at the Minkowski initial data $(g_{\mathbb{E}^n}, 0)$ one obtains asymptotic conserved quantities. Recall that $D\Phi^*_{(g_{\mathbb{E}^n},0)}(N,X) = (0,0)$ for N a constant plus a linear combination of Cartesian coordinate functions x^i, and X a Euclidean Killing field (a linear combination of generators of translations and rotations). Now, if $D\Phi^*_{(g_{\mathbb{E}^n},0)}(N,X) = (0,0)$, then (with the dot product below induced by the Euclidean metric) we have, by integration by parts,

$$\int_{\{r_0 \le |x| \le r\}} (N, X) \cdot D\Phi_{(g_{\mathbb{E}^n}, 0)}(h, \pi) \, dx$$

$$= \int_{\{r_0 \le |x| \le r\}} \left(N \sum_{i,j=1}^{n} (h_{ij,i} - h_{ii,j})_{,j} + \sum_{i,j=1}^{n} X^i \pi_{ij,j} \right) dx = \mathcal{B}(r) - \mathcal{B}(r_0),$$

where $\mathcal{B}(r)$ is given by

$$\int_{\{|x|=r\}} \sum_{j=1}^{n} \left(\sum_{i=1}^{n} \pi_{ij} X^i + N \sum_{i=1}^{n} (h_{ij,i} - h_{ii,j}) - \sum_{i=1}^{n} (N_{,i} h_{ij} - N_{,j} h_{ii}) \right) v^j \, d\sigma.$$

Here we used the fact that $X^i = X_i$ in Cartesian coordinates, so the components of the Lie derivative $L_X g_{\mathbb{E}^n}$ are given by $(L_X g_{\mathbb{E}^n})_{ij} = X_{i,j} + X_{j,i} = X^i_{,j} + X^j_{,i}$, which vanishes for a Euclidean Killing field. Therefore, by symmetry of π, $\sum_{i,j=1}^{n} X^i_{,j} \pi_{ij} = \sum_{i,j=1}^{n} \frac{1}{2}(X^i_{,j} + X^j_{,i}) \pi_{ij} = 0$. We also used that

$$0 = L^*_{g_{\mathbb{E}^n}} N = h \cdot (-(\Delta_{g_{\mathbb{E}^n}} N) g_{\mathbb{E}^n} + \text{Hess}_{g_{\mathbb{E}^n}} N) = \sum_{i,j=1}^{n} (-N_{,jj} h_{ii} + N_{,ij} h_{ij}).$$

The boundary integral for E comes from $(N, X) = (1, 0)$, while the integral for P_i comes from $(N, X) = (0, \partial/\partial x^i)$. If $(g, \pi) = (g_{\mathbb{E}^n} + h, \pi)$ is asymptotically flat, then in asymptotically flat coordinates x on an asymptotic end, $\Phi(g, \pi) = \Phi(g_{\mathbb{E}^n}, 0) + D\Phi_{(g_{\mathbb{E}^n}, 0)}(h, \pi) + O(|x|^{-2q-2}) = D\Phi_{(g_{\mathbb{E}^n}, 0)}(h, \pi) + O(|x|^{-2q-2})$, where the error term comes from a sum of terms of the algebraic forms $h * \partial^2 h$, $\partial h * \partial h$, $\pi * \pi$, $\partial h * \pi$ and $h * \partial \pi$, the latter term not needed when π is treated a $(2, 0)$-tensor. Here "$*$" means a linear combination, with smooth bounded coefficients, of terms quadratic in the components (or their partials) of h and π, as indicated. Thus if (g, π) is a solution to the vacuum constraint $\Phi(g, \pi) = 0$, then $D\Phi_{(g_{\mathbb{E}^n}, 0)}(h, \pi) = O(|x|^{-2q-2})$.

Exercise 7-38. Verify the stated form of the error term in the expansion of $\Phi(g, \pi)$ above, and confirm the order estimate.

These kernel elements of $D\Phi^*_{(g_{\mathbb{E}^n}, 0)}$ correspond to symmetries of the Minkowski spacetime, and so represent asymptotic symmetries of an asymptotically flat spacetime. Indeed $(N, X) = (1, 0)$ corresponds to time translation generated by $\partial/\partial t$, and $(N, X) = (0, \partial/\partial x^i)$ corresponds to spatial translation, and they are respectively associated to conserved quantities which are the energy and linear momentum.

We now consider the other remaining basic kernel elements of $D\Phi^*_{(g_{\mathbb{E}^n},0)}$. For instance, suppose $(N, X) = (x^k, 0)$, then

$$\mathcal{B}(r) = \int_{\{|x|=r\}} \left(x^k \sum_{i,j=1}^{n} (h_{ij,i} - h_{ii,j}) - \sum_{i,j=1}^{n} (\delta_i^k h_{ij} - \delta_j^k h_{ii}) \right) v^j \, d\sigma. \quad (7.2.4)$$

This integral need not converge as $r \to \infty$ for general asymptotically flat initial data, even for solutions of the vacuum constraints. However, when there are asymptotically flat coordinates for which the data enjoys sufficient asymptotic parity, then the above surface integral will converge. We illustrate with an example.

Exercise 7-39. Suppose $g = u^{\frac{4}{n-2}} g_{\mathbb{E}^n} = g_{\mathbb{E}^n} + h$ is harmonically flat at infinity, with expansion $u(x) = 1 + A/|x|^{n-2} + B \cdot x/|x|^n + O_\infty(|x|^{-n})$ for $|x|$ sufficiently large. Show that

$$\frac{4(n-1)}{n-2} \omega_{n-1} B^k$$

$$= \lim_{r \to \infty} \int_{\{|x|=r\}} \left(x^k \sum_{i,j=1}^{n} (h_{ij,i} - h_{ii,j}) - \sum_{i,j=1}^{n} (\delta_i^k h_{ij} - \delta_j^k h_{ii}) \right) v^j \, d\sigma. \quad (7\text{-}39a)$$

Thus for $m = 2A \neq 0$, the limit above is equal to $2(n-1)\omega_{n-1} m c^k$, where $c^k = \dfrac{B^k}{(n-2)A}$ gives the center of mass as in Exercise 7-28.

Thus for an asymptotically flat metric g with $m \neq 0$, with respect to an asymptotically flat chart for which $\lim_{r \to \infty} \mathcal{B}(r)$, with $\mathcal{B}(r)$ from (7.2.4), exists, we *define* the center of mass c^k as follows (compare (7.1.8)):

$$2(n-1)\omega_{n-1} m c^k$$

$$= \lim_{r \to \infty} \int_{\{|x|=r\}} \left(x^k \sum_{i,j=1}^{n} (h_{ij,i} - h_{ii,j}) - \sum_{i,j=1}^{n} (\delta_i^k h_{ij} - \delta_j^k h_{ii}) \right) v^j \, d\sigma. \quad (7.2.5)$$

Lastly we have the rotational Euclidean Killing fields, e.g., in dimension $n = 3$, we have three generators $Y_{(i)}$, $i = 1, 2, 3$, given by

$$Y_{(1)} := x^2 \frac{\partial}{\partial x^3} - x^3 \frac{\partial}{\partial x^2}, \quad Y_{(2)} := x^3 \frac{\partial}{\partial x^1} - x^1 \frac{\partial}{\partial x^3}, \quad Y_{(3)} := x^1 \frac{\partial}{\partial x^2} - x^2 \frac{\partial}{\partial x^1}.$$

With $(N, X) = (0, Y_{(i)})$, then, we have the surface integrals

$$\mathcal{B}(r) = \int_{\{|x|=r\}} \pi_{jk} Y_{(i)}^k v^j \, d\sigma.$$

Just as with the center of mass, when there is enough asymptotic parity symmetry in the data, the surface integrals approach a limit for $r \to \infty$, and we define the *angular momentum*

$$\mathcal{J}_i = \frac{1}{8\pi} \lim_{r\to\infty} \int_{\{|x|=r\}} \pi_{jk} Y^k_{(i)} \nu^j \, d\sigma. \tag{7.2.6}$$

(In $n \geq 3$ dimensions, 8π is replaced by $(n-1)\omega_{n-1}$.) See [161, Chapter 19] for the appearance of the angular momentum in the spacetime asymptotics, and see Exercise 7-106 for initial data asymptotics.

Asymptotic parity coordinate conditions on asymptotically flat $\big($at rate $q > \frac{n-2}{2}\big)$ initial data sufficient to guarantee convergence of the center of mass and angular momentum integrals were proposed by Regge and Teitelboim [187]: for some $\ell \in \mathbb{Z}_+$,

$$g_{ij}(x) - g_{ij}(-x) = O_{\ell+1}(|x|^{-q-1}) \tag{7.2.7}$$

$$K_{ij}(x) + K_{ij}(-x) = O_{\ell}(|x|^{-q-2}). \tag{7.2.8}$$

The parity condition on K_{ij} could be replaced by the analogue for π_{ij}. We note that the component conditions can be interpreted as relating g or K to the respective pullback of g or K under the antipodal map $x \mapsto -x$. Moreover under these conditions it is easy to see that we can replace π_{jk} with K_{jk} in (7.2.6). See Exercise 7-99 for the transformation rule for the angular momentum under boosts and translations of the center of mass, cf. Exercise 7-106, as well as [61, Appendix E] (watch sign conventions).

Exercise 7-40 (cf. [159]). From the proof of Proposition 7-33, the limit of integrals defining the energy converges if and only if $\lim_{r\to\infty} \int_{\{r_0 \leq |x| \leq r\}} R(g)\,dx$ exists. Show likewise that under condition (7.2.7) (with $\ell = 1$), the limit of integrals defining the center of mass as in (7.2.5) converges if and only if $\lim_{r\to\infty} \int_{\{r_0 \leq |x| \leq r\}} x^k R(g)\,dx$ exists, for each k.

Exercise 7-41. a. Verify the above claim that under the Regge–Teitelboim conditions, one can replace π_{jk} by K_{jk} in (7.2.6).

b. Consider the initial data on a constant t-slice in the Boyer–Lindquist coordinates used for the Kerr spacetime (2.4.6). Introduce asymptotically flat coordinates $x = r\cos\theta\sin\phi$, $y = r\sin\theta\sin\phi$, $z = r\cos\phi$, and show that the shift vector is given by

$$X^1 = \frac{2yma}{r\rho^2}, \quad X^2 = -\frac{2xma}{r\rho^2}, \quad X^3 = 0.$$

c. Show that this initial data set satisfies the Regge–Teitelboim conditions (7.2.7)–(7.2.8), so that the angular momentum flux integrals converge. To do this, you might use the ADM equation (5.3.4). Show that the nonzero component of angular momentum is given by $\mathcal{J}_3 = am$.

Remark 7-42. The Kerr spacetime can also be expressed in *Kerr–Schild form* as

$$g_{\mu\nu} = \eta_{\mu\nu} + \frac{2m\tilde{r}^3}{\tilde{r}^4 + a^2 z^2}\theta_\mu\theta_\nu,$$

where $\eta = -dt^2 + dx^2 + dy^2 + dz^2$ is the Minkowski metric, and

$$\theta_\mu dx^\mu = dx^0 - \frac{1}{\tilde{r}^2 + a^2}\left(\tilde{r}(x\,dx + y\,dy) + a(x\,dy - y\,dx)\right) - \frac{z}{\tilde{r}}dz,$$

with $\tilde{r}$ defined implicitly as the solution of the equation

$$\tilde{r}^4 - \tilde{r}^2(x^2 + y^2 + z^2 - a^2) - a^2 z^2 = 0.$$

See [23, (37)] for the second fundamental form of the constant t-slice in these coordinates, where it is shown that the leading order terms possess a form which allows the angular momentum flux integrals to converge.

7.2.2. *Weighted spaces.* We will now introduce weighted function spaces that are designed to capture growth or decay rates of functions and tensors near infinity. One can generalize the notion of asymptotically flat initial data sets in terms of these spaces. Moreover, we will want to utilize the properties of the Laplace operator between such weighted spaces in order to achieve certain kinds of conformal deformations on asymptotically flat manifolds. In particular we will construct certain deformations of the scalar curvature on asymptotically flat manifolds, with sufficient control on the change in the ADM mass induced by such conformal transformations. We remark that in their proof of the Riemannian positive mass theorem, Schoen and Yau [199] prove the necessary existence and expansion of solutions to the PDE giving the required deformations of scalar curvature directly, via a Green's function expansion. As the theory of weighted spaces has now been sufficiently developed and applied to the study of ADM mass (cf. e.g., [15]), we will discuss the definitions, fundamental results, and basic ideas behind the proofs.

Basic definitions and properties. We now introduce the basic definitions of weighted Sobolev and Hölder spaces on $\mathbb{R}^n$. Let $\sigma \geq 1$ be a smooth function identical to $r = |x|$ for $|x| \geq 2$. For weighted Sobolev spaces, we start with the $L^p_{-\tau}$-norm ($1 \leq p < \infty$):

$$\|u\|^p_{L^p_{-\tau}} = \int_{\mathbb{R}^n} |u|^p \sigma^{\tau p - n}\,dx = \int_{\mathbb{R}^n} (|u|\sigma^\tau)^p \sigma^{-n}\,dx.$$

For $\tau > 0$, the exponent on the weight factor is chosen to encourage decay of u at infinity (and also to rescale the volume measure). The norm

$$\|u\|_{W^{k,p}_{-\tau}} = \sum_{|\gamma| \leq k} \|\partial^\gamma u\|_{L^p_{-\tau-|\gamma|}}$$

determines the spaces $W^{k,p}_{-\tau}(\mathbb{R}^n)$, considered as the closure of the space of smooth compactly supported functions in the given norm, or equivalently defined in terms of weak derivatives. Note that $W^{0,p}_{-\tau}(\mathbb{R}^n) = L^p_{-\tau}(\mathbb{R}^n)$, while $L^p(\mathbb{R}^n) = L^p_{-n/p}(\mathbb{R}^n)$, and moreover for $\ell := |\gamma| \leq k$, the linear map $\partial^\gamma : W^{k,p}_{-\tau}(\mathbb{R}^n) \to W^{k-\ell,p}_{-\tau-\ell}(\mathbb{R}^n)$ is bounded. The spaces $W^{k,\infty}_{-\tau}$ can be defined in the natural way starting from $\|u\|_{L^\infty_{-\tau}} = \|\sigma^\tau u\|_{L^\infty}$, so that $\|u\|_{W^{k,\infty}_{-\tau}} = \sum_{|\gamma| \leq k} \|\partial^\gamma u\|_{L^\infty_{-\tau-|\gamma|}}$. Finally, the weighting conventions are not universal; we have chosen to follow those in [15].

There are a number of basic properties of these spaces that we will need, some of which will be given as exercises, which may be interspersed at appropriate points throughout the section. For example, the next exercise is basic but essential.

Exercise 7-43. Show that if $1 \leq p \leq q$ and $\tau_1 < \tau_2$, then $L^q_{-\tau_2}(\mathbb{R}^n) \subset L^p_{-\tau_1}(\mathbb{R}^n)$, and in fact there is a constant $C > 0$ such that $\|u\|_{L^p_{-\tau_1}} \leq C \|u\|_{L^q_{-\tau_2}}$ for all $u \in L^q_{-\tau_2}(\mathbb{R}^n)$. Prove the following weighted version of *Hölder's inequality*: if $p, q, r \geq 1$ and $\frac{1}{p} = \frac{1}{q} + \frac{1}{r}$, then if $u \in L^q_{-\tau_1}(\mathbb{R}^n)$ and $v \in L^r_{-\tau_2}(\mathbb{R}^n)$, then $uv \in L^p_{-(\tau_1+\tau_2)}(\mathbb{R}^n)$ and

$$\|uv\|_{L^p_{-(\tau_1+\tau_2)}} \leq \|u\|_{L^q_{-\tau_1}} \|v\|_{L^r_{-\tau_2}}. \tag{7-43a}$$

We can identify the dual spaces to $L^p_{-\tau}(\mathbb{R}^n)$ via a dual pairing, as in the next exercise.

Exercise 7-44. Suppose that $p > 1$ and $q > 1$ satisfy $\frac{1}{p} + \frac{1}{q} = 1$. Using the dual pairing (Riesz representation) between $L^p(dx)$ and $L^q(dx)$, identify the dual space of $L^p_{-\tau}(\mathbb{R}^n)$ as $L^q_{\tau-n}(\mathbb{R}^n)$ by considering a pairing $(u, v) \in L^p_{-\tau}(\mathbb{R}^n) \times L^q_{\tau-n}(\mathbb{R}^n) \mapsto \int_{\mathbb{R}^n} uv \, dx$.

In contrast to integral control from the weighted Sobolev norms, one can use weighted Hölder norms for direct pointwise control. Indeed, let $\tau \in \mathbb{R}$, $\alpha \in (0, 1]$, and let $\sigma(x, y) = \min(\sigma(x), \sigma(y))$. We define a weighted Hölder seminorm by

$$[f]_{\alpha, -\tau} = \sup_{x \neq y} \sigma(x, y)^{\alpha+\tau} \frac{|f(x) - f(y)|}{|x - y|^\alpha},$$

and a weighted norm for C^k-functions f by

$$\|f\|_{C^k_{-\tau}} = \sum_{|\gamma| \leq k} \sup_{x \in \mathbb{R}^n} \sigma(x)^{\tau+|\gamma|} |\partial^\gamma f(x)| = \|f\|_{W^{k,\infty}_{-\tau}}.$$

The Banach space $C^k_{-\tau}(\mathbb{R}^n)$ is comprised of those C^k-functions for which the above norm is finite, while the corresponding weighted Hölder spaces $C^{k,\alpha}_{-\tau}(\mathbb{R}^n)$ are given by those functions $f \in C^{k,\alpha}_{\text{loc}}(\mathbb{R}^n)$ such that the norm

$$\|f\|_{C^{k,\alpha}_{-\tau}} := \|f\|_{C^k_{-\tau}} + \sum_{|\gamma|=k} [\partial^\gamma f]_{\alpha,-\tau-k}$$

is finite. (This norm is equivalent to that defined in [15].) For $\ell := |\gamma| \le k$, $\partial^\gamma : C^{k,\alpha}_{-\tau}(\mathbb{R}^n) \to C^{k-\ell,\alpha}_{-\tau-\ell}(\mathbb{R}^n)$ is a bounded linear map. Likewise, for $\tau' < \tau$ and any $p \ge 1$ there is a bounded inclusion map $C^k_{-\tau}(\mathbb{R}^n) \hookrightarrow W^{k,p}_{-\tau'}(\mathbb{R}^n)$.

We have a natural extension of order notation as follows: for $\alpha \in (0, 1]$ and $\ell \in \mathbb{Z}_+ \cup \{0\}$, the notation $f = O_{\ell,\alpha}(|x|^s)$ means $f = O_\ell(|x|^s)$ as in Definition 7-5, and furthermore there is a $C > 0$ such that for all multi-indices γ with $|\gamma| = \ell$, and for all $|x|$ sufficiently large,

$$\sup_{0<|y-x|<\frac{|x|}{2}} \frac{|\partial^\gamma f(y) - \partial^\gamma f(x)|}{|y - x|^\alpha} \le C|x|^{s-\ell-\alpha}.$$

It is straightforward to check that $C^{\ell,\alpha}_{-\tau}(\mathbb{R}^n)$ is comprised of those $C^{\ell,\alpha}_{\text{loc}}(\mathbb{R}^n)$ functions which are also of order $O_{\ell,\alpha}(|x|^{-\tau})$.

Exercise 7-45. Suppose k is a nonnegative integer, $\alpha \in (0, 1]$, and $\tau_1, \tau_2 \in \mathbb{R}$. Show there is a $C > 0$ such that $u \in C^{k,\alpha}_{-\tau_1}(\mathbb{R}^n)$ and $v \in C^{k,\alpha}_{-\tau_2}(\mathbb{R}^n)$ implies that $uv \in C^{k,\alpha}_{-(\tau_1+\tau_2)}(\mathbb{R}^n)$ and

$$\|uv\|_{C^{k,\alpha}_{-(\tau_1+\tau_2)}} \le C \|u\|_{C^{k,\alpha}_{-\tau_1}} \|v\|_{C^{k,\alpha}_{-\tau_2}}.$$

If in addition $0 < \beta \le \alpha$ and $\tau_1 \le \tau_2$, prove that $C^{k,\alpha}_{-\tau_2}(\mathbb{R}^n) \subset C^{k,\beta}_{-\tau_1}(\mathbb{R}^n)$ is a continuous inclusion, i.e., there is a constant $C > 0$ such that for all $u \in C^{k,\alpha}_{-\tau_2}(\mathbb{R}^n)$,

$$\|u\|_{C^{k,\beta}_{-\tau_1}} \le C \|u\|_{C^{k,\alpha}_{-\tau_2}}.$$

Before moving on, we note that for tensor fields on $\mathbb{R}^n$, one may use Cartesian components to extend the above definitions to weighted spaces of tensor fields, as well as the embeddings between the various spaces to which we will turn next.

7.2.2.1. *Sobolev embeddings.* There are classical inequalities associated with *Sobolev embeddings* (cf., e.g., [2; 86; 107; 144]), for which there are weighted analogues. We introduced certain such embeddings in Section 6.1, and used them in Section 6.1.10. We will recall a suite of such embeddings here, and then sketch how to obtain weighted analogues. The domains Ω to which these theorems apply, and to which we restrict our discussion here, include balls and annuli such as $A_R = \{x \in \mathbb{R}^n : R < |x| < 2R\}$; these suffice for our purposes, so we may assume that Ω is a ball or an annulus. For more details, please refer to the above references, to which we defer for the proofs of these embeddings.

Embeddings of $W^{1,p}(\Omega)$. We begin with embeddings of $W^{1,p}(\Omega)$, $1 \le p < \infty$. If $p > n$, and if $0 < \alpha \le 1 - \frac{n}{p}$, there is a constant $C > 0$ such that for any element of $W^{1,p}(\Omega)$, we can choose a representative $u \in C^{0,\alpha}(\overline{\Omega})$, and moreover, $\|u\|_{C^{0,\alpha}(\overline{\Omega})} \le C\|u\|_{W^{1,p}(\Omega)}$. Thus we have a continuous embedding $W^{1,p}(\Omega) \hookrightarrow C^{0,\alpha}(\overline{\Omega})$. In other words, as we used earlier in Section 6.1.10, suitable integral properties of a function and its derivative imply pointwise estimates.

There are also embeddings when $1 \le p \le n$. In case $1 \le p < n$, if we let $p^* = \frac{np}{n-p}$, we have the inequality $\|u\|_{L^{p^*}(\Omega)} \le C\|u\|_{W^{1,p}(\Omega)}$ and the associated embedding $W^{1,p}(\Omega) \hookrightarrow L^{p^*}(\Omega)$. We could use interpolation of the norms to get $\|u\|_{L^q(\Omega)} \le C\|u\|_{W^{1,p}(\Omega)}$ along with the associated embedding $W^{1,p}(\Omega) \hookrightarrow L^q(\Omega)$ for $p \le q \le p^*$. In the borderline case $p = n$ (see [2; 144]), there is an inequality $\|u\|_{L^q(\Omega)} \le C\|u\|_{W^{1,n}(\Omega)}$ for $n \le q < \infty$, establishing the associated embedding $W^{1,n}(\Omega) \hookrightarrow L^q(\Omega)$. We make the trivial observation that since Ω has finite volume (being a ball or an annulus), Hölder's inequality immediately allows us to extend the above inequality and embedding into $L^q(\Omega)$ with $1 \le q < p$.

Embeddings of $W^{k,p}_{-\tau}(\Omega)$. We now consider how to establish weighted analogues for the above embeddings. For example, if $n < p < \infty$, a function $u \in W^{1,p}_{-\tau}(\mathbb{R}^n)$ can be represented by a function in $C^{0,\alpha}_{-\tau}(\mathbb{R}^n)$, where $0 < \alpha \le 1 - \frac{n}{p}$, and moreover there is a constant $C > 0$ such that for all $u \in W^{1,p}_{-\tau}(\mathbb{R}^n)$,

$$\|u\|_{C^{0,\alpha}_{-\tau}} \le C\|u\|_{W^{1,p}_{-\tau}}. \tag{7.2.9}$$

To prove this estimate, we can consider for $R > 0$ the rescaled function $u_R(x) = u(Rx)$, and note that if $A_R = \{x \in \mathbb{R}^n : R < |x| < 2R\}$, then $\|u\|_{C^{0,\alpha}_{-\tau}(A_R)}$ (defined as above but using $x, y \in A_R$) and $R^\tau\|u_R\|_{C^{0,\alpha}(\overline{A}_1)}$ are equivalent for $R \ge 1$: there is a constant $C > 1$ such that for all $R \ge 1$ and all $u \in C^{0,\alpha}_{-\tau}(\mathbb{R}^n)$,

$$C^{-1}\|u\|_{C^{0,\alpha}_{-\tau}(A_R)} \le R^\tau\|u_R\|_{C^{0,\alpha}(\overline{A}_1)} \le C\|u\|_{C^{0,\alpha}_{-\tau}(A_R)}, \tag{7.2.10}$$

and similarly for all $R \ge 1$ and for all $u \in W^{1,p}_{-\tau}(\mathbb{R}^n)$,

$$C^{-1}\|u\|_{W^{1,p}_{-\tau}(A_R)} \le R^\tau\|u_R\|_{W^{1,p}(A_1)} \le C\|u\|_{W^{1,p}_{-\tau}(A_R)} \tag{7.2.11}$$

(where the weighted Sobolev norms on A_R are defined as above but only integrating over A_R).

Exercise 7-46. Verify the scaling relationships (7.2.10)–(7.2.11), and then deduce the corresponding weighted estimate (7.2.9).

Exercise 7-47 (cf. [15]). If $u \in C^{0,\alpha}_{-\tau}(\mathbb{R}^n)$, then clearly $u(x) = O(|x|^{-\tau})$. Show that if $p > n$ and $u \in W^{1,p}_{-\tau}(\mathbb{R}^n)$, then in fact $u(x) = o(|x|^{-\tau})$ as $|x| \to \infty$, i.e., $\lim_{|x|\to\infty} |x|^\tau u(x) = 0$.

We can likewise establish embeddings $W^{1,p}_{-\tau}(\mathbb{R}^n) \hookrightarrow L^q_{-\tau}(\mathbb{R}^n)$ for $1 \leq p < n$ and $p \leq q \leq p^* = \frac{np}{n-p}$; in the borderline case $p = n$, there is a continuous embedding $W^{1,n}_{-\tau}(\mathbb{R}^n) \hookrightarrow L^q_{-\tau}(\mathbb{R}^n)$ for any $n \leq q < \infty$. One proves these as in [15], by using the inequality $\|u\|_{L^q(A_1)} \leq C\|u\|_{W^{1,p}(A_1)}$ associated to the relevant embedding of $W^{1,p}(A_1)$ into $L^q(A_1)$ and rescaling as above to prove $\|u\|_{L^q_{-\tau}} \leq C\|u\|_{W^{1,p}_{-\tau}}$. Indeed there are constants C_1, C_2, C_3 such that for $R \geq 1$ and for $u \in W^{1,p}_{-\tau}(\mathbb{R}^n)$, defining $u_R(x) = u(Rx)$ as above,

$$\|u\|_{L^q_{-\tau}(A_R)} \leq C_1 R^\tau \|u_R\|_{L^q(A_1)} \leq C_2 R^\tau \|u_R\|_{W^{1,p}(A_1)} \leq C_3 \|u\|_{W^{1,p}_{-\tau}(A_R)}.$$

We now proceed as in the following exercise.

Exercise 7-48 (cf. [15]). By writing the weighted Sobolev norm in terms of a series over nonoverlapping annuli, along with scaling, prove that in case $1 \leq p \leq n$, and for $p \leq q < \infty$ (and $q \leq p^*$ in case $1 \leq p < n$), there is a $C > 0$ such that for all $u \in W^{1,p}_{-\tau}(\mathbb{R}^n)$, the following estimate holds: $\|u\|_{L^q_{-\tau}} \leq C\|u\|_{W^{1,p}_{-\tau}}$. For $p > n$, this holds with $q = \infty$ by (7.2.9); show this inequality also holds for $n < p \leq q < \infty$, first using the Sobolev embedding of $W^{1,p}(A_1)$ to $C^0(\overline{A}_1)$ to establish a continuous embedding of $W^{1,p}(A_1)$ into $L^r(A_1)$, for $r \geq 1$.

There are analogous direct relations between embeddings for $W^{k,p}(\Omega)$ and $W^{k,p}_{-\tau}(\mathbb{R}^n)$ for $k \geq 1$; we state them for the weighted spaces. For $kp \leq n$ we have embeddings into integral spaces: for $kp < n$, $W^{k,p}_{-\tau}(\mathbb{R}^n)$ embeds in $L^q_{-\tau}(\mathbb{R}^n)$ for $p \leq q \leq \frac{np}{n-kp}$; for $kp = n$, $W^{k,p}_{-\tau}(\mathbb{R}^n)$ embeds in $L^q_{-\tau}(\mathbb{R}^n)$ for $p \leq q < \infty$. For $kp > n \geq (k-1)p$, we have embeddings into Hölder spaces: for $kp > n > (k-1)p$, and $0 < \alpha \leq k - \frac{n}{p} < 1$, $W^{k,p}_{-\tau}(\mathbb{R}^n)$ embeds into $C^{0,\alpha}_{-\tau}(\mathbb{R}^n)$; for $n = (k-1)p$, $W^{k,p}_{-\tau}(\mathbb{R}^n)$ embeds into $C^{0,\alpha}_{-\tau}(\mathbb{R}^n)$ for $0 < \alpha < 1$. Moreover, if $n < kp < \infty$ and $u \in W^{k,p}_{-\tau}(\mathbb{R}^n)$, then $u(x) = o(|x|^{-\tau})$, just as in Exercise 7-47. We have extensions to higher regularity embeddings; for example, for $\ell \in \mathbb{Z}_+$, $(k-1)p \leq n < kp$ and corresponding α as above, $W^{k+\ell,p}_{-\tau}(\mathbb{R}^n)$ embeds into $C^{\ell,\alpha}_{-\tau}(\mathbb{R}^n)$.

Compact embeddings. Many of the above embeddings are *compact*, the definition of which we recalled in Section 6.1.

For example, recall that we obtain some compact embeddings by employing the Ascoli–Arzelà theorem, a direct corollary of which is the following: if Ω is a bounded domain, then the space $C^{0,\alpha}(\overline{\Omega})$ $(0 < \alpha \leq 1)$ is compactly included in $C^0(\overline{\Omega})$. The Hölder continuity yields the required equicontinuity criterion for Ascoli–Arzelà. Combining this observation with just a little work, you can obtain the following.

Exercise 7-49. Show that for a bounded domain Ω, $C^{0,\alpha}(\overline{\Omega})$ is compactly included in $C^{0,\beta}(\overline{\Omega})$, for $0 < \beta < \alpha \leq 1$.

For Ω a convex, bounded domain, the mean value theorem can be used together with Ascoli–Arzelà to conclude that $C^1(\overline{\Omega})$ is compactly included in $C^{0,\beta}(\overline{\Omega})$ for $0 < \beta < 1$, and as well in $C^0(\overline{\Omega})$; likewise for k a nonnegative integer, $C^{k+1}(\overline{\Omega})$ is compactly included in $C^k(\overline{\Omega})$ and in $C^{k,\beta}(\overline{\Omega})$ for $0 < \beta < 1$, and moreover, $C^{k,\alpha}(\overline{\Omega})$ is compactly included $C^k(\overline{\Omega})$, and in fact in $C^{k,\beta}(\overline{\Omega})$ for $0 < \beta < \alpha \leq 1$. These results extend to compact smooth manifolds-with-boundary, such as an annular region, by a simple covering argument.

There are also compact embeddings between certain Sobolev spaces, given by the *Rellich–Kondrachov theorem* [2; 86; 107; 144]. For an example of such an embedding, for suitable bounded domains Ω, such as a ball or an annulus, the inclusion $W^{k,p}(\Omega) \hookrightarrow W^{j,p}(\Omega)$ for $0 \leq j < k$, $1 \leq p < \infty$, is compact. For a proof in the $p = 2$ case (*Rellich's lemma*) using the Fourier transform and Ascoli–Arzelà, see, e.g., [139]. Some of the Sobolev embeddings mentioned in the preceding section are compact for suitable such domains Ω, for example, for $kp > n$, with $(k-1)p \leq n < kp$, and for α with $0 < \alpha < k - \frac{n}{p}$, then for $\ell \in \mathbb{Z}_+$, $W^{k+\ell,p}(\Omega)$ embeds *compactly* into $C^{\ell,\alpha}(\overline{\Omega})$ (and hence compactly into $C^\ell(\overline{\Omega})$ and $W^{\ell,q}(\Omega)$ for $q \geq 1$).

Lemma 7-50. *For $1 \leq p < \infty$, $k > j \geq 1$ and $\mu > 0$, $W^{k,p}_{-\tau-\mu}(\mathbb{R}^n)$ is compactly included in $W^{j,p}_{-\tau}(\mathbb{R}^n)$.*

Exercise 7-51. Prove the preceding lemma, employing Rellich–Kondrachov and a diagonal argument.

Exercise 7-52. In a similar manner as the preceding, prove that $C^{0,\alpha}_{-\tau-\mu}(\mathbb{R}^n)$ is compactly included in $C^{0,\beta}_{-\tau}(\mathbb{R}^n)$, for $\mu > 0$ and $0 < \beta < \alpha \leq 1$, and that $C^1_{-\tau-\mu}(\mathbb{R}^n)$ is compactly included in $C^{0,\beta}_{-\tau}(\mathbb{R}^n)$ for $0 < \beta < 1$. Formulate higher regularity versions as well.

7.2.2.2. *Weighted spaces and asymptotically flat manifolds.* One can extend the definitions of weighted Sobolev spaces to an asymptotically flat manifold (M, g), using asymptotically flat charts for each of the finitely many ends whose union is $M \setminus \mathcal{C}$, and a suitable finite covering by charts of the compact set $\mathcal{C} \subset M$; equivalently, one could define these spaces using the induced connection and volume measure from a smooth background metric $\mathring{g}$ with $\mathring{g}_{ij} = \delta_{ij}$ in the asymptotically flat charts. Similar considerations apply for defining weighted Hölder spaces on asymptotically flat manifolds, by using a suitable covering by charts, and computing Hölder quotients for components, or using parallel transport. We often focus on an asymptotically flat end $(\mathcal{E}, g)$, and can consider weighted spaces defined on the end. We will sometimes omit the domain in the notation for a weighted function space, for example letting $W^{k,p}_{-\tau} = W^{k,p}_{-\tau}(M, g)$.

Suppose (M, g) is asymptotically flat with rate $q > 0$ and order $\ell + 1$, $\ell \in \mathbb{Z}_+$. On each end $\mathcal{E}$, the component functions satisfy $(g_{ij} - \delta_{ij}) \in W_{-\mu}^{\ell+1,p'} \cap C_{-q}^{\ell,\alpha}$ for $0 < \alpha \leq 1$, $\mu < q$ and $1 \leq p' < \infty$; in this discussion we restrict to $p' < \infty$ without further comment. In fact, asymptotic flatness is sometimes *defined* by requiring the components $(g_{ij} - \delta_{ij})$ to lie in a suitable weighted space built using the chosen asymptotic chart on each end (equivalently, one might define weighted spaces in terms of a smooth background metric $\mathring{g}$ on M as above, and require $g - \mathring{g}$ to lie in a weighted space); in addition, if this does not already follow from the weighted space condition, one often requires that g_{ij} and δ_{ij} give uniformly equivalent metrics on each end; in some circumstances, there is also some overarching regularity assumption that g be (sufficiently) smooth, e.g., $C_{\mathrm{loc}}^{k,\alpha}$ for some nonnegative integer k and $0 < \alpha \leq 1$. To illustrate, suppose $(g_{ij} - \delta_{ij}) \in W_{-\mu}^{1,p'}$ for $\mu > 0$ and $p' > n \geq 3$; then by Sobolev embedding, $(g_{ij} - \delta_{ij}) = o(|x|^{-\mu})$, so the metric components $g_{ij}(x)$ converge uniformly to δ_{ij} as $|x| \to \infty$.

The bottom line, of course, is that one should always state their assumptions precisely when using the term "asymptotically flat", whether formulated in terms of weighted spaces or in terms of Definition 7-29, as we strive to do in what follows. In any case, the regularity and decay assumptions on g, and in particular on the connection coefficients (equivalently on $\mathring{\nabla}(g - \mathring{g}) = \mathring{\nabla}g$) may restrict the spaces between which differential operators associated to g act continuously. One might naturally propose using the induced connection and volume measure from g to *define* the weighted spaces, noting the decay of the connection coefficients would be controlled by an asymptotic assumption on $\nabla(g - \mathring{g}) = -\nabla \mathring{g}$. Analogous to what we mentioned just above, the regularity and decay assumptions would impose a limit (again coming from the connection coefficients) on k in defining, for instance, $W_{-\tau}^{k,p}$. We will now discuss two basic calculations using the bookkeeping on the regularity and weights for an asymptotically flat metric, the first involving the ADM mass, and the second involving the Laplace operator.

Suppose we take $(g_{ij} - \delta_{ij}) \in W_{-\mu}^{2,p'}$ on each end $\mathcal{E}$, for $p' > n \geq 3$ and $\mu \geq \frac{n-2}{2}$, with $R(g) \in L^1(M, dv_g)$ (the *mass decay conditions* in [15]). Then one can expand the scalar curvature in asymptotic coordinates, and use the Sobolev embedding to show the ADM mass is defined. We refer to the proof of Proposition 4.1 in [15] to handle the borderline case, while noting here that in case $\mu > \frac{n-2}{2}$, one can expand simply as in the proof of Proposition 7-31. To handle the error terms, one notes, for example, that since $g_{ij,k} \in W_{-\mu-1}^{1,p'}$, we have from Sobolev embedding $g_{ij,k}(x) = o(|x|^{-\mu-1})$, so it is easy to check

$\partial g \in L^2(\mathcal{E})$ (the same holds at the borderline, using the *proof* of the Sobolev embedding). To complete the proof that the ADM mass definition extends to this setting, proceed as in the following exercise.

Exercise 7-53. Let $\mu > \frac{n-2}{2}$. Use $(g_{ij} - \delta_{ij}) = o(|x|^{-\mu})$ along with $g_{ij,k\ell} \in L^{p'}_{-\mu-2}$ and Hölder's inequality to deduce that $(g_{ij} - \delta_{ij})g_{k\ell,ms} \in L^1(\mathcal{E})$. From here, conclude that $(g_{ij} - \delta_{ij})\Gamma^k_{\ell m,s} \in L^1(\mathcal{E})$; then, expanding as in the proof of Proposition 7-31 and using that $R(g)$ is integrable, show that the mass is defined.

Laplacian on weighted spaces. We now turn to the Fredholm theory for the Laplace operator on functions in weighted spaces. In preparation for this, we note that since the operator Δ_g involves the metric g, the regularity and decay of the components of g in an asymptotic chart are reflected in the coefficients of the operator Δ_g and the weighted spaces between which it operates. It is simple to keep track of this in case the decay conditions are pointwise conditions; we illustrate more generally with the following lemma.

Lemma 7-54. *Suppose (M, g) is asymptotically flat, with $(g_{ij} - \delta_{ij}) \in W^{1,p'}_{-\mu}$ for $\mu > 0$ and $p' > n \geq 3$ on each asymptotic end. Then, for $1 \leq p \leq p'$,*

$$\Delta_g : W^{2,p}_{-\tau} \to W^{0,p}_{-\tau-2}$$

is a bounded linear map. Moreover, $(\Delta - \Delta_g) : W^{2,p}_{-\tau} \to W^{0,p}_{-\tau-\mu-2}$ is also bounded on each end, or equivalently, $(\mathring{\Delta} - \Delta_g) : W^{2,p}_{-\tau} \to W^{0,p}_{-\tau-\mu-2}$ is bounded, where $\mathring{\Delta}$ is the Laplacian for a smooth background metric $\mathring{g}$ as above.

We remark that when g is asymptotically flat of rate $q > 0$, p' can be taken as large as desired, so there would be no restriction on $p \in [1, \infty)$ in the above.

Proof. The main step is to show that for $w \in W^{2,p}_{-\tau}$, $\Gamma^k_{ij}\,\partial w/\partial x^k \in W^{0,p}_{-\tau-\mu-2}$, with a suitable bound on its norm. Note that $1 \leq p < p'$ implies $p < \frac{p'p}{p'-p}$; if $1 \leq p < n$ as well, then $\frac{p'p}{p'-p} < p^* = \frac{np}{n-p}$. Thus when $1 \leq p < p'$, we can apply Sobolev embedding to conclude that for $w \in W^{2,p}_{-\tau}$, $\partial w/\partial x^j \in W^{1,p}_{-1-\tau} \hookrightarrow W^{0,q}_{-1-\tau}$ for $q = \frac{p'p}{p'-p}$. Since metric components are uniformly bounded near infinity in any end by Sobolev embedding, the Christoffel symbols of g in these coordinates satisfy $\Gamma^k_{ij} \in W^{0,p'}_{-\mu-1}$, and we can apply Hölder's inequality (7-43a) (p. 259) to conclude $\Gamma^k_{ij}\,\partial w/\partial x^k \in W^{0,p}_{-\tau-\mu-2}$. To handle the case $p = p'$, we use the Sobolev embedding $\partial w/\partial x^j \in W^{1,p}_{-1-\tau} \hookrightarrow C^{0,\alpha}_{-1-\tau}$ for $p > n$, where $\alpha = 1 - \frac{n}{p}$, to conclude $\Gamma^k_{ij}\,\partial w/\partial x^k \in W^{0,p}_{-\tau-\mu-2}$.

By the Sobolev embedding, $(g_{ij} - \delta_{ij}) = o(|x|^{-\mu})$, and so for $w \in W^{2,p}_{-\tau}$,

$$\Delta_g w = g^{ij}(w_{,ij} - \Gamma^k_{ij}w_{,k}) \in W^{0,p}_{-\tau-2}$$

$$(\Delta - \Delta_g)(w) = (\delta^{ij} - g^{ij})w_{,ij} + g^{ij}\Gamma^k_{ij}w_{,k} \in W^{0,p}_{-\tau-\mu-2}. \qquad \square$$

Exercise 7-55. Let $k \geq 2$, and suppose $(g_{ij} - \delta_{ij}) \in W^{k-1,p'}_{-\mu}$ for $\mu > 0$ and $p' > n \geq 3$. Prove that for $2 \leq k' \leq k$ and $1 \leq p \leq p'$, $\Delta_g : W^{k',p}_{-\tau} \to W^{k'-2,p}_{-\tau-2}$ and $(\mathring{\Delta} - \Delta_g) : W^{k',p}_{-\tau} \to W^{k'-2,p}_{-\tau-\mu-2}$ are bounded linear maps. To analyze the term $\Gamma^m_{ij} \partial w / \partial x^m$, you can employ the product rule together with methods used above. If instead $(g_{ij} - \delta_{ij}) \in C^{k-1,\alpha}_{-\mu}$ for $\mu > 0$ and $\alpha \in (0, 1]$, prove the analogous statements for $\Delta_g : C^{k',\beta}_{-\tau} \to C^{k'-2,\beta}_{-\tau-2}$ and $\mathring{\Delta} - \Delta_g : C^{k',\beta}_{-\tau} \to C^{k'-2,\beta}_{-\tau-\mu-2}$ in weighted Hölder spaces, with $0 < \beta \leq \alpha$ and $2 \leq k' \leq k$. Compare Exercise 7-45.

We will now recall from [15; 155; 211] some fundamental results regarding the Laplacian acting on scalar-valued functions in weighted spaces; see also [39; 44; 45; 168]. We will mostly use these when solving for conformal deformations of asymptotically flat metrics in the next section. You might review the definition of a Fredholm operator from Section 6.1.

The Euclidean Laplacian. We start by revisiting the Euclidean Laplacian Δ on $\mathbb{R}^n$ ($n \geq 3$). Suppose $\Delta u = 0$ on $\mathbb{R}^n$. By the maximum principle, if u goes to 0 at infinity, then u is identically zero. By Proposition 7-14 (Liouville's theorem), if u is bounded then u must be constant. If $p \in [1, \infty)$ and $\tau \geq 0$, then if $u \in L^p_{-\tau}(\mathbb{R}^n)$ is harmonic, then an application of the mean value property shows that u vanishes identically (Exercise 7-102, cf. Exercise 7-15). On the other hand, by [15, Corollary 1.9], if $1 < p < \infty$, and $\tau < 0$, then u must be a harmonic polynomial lying in $L^p_{-\tau}(\mathbb{R}^n)$, so that its degree must be less than $-\tau$; as a corollary, for $\tau \leq 0$, a harmonic function $u \in C^0_{-\tau}(\mathbb{R}^n)$ must be a harmonic polynomial of degree at most $-\tau$. For the rest of the section, we take $1 < p < \infty$ (with or without comment).

For $p > 1$ and any τ, then, $\Delta : W^{2,p}_{-\tau}(\mathbb{R}^n) \to L^p_{-\tau-2}(\mathbb{R}^n)$ has finite-dimensional kernel. Recalling what we have seen in Section 7.1.2.3, the set

$$\{\ell \in \mathbb{Z} : \ell \leq 2-n \text{ or } \ell \geq 0\} = \{\dots, -1-n, -n, 1-n, 2-n, 0, 1, 2, 3, \dots\}$$

is comprised of growth rates of functions in weighted spaces which are Euclidean harmonic functions in a neighborhood of infinity. These numbers form a set of *exceptional weights* for the function theory of the Laplace operator, the complement of which gives the set of *nonexceptional weights*. The connection between the exceptional weights as growth rates for harmonic functions and the function theory for the Laplacian is laid bare in the following theorem. For m a nonnegative integer, let $\mathcal{K}_m$ be the space of harmonic polynomials of degree at most m.

Theorem 7-56. *Suppose $p > 1$. For $-\tau$ exceptional, $\Delta : W^{2,p}_{-\tau}(\mathbb{R}^n) \to L^p_{-\tau-2}(\mathbb{R}^n)$ does not have closed range. If $-\tau$ is nonexceptional, this operator is Fredholm,* *and its index depends only on the interval inside the nonexceptional set that contains $-\tau$. More explicitly, if m is a nonnegative integer, the operator $\Delta : W^{2,p}_{-\tau}(\mathbb{R}^n) \to L^p_{-\tau-2}(\mathbb{R}^n)$ is surjective with kernel $\mathcal{K}_m$ if $m < -\tau < m+1$, while for $(n-2)+m < \tau < (n-2)+m+1$ the operator has trivial kernel, and closed range*

$$\Delta(W^{2,p}_{-\tau}(\mathbb{R}^n)) = \left\{ f \in L^p_{-\tau-2}(\mathbb{R}^n) : \int_{\mathbb{R}^n} f(x)v(x)\,dx = 0 \text{ for all } v \in \mathcal{K}_m \right\}.$$

For $0 < \tau < n-2$, $\Delta : W^{2,p}_{-\tau}(\mathbb{R}^n) \to L^p_{-\tau-2}(\mathbb{R}^n)$ is an isomorphism.

The theorem is proved in [155]; see also [168; 15]. (We note that the weighting conventions in [155] differ from [15]. For comparison, the weighted Sobolev spaces $M^p_{s,\delta}(\mathbb{R}^n)$ as defined in [155] are such that $W^{k,p}_{-\tau}(\mathbb{R}^n) = M^p_{k,\delta}(\mathbb{R}^n)$ for $\tau = \delta + \frac{n}{p}$, and so the Laplace operator $\Delta : M^p_{k,\delta}(\mathbb{R}^n) \to M^p_{k-2,\delta+2}(\mathbb{R}^n)$ is a bounded linear map.) We will not go into full details of the proof of Theorem 7-56 here, but we will make some observations and comments.

Comments on the proof. We discussed the kernel above, so we turn to the range $\Delta(W^{2,p}_{-\tau}(\mathbb{R}^n)) \subset L^p_{-\tau-2}(\mathbb{R}^n)$. Recall that the dual space of $L^p_{-\tau-2}(\mathbb{R}^n)$ can be identified with $L^q_{\tau+2-n}(\mathbb{R}^n)$, where $\frac{1}{p} + \frac{1}{q} = 1$ (Exercise 7-44), as follows: for $v \in L^q_{\tau+2-n}(\mathbb{R}^n)$, $\ell_v(w) = \int_{\mathbb{R}^n} v(x)w(x)\,dx$ defines a bounded linear functional on $L^p_{-\tau-2}(\mathbb{R}^n)$; as such, ℓ_v has a closed nullspace, which is codimension-one for nontrivial v. Now suppose $v \in L^q_{\tau+2-n}(\mathbb{R}^n)$ annihilates the range $\Delta(W^{2,p}_{-\tau}(\mathbb{R}^n))$, i.e., for all $u \in W^{2,p}_{-\tau}(\mathbb{R}^n)$, $\int_{\mathbb{R}^n} v\,\Delta u\,dx = 0$; in particular this holds for all smooth, compactly supported u, and hence $\Delta v = 0$ distributionally, so that v is a smooth harmonic function. In fact, by density, $v \in L^q_{\tau+2-n}(\mathbb{R}^n)$ annihilates $\Delta(W^{2,p}_{-\tau}(\mathbb{R}^n))$ if and only if it annihilates $\Delta(C^\infty_c(\mathbb{R}^n))$. As we have seen, if $v \in L^q_{\tau+2-n}(\mathbb{R}^n)$ is a harmonic function, then v must vanish identically in case $\tau + 2 - n \le 0$, while in case $\tau > n-2$, $v \in \mathcal{K}_m$, where m is the nonnegative integer such that $m < \tau + 2 - n \le (m+1)$. Since for such m, $\mathcal{K}_m \subset L^q_{\tau+2-n}(\mathbb{R}^n)$, we see that the range $\Delta(W^{2,p}_{\tau}(\mathbb{R}^n))$ is contained in the closed subspace

$$^\perp\mathcal{K}_m := \bigcap_{v \in \mathcal{K}_m} \ell_v^{-1}(0) = \left\{ f \in L^p_{-\tau-2}(\mathbb{R}^n) : \int_{\mathbb{R}^n} f(x)v(x)\,dx = 0 \text{ for all } v \in \mathcal{K}_m \right\}.$$

By the Hahn–Banach theorem, we conclude $\Delta(W^{2,p}_{-\tau}(\mathbb{R}^n))$ is dense in $L^p_{-\tau-2}(\mathbb{R}^n)$ for $\tau \le n-2$, while in case $\tau > n-2$, with the nonnegative integer m as above, the closure of $\Delta(W^{2,p}_{-\tau}(\mathbb{R}^n))$ in $L^p_{-\tau-2}(\mathbb{R}^n)$ is precisely $^\perp\mathcal{K}_m$, a space with finite codimension ($\mathcal{K}_m$ is finite-dimensional).

In summary, then, the main issue for deriving the Fredholm property is establishing closed range. By the above discussion, the range of $\Delta : W^{2,p}_{-\tau}(\mathbb{R}^n) \to L^p_{-\tau-2}(\mathbb{R}^n)$ will be as stated in Theorem 7-56, for all τ for which the range is closed. That said, to establish closed range for $-\tau < 0$ nonexceptional, one may proceed directly by solving the Poisson equation $\Delta u = f$, for f in the appropriate space, either for $f \in L^p_{-\tau-2}(\mathbb{R}^n)$ in case $0 < \tau < n - 2$, or for $f \in {}^\perp \mathcal{K}_m \subset L^p_{-\tau-2}(\mathbb{R}^n)$ when $\tau > n - 2$; we will say a few words about this now. Once this is established, one may apply a duality argument to handle the case $-\tau > 0$ nonexceptional, cf. [155].

For $p > 1$, let $p' = \frac{p}{p-1}$, and suppose a and b satisfy $a + b > 0$. Consider the integral kernel $K'(x, y) = |x|^{-a} |x - y|^{a+b-n} |y|^{-b}$, and the related kernel $\widetilde{K}(x, y) = (1 + |x|)^{-a} |x - y|^{a+b-n} (1 + |y|)^{-b}$. A workhorse lemma for the analysis of the Laplacian Δ on weighted spaces is that $T'(u)(x) = \int_{\mathbb{R}^n} K'(x, y) u(y)\, dy$ defines a bounded linear operator T' on $L^p(\mathbb{R}^n)$ if and only if $a < \frac{n}{p}$ and $b < \frac{n}{p'}$. For a proof, see [168, Lemma 2.1]. See also [155, Lemma A] for other references; as noted there, a straightforward reformulation is that $\widetilde{T}(u)(x) = \int_{\mathbb{R}^n} \widetilde{K}(x, y) u(y)\, dy$ defines a bounded linear operator on $L^p(\mathbb{R}^n)$ if and only if $a < \frac{n}{p}$ and $b < \frac{n}{p'}$. We address the necessity of the condition on a and b in Exercise 7-103.

Let $c_n = 1/((2 - n)\omega_{n-1})$. From our discussion of the Poisson equation and the role of $c_n |x - y|^{2-n}$ as the fundamental solution for Δ, it might not be surprising that the above integral kernels play a role in the analysis, at least for $a + b = 2$, which is the case stated in [15, Lemma 1.8]. If we let $a = -\tau + \frac{n}{p}$ and $b = \tau + 2 - \frac{n}{p}$, then the conditions $a < \frac{n}{p}$ and $b < \frac{n}{p'}$ translate into $0 < \tau < n - 2$. For τ in this range, the mapping $f \mapsto N_f$, where N_f is the Newtonian potential $N_f(x) = \int_{\mathbb{R}^n} c_n |x - y|^{2-n} f(y)\, dy$, defines a bounded linear operator $N : L^p_{-\tau-2}(\mathbb{R}^n) \to L^p_{-\tau}(\mathbb{R}^n)$. To see this, we let

$$\tilde{f}(y) := f(y)(1 + |y|)^{\tau+2-\frac{n}{p}} = f(y)(1 + |y|)^b,$$

and note that $f \in L^p_{-\tau-2}(\mathbb{R}^n)$ if and only if $\tilde{f} \in L^p(\mathbb{R}^n)$, with $C^{-1} \|f\|_{L^p_{-\tau-2}} \leq \|\tilde{f}\|_{L^p} \leq C \|f\|_{L^p_{-\tau-2}}$ for some constant $C > 1$. Since

$$N_f(x) = (1 + |x|)^a \widetilde{T}(\tilde{f})(x) = (1 + |x|)^{-\tau+\frac{n}{p}} \widetilde{T}(\tilde{f})(x),$$

we conclude that $N_f \in L^p_{-\tau}(\mathbb{R}^n)$ and that N is bounded as claimed. We will recall a weighted analogue of elliptic regularity in Proposition 7-61, from which we conclude $N_f \in W^{2,p}_{-\tau}(\mathbb{R}^n)$, as desired. This gives us the theorem in the isomorphism range $0 < \tau < n - 2$, which is the case we will need.

That said, it is instructive to make some comments about the other weight ranges. We recall that for $\tau_1 < \tau_2$, $L^p_{-\tau_2}(\mathbb{R}^n) \subset L^p_{-\tau_1}(\mathbb{R}^n)$, so that in particular, for $\tau \geq n - 2$, $L^p_{-\tau}(\mathbb{R}^n) \subset \bigcap_{0 < \sigma < n-2} L^p_{-\sigma}(\mathbb{R}^n)$. For $\tau \geq n - 2$, it follows from the preceding discussion that $f \in L^p_{-\tau-2}(\mathbb{R}^n)$ implies $f = \Delta u$ for $u = N_f \in$

$\bigcap_{0<\sigma<n-2} W^{2,p}_{-\sigma}(\mathbb{R}^n)$. This space contains $W^{2,p}_{-\tau}(\mathbb{R}^n)$, but it is certainly possible that N_f is not in $W^{2,p}_{-\tau}(\mathbb{R}^n)$, as in the following example.

Example 7-57. Let φ be a smooth function which is zero in a neighborhood of the origin, and is identically one outside a compact set. Suppose v is a nontrivial homogeneous harmonic polynomial with degree m and Kelvin transform $K[v]$ (Section 7.1.2.3). We let $f = \Delta u$ with $u(x) = \varphi(x) K[v](x)$; for example if $m = 0$ and $v = 1$, we have $u(x) = \varphi(x)|x|^{2-n}$. We have $u \in \bigcap_{0<\sigma<m+n-2} W^{2,p}_{-\sigma}(\mathbb{R}^n)$ but $u \notin W^{2,p}_{-\tau}(\mathbb{R}^n)$ if $\tau \geq m + n - 2$. Clearly $f \in L^p_{-\tau-2}(\mathbb{R}^n)$ for any τ (as f vanishes near infinity), and so $(u - N_f)$ is harmonic and lies in $L^p_{-\sigma}(\mathbb{R}^n)$ for some $\sigma > 0$. Hence $N_f = u \notin W^{2,p}_{-(m+n-2)}(\mathbb{R}^n)$.

Exercise 7-58. Continuing the previous example, show that, whereas for $\tau > m+n-2$, $\Delta(W^{2,p}_{-\tau}(\mathbb{R}^n))$ is annihilated by $\mathcal{K}_m$ under the dual pairing, a simple application of Green's identity, together with the fact that $K[v](x) = |x|^{2-n-2m}v(x)$, shows that $\int_{\mathbb{R}^n} v(x) f(x)\, dx \neq 0$.

To prove Theorem 7-56 it remains to show that $N_f \in W^{2,p}_{-\tau}(\mathbb{R}^n)$ for all f in $^\perp\mathcal{K}_m \subset L^p_{-\tau-2}(\mathbb{R}^n)$, with $m + (n-2) < \tau < m+1+(n-2)$ and m a nonnegative integer. As you might imagine by now, there is an elegant way to do this using the spherical harmonic expansion of $c_n |x-y|^{2-n}$, which for $|x| > |y|$ and $\hat{x} = x|x|^{-1}$ we write as

$$c_n |x - y|^{2-n} = c_n |x|^{2-n} \sum_{k=0}^{\infty} h_k(\hat{x}, y)|x|^{-k},$$

where for fixed $\hat{x} \in \mathbb{S}^{n-1}$, $h_k(\hat{x}, y)$ is a homogeneous harmonic polynomial of degree k with respect to y. Let $0 < \psi \leq 1$ be a continuous function with $\psi(x) = |x|^{-1}$ for $|x| \geq 2$. For m a nonnegative integer, we consider the functions

$$K'_m(x, y) = c_n |x - y|^{2-n} - c_n |x|^{2-n} \sum_{k=0}^{m} h_k(\hat{x}, y)|x|^{-k},$$

$$K_m(x, y) = c_n |x - y|^{2-n} - c_n(\psi(x))^{n-2} \sum_{k=0}^{m} h_k(\hat{x}, y)(\psi(x))^{k}.$$

K'_m is used in [15], which works with weighted spaces defined on punctured Euclidean space, whereas [155] applies the integral kernel K_m in weighted spaces on $\mathbb{R}^n$, as the use of $\psi(x)$ removes the singularity at $x = 0$ in the sum. Using the workhorse lemma on $\tilde{T}$ cited above, with elementary estimates of K_m, one shows that $K_m : L^p_{-\tau-2}(\mathbb{R}^n) \to L^p_{-\tau}(\mathbb{R}^n)$ is bounded for $m+(n-2) < \tau < m+1+(n-2)$. Of course, if $f \in {}^\perp\mathcal{K}_m \subset L^p_{-\tau-2}(\mathbb{R}^n)$, then $\int_{\mathbb{R}^n} K_m(x, y) f(y)\, dy = N_f(x)$, so that $N_f \in L^p_{-\tau}(\mathbb{R}^n)$, with $\Delta N_f = f \in L^p_{-\tau-2}(\mathbb{R}^n)$. Just as above, one can conclude in fact $N_f \in W^{2,p}_{-\tau}(\mathbb{R}^n)$, as desired.

That the range is not closed for $-\tau$ exceptional is addressed in Exercise 7-104. For more details and the remainder of the proof, see [15; 155]. $\square$

Remark 7-59. Using the regularity statement in Proposition 7-61 below, one can conclude that $\Delta : W^{k,p}_{-\tau}(\mathbb{R}^n) \to W^{k-2,p}_{-\tau-2}(\mathbb{R}^n)$ is Fredholm if $-\tau$ is nonexceptional, $p > 1$, and $k \geq 2$. The map is surjective for $-\tau$ nonexceptional and $\tau < n - 2$, while for m a nonnegative integer and $(n-2) + m < \tau < (n-2) + m + 1$, $\Delta(W^{k,p}_{-\tau}(\mathbb{R}^n)) = {}^{\perp}\mathcal{K}_m \cap W^{k-2,p}_{-\tau-2}(\mathbb{R}^n)$; the kernel is the same as in the $k = 2$ case. The analogous Fredholm properties for $\Delta : C^{k,\alpha}_{-\tau}(\mathbb{R}^n) \to C^{k-2,\alpha}_{-\tau-2}(\mathbb{R}^n)$, with $k \geq 2, 0 < \alpha < 1$ and $-\tau$ nonexceptional, can be established directly using the expansion of the Newtonian potential, along with weighted Schauder estimates and regularity, as in [157].

Exercise 7-60. Combine the expansion result in [157] with the Fredholm properties in weighted Sobolev spaces to obtain the Fredholm properties in weighted Hölder spaces. Begin by letting $\tau' < \tau$ be slightly less than τ, so that $-\tau'$ is nonexceptional, and letting $p > 1$ be such that the Sobolev embedding $W^{2,p}_{-\tau'}(\mathbb{R}^n) \hookrightarrow C^{0,\alpha}_{-\tau'}(\mathbb{R}^n)$ holds; note that $C^{k,\alpha}_{-\tau}(\mathbb{R}^n) \hookrightarrow W^{k,p}_{-\tau'}(\mathbb{R}^n)$. Use the weighted Hölder estimate and regularity as in Proposition 7-63 (cf. [157, Theorem 1]) to get that for $f \in \Delta(W^{k,p}_{-\tau'}(\mathbb{R}^n)) \cap C^{k-2,\alpha}_{-\tau-2}(\mathbb{R}^n)$, there is $w \in C^{k,\alpha}_{-\tau'}(\mathbb{R}^n)$ with $\Delta w = f$. Use [157, Theorem 2] to conclude that $w \in C^{k,\alpha}_{-\tau}(\mathbb{R}^n)$. From here one can easily find the image of the map $\Delta : C^{k,\alpha}_{-\tau}(\mathbb{R}^n) \to C^{k-2,\alpha}_{-\tau-2}(\mathbb{R}^n)$, and conclude that the map is Fredholm.

The Laplacian Δ_g on asymptotically flat (M, g). The extension of the Fredholm property from Theorem 7-56 to the Laplacian Δ_g on an asymptotically flat manifold (M, g), namely that $\Delta_g : W^{2,p}_{-\tau} \to L^p_{-\tau-2}$ is Fredholm for $-\tau$ nonexceptional (and more generally for operators which are asymptotic to Δ in a suitable sense), is established in [15, Proposition 1.14, Proposition 2.2]; see also [39]. For our purposes, the following fundamental weighted elliptic estimate (7.2.12) and regularity result for Δ_g, and the weighted Hölder version in Proposition 7-63, will suffice; compare Lemma 7-54. We emphasize the stronger estimate (7.2.13) that holds in the weight range for which Δ_g is an isomorphism.

Proposition 7-61. *Let $\tau \in \mathbb{R}, \mu > 0, k \geq 2, p' > n \geq 3$ and $1 < p \leq p'$. Suppose (M^n, g) is asymptotically flat, admitting coordinate charts for each end in which $(g_{ij} - \delta_{ij}) \in W^{k-1,p'}_{-\mu}$. The bounded linear operator $\Delta_g : W^{k,p}_{-\tau} \to W^{k-2,p}_{-\tau-2}$ satisfies the following a priori weighted elliptic estimate: for some $C > 0$ and for $w \in W^{k,p}_{-\tau}$,*

$$\|w\|_{W^{k,p}_{-\tau}} \leq C \left(\|\Delta_g w\|_{W^{k-2,p}_{-\tau-2}} + \|w\|_{L^p_{-\tau}} \right). \tag{7.2.12}$$

Moreover, if $u \in L^p_{-\tau}$ and $\Delta_g u \in W^{k-2,p}_{-\tau-2}$, then $u \in W^{k,p}_{-\tau}$.

If in addition $\tau \in (0, n-2)$, then $\Delta_g : W^{k,p}_{-\tau} \to W^{k-2,p}_{-\tau-2}$ is an isomorphism, so in particular, there is a $C > 0$ such that for all $w \in W^{k,p}_{-\tau}$,

$$\|w\|_{W^{k,p}_{-\tau}} \leq C \|\Delta_g w\|_{W^{k-2,p}_{-\tau-2}}. \tag{7.2.13}$$

Thus, if $0 < \tau_1 < \tau_2 < n-2$, with $u \in L^p_{-\tau_1}$ and $\Delta_g u \in W^{k-2,p}_{-\tau_2-2}$, then $u \in W^{k,p}_{-\tau_2}$.

Remark 7-62. Aside from the various parameters in the proposition, the constant C in (7.2.12) depends on the ellipticity constant $\lambda > 0$ such that $\lambda|\xi|^2 \leq g^{ij}\xi_i\xi_j \leq \lambda^{-1}|\xi|^2$ and on the norm $\|g_{ij} - \delta_{ij}\|_{W^{k-1,p'}_{-\mu}}$ (defined using a suitable covering), as reflected in the coefficients of Δ_g. We could also write this in terms of the analogous norm of the difference $g - \mathring{g}$ for a background metric $\mathring{g}$ which in each asymptotic chart is given by $\mathring{g}_{ij} = \delta_{ij}$. Thus the constant C can be chosen uniformly across a set of metrics, each asymptotically flat with respect to the same asymptotically flat coordinates and suitably controlled in $W^{k-1,p'}_{-\mu}$. The same is true for the constant in (7.2.13); see [15, Propositions 1.6 and 1.11, Corollary 1.16].

Outline of the proof of the proposition. To go from Δ to Δ_g, we have to modify the operator, and in addition, for a general asymptotically flat manifold, the topology can be more complicated; for example, there may be more than one asymptotically flat end. The proof of the *a priori* estimates and regularity employs the standard interior elliptic estimate and regularity combined with a scaling argument akin to ones we have used earlier. There is however a key difference between the asymptotic setting and that of a closed manifold, with respect to the Fredholm property. In the latter case, from the basic elliptic estimate, one concludes the Fredholm property of the elliptic operator as in Exercise 6-25, by using an estimate of the form $\|u\|_{W^{k,p}} \leq C(\|Lu\|_{W^{k-m,p}} + \|u\|_{L^p})$, where L has order m and $k \geq m$, say. In fact, one couples this estimate along with the compactness of the embedding $W^{k,p}(M) \hookrightarrow L^p(M)$ to conclude the kernel of L is finite-dimensional and the range of $L : W^{k,p}(M) \to W^{k-m,p}(M)$ is closed. This same strategy fails in the weighted case, since the weighted compactness results discussed earlier require the weights to *change*, e.g., for $\tau_1 < \tau_2$, the embedding $W^{1,p}_{-\tau_2} \hookrightarrow L^p_{-\tau_1}$ is compact. In [15] the Fredholm property is established via a "scale-broken" estimate for nonexceptional weights, [15, Theorem 1.10], in which the lower order term $\|w\|_{L^p_{-\tau}}$ in (7.2.12) is replaced by $\|w\|_{L^p(\Omega)}$, where Ω is the compactly contained domain whose complement is a union of the exterior regions of the form $\{x| \geq R\}$ in each asymptotic end, for sufficiently large R. The proof of the estimate works by decomposing functions into pieces each supported on an asymptotic end, plus a compactly supported piece. On the ends, where Δ_g is appropriately close to Δ, one can use an injectivity estimate for Δ at nonexceptional weights, and then patch these together with interior estimates on the compactly supported piece. The key for the Fredholm property is that the restriction $W^{1,p}_{-\tau}(M) \hookrightarrow L^p(\Omega)$ *is compact* by Rellich–Kondrachov on Ω. We will use this form of the estimate again in the proof of Proposition 7-73. $\square$

We turn to a result analogous to Proposition 7-61 for Hölder spaces, and we include a formulation of an isomorphism statement in the borderline case $\tau = n - 2$, following [211].

Proposition 7-63. *Let $n \geq 3$, $\tau \in \mathbb{R}$, $\mu > 0$, $\alpha \in (0, 1)$, and $k \geq 2$. Suppose (M^n, g) is asymptotically flat, admitting coordinate charts for each end in which $(g_{ij} - \delta_{ij}) \in C^{k-1,\alpha}_{-\mu}$. Then $\Delta_g : C^{k,\alpha}_{-\tau} \to C^{k-2,\alpha}_{-\tau-2}$ is a bounded linear operator satisfying the following a priori weighted elliptic estimate: there is a $C > 0$ such that for all $w \in C^{k,\alpha}_{-\tau}$,*

$$\|w\|_{C^{k,\alpha}_{-\tau}} \leq C(\|\Delta_g w\|_{C^{k-2,\alpha}_{-\tau-2}} + \|w\|_{C^0_{-\tau}}). \tag{7.2.14}$$

Moreover, if $w \in C^0_{-\tau}$ and $\Delta_g w \in C^{k-2,\alpha}_{-\tau-2}$, then $w \in C^{k,\alpha}_{-\tau}$.

If in addition, $\tau \in (0, n-2)$, then $\Delta_g : C^{k,\alpha}_{-\tau} \to C^{k-2,\alpha}_{-\tau-2}$ is an isomorphism, so in particular, there is a $C > 0$ such that for all $w \in C^{k,\alpha}_{-\tau}$,

$$\|w\|_{C^{k,\alpha}_{-\tau}} \leq C\|\Delta_g w\|_{C^{k-2,\alpha}_{-\tau-2}}. \tag{7.2.15}$$

Thus, if $0 < \tau_1 < \tau_2 < n - 2$, with $u \in C^0_{-\tau_1}$ and $\Delta_g u \in C^{k-2,\alpha}_{-\tau_2-2}$, then $u \in C^{k,\alpha}_{-\tau_2}$.

Finally, suppose $\tau > 0$. If $w \in C^0_{-\tau}$ and $\Delta_g w \in C^{k-2,\alpha}_{-n} \cap L^1$, then $w \in C^{k,\alpha}_{2-n}$. Moreover, there is a $C > 0$ such that for all $w \in C^{k,\alpha}_{2-n}$,

$$\|w\|_{C^{k,\alpha}_{2-n}} \leq C(\|\Delta_g w\|_{C^{k-2,\alpha}_{-n}} + \|\Delta_g w\|_{L^1}). \tag{7.2.16}$$

Given any $f \in C^{k-2,\alpha}_{-n} \cap L^1$, there is a unique $w \in C^{k,\alpha}_{2-n}$ such that $\Delta_g w = f$.

Comments on the proof. As with the Sobolev case, (7.2.14) is established using interior estimates and scaling. Given what we saw earlier about the role of the relevant integral kernels in the analysis of Δ on weighted Sobolev spaces, it should not be surprising that the injectivity estimates (7.2.15) and (7.2.16) can be obtained using the fundamental solution, as we will indicate shortly for the Euclidean case. Thus to establish these two estimates in general, one approach is to show there is a fundamental solution for Δ_g with the same asymptotics as that for Δ, and compute with it as we will indicate below for the Euclidean metric, cf. [211].

On the proof of (7.2.15) *and* (7.2.16) *for* Δ. Recall that the Newtonian potential is defined by $N_f(x) = \int_{\mathbb{R}^n} c_n |x - y|^{2-n} f(y)\, dy$.

Lemma 7-64. *For $0 < \tau < n - 2$, there is a constant $C > 0$ such that for all $f \in C^0_{-\tau-2}(\mathbb{R}^n)$, it follows that $f(y)|x - y|^{2-n} \in L^1(\mathbb{R}^n, dy)$, and in fact $N_f \in C^0_{-\tau}(\mathbb{R}^n)$ with $\|N_f\|_{C^0_{-\tau}} \le C \|f\|_{C^0_{-\tau-2}}$.*

Exercise 7-65. Prove the preceding lemma. Note that this claim involves deriving a supremum estimate and establishing continuity of N_f. To show existence and continuity of N_f, you might break up the integral for $N_f(x)$ into two parts centered around x; continuity will follow from an application of dominated convergence. To show the $C^0_{-\tau}$-estimate, for $|x| > 0$, you might break the integral into three parts, integrating over $\left\{ y : |x - y| \le \frac{|x|}{2} \right\}$, $\left\{ y : \frac{|x|}{2} \le |x - y| \le 2|x| \right\}$, and $\{ y : |x - y| \ge 2|x| \}$. Remark where you use the conditions $0 < \tau$ and $\tau < n - 2$.

The case $\tau = n - 2$ is handled similarly, as in the next lemma.

Lemma 7-66. *There is a constant $C > 0$ such that for all $f \in C^0_{-n}(\mathbb{R}^n) \cap L^1(\mathbb{R}^n)$, it follows that $f(y)|x - y|^{2-n} \in L^1(\mathbb{R}^n, dy)$, and in fact $N_f \in C^0_{2-n}(\mathbb{R}^n)$ with $\|N_f\|_{C^0_{2-n}} < C(\|f\|_{C^0_{-n}} + \|f\|_{L^1})$.*

Exercise 7-67. Prove the preceding lemma. To do this, for $|x| > 0$, you might break the integral into two parts: over $\left\{ y : |x - y| \le \frac{|x|}{2} \right\}$ and $\left\{ y : |x - y| \ge \frac{|x|}{2} \right\}$. Continuity of N_f can be shown with a dominated convergence argument, or by showing N_f is differentiable.

With these lemmas in hand, what remains to be shown is that if $0 < \alpha < 1$, $0 < \tau < n-2$ and $f \in C^{k-2,\alpha}_{-\tau-2}(\mathbb{R}^n)$, or if $\tau = n-2$ and $f \in C^{k-2,\alpha}_{-n}(\mathbb{R}^n) \cap L^1(\mathbb{R}^n)$, then $N_f \in C^{k,\alpha}_{-\tau}(\mathbb{R}^n)$ and $\Delta N_f = f$. A straightforward approach is to compute the partial derivatives of N_f directly, using the Hölder continuity of f when computing the second partials of N_f, as in [107, Chapter 4], for example, to conclude $\Delta N_f = f$. From here, the claim follows from the regularity result one gets from the basic estimate (7.2.14). A slight modification of this approach is as follows, assuming you know $\Delta N_\psi - \varphi$ for $\varphi \subset C^{0,\alpha}_c(\mathbb{R}^n)$. For $R \ge 1$ we let $\psi_R(x) = \psi(|x|/R)$, where $\psi : \mathbb{R} \to \mathbb{R}$ is a smooth function with $0 \le \psi \le 1$, with $\psi(t) = 1$ for $t \le 1$, and $\psi(t) = 0$ for $t \ge 2$. Let $f_R = \psi_R f \in C^{k-2,\alpha}_c(\mathbb{R}^n)$. Then $\Delta N_{f_R} = f_R$, and f_R converges to f in $C^{k-2,\alpha}_{-\tau'-2}(\mathbb{R}^n)$ for $0 < \tau' < \tau$. By applying Lemma 7-64 to the difference $f - f_R$, we have that N_{f_R} converges to N_f in $C^0_{-\tau'}(\mathbb{R})$; using (7.2.14) we conclude that N_{f_R} is Cauchy in $C^{k,\alpha}_{-\tau'}(\mathbb{R}^n)$ as $R \nearrow \infty$, so that the limit N_f is in $C^{k,\alpha}_{-\tau'}(\mathbb{R}^n)$, and $\Delta N_f = f$ as desired. Since $N_f \in C^0_{-\tau}(\mathbb{R}^n)$, we conclude from Proposition 7-63 that $N_f \in C^{k,\alpha}_{-\tau}(\mathbb{R}^n)$. From here the isomorphism statements are immediate. $\square$

Before moving on, we urge the reader to analyze the proofs of the preceding lemmas in case $f \in C^0_{-\tau-2}(\mathbb{R}^n)$ for $\tau > n - 2$. Note for such f, we have

$f \in L^1(\mathbb{R}^n)$ as well, but we cannot conclude $N_f \in C^0_{-\tau}(\mathbb{R}^n)$. Where does the obstruction appear in the proofs of the lemmas? We see that with $\tau > n-2$, for general $f \in C^{0,\alpha}_{-\tau-2}(\mathbb{R}^n)$, one can conclude $N_f \in C^{2,\alpha}_{2-n}(\mathbb{R}^n)$ and $\Delta N_f = f$, but we cannot conclude $N_f \in C^{2,\alpha}_{-\tau}(\mathbb{R}^n)$. The reader should keep in mind that $|x|^{2-n}$ is harmonic on $\mathbb{R}^n \setminus \{0\}$, and gives the leading order rate of decay for functions which are harmonic at infinity in Euclidean space; as such, you can let w be a smooth function which is equal to $|x|^{2-n}$ outside a ball, and $f = \Delta w$ is any of the weighted spaces under discussion. If $\Delta v = f$ with $\lim_{|x|\to\infty} v(x) = 0$, then $w = v$ by the maximum principle applied to the harmonic function $w - v$. For more on the asymptotics of solutions of the Poisson equation in weighted Hölder spaces, please see [157]. $\square$

The isomorphism result is often applied as follows:

Corollary 7-68. *Suppose $0 < \tau < n-2$, $\mu > 0$, $k \geq 2$, $p' > n \geq 3$ and $1 < p \leq p'$. Suppose (M^n, g) is asymptotically flat, admitting coordinate charts for each end in which $(g_{ij} - \delta_{ij}) \in W^{k-1,p'}_{-\mu}$. Furthermore, suppose $v > 2$ and $h \in W^{k-2,\infty}_{-v}$. If the operator $\Delta_g - h : W^{k,p}_{-\tau} \to W^{k-2,p}_{-\tau-2}$ is injective, such as is the case for $\|h\|_{W^{k-2,\infty}_{-v}}$ sufficiently small, then it is an isomorphism, and so there is a $C > 0$ such that, for all $w \in W^{k,p}_{-\tau}$,*

$$\|w\|_{W^{k,p}_{-\tau}} \leq C \|(\Delta_g - h)w\|_{W^{k-2,p}_{-\tau-2}}. \tag{7.2.17}$$

Proof. We show that the operator $\Delta_g - h$ is a compact perturbation of an isomorphism, from which we can conclude it is Fredholm of index zero. Indeed, for $u \in W^{k-2,p}_{-\tau}$ and $h \in W^{k-2,\infty}_{-v}$, a simple argument using the product rule gives

$$\|hu\|_{W^{k-2,p}_{-\tau-2}} \leq C\|h\|_{W^{k-2,\infty}_{-v}} \|u\|_{W^{k-2,p}_{-\tau-2+v}} \leq C\|h\|_{W^{k-2,\infty}_{-v}} \|u\|_{W^{k-2,p}_{-\tau}}. \tag{7.2.18}$$

In particular, then, the function T_h given by $T_h(u) = hu$ gives a bounded linear operator $T_h : W^{k,p}_{-\tau} \to W^{k-2,p}_{-\tau-2}$. We show that T_h is compact. Consider a bounded sequence in $W^{k,p}_{-\tau}$; by Lemma 7-50, we can assume by taking a subsequence that it converges in $W^{k-2,p}_{-\tau+v-2}$. Let u_i be such a subsequence. By applying the first inequality in (7.2.18) with $u = u_i - u_j$, we see that $T_h(u_i) = hu_i$ is Cauchy, hence convergent, in $W^{k-2,p}_{-\tau-2}$.

If $\|h\|_{W^{k-2,\infty}_{-v}}$ is sufficiently small, then by (7.2.18) the operator $\Delta_g - h$ is a small perturbation of an isomorphism, and hence is an isomorphism. $\square$

There is of course an analogous statement in Hölder spaces.

Corollary 7-69. *Let $n \geq 3$, $\mu > 0$, $\alpha \in (0, 1)$, and $k \geq 2$. Suppose (M^n, g) is asymptotically flat, admitting coordinate charts for each end in which $(g_{ij}-\delta_{ij}) \in C^{k-1,\alpha}_{-\mu}$. Suppose furthermore that $v > 2$ and $h \in C^{k-2,\alpha}_{-v}$. If $0 < \tau < n-2$ and if* *the operator $\Delta_g - h : C^{k,\alpha}_{-\tau} \to C^{k-2,\alpha}_{-\tau-2}$ is injective, as for example when $\|h\|_{C^{k-2,\alpha}_{-\nu}}$ is sufficiently small, then it is an isomorphism, so that there is a $C > 0$ such that for all $w \in C^{k,\alpha}_{-\tau}$,*

$$\|w\|_{C^{k,\alpha}_{-\tau}} \leq C \|(\Delta_g - h)w\|_{C^{k-2,\alpha}_{-\tau-2}}. \tag{7.2.19}$$

Now suppose that the only solution w of $(\Delta_g - h)w = 0$ with $w \in C^{k,\alpha}_{2-n}$ and $\Delta_g w \in L^1$ is the trivial solution $w = 0$. Then for any $f \in C^{k-2,\alpha}_{-n} \cap L^1$, there is a unique $w \in C^{k,\alpha}_{2-n}$ such that $(\Delta_g - h)w = f$. There is a constant $C > 0$ such that for all $w \in C^{k,\alpha}_{2-n}$

$$\|w\|_{C^{k,\alpha}_{2-n}} \leq C(\|(\Delta_g - h)w\|_{C^{k-2,\alpha}_{-n}} + \|(\Delta_g - h)w\|_{L^1}). \tag{7.2.20}$$

The proof is similar to that of the preceding corollary, and is sketched in Exercise 7-105.

Asymptotic expansion. Before we move back to geometry and the Einstein constraint equations, we will discuss a partial expansion near infinity for solutions of a Poisson equation in weighted spaces; cf. [15, Theorem 1.17]. The proof, which we recall below, uses the fact that Δ_g is asymptotic to Δ in an appropriate sense (the coefficients of $(\Delta - \Delta_g)$ decay near infinity in any asymptotic end), where Δ is the Euclidean Laplacian, along with the spherical harmonic expansion of Euclidean harmonic functions discussed earlier (p. 244).

Proposition 7-70. *Suppose $\mathcal{E}$ is an asymptotically flat end in (M^n, g), admitting asymptotic coordinates x in which $(g_{ij} - \delta_{ij}) \in W^{1,p'}_{-\mu}(\mathcal{E})$ for $\mu > 0$ and $p' > n \geq 3$. If $\frac{n}{2} < p \leq p'$, $\tau \in (0, n-2)$, and if $v \in W^{0,p}_{-\tau}(\mathcal{E})$ satisfies $\Delta_g v = f \in W^{0,p}_{-\beta}(\mathcal{E})$ for some $\beta > n$, there is a constant A and a $\gamma > 0$ such that*

$$v(x) = \frac{A}{|x|^{n-2}} + o(|x|^{-(n-2)-\gamma}).$$

Proof. By multiplying v by a cutoff function which is supported in $\mathcal{E}$ and is identically one near infinity, if necessary, we may assume that we are working on an asymptotically flat metric on $\mathbb{R}^n$. By Proposition 7-61, we have $v \in W^{2,p}_{-\tau}$.

We choose $\delta > 0$ small enough so that $\delta \leq \mu$, $0 < \delta + \tau < n - 2$ and $\delta < \beta - n$. We can further restrict δ so that $-(a\delta + \tau)$ is nonexceptional for all $a \in \mathbb{Z}_+$. Therefore $\Delta : W^{2,p}_{a\delta - \tau} \to W^{0,p}_{-a\delta-\tau-2}$ is Fredholm, and there is an $\ell \in \mathbb{Z}_+$ such that $\ell\delta + \tau + 2 < n < (\ell+1)\delta + \tau + 2 < \beta$. Since $\delta \leq \mu$, by Lemma 7-54, we have $\Delta v = (\Delta - \Delta_g)(v) + f$, with $(\Delta - \Delta_g)(v) \in W^{0,p}_{-\delta-\tau-2}$.

We also observe that $f \in W^{0,p}_{-\beta} \subset W^{0,p}_{-(\ell+1)\delta-\tau-2} \subset W^{0,p}_{-\delta-\tau-2}$. Thus $\Delta v \in W^{0,p}_{-\delta-\tau-2}$, and so $v \in W^{2,p}_{-\delta-\tau}$, from which we again use $\delta \leq \mu$ to conclude that

$(\Delta - \Delta_g)(v) \in W^{0,p}_{-2\delta-\tau-2}$. Hence $\Delta v \in W^{0,p}_{-2\delta-\tau-2}$, and so if $\ell \geq 2$, we have $v \in W^{2,p}_{-2\delta-\tau}$.

We can continue this way until we get to $\Delta v \in W^{0,p}_{-(\ell+1)\delta-\tau-2}$. By the Fredholm property, $\Delta : W^{2,p}_{-(\ell+1)\delta-\tau} \to W^{0,p}_{-(\ell+1)\delta-\tau-2}$ has finite codimension, which together with the density of smooth functions with compact support, implies there is a finite-dimensional subspace $S \subset C^\infty_c$ with $\Delta(W^{2,p}_{-(\ell+1)\delta-\tau}) \oplus S = W^{0,p}_{-(\ell+1)\delta-\tau-2}$. Thus we conclude there is an $R > 0$ and a $w \in W^{2,p}_{-(\ell+1)\delta-\tau}$ such that $\Delta(v-w) = 0$ on $\mathcal{E}_R := \{|x| > R\}$. As v and w both decay, $v - w$ is harmonic at infinity and so as in Section 7.1.2.4, $v - w$ admits a spherical harmonic expansion in $\mathcal{E}_R$, with the first term in the expansion of the form $A/|x|^{n-2}$ for some constant A.

Now we apply Sobolev embedding (using $p > \frac{n}{2}$) to complete the argument. Note that since $A/|x|^{n-2} \in W^{2,p}_{-\ell\delta-\tau}(\mathcal{E}_R) \setminus W^{2,p}_{-(\ell+1)\delta-\tau}(\mathcal{E}_R)$ if $A \neq 0$, we cannot in general go further due to the degree of the leading harmonic. $\qquad\square$

Of particular importance in applying this below is the ability to differentiate the expansion and improve decay by a power of $|x|^{-1}$ for each successive partial derivative. We can formulate the above proposition in appropriate higher regularity weighted spaces, so that the error term and some number of its derivatives enjoy suitable decay. The pointwise estimates will come via Sobolev embedding.

Exercise 7-71. a. Assume that $p > n$ in Proposition 7-70. Prove that $v(x) = A/|x|^{n-2} + O_1(|x|^{-(n-2)-\gamma})$ for some A and $\gamma > 0$.

b. Formulate and prove an analogue of Proposition 7-70 whose conclusion is that $v(x) = A/|x|^{n-2} + O_2(|x|^{-(n-2)-\gamma})$ for some A and $\gamma > 0$.

We can also get pointwise estimates using weighted Hölder spaces and Schauder estimates from Proposition 7-63; see [157, Theorem 2].

Proposition 7-72. *Suppose $\mathcal{E}$ is an asymptotically flat end in (M^n, g), admitting asymptotic coordinates x in which $(g_{ij} - \delta_{ij}) \in C^{k-1,\alpha}_{-\mu}(\mathcal{E})$ for some $\mu > 0$ and $k \geq 2$. If $\beta > n \geq 3$ and $\tau \in (0, n-2)$, and if $v \in C^0_{-\tau}(\mathcal{E})$ and $\Delta_g v \in C^{k-2,\alpha}_{-\beta}(\mathcal{E})$, then there is a constant A and $\gamma > 0$ such that*

$$v(x) = \frac{A}{|x|^{n-2}} + O_{k,\alpha}(|x|^{-(n-2)-\gamma}).$$

Sketch of proof. Argue as in the proof of Proposition 7-70, with δ as chosen there. Then $\Delta v \in C^{k-2,\alpha}_{-\delta-\tau-2}$. Recall Remark 7-59. $\qquad\square$

7.2.3. Application to scalar curvature deformation. We now develop some controlled scalar curvature deformation results, using conformal and nonconformal techniques, on asymptotically flat manifolds.

Suppose M admits an asymptotically flat metric, i.e., M can be obtained from a closed and connected manifold by removing a finite nonempty set of points. Then there is a (smooth) metric $\mathring{g}$ on M for which $\mathring{g}_{ij} = \delta_{ij}$ in each of the end charts. For $kp > n$ and $\tau > 0$, we let $\mathcal{M}_{-\tau}^{k,p}$ denote the space of metrics on M for which, relative to the family of charts for the ends, $g_{ij} - \delta_{ij} \in W_{-\tau}^{k,p}$ on each end. By Sobolev embedding, if $g \in \mathcal{M}_{-\tau}^{k,p}$, there is $\varepsilon > 0$ such that for $h \in W_{-\tau}^{k,p}$ a symmetric two-tensor field with $\|h\|_{W_{-\tau}^{k,p}} < \varepsilon$ (norm taken with respect to g or $\mathring{g}$), then $g + h \in \mathcal{M}_{-\tau}^{k,p}$.

Suppose $k \geq 2$. If furthermore $k > \frac{n}{p}$, judicious use of the Sobolev embedding shows the scalar curvature map is smooth as a map from the set of metrics $\mathcal{M}_{-\tau}^{k,p}$ to $W_{-\tau-2}^{k-2,p}$. Indeed, the scalar curvature is given by

$$R(g) = g^{jk}(\Gamma_{jk,\ell}^{\ell} - \Gamma_{\ell k,j}^{\ell} + \Gamma_{jk}^{m}\Gamma_{\ell m}^{\ell} - \Gamma_{\ell k}^{m}\Gamma_{jm}^{\ell}).$$

Thus it suffices to prove that the multiplication maps $W_{-\tau}^{k,p} \times W_{-\tau}^{k,p} \to W_{-\tau}^{k,p}$, $W_{-\tau}^{k,p} \times W_{-\tau-2}^{k-2,p} \to W_{-\tau-2}^{k-2,p}$ and $W_{-\tau-1}^{k-1,p} \times W_{-\tau-1}^{k-1,p} \to W_{-\tau-2}^{k-2,p}$ are continuous maps. We leave this as an exercise for the reader, as these are weighted versions of the analogous statements analyzed in the proof of Proposition 6-54.

We now state an asymptotic version of Theorem 6-52; cf. [89].

Proposition 7-73 (Fischer–Marsden). *Suppose $n \geq 3$, $p > 1$ and $k \geq 2$ satisfy $kp > n$, and suppose $\tau \in (0, n-2)$. The scalar curvature map $R : \mathcal{M}_{-\tau}^{k,p}(\mathbb{R}^n) \to W_{-\tau-2}^{k-2,p}(\mathbb{R}^n)$ maps a neighborhood of the Euclidean metric onto a neighborhood of the zero function. Moreover, any $h_0 \in W_{-\tau}^{k,p}(\mathbb{R}^n)$ which satisfies $DR_{g_{\mathbb{E}^n}}(h_0) = 0$ is tangent to a path of metrics $g(t) \in \mathcal{M}_{-\tau}^{k,p}(\mathbb{R}^n)$ with $R(g(t)) = 0$.*

Proof. Let $L = DR_{g_{\mathbb{E}^n}}$, so that $Lh = -\Delta(\operatorname{tr} h) + \operatorname{div}\operatorname{div} h$. We show that $L : W_{-\tau}^{k,p}(\mathbb{R}^n) \to W_{-\tau-2}^{k-2,p}(\mathbb{R}^n)$ is surjective. This is a plausible claim, since L^* is easily seen to be injective on weighted spaces which promote decay at infinity. Indeed, if we trace $0 = L^*N = -(\Delta N)g_{\mathbb{E}^n} + \operatorname{Hess} N$, we obtain $\Delta N = 0$, and thus $\operatorname{Hess} N = 0$, so that $N \subset \operatorname{span}\{1, x^1, \ldots, x^n\}$.

Recall that for $\tau \in (0, n-2)$, $\Delta : W_{-\tau}^{k,p}(\mathbb{R}^n) \to W_{-\tau-2}^{k-2,p}(\mathbb{R}^n)$ is an isomorphism. With the identity $L(wg_{\mathbb{E}^n}) = -(n-1)\Delta w$ in mind, given any $f \in W_{-\tau-2}^{k-2,p}(\mathbb{R}^n)$, we let $w = -\frac{1}{n-1}\Delta^{-1}f \in W_{-\tau}^{k,p}(\mathbb{R}^n)$. Then $h := wg_{\mathbb{E}^n} \in W_{-\tau}^{k,p}(\mathbb{R}^n)$ satisfies $Lh = f$, as desired.

It follows that the space of symmetric tensor fields has a splitting $W_{-\tau}^{k,p}(\mathbb{R}^n) = \{wg_{\mathbb{E}^n} : w \in W_{-\tau}^{k,p}(\mathbb{R}^n)\} \oplus \{h_0 \in W_{-\tau}^{k,p}(\mathbb{R}^n) : Lh_0 = 0\}$. It is easy to see that these subspaces are both closed. Since the scalar curvature map is smooth, we can conclude from here by the implicit function theorem. $\square$

The directions transverse to the kernel in the above proof are tangent to conformal deformations. For a splitting analogous to (6.3.2), see Exercise 7-109.

Exercise 7-74. Let $L = DR_{g_{\mathbb{E}^n}}$ as above.

a. Show directly (and in one line) that if the symmetric $(0,2)$-tensor h has compact support, and if $Lh \geq 0$, then $Lh = 0$.

b. Show by an elementary construction that there exists an infinite-dimensional space of smooth symmetric TT tensors (trace-free, divergence-free) on $(\mathbb{R}^3, g_{\mathbb{E}^3})$ *with compact support*. Such tensors automatically satisfy $Lh = 0$. (The existence follows also from more general results of [24].)

As discussed following Corollary 6-53, the only metrics on $\mathbb{T}^n$ with nonnegative scalar curvature are flat. If h is a smooth symmetric $(0,2)$-tensor on $\mathbb{R}^n$ with $Lh = 0$ (compactly supported, or in a suitable weighted space as above), Proposition 7-73 yields a path of scalar-flat metrics $g(t)$ tangent to h at $g_{\mathbb{E}^n}$. Even if h has compact support, then, we see that if $g(t)$ is not isometric to the Euclidean metric, then it cannot be Euclidean near infinity either, so the perturbation $g(t) - g_{\mathbb{E}^n}$ is not compactly supported (else we could use $g(t)$ to construct a non-flat, scalar-flat quotient $\mathbb{T}^n$). In the next proposition, we show the analogue of the local surjectivity result in Proposition 7-73 holds near any asymptotically flat metric g, and again, we note that from the conformal deformation used, the metric perturbation $\gamma - g$ so obtained will not in general have compact support.

Proposition 7-75. *Let $p > 1$ and $k \geq 2$ be such that $kp > n \geq 3$, and consider a (smooth) asymptotically flat manifold (M^n, g) (of rate $q > 0$ and order k). For $\tau \in (0, n-2)$ with $\tau < q$, there is an $\varepsilon > 0$ such that if $\|S\|_{W^{k-2,p}_{-\tau-2}} < \varepsilon$, there is a function $u > 0$ with $u(x) \to 1$ as $|x| \to \infty$, and a smooth symmetric $(0,2)$-tensor h compactly supported in M, such that $\gamma := u^{\frac{4}{n-2}} g + h$ is a metric and $R(\gamma) = R(g) + S$. If $q + \tau > n - 2$, and $S \in W^{0,p'}_{-\delta'}$ for some $p' > \frac{n}{2}$ and $\delta' > n$, then u admits an expansion $u(x) = 1 + A/|x|^{n-2} + O(|x|^{-\beta})$ in any end, for some $\beta > n - 2$.*

Proof. Let $c_n = \frac{n-2}{4(n-1)}$. Let T be the nonlinear operator defined on the open subset of $W^{k,p}_{-\tau}$ where $v > -1$ by

$$T(v) = R((1+v)^{\frac{4}{n-2}} g) = (1+v)^{-\frac{n+2}{n-2}}\left(R(g)(1+v) - c_n^{-1}\Delta_g(1+v)\right).$$

T is a smooth map to $W^{k-2,p}_{-\tau-2}$, by arguments similar to those showing that the scalar curvature map is smooth. Of course $T(0) = R(g)$, and the linearization

$DT|_0$ at $v = 0$ is given by

$$DT|_0(w) = R(g)w - c_n^{-1}\Delta_g w - \tfrac{n+2}{n-2}w R(g) = -\big(c_n^{-1}\Delta_g w + \tfrac{4}{n-2}w R(g)\big).$$

For S suitably small, we would like to solve $T(v) = R(g) + S$, for v small, using the inverse function theorem. Let $2 \le k' \le k$. By the asymptotics of g, $w \mapsto w R(g)$ gives a bounded linear map from $W^{k',p}_{-\tau}$ to $W^{k'-2,p}_{-\tau-q-2}$, which is readily seen to be compact as a map to $W^{k'-2,p}_{-\tau-2}$ by Rellich's lemma (Lemma 7-50), as in the proof of Corollary 7-68. Thus, $DT|_0 : W^{k',p}_{-\tau} \to W^{k'-2,p}_{-\tau-2}$ is a Fredholm operator of zero index, as a compact perturbation of the isomorphism $-c_n^{-1}\Delta_g$.

We use a method from [69] to find h. We first show that the linearization $L_g := DR_g : W^{2,p}_{-\tau} \to W^{0,p}_{-\tau-2}$ of the scalar curvature operator is surjective, for $\tau \in (0, n-2)$. The range of L_g contains $DT|_0(W^{2,p}_{-\tau})$, and so it has finite codimension, and hence is closed. Suppose a dual element $f \in W^{0,q'}_{\tau+2-n}$, $q' = \frac{p}{p-1}$, annihilates the range of L_g. Then $0 = L_g^* f = -(\Delta_g f)g + \mathrm{Hess}_g f - f\,\mathrm{Ric}(g)$. By elliptic regularity, f is smooth and as we proved earlier (see Section 2.4.5), if f were nontrivial, $R(g)$ would be constant (since M is connected), so that by asymptotic flatness, $R(g) = 0$. However in case $R(g)$ vanishes identically, we see $\Delta_g f = 0$, and so $f = 0$ (since $0 < n - 2 - \tau < n - 2$). Thus in any case we have $f = 0$, showing that L_g is surjective as desired.

By the density of C_c^∞, we let $\omega_1, \ldots, \omega_\ell$ be compactly supported smooth $(0, 2)$-tensors for which $\mathrm{span}\{L_g\omega_1, \ldots, L_g\omega_\ell\}$ is a complementary subspace to $DT|_0(W^{2,p}_{-\tau})$ in $W^{0,p}_{-\tau-2}$. This span is easily seen to form a complementary subspace to $DT|_0(W^{k,p}_{-\tau})$ in $W^{k-2,p}_{-\tau-2}$, by elliptic regularity (Proposition 7-61) applied to $DT|_0(w) + \sum_{i=1}^\ell c_i L_g\omega_i \in W^{k-2,p}_{-\tau-2}$. (Alternatively, since a distributional solution of $L_g^* f = 0$ is smooth, essentially the same argument as above shows $L_g := DR_g : W^{k,p}_{-\tau} \to W^{k-2,p}_{-\tau-2}$ is surjective.)

We let X be a complementary space to the kernel of $DT|_0$, and we define, for suitably chosen neighborhoods Y of $0 \in X$ and Z of $0 \in \mathrm{span}\{\omega_1, \ldots, \omega_\ell\}$, the map $\overline{T} : Y \oplus Z \to W^{k-2,p}_{-\tau-2}$ by $\overline{T}(v, h) = R((1 + v)^{\frac{4}{n-2}}g + h)$. Note that $\overline{T}(v, 0) = T(v)$, so that $D\overline{T}|_{(0,0)}(w, 0) = DT|_0(w)$, while $D\overline{T}|_{(0,0)}(0, \omega) = L_g\omega$. Thus by design, $\overline{T}$ is a smooth map for which $D\overline{T}|_{(0,0)}$ is an isomorphism. From here, the result follows from the inverse function theorem, and the standard asymptotic expansion follows as in Proposition 7-70 (see Exercise 7-108). $\qquad\square$

This result gives us good control on u and h for small S; however, for our purposes in studying the ADM energy, the above result does not suffice, since we do not get control on the sign of A. As we will see, for suitably small $R(g)$, or for $R(g) \ge 0$, we can avoid the difficulty encountered in the above proof, which accounts for the possibility that the linearization is not invertible. Note that when we work in the smooth setting and carry out constructions like that above, we should analyze the regularity of the result. In the preceding, for example, one can readily show that if S were smooth, then the solution v would be smooth outside the support of h. To guarantee this inside the support of h, one could increase k if needed and use Sobolev embedding (to control the nonlinear terms) and elliptic bootstrapping.

We now consider what happens to the ADM energy if we conformally transform away the matter density in a time-symmetric asymptotically flat initial data set. In the time-symmetric setting, the ADM linear momentum vanishes, and we will let $m(g)$ be the ADM mass (energy) for a chosen end of an asymptotically flat (M^n, g).

Proposition 7-76. *Suppose (M^n, g), $n \geq 3$, is asymptotically flat at rate $q > \frac{n-2}{2}$ and order $k+1 \geq 3$, with nonnegative scalar curvature $R(g) \in L^1(M)$. There is a conformally related metric $\bar{g} = u^{\frac{4}{n-2}} g$ which is asymptotically flat (of order k) and has $R(\bar{g}) = 0$, with $m(\bar{g}) \leq m(g)$ (in each end).*

Proof. Recall the conformal Laplacian $\mathscr{L}_g = \Delta_g - \frac{n-2}{4(n-1)} R(g)$, a self-adjoint linear operator. By (6.1.8), for $u > 0$ and $\bar{g} = u^{\frac{4}{n-2}} g$, $R(\bar{g})$ vanishes if and only if $\mathscr{L}_g u = 0$. This is a linear equation, and we study $\mathscr{L}_g$ acting on a weighted function space.

Consider $\tau \geq \frac{n-2}{2}$, and suppose that $v \in C^{k,\alpha}_{-\tau}$ solves $\mathscr{L}_g v = 0$. We have $R(g)v \in C^{k-2,\alpha}_{-q-2-\tau}$ and $q + \tau + 2 > n$. From Proposition 7-72, we see that $v \, \partial v / \partial x^j = O(|x|^{-2n+3})$ in asymptotic coordinates x, so that since $n \geq 3$, $\lim_{r \to \infty} \int_{\{|x|=r\}} v \, (\partial v/\partial v) \, d\sigma = 0$, and hence $\lim_{r \to \infty} \int_{\{|x|=r\}} v \, (\partial v/\partial v_g) \, d\sigma_g = 0$ by Exercise 7-35. Integrating by parts we have

$$-\int_M |\nabla_g v|^2 \, dv_g = \frac{n-2}{4(n-1)} \int_M R(g) v^2 \, dv_g \geq 0.$$

Thus v is a constant, and must vanish by the decay of v. (While this follows more generally for $\tau > 0$ by the maximum principle, the argument above is also useful; cf. Remark 7-78.)

We take $\tau \leq q$ with $\tau \in \left(\frac{n-2}{2}, n-2 \right)$, and apply Corollary 7-69 to the operator $\mathscr{L}_g$: we solve $\mathscr{L}_g v = \frac{n-2}{4(n-1)} R(g)$ uniquely for $v \in C^{k,\alpha}_{-\tau}$, and then let $u = 1 + v$. We see that $u \to 1$ on approach to infinity in any end, and that $\mathscr{L}_g u = 0$, and thus u is smooth. We only need to show that $u > 0$ to have an honest conformal factor and the asymptotically flat metric $\bar{g}$ (of order k) with zero scalar curvature. By the weak maximum principle, $0 \leq u \leq 1$, and by the strong maximum principle (or the Harnack inequality), $u > 0$.

Suppose first that $R(g) \in C^{0,\alpha}_{-\beta}$ for some $\beta > n$ (such as would hold in case (g, K) satisfies the vacuum constraints). Then, since $u \leq 1$, we see that

$A \leq 0$ in the expansion from Proposition 7-72, so that by Lemma 7-37, we have $m(\bar{g}) = m(g) + 2A \leq m(g)$. For more general $R(g)$, the mass estimate follows in the case of one end by showing that

$$m(\bar{g}) - m(g) = -\frac{1}{2(n-1)\omega_{n-1}} \int_M R(g)u \, dv_g, \qquad (7.2.21)$$

which can be derived by an expansion similar to that in the proof of Lemma 7-37, along with integration by parts as in the following remark. For the case of multiple ends, see, e.g., [142, pp. 88–89]. $\qquad \square$

Remark 7-77. For a nontrivial case of the preceding proposition (meaning that $R(g)$ does not vanish identically), the mass of each end *strictly* decreases. When there is a single asymptotic end, this follows by (7.2.21), whereas in the case of multiple ends, we can conclude from here that the mass of at least one end strictly decreases.

Consider the case $R(g) \in C^{0,\alpha}_{-\beta}$, $\beta > n$, when the conformal factor u admits an expansion. In the case of one asymptotic end, we find, using Exercise 7-35 to switch to the Euclidean surface measure,

$$\int_M \frac{n-2}{4(n-1)} R(g)u \, dv_g = \int_M \Delta_g u \, dv_g = \lim_{r \to \infty} \int_{\{|x|=r\}} \frac{\partial u}{\partial v_g} \, d\sigma_g$$

$$= \lim_{r \to \infty} \int_{\{|x|=r\}} \frac{\partial u}{\partial v} \, d\sigma = -(n-2)\omega_{n-1} A.$$

In the case of multiple ends, one could isolate one end at a time, by considering a comparison principle on a manifold-with-boundary which contains only one asymptotic end, the chosen asymptotic end, see Exercise 7-94. The exercise invokes a solution to a certain Dirichlet boundary value problem with asymptotic conditions, and we note that one can modify the theory of weighted spaces to allow boundary, see e.g., [152]. In fact we remark that Schoen and Yau [199] give a direct proof for the existence and expansion of solutions of the analogue *Neumann* boundary value problem.

Note that the above integration by parts formula would hold for a conformal factor u satisfying the PDE $\mathscr{L}_g u = 0$ on a manifold with compact boundary and one asymptotic end, such that u goes to 1 at infinity in the end, and with $\partial u/\partial v = 0$ on the boundary. One could apply this together with (7.2.21) to handle the multiple end case; cf. [142, pp. 88–89].

Remark 7-78. We adapt an argument from [199] to show that the conformal Laplacian $\mathscr{L}_g : W^{2,p}_{-\tau} \to W^{0,p}_{-\tau-2}$ (where $\tau \in (0, n-2)$, $p > \frac{n}{2}$) is invertible for g asymptotically flat with $R(g) \geq 0$, or in fact, whenever $R(g)$ has suitably small negative part $R(g)^-$. The operator $\mathscr{L}_g$ is Fredholm of index zero, since it is a compact perturbation of the Laplacian. For any w in the kernel, we obtain by integration by parts, using the decay of w and the Hölder inequality (Exercise 7-95), that

$$\|\nabla w\|^2_{L^2(dv_g)} \leq c \|R(g)^-\|_{L^{n/2}(dv_g)} \|w\|^2_{L^{2^*}(dv_g)},$$

where $2^* = \frac{2n}{n-2}$ is the Sobolev conjugate exponent to 2, in dimension $n > 2$. For $R(g)^-$ small in $L^{n/2}$, we see that w must be zero by the Sobolev inequality: $\|w\|_{L^{2^*}(dv_g)} \leq C\|\nabla w\|_{L^2(dv_g)}$, and therefore $\mathcal{L}_g : W^{2,p}_{-\tau} \to W^{0,p}_{-\tau-2}$ is an isomorphism.

For analyzing the question of the sign of the ADM mass of time-symmetric initial data sets with the dominant energy condition, it thus suffices to study the vacuum case. In the next section we reduce further to the case of harmonically flat asymptotics.

7.3. Harmonically flat asymptotics

Schoen and Yau [202] showed that asymptotically flat metrics with zero scalar curvature can be approximated by metrics which are harmonically flat on the asymptotic ends. We sketch the proof of a modestly strengthened statement, slightly modifying and improving the treatment from [68]. In this section and the next, asymptotically flat metrics have a rate $q > \frac{n-2}{2}$.

For the next proposition, we let $X^2_{-\tau}$ be either $C^{2,\alpha}_{-\tau}(M)$ (only if the order is at least 3) or $W^{2,p}_{-\tau}(M)$, with $p > \frac{n}{2}$, $\frac{n-2}{2} < \tau < \min(q, n-2)$, $0 < \alpha < 1$. Likewise $X^0_{-\tau-2}$ stands for either $C^{0,\alpha}_{-\tau-2}(M)$ or $W^{0,p}_{-\tau-2}(M)$, as appropriate.

Proposition 7-79 (Schoen–Yau). *Suppose (M, g) is asymptotically flat with nonnegative $R(g) \in L^1(M)$. For any $\varepsilon > 0$, there is a metric $\bar{g}$ within ε of g in $X^2_{-\tau}$, with $R(\bar{g}) \geq 0$, and which is harmonically flat near infinity in each end, with $|m(g) - m(\bar{g})| < \varepsilon$. There is also a metric $\tilde{g}$ with $R(\tilde{g}) = 0$ which is harmonically flat near infinity in each end, with $m(\tilde{g}) < m(g) + \varepsilon$.*

Proof. By Proposition 7-76, the second claim follows from the first, the proof of which will in fact yield $R(\bar{g}) = 0$ in the case $R(g) = 0$ to start.

Let $0 \leq \psi \leq 1$ be a smooth cutoff function so that $\psi(t) = 1$ for $t < 1$ and $\psi(t) = 0$ for $t > 2$. On each end we choose asymptotically flat coordinates defined for $|x| > 1$. For $\theta > 1$, let $\psi_\theta(x) = \psi(|x|\theta^{-1})$; ψ_θ extends smoothly from the ends to all of M. Now consider the metric $g_\theta(x) = \psi_\theta(x)g(x) + (1 - \psi_\theta(x))g_{\mathbb{E}^n}(x)$. This metric is identical to the Euclidean metric for $|x| > 2\theta$, and (as $q > \tau$) g_θ can be made arbitrarily close to g in $X^2_{-\tau}$, for θ sufficiently large. Note that $R(g_\theta) \geq 0$ except possibly for $\theta < |x| < 2\theta$, and we will use a conformal deformation to impose nonnegative scalar curvature.

From (6.1.8) we have

$$R(u^{\frac{4}{n-2}} g_\theta) = -\frac{4(n-1)}{n-2} u^{-\frac{n+2}{n-2}} \Big(\Delta_{g_\theta} u - \frac{n-2}{4(n-1)} R(g_\theta) u \Big).$$

We want to impose $R(u^{\frac{4}{n-2}} g_\theta) = \psi_\theta u^{-\frac{4}{n-2}} R(g)$, i.e.,

$$\Delta_{g_\theta} u - \frac{n-2}{4(n-1)} \big(R(g_\theta) - \psi_\theta R(g) \big) u = 0.$$

With $u = 1 + v$ this becomes

$$\Delta_{g_\theta} v - \frac{n-2}{4(n-1)} \big(R(g_\theta) - \psi_\theta R(g) \big) v = \frac{n-2}{4(n-1)} \big(R(g_\theta) - \psi_\theta R(g) \big). \quad (7.3.1)$$

Note that $h_\theta := \frac{n-2}{4(n-1)} \big(R(g_\theta) - \psi_\theta R(g) \big)$ vanishes unless $\theta < |x| < 2\theta$, with $|h_\theta| \leq C\theta^{-2-q}$. In fact since $q > \tau$, for sufficiently large θ, h_θ has *small* norm in $X^0_{-\tau-2}$. Thus, $\Delta_{g_\theta} - h_\theta$ is a small perturbation of the invertible operator $\Delta_g : X^2_{-\tau} \to X^0_{-\tau-2}$, and thus it is an isomorphism, with inverse bounded uniformly in θ, cf. Corollaries 7-68 and 7-69; in particular, the constant in (7.2.17) or (7.2.19) can be chosen uniformly in h suitably small. Thus for θ large enough, we can solve (7.3.1), and the solution v_θ will be smooth and close to zero in $X^2_{-\tau}$, so that $u_\theta = 1 + v_\theta > 0$, while $\bar{g}_\theta = u_\theta^{4/(n-2)} g_\theta$ is close in the weighted norm to g_θ and hence to g, and is harmonically flat near infinity in each end.

Upon further restricting to $\tau > \frac{n-2}{2}$, the masses will also be close for large θ. We will sketch the idea of the proof following [202; 197] using the Hölder space for v_θ, cf. [78; 119] for the Sobolev case. From (7.2.2) and Propositions 7-33 and 7-31, we know that $m(g)$ equals

$$\frac{1}{2(n-1)\omega_{n-1}} \Big(\int_{\{|x|=r_1\}} \sum_{i,j=1}^{n} (g_{ij,i} - g_{ii,j}) \nu^j \, d\sigma + \int_{\{r_1 < |x|\}} (R(g) + \mathcal{Q}_g) \, dx \Big),$$

where we have set $\sum_{i,j=1}^{n} (g_{ij,ij} - g_{ii,jj}) = R(g) + \mathcal{Q}_g$.

From the expansion of the scalar curvature (Proposition 7-31; cf. Exercise 7-38) we have $\mathcal{Q}_g(x) = O(|x|^{-2q-2})$, so that $R(g) + \mathcal{Q}_g$ is integrable on the end. Given $\varepsilon > 0$, we can bound the second integral by $\frac{\varepsilon}{3}$ by choosing r_1 sufficiently large. We replace g by $\bar{g}_\theta$ in the integrals above to compute $m(\bar{g}_\theta)$. By our control on v_θ in $C^{2,\alpha}_{-\tau}(M)$, and hence on $\bar{g}_\theta$, we can uniformly estimate $R(\bar{g}_\theta) = \psi_\theta u_\theta^{-4/(n-2)} R(g)$, and we can see that $|\mathcal{Q}_{\bar{g}_\theta}(x)| \leq C|x|^{-2\tau-2}$, where C is uniform in θ large. Thus we can bound the corresponding integral for $\bar{g}_\theta$ by $\frac{\varepsilon}{3}$ by choosing an r_1 large enough, *uniformly* for $\theta \geq \theta_0$ for some sufficiently large

θ_0. By using (7.2.19), we see that for θ sufficiently large, h_θ and hence v_θ and ∂v_θ are small enough that

$$\left| \int_{\{|x|=r_1\}} \left(\sum_{i,j=1}^{n} (g_{ij,i} - g_{ii,j}) - \sum_{i,j=1}^{n} ((\bar{g}_\theta)_{ij,i} - (\bar{g}_\theta)_{ii,j}) \right) v^j \, d\sigma \right| < \frac{\varepsilon}{3}. \qquad \square$$

Bray [25] observed that the Schoen–Yau approximation could be modified to produce ends that are *precisely* Schwarzschild, while preserving *nonnegative* scalar curvature. To present his argument, we first recall a standard setup for mollification. Let $\phi : \mathbb{R} \to \mathbb{R}$ be a smooth, nonnegative function with support $[-1, 1]$, which is constant near the origin. Let $\varphi(x) = \phi(|x|)$ be the associated smooth, rotationally symmetric bump function supported in the unit ball, and by scaling we may assume $\int_{\mathbb{R}^n} \varphi(x) \, dx = 1$. Consider the approximate identity $\varphi_\sigma(x) = \sigma^{-n} \varphi(\frac{x}{\sigma})$ for $\sigma \downarrow 0$; note that φ_σ has unit integral, and is supported on $\{|x| \le \sigma\}$. We can use φ_σ to mollify functions by convolution: $(\varphi_\sigma * w)(x) = \int_{\mathbb{R}^n} \varphi_\sigma(y) w(x - y) \, dy$.

We now state and prove Bray's approximation. The closeness of the metrics g and $\tilde{g}$ can be measured in a weighted norm as above, or, as Bray states it, as an *ε-quasi-isometry*, meaning that for all nonzero $v \in TM$, the ratio $g(v, v)/\tilde{g}(v, v)$ lies in $(e^{-\varepsilon}, e^{\varepsilon})$.

Proposition 7-80 (Bray). *Suppose (M, g) is asymptotically flat end with non-negative $R(g) \in L^1(M)$. For any $\varepsilon > 0$, there is a metric $\tilde{g}$ with $R(\tilde{g}) \ge 0$, which is ε-close to g, which is isometric to a Riemannian Schwarzschild metric near infinity in each end of M, and for which $|m(g) - m(\tilde{g})| < \varepsilon$.*

Proof. By applying Proposition 7-79, we may modify the metric so that each end is harmonically flat. For any chosen end $\mathcal{E}$, we may choose asymptotic coordinates along with an $r_0 > 0$, so that for $|x| > r_0$, the metric has the form $g_{ij}(x) = u(x)^{\frac{4}{n-2}} \delta_{ij}$, with $\Delta u = 0$, and $u(x) = 1 + \frac{1}{2} m(g)/|x|^{n-2} + O_\infty(|x|^{-(n-1)})$. We will handle each end individually, so we do not introduce notation to distinguish the mass on different ends. Now for any $R > r_0$ and $\delta > 0$, consider the harmonic function $\tilde{u}(x) = C_1 + C_2/|x|^{n-2}$, with C_1 and C_2 chosen so that

$$C_1 + \frac{C_2}{R^{n-2}} = \max_{\{|x|=R\}} u(x) + \delta \quad \text{and} \quad C_1 + \frac{C_2}{(2R)^{n-2}} = \min_{\{|x|=2R\}} u(x) - \delta.$$

By the choice of C_1 and C_2, the function w defined by

$$w(x) = \begin{cases} u(x) & \text{if } |x| < R, \\ \min(u(x), \tilde{u}(x)) & \text{if } R \le |x| \le 2R, \\ \tilde{u}(x) & \text{if } |x| > 2R \end{cases}$$

is continuous. The expansion of u also implies that, for any $\eta > 0$, if we take R sufficiently large and δR^{n-2} sufficiently small, then $|C_1 - 1| < \eta$ and $\left|C_2 - \frac{1}{2}m(g)\right| < \eta$. Since u and $\tilde{u}$ are harmonic, and the minimum of harmonic functions is (weakly) superharmonic, we have that w is superharmonic. If we convolve w with a spherically symmetric mollifier φ_σ supported in $\{y : |y| \leq \sigma\}$ $(0 < \sigma < R - r_0)$, we produce a smooth, superharmonic function $\tilde{w} = \varphi_\sigma * w$ which satisfies, by the mean value property, $\tilde{w}(x) = u(x)$ for $|x| < R - \sigma$, and $\tilde{w}(x) = \tilde{u}(x)$ for $|x| > 2R + \sigma$. Thus if we let $\tilde{g}_{ij} = \tilde{w}^{\frac{4}{n-2}}\delta_{ij}$ on $\mathcal{E}$, then $\tilde{g}$ agrees with g for $r_0 < |x| < R - \sigma$, and $\tilde{g}$ is precisely Schwarzschild on $|x| > 2R + \sigma$, with mass given by (see Exercise 7-36) $m(\tilde{g}) = 2C_1 C_2 \approx m(g)$. $\qquad\square$

We emphasize that applied to an end which is already harmonically flat, this construction is local to the end.

7.4. On the positive mass theorem

As we saw earlier, to each end in an asymptotically flat initial data set with integrable mass and momentum densities (ρ, J), we can associate an energy E and linear momentum vector P, which together form the ADM energy-momentum vector for the asymptotic end. The constraint equations have the form $\Phi(g, \pi) = (2\kappa\rho, \kappa J)$ (we take $\Lambda = 0$). Without a model for, or condition on, (ρ, J), any (g, π) satisfies this system by definition, and indeed by simple patching one can construct asymptotically flat initial data sets with negative energy. We want a condition that will imply that the ADM energy-momentum vector is future-pointing causal, i.e., $E \geq |P|$ (with $c = 1$), in which case we define the ADM mass as $m = \sqrt{E^2 - |P|^2}$. The *positive mass theorem* asserts that indeed $E \geq |P|$ for an asymptotically flat initial data set (M, g, π) with the dominant energy condition in the form $\rho \geq |J|_g$ (cf. Section 5.2), while the *positive energy theorem* asserts that $E \geq 0$.

Sometimes these statements are called the *spacetime* positive mass theorem and *spacetime* positive energy theorem, in contrast to the situation where the object of interest is an asymptotically flat manifold (M, g). Indeed, in the time-symmetric case $(K = 0$, i.e., $\pi = 0)$, we have $P = 0$, and so we often write $E = m$, which we note can be defined for any asymptotically flat (M, g) with $R(g) \in L^1(M)$. The dominant energy condition in the time-symmetric case, or more generally in the maximal $(\mathrm{tr}_g K = 0)$ case, reduces to $R(g) \geq 0$. The *Riemannian* positive mass theorem for an asymptotically flat (M, g) with $R(g) \geq 0$ is then $m \geq 0$.

In each situation above, the theorem is really comprised of an inequality, together with a *rigidity* statement which characterizes the ground state. In the positive energy theorem, if we find $E = 0$, we would like to conclude our initial data is a slice in Minkowski spacetime, and in the Riemannian positive mass theorem we would like the conclude if $m = 0$, then g is Euclidean. Finally in the general case if we find $E = |P|$, we would again like to conclude the initial data is a slice in Minkowski spacetime.

The inequalities and rigidity statements have a long and storied history which we will not attempt to detail here, but we give some references for further exploration. The Riemannian positive mass theorem in dimensions less than eight was established in the seminal works of Schoen and Yau [199; 201; 202], cf. [197]. The extension to all dimensions has recently appeared in [205], where one can also find references to an approach in a series of papers by Lohkamp. The positive energy theorem in dimension three was established in [203], the methods of which have been extended to dimensions less than eight by Eichmair [77]. The inequality in the positive mass theorem for dimensions less than eight was proven in [78], to which we refer the reader for further background and discussion, while the corresponding rigidity statement recently appeared in [120].

We also want to point out that in the case M is *spin*, there is an approach to the positive mass theorem discovered by Witten [224], cf. [15; 175].

On the proof of the Riemannian PMT. We now present a useful observation due to Lohkamp. We give a somewhat simpler proof than in [146], using Bray's Proposition 7-80.

Proposition 7-81 (Lohkamp). *Suppose the end $(\mathcal{E}, g)$ is harmonically flat with negative mass. Then there is a metric on $\mathcal{E}$ which has nonnegative scalar curvature (and which is not identically zero), which agrees with g near $\partial\mathcal{E}$ and which is flat outside a compact set.*

Proof. By applying the construction in the proof of Proposition 7-80, we may assume that in appropriate asymptotically flat coordinates for $\mathcal{E}$, we have

$$g_{ij}(x) = \left(1 + \frac{m}{2|x|^{n-2}}\right)^{\frac{4}{n-2}} \delta_{ij},$$

with $m < 0$. For $R > -\frac{m}{2}$ large enough that $\mathcal{E}$ contains a neighborhood of $\{|x| = R\}$ in coordinates, we consider the positive continuous function U on $\mathcal{E}$ given by

$$U(x) = \min\left(1 + \frac{m}{2|x|^{n-2}}, \ 1 + \frac{m}{2R^{n-2}}\right) = \begin{cases} 1 + \dfrac{m}{2|x|^{n-2}} & \text{if } |x| \leq R, \\ 1 + \dfrac{m}{2R^{n-2}} & \text{if } |x| \geq R; \end{cases}$$

note where $m < 0$ has been used. Since U is the minimum of two (Euclidean) harmonic functions, it is superharmonic. We use a spherically symmetric mollifier

φ_ε as above to mollify U: let $\widetilde{U} = (\varphi_\varepsilon * U)$. $\widetilde{U}$ is well-defined on $\mathcal{E}$ for $\varepsilon > 0$ small enough, and it is smooth and positive. Indeed for $|x| < R - \varepsilon$, $\widetilde{U}(x) = u(x)$, by the mean value property of harmonic functions, and for $|x| > R + \varepsilon$, $\widetilde{U}(x) = 1 + \frac{1}{2}m/R^{n-2}$ is a positive constant. The mollification preserves superharmonicity, and thus, as it is clear that $\widetilde{U}$ is not harmonic on all of $\mathcal{E}$, there is a region where $\Delta \widetilde{U} < 0$. Using the chosen coordinates, the metric $\tilde{g}_{ij} = \widetilde{U}^{\frac{4}{n-2}} \delta_{ij}$ has nonnegative scalar curvature *which is not identically zero*, and it is a flat metric on $\{|x| > R + \varepsilon\}$. $\qquad\square$

A basic version of the positive mass theorem follows as a corollary of this result, using the Schoen–Yau topological obstruction to positive scalar curvature.

Theorem 7-82 (Riemannian positive mass theorem). *Suppose (M, g) is an asymptotically flat three-manifold with nonnegative scalar curvature $R(g)$. Then the ADM mass of any end is nonnegative. If the mass of any end is zero, then (M, g) is isometric to $(\mathbb{R}^3, g_{\mathbb{E}^3})$.*

Proof. From Proposition 7-79, we may assume that g is harmonically flat at the ends, and while nonnegative scalar curvature would suffice for the argument, we can also arrange $R(g) = 0$. (From the proof of Proposition 7-33, if $R(g) \geq 0$ is not integrable on an end, the mass of that end is $+\infty$.) In fact, we will now further reduce to the case where there is only one asymptotically flat end.

If M has more than one end, choose one, say $\mathcal{E}$, and let u be a harmonic function, $\Delta_g u = 0$, with $u(x) \to 1$ as $|x| \to \infty$ in $\mathcal{E}$, and $u(x) \to 0$ as $|x| \to \infty$ in the other ends. To find such a u, we fix $w \in C^\infty(M)$ with $w = 1$ near infinity in the end $\mathcal{E}$, and $w = 0$ near infinity in the other ends. Then $\Delta_g w \in C_c^\infty(M)$, and so we can solve $\Delta_g v = -\Delta_g w$, for v in a weighted space (so that v decays to zero in each end). Let $u = v + w$. By the maximum principle, $0 < u < 1$ on M.

Since $R(g) = 0$, the metric $u^4 g$ has vanishing scalar curvature, and near infinity in any end, it can be written $u^4 g = U^4 g_{\mathbb{E}^3}$ for $U > 0$, where $\Delta U = 0$. In the ends other than $\mathcal{E}$, U tends to zero at infinity, and so if we write U in spherical harmonics in these ends, we have $U(x) = c/|x| + O_\infty(1/|x|^2)$. The higher spherical harmonics are not everywhere-positive, so that $c > 0$. A simple calculation using the Kelvin transform $x \mapsto x/|x|^2$, shows that $(M, u^4 g)$ can be completed to a smooth asymptotically flat manifold $(\overline{M}, \bar{g})$, with each end apart from $\mathcal{E}$ compactified with an additional point. Since $u < 1$, the mass $m(\bar{g})$ is no more than that of $(\mathcal{E}, g)$, as in the proof of Proposition 7-76.

If the ADM mass of $(\mathcal{E}, g)$ were negative, then the ADM mass of $(\overline{M}, \bar{g})$ would also be negative. Now we apply the preceding proposition to assert the existence of a metric on $\overline{M}$ with nonnegative scalar curvature which is flat outside a compact set. We can thus consider a region $W \subset \overline{M}$ such that

$\partial W = \{x : |x^i| = r_0, \ i = 1, 2, 3\}$ is a large cube in the region where the metric is flat. The metric thus descends to a metric with nonnegative (not identically zero) scalar curvature on the quotient manifold $\check{M}$ obtained by identifying the opposite coordinate faces in pairs. $\check{M}$ can be expressed as a connected-sum of a closed three-manifold with the torus $\mathbb{T}^3$, and in particular there is a copy of $\mathbb{Z} \oplus \mathbb{Z}$ in $\pi_1(\check{M})$. We clearly have a contradiction to the Schoen–Yau obstruction to positive scalar curvature (Theorem 6-56), in the case when M, and hence $\check{M}$ is orientable. In the non-orientable case, one can either apply Exercise 7-93 and replace (M, g) by an asymptotically flat orientable double cover, and argue as above from here, or one can argue that an orientable double cover $\widehat{M}$ of $\check{M}$ has a copy of $\mathbb{Z} \oplus \mathbb{Z}$ inside $\pi_1(\widehat{M})$, as $\pi_1(\widehat{M})$ injects under the covering map as an index two subgroup in $\pi_1(\check{M})$.

We address the rigidity case below. $\hfill\square$

The minimal hypersurface proof of the positive mass theorem given by Schoen and Yau in [199; 197] does not use Theorem 6-56 directly, but the proofs share some of the same ideas (and note that more recent work of Schoen and Yau [205] does employ a compactification argument in higher dimensions). A simple calculation in asymptotically flat coordinates for a metric $g_{ij}(x) = (1 + 2m/|x|)\delta_{ij} + O_1(|x|^{-2})$ on a three-manifold yields the Christoffel symbol $\Gamma_{ij}^3 = (mx^3/|x|^3)\delta_{ij} + O(|x|^{-3})$, for $i, j = 1, 2$. By tracing over $i, j = 1, 2$ (and estimating the difference between $\partial/\partial x^3$ and the unit normal), we can conclude that if the mass were negative, the coordinate hyperplanes $x^3 = \pm\xi$ for large enough $\xi > 0$ would have mean curvature vector $\boldsymbol{H}$ with $g(\boldsymbol{H}, \pm\partial/\partial x^3) < 0$; cf. Exercise X-5 where the mean curvature calculation is done for the Schwarzschild metric, of which g is a perturbation near infinity. These hyperplanes can thus be used as barriers for finding a stable minimal hypersurface asymptotic to a plane, by solving a Plateau problem on large cylinders with axis along the x^3-direction. The stability inequality and Gauss–Bonnet yield a contradiction, in a similar manner as in the proof of Proposition 6-57; cf. [197]. We remark that in [199] Schoen and Yau consider asymptotically flat manifolds-with-boundary. The mean curvature vector along all boundary components should point inward, so that each boundary component also serves as a barrier. Indeed, the negative mass Schwarzschild metric is incomplete, and if you excise a ball centered around the singularity, you obtain an asymptotically flat manifold-with-boundary and vanishing scalar curvature, but the mean curvature vector along the boundary sphere points outward.

On the proof of rigidity. We now address the rigidity statement: if the mass of any end vanishes, then (M^3, g) is isometric to $(\mathbb{R}^3, g_{\mathbb{E}^3})$; in particular, the vanishing of the mass rules out nontrivial topology in the manifold. We compare to a simple analogue in dimension $n \geq 2$ as remarked following Proposition 7-73, by compactification to $\mathbb{T}^n$: any metric g on $\mathbb{R}^n$ with nonnegative scalar curvature and which outside a compact set is isometric to the Euclidean metric is actually isometric to $g_{\mathbb{E}^n}$. (Thus, for $n \geq 3$, the argument of the preceding section implies $m \geq 0$ for $(\mathbb{R}^n, g)$ asymptotically flat with nonnegative scalar curvature.)

We indicate a rigidity proof in the next exercise, given the Riemannian positive mass theorem in higher dimensions [197; 205]. The proof below uses volume comparison; for a proof using harmonic coordinates, see [196, Proposition 2], and for further discussion, see [142].

Exercise 7-83. Suppose (M^n, g), $n \geq 3$, is asymptotically flat with nonnegative scalar curvature and vanishing ADM mass. By Remark 7-77 (see Exercise 7-95), g has vanishing scalar curvature. Suppose M has a single asymptotic end.

Suppose h is a smooth, symmetric $(0, 2)$-tensor with compact support in M. Let $\gamma_\epsilon = g + \epsilon h$, which is a metric for $|\epsilon|$ small. By Corollary 7-69 and Proposition 7-72, there is $u_\epsilon = 1 + v_\epsilon > 0$ so that $g_\epsilon := u_\epsilon^{4/(n-2)} \gamma_\epsilon$ has $R(g_\epsilon) = 0$, with $v_\epsilon \subset C_{-\tau}^{2,\alpha}(M)$, so that g_ϵ is asymptotically flat, and $u_\epsilon(x) = 1 + \frac{1}{2}m(\epsilon)/|x|^{n-2} + O_2(|x|^{-(n-2)-\gamma})$, for some $\gamma > 0$. Lemma 7-37 and the hypothesis on g imply that $m(g_\epsilon) = m(\epsilon)$, and as in Remark 7-77, we have

$$m(\epsilon) = -\frac{1}{2(n-1)\omega_{n-1}} \int_M R(\gamma_\epsilon) u_\epsilon \, dv_{\gamma_\epsilon}.$$

a. Argue that the map $\epsilon \mapsto v_\epsilon \in C_{-\tau}^{2,\alpha}(M)$ is differentiable in ϵ, then obtain an identity from computing $m'(0)$. (Differentiability at $\epsilon = 0$ suffices for this.)

b. Let ζ be any compactly supported smooth bump function. Let $h = \zeta \operatorname{Ric}(g)$ in the above. Conclude that $\operatorname{Ric}(g) = 0$.

c. Use the Bishop–Gromov volume comparison theorem to conclude that (M^n, g) is isometric to Euclidean space, cf. [182, Chapter 9, Exercise 5].

We remark that one way to handle multiple ends is to proceed following an argument in [199]: given M as above with an end $\mathcal{E}$ with vanishing mass, namely, let $N \subset M$ be an asymptotically flat manifold-with-boundary containing one asymptotic end $\mathcal{E}$, and with boundary given by a union of large coordinate spheres from the remaining ends, chosen so large that the mean curvature vector of these spheres points *into* N. Now run the argument indicated in the exercise, solving for the conformal factor u_ϵ which solves the same PDE but has the *Neumann* boundary condition; see Remark 7-77. The conformal metric $g_\epsilon = u_\epsilon^{4/(n-2)} \gamma_\epsilon$ will be asymptotically flat with vanishing scalar curvature, and with mean curvature vector still pointing into N along ∂N, by Exercise X-5. Thus by Schoen–Yau

[199], the mass of g_ϵ must be nonnegative. From here you can conclude N is Ricci-flat, and then that M is Ricci-flat, and conclude as above.

We see that the decay rate to the Euclidean metric of an asymptotically flat metric with nonnegative scalar curvature is constrained by the mass: any such metric which admits asymptotically flat coordinates with rate $q > n - 2$ must in fact be flat. We point out that this holds for complete asymptotically flat manifolds. One can easily write down a nontrivial harmonically flat end with zero mass, but just as with analogous negative mass examples, that end cannot be completed to an asymptotically flat manifold with nonnegative scalar curvature.

7.5. Localized scalar curvature deformation and asymptotics

In the final section of this chapter we will address the following question [204, p. 371]: are there nontrivial asymptotically flat solutions of the vacuum Einstein constraint equations on $\mathbb{R}^n$ which reduce to a Riemannian Schwarzschild metric outside a compact set? Here we put in the condition *nontrivial* to rule out the simple case of the Euclidean metric, with vanishing ADM mass. We saw in Proposition 7-80 that if the vacuum condition is removed and only the dominant energy condition in the form $R(g) \geq 0$ is enforced, then the answer to the question is *yes*, there are lots of such solutions; this sufficed for the proof of the Riemannian positive mass theorem. In the purely Riemannian case of the vacuum constraints (time-symmetric, $K = 0$), the Hamiltonian constraint is simply $R(g) = 0$. If we consider conformal methods to construct solutions to the constraints, unique continuation for harmonic functions suggests that maybe the answer is no: maybe the condition of vanishing scalar curvature and being identical to Schwarzschild near infinity is a rigid condition, cf. Example 7-19.

The strategy we will outline here is to investigate the question using a localized version of the Fischer–Marsden scalar curvature deformation. The idea is to patch together an asymptotically flat metric with zero scalar curvature to a Schwarzschild end, using a smooth cutoff function. In an annular transition region the scalar curvature may fail to vanish, and we seek to reimpose the scalar curvature constraint by considering *localized* deformations whose support, unlike in the case of conformal deformations in general, does not extend to the Schwarzschild end.

Thus as a first step, we have to address to what extent the metric deformation in the Fischer–Marsden theorem, which has been discussed earlier both in the closed case (Theorem 6-52) and near the Euclidean metric in the asymptotically flat case (Proposition 7-73), can be localized. We state a sufficient version of a localized scalar curvature deformation (cf. [66; 70]). We introduce a term before doing so: if M is a smooth manifold, an open subset $\Omega \subset M$ is a *precompact smooth domain in* M provided the closure $\overline{\Omega} \subset M$ is a compact manifold-with-boundary embedded in M via inclusion, with manifold interior Ω and smooth boundary $\partial \Omega$, which is then a smooth embedded submanifold (hypersurface) in M. Let $L_g = DR_g$, the linearization of the scalar curvature map at g.

Theorem 7-84. *Let* $0 < \alpha < 1$. *Suppose* (M, g) *is a smooth Riemannian manifold, and* $\Omega \subset M$ *is a precompact smooth domain in* M, *so that* L_g^* *has trivial kernel on* Ω. *For any open* Ω_0 *compactly contained in* Ω, *there is* $\varepsilon_0 > 0$ *and* $C_0 > 0$ *such that for any* $S \in C_c^\infty(\Omega_0)$ *with* $\|S\|_{C^{0,\alpha}} < \varepsilon_0$, *there is a smooth symmetric* $(0, 2)$-*tensor* h *on* M, *with support contained in* Ω, *with* $\|h\|_{C^{2,\alpha}} \le C_0 \|S\|_{C^{0,\alpha}}$ *and* $R(g + h) = R(g) + S$.

A number of remarks are in order. First, this is a localized scalar curvature deformation: it is a perturbation result under a nondegeneracy condition, with a localization of the support of the deformation. We note the nondegeneracy condition states precisely that (Ω, g) does not support any nontrivial static potentials (cf. Section 2.4.5). Just as with Theorem 6-52, the need for some nondegeneracy condition is illustrated at a flat metric, which is static: there is no non-flat metric on $\mathbb{R}^n$ with nonnegative scalar curvature which is Euclidean outside a compact set, hence there certainly cannot be a localized deformation of the type in the above theorem about this metric. For other rigidity results for static metrics, see [34; 185]. For the purpose of comparison and inspiration, we mention that Lohkamp showed [146] that there is no obstruction to deforming the scalar curvature *downward* in a compact subdomain. As we noted above, conformal deformations appear ill-suited for localized deformation results, and in fact Yuan [230] shows certain localized deformations *must* leave the conformal class. We note that the round sphere is static (Exercise 2-43), but the spherical case is different than the case of Euclidean space or the flat torus. Indeed, Brendle, Marques and Neves [35] constructed a metric on the sphere $\mathbb{S}^n$ with scalar curvature $R(g) \ge n(n-1)$, with strict inequality on a nonempty subset, and for which g agrees with the round unit sphere metric in a neighborhood of a closed hemisphere, thereby giving a counterexample to the *Min-Oo conjecture*.

We have seen that L_g^* is overdetermined-elliptic, and there are results that indicate that the presence of nontrivial static potentials (kernel elements of L_g^*) is non-generic; for example, by taking the divergence of $L_g^* f = 0$ and using the Ricci formula, it follows that a connected open set with a nontrivial static potential must have constant scalar curvature (Corollary 2-45, Exercise 2-52a.; see also [19]). Whereas, say, the Laplacian would have an infinite-dimensional kernel parametrized by boundary values, if Ω is an n-dimensional connected domain, the kernel of L_g^* has dimension at most $n+1$, since the equation $L_g^* f = 0$ induces an ODE along geodesics emanating from a point; from this we also see that on a connected manifold, no nontrivial static potential can vanish on an open set (Proposition 2-44). We can even consider weak solutions to $L_g^* f = 0$, but by elliptic regularity (assuming g is smooth), we will obtain that f is smooth, and in fact, it is a simple exercise to use the induced ODE to show that a static potential extends smoothly to the boundary [66; 70].

In the remaining sections, we will give an idea of the proof of Theorem 7-84 and then answer the question posed at the start of the section in the form of Theorem 7-89. We will focus on the linear theory, while leaving a number of technical details to the references. Also, the discussion will be somewhat breezy, as the goal is to map out the big ideas of the proofs.

7.5.1. *On the proof of Theorem 7-84.* We now take a precompact smooth domain Ω in a Riemannian manifold (M, g),[6] and assume that L_g^* has trivial kernel on Ω. For functions $S \in C_c^\infty(\Omega)$ sufficiently small, we will construct a smooth metric $g + h$ on M, with h extending smoothly by zero across $\partial\Omega$, so that $R(g + h) = R(g) + S$. We briefly remark that this paraphrasing of Theorem 7-84 is correct in spirit, but is not precise enough in regards to the smallness condition on S. As we will see in the outline of the proof, certain weighted spaces are used to control the support of the deformation h. One can state a version of Theorem 7-84 in such spaces, cf. [70]. Because of the nature of the weights, when formulating the theorem in simple terms as stated above, the ε_0 depends on Ω_0 (and in particular on the distance $d(\Omega_0, \partial\Omega)$). In applications one might indeed be working on a fixed subdomain Ω_0, in which case this paraphrasing is reasonably faithful. Since L_g^* has trivial kernel on each component of Ω, we will assume that Ω is connected with nonempty boundary (which may or may not be connected).

The approach we take follows the spirit of the proof of Theorem 6-52. The trivial kernel condition should translate into a surjectivity property for L_g, and then a local surjectivity result for the nonlinear operator. The way we proved the linearized surjectivity above was through a Hodge splitting, which we chose to prove variationally; that variational approach can be adapted suitably for Theorem 7-84. At this point, in contrast with the case of a closed manifold, we do not have a functional framework to simply invoke the inverse function theorem. That said, the idea of the proof of the implicit/inverse function theorem is to solve the nonlinear problem by iterating linear corrections (Newton–Picard

---

[6]Actually, the proof does not require M extending $\overline{\Omega}$, so we could formulate the problem on the manifold-with-boundary $(\overline{\Omega}, g)$.

iteration). For the proof of Theorem 7-84, we run the iteration by hand and show it converges to a solution.

To do this, we employ ideas from our analysis in the previous chapter. First, we get the elliptic estimate used to make the variational method for establishing linear surjectivity succeed. We then iterate the process of linear corrections, using interior estimates to control the sequence of approximate solutions, and establish convergence. At this point, this should sound reasonable and somewhat unremarkable. That said, we have not yet indicated how we will impose the decay of the deformation tensor h for which we solve. We will define weighted spaces to accomplish this, and broach some of the technical hurdles involved and how these are addressed in the analysis.

The linearized problem. We used elliptic estimates to serve as the foundation for much of the elliptic PDE theory that we have employed. The elliptic estimates in Section 6.1 were on closed manifolds, or the related interior estimates. If you study, say, the Laplace operator on a bounded domain, the relevant elliptic estimate will have to include a term for the boundary data. In our present setting, we do not want to solve a boundary value problem *per se*, because we want our deformation tensor to vanish on $\partial\Omega$, along with derivatives (a finite number, or all derivatives, depending on how regular we want h to be across $\partial\Omega$). The operator L_g^* is overdetermined-elliptic, and the structure of the operator allows us to get an absolute elliptic estimate on any domain Ω, *without boundary terms*. This is easy for this operator, because $\mathrm{tr}_g(L_g^* f) = -(n-1)\Delta_g f - f R(g)$, so that $\mathrm{Hess}_g\, f = L_g^* f - \frac{1}{n-1}\big(\mathrm{tr}_g L_g^* f + f R(g)\big)g + f\,\mathrm{Ric}(g)$. So on any domain Ω where the Ricci curvature is bounded, say on a compactly contained domain, we immediately have a constant C such that for all f,

$$\|f\|_{H^2(\Omega)} \leq C(\|L_g^* f\|_{L^2(\Omega)} + \|f\|_{H^1(\Omega)}). \tag{7.5.1}$$

(For a different proof, see [66, Appendix].)

Suppose Ω is a precompact smooth domain, so we can apply Rellich compactness to the inclusion $H^2(\Omega) \hookrightarrow H^1(\Omega)$. If L_g^* has trivial kernel on Ω, we can then promote the estimate to

$$\|f\|_{H^2(\Omega)} \leq C\|L_g^* f\|_{L^2(\Omega)}, \tag{7.5.2}$$

with possibly a different constant; cf. Proposition 6-15.

We introduce a weight function to help achieve the desired support of the deformation tensor h. Given $0 < \tau_0 < \tau_1$, we let $\tilde\rho : (0, +\infty) \to \mathbb{R}$ be a smooth function that satisfies $\tilde\rho(t) = 1$ for $t \geq \tau_1$, and for $0 < t < \tau_0$, either $\tilde\rho(t) = t^N$ for some $N > 0$, or $\tilde\rho(t) = e^{-1/t}$, and finally for all $t > 0$, $\tilde\rho'(t) \geq 0$. Let

$\mathring{g}$ be a smooth background metric on $\overline{\Omega}$. The weight function in Ω is then $\rho(x) = \tilde{\rho}(d(x))$, where $d(x) = d_{\mathring{g}}(x, \partial\Omega) = \inf_{y \in \partial\Omega} d_{\mathring{g}}(x, y)$. Since $\partial\Omega$ is a compact, embedded smooth hypersurface, d is smooth in a neighborhood of $\partial\Omega$ in $\overline{\Omega}$, with nowhere-vanishing differential; i.e., d is a *defining function* for the boundary. Thus for small ϵ, the level sets $\{x \in \Omega : d(x) = \epsilon\}$ are smooth, and by choosing τ_1 small, the weight function ρ is smooth in $\overline{\Omega}$ as well. Such ρ will be used to achieve the decay of h at the boundary, and hence the smoothness of the extension of h by zero outside of $\overline{\Omega}$, where h will be our constructed solution to $R(g + h) = R(g) + S$. The power weight suffices for finite regularity solutions (using sufficiently large N), and the exponential weight can be used for smooth solutions.

The linearized problem we want to solve is $L_g h = S$. We will solve for h of the form $h = \rho L_g^* u$; you may recall in the closed setting the range of L_g is indeed the range of $L_g L_g^* \sim (n-1)\Delta_g^2$. In the situation here, the equation $L_g \rho L_g^* u = S$ is not strictly elliptic. To fix this, we rewrite the equation as $\rho^{-1} L_g \rho L_g^* u = \rho^{-1} S$, which is strictly elliptic, but there are a couple points to note. First, the coefficients of the lower-order terms will be on the order of some power of d^{-1}. As it turns out, one can actually handle this in the elliptic estimates by suitable scaling [70, Appendix A]; the interested reader can compare this to the formulation of the interior Schauder estimates in [107, Chapter 6]. Secondly, the right-hand side $\rho^{-1} S$ needs to be reasonably well-behaved, for example, S can compactly supported in the interior Ω, or more generally S must decay fast enough at the boundary. Finally, as we will see presently, the solution u need not be *a priori* bounded on Ω, but will behave well enough so that $\rho L_g^* u$ decays suitably at the boundary.

To solve $L_g h = S$ variationally, we define $\mathcal{G}(u) = \int_\Omega \left(\frac{1}{2}\rho|L_g^* u|^2 - uS\right) dv_g$. We will identify a suitable space on which to minimize $\mathcal{G}$, such that if u is in the space and v is smooth and compactly supported, then $u + tv$ is in the space for any t. If u is critical for $\mathcal{G}$ under such perturbations, the Euler–Lagrange equation is derived from $0 = \frac{d}{dt}\big|_{t=0} \mathcal{G}(u + tv) = \int_\Omega (\langle \rho L_g^* u, L_g^* v \rangle - Sv) dv_g$. This is a weak formulation of $L_g \rho L_g^* u = S$, as desired.

In order to get the functional framework to produce a minimizer, we need to promote (7.5.2) to a weighted estimate. To do this, we define certain weighted L^2-Sobolev spaces. The definition we will give is used in [66; 70], as it is simple and suffices for our purposes here; in [67] we compare these spaces to another suite of weighted L^2-spaces that are naturally defined.

We say $u \in L^2_\rho(\Omega)$ if $\|u\|^2_{L^2_\rho(\Omega)} = \int_\Omega |u|^2 \rho \, dv_g$ is finite, and we define $H^k_\rho(\Omega)$ similarly for any nonnegative integer k, where $\|u\|^2_{H^k_\rho(\Omega)} = \sum_{\ell=0}^k \|\nabla^\ell u\|^2_{L^2_\rho(\Omega)}$; we note $H^0_\rho(\Omega) = L^2_\rho(\Omega)$. It is a simple exercise to show that $H^k_\rho(\Omega)$ is a Hilbert space. Furthermore, since ρ is bounded, we have by [69, Lemma 2.1] that $C^\infty(\overline{\Omega})$ is dense in $H^k_\rho(\Omega)$; this can be useful in establishing certain relations or estimates, by proving on the dense subset and then taking limits. As a set, $H^k_\rho(\Omega)$ is the same even if the metric g, and hence the norm, changes, and we will usually suppress the metric from the notation. We remark that, considering the behavior of the weight function near the boundary, it would be natural to define the weighted spaces with a different weighting on different order derivatives, cf. [67; 61].

We state the key estimate: with $\overline{\Omega}$ a compact smooth manifold-with-boundary, assuming L^*_g has no kernel on the manifold interior Ω, there is a constant $C > 0$ such that for all $f \in H^2_\rho(\Omega)$

$$\|f\|_{H^2_\rho(\Omega)} \le C \|L^*_g f\|_{L^2_\rho(\Omega)}. \tag{7.5.3}$$

We sketch the proof in the form of two exercises.

Exercise 7-85. Prove that with $\Omega_\epsilon = \{x \in \Omega : d(x) > \epsilon\}$, there is a $C > 0$ such that for all small enough $\epsilon > 0$, $\|f\|_{H^2(\Omega_\epsilon)} \le C \|L^*_g f\|_{L^2(\Omega_\epsilon)}$ holds. The argument can be carried out by contradiction, assuming a sequence $\epsilon_j \searrow 0$ and $f_j \in H^2(\Omega_{\epsilon_j})$ with $\|f_j\|_{H^2(\Omega_{\epsilon_j})} > j \|L^*_g f_j\|_{L^2(\Omega_{\epsilon_j})}$. One can extend f_j to $\tilde{f}_j \in H^2(\Omega)$ with $\|\tilde{f}_j\|_{H^2(\Omega)} \le D \|f_j\|_{H^2(\Omega_{\epsilon_j})}$, with a constant D uniform in j, then normalize to $\|\tilde{f}_j\|_{H^1(\Omega)} = 1$. Using (7.5.1) on Ω_{ϵ_j} and an application of Rellich will yield the contradiction, cf. [66; 70].

The next exercise indicates how to get from here to (7.5.3).

Exercise 7-86. Establish (7.5.3) as follows. Since $\tilde{\rho}' \ge 0$, we have

$$\tilde{\rho}'(\epsilon) \|f\|^2_{H^2(\Omega_\epsilon)} \le C^2 \tilde{\rho}'(\epsilon) \|L^*_g f\|^2_{L^2(\Omega_\epsilon)}.$$

Foliate a neighborhood of $\partial\Omega$ with smooth level sets of d, containing $\{x \in \overline{\Omega} : 0 \le d(x) \le d_1\}$ for some small $d_1 > 0$. Let $C_0 = \int_0^{d_1} \tilde{\rho}'(t) \, dt = \tilde{\rho}(d_1) > 0$. Use the co-area formula to argue $\int_0^{d_1} \tilde{\rho}'(\epsilon) \|f\|^2_{H^2(\Omega_\epsilon \setminus \overline{\Omega}_{d_1})} \, d\epsilon = \|f\|^2_{H^2_\rho(\Omega \setminus \overline{\Omega}_{d_1})}$.

For the full constraint operator the analogous weighted estimates for $D\Psi^*_{(g,\pi)}$ hold, but are somewhat harder to establish, without having an analogue of the absolute (unweighted) estimate (7.5.2), cf. [69; 67; 61].

We now see that (7.5.3) gives us a coercivity bound on the functional $\mathcal{G}$: assuming the kernel of L^*_g on Ω is trivial and setting $\|S\|^2_{L^2_{\rho^{-1}}(\Omega)} = \int_\Omega |S|^2 \rho^{-1} dv_g$, we obtain for $u \in H^2_\rho(\Omega)$

$$\mathcal{G}(u) \geq C' \|u\|^2_{H^2_\rho(\Omega)} - \|u\|_{L^2_\rho(\Omega)} \|S\|_{L^2_{\rho-1}(\Omega)} \geq C' \|u\|^2_{H^2_\rho(\Omega)} - \|u\|_{H^2_\rho(\Omega)} \|S\|_{L^2_{\rho-1}(\Omega)}.$$

In applying Cauchy–Schwarz, we incorporated the weight on u, and thus we had to incorporate a dual weight on S. We already saw above the indication that S will have to decay suitably near the boundary.

By applying standard functional analysis just like in the proof of the Hodge decomposition in Section 6.1, one can prove the existence of a minimizer u for $\mathcal{G}$ on $H^2_\rho(\Omega)$. We note there is a *unique* minimizer, as a corollary of the following exercise.

Exercise 7-87. Under the trivial kernel condition in the setting above, show that for $u_1 \neq u_2 \in H^2_\rho(\Omega)$, the function $G(s) = \mathcal{G}((1-s)u_1 + s(u_2))$ is strictly convex.

For this unique minimizer u, we note that $\mathcal{G}(u) \leq \mathcal{G}(0) = 0$. This implies

$$\|\rho L^*_g u\|^2_{L^2_{\rho-1}(\Omega)} = \|L^*_g u\|^2_{L^2_\rho(\Omega)} \leq 2\|S\|_{L^2_{\rho-1}(\Omega)} \|u\|_{L^2_\rho(\Omega)}.$$

Now we can apply (7.5.3) to infer $\|\rho L^*_g u\|_{L^2_{\rho-1}(\Omega)} \leq C' \|u\|_{H^2_\rho(\Omega)} \leq C'' \|S\|_{L^2_{\rho-1}(\Omega)}$.

We summarize what we have discussed.

Proposition 7-88. *Let $0 < \alpha < 1$. Suppose (M, g) is a smooth Riemannian manifold, and $\Omega \subset M$ is a precompact smooth domain in M, so that L^*_g has trivial kernel on Ω. There are positive constants C' and C'' such that for any $S \in L^2_{\rho-1}(\Omega)$, there is unique $u \in H^2_\rho(\Omega)$ such that for $h = \rho L^*_g u$, we have $L_g h = S$, with the following estimate: $\|h\|_{L^2_{\rho-1}(\Omega)} \leq C' \|u\|_{H^2_\rho(\Omega)} \leq C'' \|S\|_{L^2_{\rho-1}(\Omega)}$.*

On the nonlinear problem. If we assume S and g in the preceding proposition are smooth, then by elliptic regularity, u, and hence h, will be smooth in Ω as well. What remains to be understood for the proof of Theorem 7-84 is for which S we can infer that h decays sufficiently fast on approach to the boundary that we can extend it by zero to (and past) the boundary smoothly. We also expect that if we choose S suitably small, we will be able to control h to be pointwise small, so that in particular $g + h$ will be a metric. In this case we see that $R(g+h) = R(g) + L_g h + Q_g h = R(g) + S + Q_g h$, where $Q_g h$ is the Taylor remainder in the expansion of the scalar curvature map. Solving the linearized problem results in an approximate solution where the error is quadratic in the perturbation, after which one expects to be able to solve the nonlinear equation with a Newton–Picard-type iteration.

The pointwise estimates, including the smallness condition and the behavior near the boundary, are carried out with interior Schauder estimates and scaling.

This can all be done, in the spirit of [66] (with correction to the norm needed on S as noted in [69], and cf. [70] for a cleaner formulation). When S has compact support, if we assume it is $C^{0,\alpha}$-small, then it will also be small in the required weighted norm, depending on the support (in particular, on a lower bound of ρ on the support of S), which explains the way in which Theorem 7-84 is stated. We note here that given k, by taking a power weight $\rho(x) = (d(x))^N$ near the boundary, for N sufficiently large we can make h extend by zero in a C^k fashion across the boundary; an exponential weight can be used to produce a C^∞-solution. The solution just constructed is supported on $\overline{\Omega}$; we can arrange the support to lie strictly in Ω by running the above construction, replacing Ω by $\Omega' = \Omega_{\epsilon'}$, for $\epsilon' > 0$ sufficiently small.

There are numerous technical issues remaining in the proof of Theorem 7-84, which we will not go into here, hoping we have at least given the spirit of the argument. Instead we spend the remaining section in this chapter discussing what modifications need to be made to address the question posed at the start of the section: can we construct a zero scalar curvature metric by gluing an asymptotically flat metric of zero scalar curvature to a Schwarzschild metric and doing a localized perturbation to make the scalar curvature vanish?

7.5.2. Asymptotic gluing construction. We now state a result that answers the question posed at the start of the section. For simplicity we continue to work with metrics that are smooth, and for asymptotically flat metrics we require the rate $q > \frac{n-2}{2}$, with order $\ell \geq 3$, though we can replace this with a weighted $C^{2,\alpha}_{-q}$-assumption; in any case, we can use this weighted norm to measure the closeness of two such metrics as in the next theorem.

Theorem 7-89. *Suppose $(\mathcal{E}, g)$ is an asymptotically flat end with vanishing scalar curvature, nonzero ADM mass $m(g)$, and asymptotically flat coordinates x. There is $\theta_0 > 0$ so that for all $\theta \geq \theta_0$, there is a metric $\bar{g}$ on $\mathcal{E}$ with $R(\bar{g}) = 0$, $\bar{g} = g$ for $|x| \leq \theta$, and $\bar{g}$ is a Schwarzschild metric for $|x| \geq 2\theta$. Given $\varepsilon > 0$, for large enough θ, $\bar{g}$ is ε-close to g, and $|m(\bar{g}) - m(g)| < \varepsilon$.*

The theorem applies to ends of negative or positive mass, while if $\mathcal{E} \subset M$ is an asymptotic end in M, and since the construction is local to the end, $\bar{g}$ extends smoothly to all of M. The remainder of the chapter will be spent illustrating the steps in the proof. The basic strategy is simple to lay out, but a number of technical issues will make the discussion in parts a bit cumbersome.

With respect to the asymptotic coordinates x, we let

$$\mathring{g}_{m,c}(x) = \left(1 + \frac{m}{2|x-c|^{n-2}}\right)^{\frac{4}{n-2}} g_{\mathbb{E}^n}$$

be the Riemannian Schwarzschild metric of mass m and center c. Let $0 \le \psi \le 1$ be a smooth cutoff function so that $\psi(t) = 1$ for $t < 1$ and $\psi(t) = 0$ for $t > 2$. For $\theta \ge r_0 > 1$, let $\psi_\theta(x) = \psi(|x|\theta^{-1})$. Now consider the metric $g_\theta(x) = \psi_\theta(x)g(x) + (1 - \psi_\theta(x))\mathring{g}_{m,c}(x)$. We assume that $|c| \le \theta/2$, so that $|x - c| \ge \theta/2$ on $|x| \ge \theta$; in particular the metric is well-defined on $\mathcal{E}$ (for θ large enough in case $m < 0$).

The scalar curvature $R(g_\theta)$ is supported in the annulus $\overline{A}_\theta$ given by $\theta \le |x| \le 2\theta$. Since the metric g is asymptotically flat, the scalar curvature $R(g_\theta)$ goes to zero as $\theta \to \infty$; in fact, a simple exercise shows that $R(g_\theta) = O(\theta^{-q-2})$. Here and below, it is important that such estimates hold uniformly for c as above and for m in a bounded set. In terms of the localized scalar curvature deformation result discussed above, the direction seems clear from here: apply a localized deformation supported in the annulus $\overline{A}_\theta$ to perturb the small scalar curvature back to zero. Upon a closer look, we find a delicate issue: namely, we have to get the size $O(\theta^{-q-2})$ of the desired perturbation to be within ε_0 from Theorem 7-84. To do this, we would like to choose θ large. Doing so changes the metric, and hence the ε_0. In fact, we really get stuck here, because as θ increases, the metric g_θ in the annulus approaches the Euclidean metric, which *is static vacuum*.

To understand this more clearly, we rescale the metric g_θ to $\tilde{g}_\theta$ on the fixed annulus A_1, via the map $\phi : A_1 \to A_\theta$ given by $\phi(x) = \theta x$: $\tilde{g}_\theta = \theta^{-2}\phi^* g_\theta$. Then $\partial_x^\beta\big((\tilde{g}_\theta)_{ij}(x) - \delta_{ij}\big) = O(\theta^{-q})$ for $0 \le |\beta| \le \ell$, so $\|R(\tilde{g}_\theta)\|_{C^{0,\alpha}(\overline{A}_1)} = O(\theta^{-q})$.

Now if g_θ does not admit any nontrivial static potentials on A_θ (actually if it did, then we could already have concluded $R(g_\theta) = 0$), then neither does $\tilde{g}_\theta$ (on A_1), in which case we have the estimate $\|f\|_{H^2_\rho(A_1)} \le C\|L^*_{\tilde{g}_\theta} f\|_{L^2_\rho(A_1)}$. The problem is that the constant C is not uniform in θ large. This is easy to see: just let $f = 1$ or $f = x^j$ for a coordinate function on A_1; the span of these functions is basically an *obstruction space* to being able to carry out our desired construction. Handling the obstruction is a two-step process: in the next section, we will discuss solving the problem transverse to the obstruction, and then finish by addressing how to correct the resulting finite-dimensional error.

The projected problem. We expect to be able to formulate function spaces so that the image of $L = L_{g_{\mathbb{E}^n}}$ should be transverse to the kernel $\mathring{\mathcal{K}}$ of $L^* = L^*_{g_{\mathbb{E}^n}}$, $\mathring{\mathcal{K}} = \mathrm{span}\{1, x^1, \ldots, x^n\}$. As motivation, consider the following exercise, for which we introduce some more notation. Let $0 \le \zeta \le 1$ be a smooth, rotationally symmetric bump function supported in A_1 and that equals 1 near $\{|x| = \frac{3}{2}\}$, say. Let $\mathcal{K} = \zeta\mathring{\mathcal{K}}$, and write $L^2(A_1, dx)$ as the orthogonal decomposition $L^2(A_1, dx) = \mathcal{K}^\perp \oplus \mathcal{K}$, and let $\overline{\Pi}$ be the projection onto the $\mathcal{K}^\perp$-factor.

Exercise 7-90. Suppose $\mathring{\sigma} \in \mathcal{K}^{\perp}$, and let $\mathring{\mathcal{G}}(u) = \int_{A_1} \left(\frac{1}{2}|L^*u|^2 - \mathring{\sigma}u\right) dx$, for $u \in H^2(A_1)$. Suppose $u \in H^2(A_1) \cap \mathcal{K}^{\perp}$ minimizes $\mathring{\mathcal{G}}$ restricted to the Banach space $H^2(A_1) \cap \mathcal{K}^{\perp}$. Argue that $u \in H^4_{\mathrm{loc}}(A_1)$ and $LL^*u - \mathring{\sigma} \in \mathcal{K}$, so that $LL^*u \in L^2(A_1, dx)$ and $\mathring{\Pi}(LL^*u) = \mathring{\sigma}$.

Even if $L^*_{\tilde{g}_\theta}$ has trivial kernel, $\mathcal{K}$ can be thought of as an *approximate kernel* for large θ, an obstruction to uniform estimates. We proceed as in Proposition 6-15: we get uniform estimates for $L^*_{\tilde{g}_\theta}$ if we work *transverse* to $\mathcal{K}$. To this end, let $^{\perp}\mathcal{K} = \left\{ u \in H^2_\rho(A_1) : \int_{A_1} u\zeta \, dx = 0 = \int_{A_1} u\zeta x^i dx, \ i = 1, \dots, n \right\}$. Since ζ is compactly supported away from the boundary of the annulus, integration against an element of $\mathcal{K}$ defines a continuous linear functional on $H^2_\rho(A_1)$, and so $^{\perp}\mathcal{K}$ is closed, which also follows from $H^2_\rho(A_1) = {}^{\perp}\mathcal{K} \oplus \mathcal{K}$. We remark that $H^2_\rho(A_1) = {}^{\perp}\mathcal{K} \oplus \mathring{\mathcal{K}}$ as well, since $\mathring{\mathcal{K}} \cap {}^{\perp}\mathcal{K} = \{0\}$, and $\mathring{\mathcal{K}}$ has the same dimension as $H^2_\rho(A_1)/{}^{\perp}\mathcal{K} \cong \mathcal{K}$. To be explicit, let $x^0 := 1$; we can decompose $u \in H^2_\rho(A_1)$ as $u = R_0(u) + P_0(u)$, where

$$P_0(u) = \sum_{j=0}^{n} \frac{\langle u, \zeta x^j \rangle_{L^2}}{\langle x^j, \zeta x^j \rangle_{L^2}} \, x^j \in \mathring{\mathcal{K}}.$$

We can also write $u = R(u) + P(u)$, with $P(u) = \sum_{j=0}^{n} \frac{\langle u, \zeta x^j \rangle_{L^2}}{\langle \zeta x^j, \zeta x^j \rangle_{L^2}} \zeta x^j \in \mathcal{K}$. It is easy to see that $R_0(u)$ and $R(u)$ are in $^{\perp}\mathcal{K}$.

Surjectivity for the linearized projected problem. Let

$$^{\perp}\mathcal{K}_\theta = \left\{ u \in H^2_\rho(A_1) : \int_{A_1} u\zeta \, dv_{\tilde{g}_\theta} = 0 = \int_{A_1} u\zeta x^i dv_{\tilde{g}_\theta}, \ i = 1, \dots, n \right\}.$$

Since $\mathring{\mathcal{K}} \cap {}^{\perp}\mathcal{K}_\theta = \{0\}$ for large θ, we have $H^2_\rho(A_1) = {}^{\perp}\mathcal{K}_\theta \oplus \mathcal{K} = {}^{\perp}\mathcal{K}_\theta \oplus \mathring{\mathcal{K}}$. Let $\mathcal{G}(u) = \int_{A_1} \left(\frac{1}{2}\rho|L^*_{\tilde{g}_\theta}u|^2 - u\sigma\right) dv_{\tilde{g}_\theta}$ for $\sigma \in L^2_{\rho^{-1}}(A_1)$. If we restrict $\mathcal{G}$ to the set $^{\perp}\mathcal{K}_\theta$, we can get a coercivity bound using an injectivity estimate transverse to the kernel by adapting the proof of (7.5.3) to obtain that there is a constant $C > 0$ such that for all θ large and $f \in {}^{\perp}\mathcal{K}_\theta$, $\|f\|_{H^2_\rho(A_1)} \leq C\|L^*_{\tilde{g}_\theta}f\|_{L^2_\rho(A_1)}$. We use this as before to produce a minimizer $u \in {}^{\perp}\mathcal{K}_\theta$ for $\mathcal{G}$ on $^{\perp}\mathcal{K}_\theta$. For any smooth, compactly supported $v \in {}^{\perp}\mathcal{K}_\theta$, we have

$$0 = \frac{d}{dt}\bigg|_{t=0} \mathcal{G}(u + tv) = \int_{A_1} \left(\langle \rho L^*_{\tilde{g}_\theta}u, L^*_{\tilde{g}_\theta}v \rangle - v\sigma\right) dv_{\tilde{g}_\theta}.$$

Since $C^\infty_c(A_1) = ({}^{\perp}\mathcal{K}_\theta \cap C^\infty_c(A_1)) \oplus \mathcal{K}$, we infer the distributional equation $L_{\tilde{g}_\theta}\rho L^*_{\tilde{g}_\theta}u - \sigma \in \mathcal{K}$. By elliptic regularity, if $\sigma \in C^\infty(A_1)$, then $u \in C^\infty(A_1)$.

Remark 7-91. We have not used anything about $\tilde{g}_\theta$ except that it is near the Euclidean metric, and so the results just obtained extend to g suitably near $g_{\mathbb{E}^n}$.

Nonlinear projected problem. We summarize what the above analysis gives us. We consider the projected operator $g \mapsto \mathring{\Pi} R(g)$, whose linearization is $\mathring{\Pi} \circ L_g$. For g near $g_{\mathbb{E}^n}$ this linear operator surjects onto $L^2_{\rho-1}(A_1) \cap \mathcal{K}^\perp$, with $\mathring{\Pi} L_g h = \mathring{\sigma} \in L^2_{\rho-1}(A_1) \cap \mathcal{K}^\perp$, where $h = \rho L_g^* u \in L^2_{\rho-1}(A_1) \cap H^2_{\mathrm{loc}}(A_1)$. Just as in Theorem 7-84 (see also Proposition 7-88), h is bounded in terms of $\mathring{\sigma}$, first in a weighted L^2-norm from the variational estimates, then in a Hölder norm by Schauder estimates to obtain regularity and decay of h at the boundary, as in [66; 70] (see also [69; 61; 67]). We do point the reader to a suggestive linear model problem in Exercise 7-110. In any case, such pointwise estimates are then used in a Newton–Picard iteration to solve the nonlinear problem $\mathring{\Pi}(R(g+h)-R(g)) = \mathring{\sigma}$ for $\mathring{\sigma} \in \mathcal{K}^\perp$ small, and suitably decaying (e.g., compactly supported). We will not go into the details of the proofs, but we indicate in the final section how to use this to accomplish the gluing to Schwarzschild near infinity.

Remark 7-92. There is one technical point that could have been raised in the proof of Theorem 7-84, but it was not needed there. In the next section, we will need to know that the solution tensor h depends continuously on $\mathring{\sigma}$ and on g. This is generally straightforward: we can proceed as in [6, Appendix A.7] and the proof of [70, Proposition 3.7], and note that we chose to use a *fixed* weight ρ which does not change with g (cf. [67, Remark 6.7]).

Handling the obstruction space. For large θ, we have $\tilde{g}_\theta$ near the Euclidean metric on the annulus, with $\|R(\tilde{g}_\theta)\|_{C^{0,\alpha}(A_1)} = O(\theta^{-q})$. In the preceding section we presented the linear theory needed to solve the problem $\mathring{\Pi}(R(\tilde{g}_\theta + h_\theta)) = 0$ for h_θ, by letting $\mathring{\sigma} = -\mathring{\Pi} R(\tilde{g}_\theta)$. As mentioned in Remark 7-92, h_θ can be constructed continuously in θ, as well as with respect to m and c, and such that it extends smoothly by zero outside A_1, together with a bound $\|h_\theta\|_{C^{2,\alpha}(A_1)} \leq C\|R(\tilde{g}_\theta)\|_{C^{0,\alpha}(A_1)} = O(\theta^{-q})$, cf. Theorem 7-84.

Thus we have solved $R(\tilde{g}_\theta + h_\theta) \in \mathcal{K}$. We now want to see how, for θ large enough, by choosing m and c appropriately, we will in fact have $R(\tilde{g}_\theta + h_\theta) = 0$; upon rescaling, we will have finished the proof of Theorem 7-89. Now since $\mathcal{K} \cap \mathcal{K}^\perp = \{0\}$, we need only arrange $\int_{A_1} x^k R(\tilde{g}_\theta + h_\theta)\, dx = 0$, for $k \in \{0, 1, \ldots, n\}$, where recall $x^0 := 1$. We proceed to indicate how to estimate these integrals. Before we do, note that near the boundary $\{|x| = 2\}$ of A_1,

$$(\tilde{g}_\theta)_{ij}(x) = \left(1 + \frac{m/\theta^{n-2}}{2|x - c/\theta|^{n-2}}\right)^{\frac{4}{n-2}} \delta_{ij},$$

which is Schwarzschild of mass m/θ^{n-2} and center c/θ. We also recall that $v^j = x^j/|x|$ and that $d\sigma$ is Euclidean surface measure.

The analysis begins with an expansion of $R(\tilde{g}_\theta + h_\theta)$, cf. Proposition 7-31:

$$R(\tilde{g}_\theta + h_\theta) = R(\tilde{g}_\theta) + L_{\tilde{g}_\theta} h_\theta + O(\theta^{-2q})$$

$$= \sum_{i,j=1}^{n} \left((\tilde{g}_\theta)_{ij,ij} - (\tilde{g}_\theta)_{ii,jj} \right) + L h_\theta + O(\theta^{-2q}).$$

Note that we replaced $L_{\tilde{g}_\theta}$ with the operator L at the Euclidean metric, up to error terms, since $\tilde{g}_\theta$ is close to the Euclidean metric. Integrating by parts, using $L^*(x^k) = 0$ and the fact that h_θ and ∂h_θ vanish at the boundary, we get

$$\int_{A_1} x^k R(\tilde{g}_\theta + h_\theta)\,dx = \int_{A_1} x^k \sum_{i,j=1}^{n} \left((\tilde{g}_\theta)_{ij,ij} - (\tilde{g}_\theta)_{ii,jj} \right) dx + O(\theta^{-2q}). \quad (7.5.4)$$

We first consider $k = 0$. Recall that $\frac{n-2}{2} < q \le n - 2$; the upper bound holds since $m(g) \ne 0$ by assumption. From the fact that g is asymptotically flat with vanishing scalar curvature, we know that $\sum_{i,j=1}^{n}(g_{ij,ij} - g_{ii,jj})(x) = O(|x|^{-2-2q})$, from which we conclude

$$\int_{\{|x| \ge \theta\}} \sum_{i,j=1}^{n} (g_{ij,ij} - g_{ii,jj})\,dx = O(\theta^{n-2-2q}),$$

and similarly for $\mathring{g}_{m,c}$ (with $q = n - 2$). Thus by (7.5.4) and (7.2.2) (see also Proposition 7-33), we have

$$\int_{A_1} R(\tilde{g}_\theta + h_\theta)\,dx$$

$$= \left(\int_{\{|x|=2\}} - \int_{\{|x|=1\}} \right) \sum_{i,j=1}^{n} ((\tilde{g}_\theta)_{ij,i} - (\tilde{g}_\theta)_{ii,j}) v^j\, d\sigma + O(\theta^{-2q})$$

$$= \theta^{2-n} \left(\int_{\{|r|=2\theta\}} \sum_{i,j=1}^{n} ((\mathring{g}_{m,c})_{ij,i} - (\mathring{g}_{m,c})_{ii,j}) v^j\, d\sigma \right.$$

$$\left. - \int_{\{|x|=\theta\}} \sum_{i,j=1}^{n} (g_{ij,i} - g_{ii,j}) v^j\, d\sigma \right) + O(\theta^{-2q})$$

$$= \theta^{2-n} \cdot \left(2(n-1)\omega_{n-1} (m - m(g)) + O(\theta^{n-2-2q}) \right). \quad (7.5.5)$$

Up to scaling, this projection integral is governed to leading order by the *mass difference* of the glued metrics.

From Corollary 7-6 and equations (7-39a) and (7.2.5) (p. 256), we know to expect that the other projection integrals should involve the *center* of mass. That said, without further assumption, g may fail to have a well-defined center of mass integral; see (7.2.7). With $h_{ij} = g_{ij} - \delta_{ij}$ and $k \in \{1, \ldots, n\}$, we have

$$\int_{\{r_0 \leq |x| \leq \theta\}} x^k \sum_{i,j=1}^{n} (g_{ij,ij} - g_{ii,jj}) \, dx$$

$$= \left(\int_{\{|x|=\theta\}} - \int_{\{|x|=r_0\}} \right) \sum_{i,j=1}^{n} \left(x^k (h_{ij,i} - h_{ii,j}) - (\delta_i^k h_{ij} - \delta_j^k h_{ii}) \right) v^j \, d\sigma.$$

Using $\sum_{i,j=1}^{n} (g_{ij,ij} - g_{ii,jj})(x) = O(|x|^{-2-2q})$ again, we see the integrand on the left is $O(|x|^{-1-2q})$. If $n - 1 \neq 2q$, then

$$\int_{r_0}^{\theta} t^{-1-2q} \cdot t^{n-1} dt = \frac{1}{n-1-2q} (\theta^{n-1-2q} - r_0^{n-1-2q}),$$

whereas if $n - 1 = 2q$ (so $q = 1$ if $n = 3$), $\int_{r_0}^{\theta} t^{-1-2q} \cdot t^{n-1} dt = \log \theta - \log r_0$. For $t > 1$, let $\gamma(t) = \max(1, t^{n-1-2q})$ in case $n - 1 \neq 2q$, and let $\gamma(t) = 1 + \log t$ for $n - 1 = 2q$. Since for the range of q we are considering, we have $n - 1 - 2q < 1$, in all cases, $\lim_{t \to \infty} \gamma(t)/t = 0$, i.e., $\gamma(t) = o(t)$. Thus we conclude there is a C such that for $\theta \geq r_0$,

$$\left| \int_{\{|x|=\theta\}} \sum_{i,j=1}^{n} \left(x^k (h_{ij,i} - h_{ii,j}) - (\delta_i^k h_{ij} - \delta_j^k h_{ii}) \right) v^j \, d\sigma \right| \leq C\gamma(\theta). \qquad (7.5.6)$$

We now compute, with $(\mathring{h}_{m,c})_{ij} = (\mathring{g}_{m,c})_{ij} - \delta_{ij}$ and $(\tilde{h}_\theta)_{ij} = (\tilde{g}_\theta)_{ij} - \delta_{ij}$, and using (7.5.4),

$$\int_{A_1} x^k R(\tilde{g}_\theta + h_\theta) \, dx$$

$$= \left(\int_{\{|x|=2\}} - \int_{\{|x|=1\}} \right) \sum_{i,j=1}^{n} \left(x^k ((\tilde{h}_\theta)_{ij,i} - (\tilde{h}_\theta)_{ii,j}) - (\delta_i^k (\tilde{h}_\theta)_{ij} - \delta_j^k (\tilde{h}_\theta)_{ii}) \right) v^j \, d\sigma$$

$$+ O(\theta^{-2q})$$

$$= \theta^{1-n} \int_{\{|x|=2\theta\}} \sum_{i,j=1}^{n} \left(x^k ((\mathring{h}_{m,c})_{ij,i} - (\mathring{h}_{m,c})_{ii,j}) - (\delta_i^k (\mathring{h}_{m,c})_{ij} - \delta_j^k (\mathring{h}_{m,c})_{ii}) \right) v^j \, d\sigma$$

$$- \theta^{1-n} \int_{\{|x|=\theta\}} \sum_{i,j=1}^{n} \left(x^k (h_{ij,i} - h_{ii,j}) - (\delta_i^k h_{ij} - \delta_j^k h_{ii}) \right) v^j \, d\sigma$$

$$+ O(\theta^{-2q}). \qquad (7.5.7)$$

Since $R(\mathring{g}_{m,c}) = 0$, we have $\sum_{i,j=1}^{n} \left((\mathring{g}_{m,c})_{ij,ij} - (\mathring{g}_{m,c})_{ii,jj} \right) = \mathcal{Q}_{\mathring{g}_{m,c}}$, where $\mathcal{Q}_g$ is defined on page 283 (cf. Proposition 7-31). Using (7-39a) (p. 256) we then obtain

$$\int_{\{|x|=2\theta\}} \sum_{i,j=1}^{n} \left(x^k((\mathring{h}_{m,c})_{ij,i} - (\mathring{h}_{m,c})_{ii,j}) - (\delta^k_i(\mathring{h}_{m,c})_{ij} - \delta^k_j(\mathring{h}_{m,c})_{ii}) \right) v^j \, d\sigma$$

$$= 2(n-1)\omega_{n-1}mc^k - \lim_{r \to \infty} \int_{\{2\theta \le |x| \le r\}} x^k \sum_{i,j=1}^{n} \left((\mathring{g}_{m,c})_{ij,ij} - (\mathring{g}_{m,c})_{ii,jj} \right) dx$$

$$= 2(n-1)\omega_{n-1}mc^k - \lim_{r \to \infty} \int_{\{2\theta \le |x| \le r\}} x^k \mathcal{Q}_{\mathring{g}_{m,c}}(x) \, dx. \tag{7.5.8}$$

While $\mathcal{Q}_{\mathring{g}_{m,c}}(x) = O(|x|^{2-2n})$, a closer analysis of $\mathcal{Q}_{\mathring{g}_{m,c}}(x)$ shows that it takes the form

$$\mathring{h}_{m,c} * \partial^2 \mathring{g}_{m,c} + \partial \mathring{g}_{m,c} * \partial \mathring{g}_{m,c} = \mathring{h}_{m,c} * \partial^2 \mathring{h}_{m,c} + \partial \mathring{h}_{m,c} * \partial \mathring{h}_{m,c},$$

which denotes a linear combination of terms of the form $(\mathring{h}_{m,c})_{ij}(\partial^2 \mathring{h}_{m,c})_{k\ell}$ and $(\partial \mathring{h}_{m,c})_{ij}(\partial \mathring{h}_{m,c})_{k\ell}$, with bounded coefficients which are rational expressions in the components of $\mathring{g}_{m,c}$. By explicit expansion, $\mathcal{Q}_{\mathring{g}_{m,c}}(x) = \mathcal{Q}^e_{\mathring{g}_{m,c}}(x) + O(|x|^{1-2n})$, where $\mathcal{Q}^e_{\mathring{g}_{m,c}}(-x) - \mathcal{Q}^e_{\mathring{g}_{m,c}}(x)$; cf. (7.2.7) and the discussion of approximate parity symmetry.

Thus we see for any $r > 2\theta$, $\int_{\{2\theta \le |x| \le r\}} x^k \mathcal{Q}^e_{\mathring{g}_{m,c}}(x) \, dx = 0$, while we also have the estimate $\int_{\{|x| \ge 2\theta\}} x^k (\mathcal{Q}_{\mathring{g}_{m,c}}(x) - \mathcal{Q}^e_{\mathring{g}_{m,c}}(x)) \, dx = O(\theta^{2-n})$. From (7.5.6)–(7.5.8), we have

$$\int_{A_1} x^k R(\tilde{g}_\theta + h_\theta) dx = \theta^{1-n} \cdot \left(2(n-1)\omega_{n-1}mc^k + O(\gamma(\theta)) \right).$$

We saw above that the rescaled Schwarzschild indeed has mass times center equal to $(m/\theta^{n-2})(c/\theta) = \theta^{1-n}mc$. With $\hat{c}^k = c^k/\theta$, recalling that $\gamma(\theta) = o(\theta)$, we obtain

$$\int_{A_1} x^k R(\tilde{g}_\theta + h_\theta) \, dx = \theta^{2-n} \cdot \left(2(n-1)\omega_{n-1}m\hat{c}^k + o(1) \right).$$

Thus even though the metric g may not have had a convergent center of mass integral, we still have the following vector identity (recall (7.5.5)):

$$\frac{\theta^{n-2}}{2(n-1)\omega_{n-1}} \left(\int_{A_1} R(\tilde{g}_\theta + h_\theta) \, dx, \int_{A_1} x R(\tilde{g}_\theta + h_\theta) \, dx \right)$$

$$=: \left(m - m(g), m\hat{c} \right) + (\xi^0(m, \hat{c}), \xi(m, \dot{c}))$$

$$= (m - m(g), m\dot{c}) + o(1), \tag{7.5.9}$$

where $(\xi^0, \xi) := (\xi^0, \xi^1, \ldots, \xi^n)$, so that each ξ^k is a continuous function of m and $\hat{c}$ (and θ).

This is spectacular: as we might have surmised as the start, we cannot expect to attach *any* Schwarzschild metric end and locally perturb to obtain vanishing scalar curvature. We have to tune it so that the mass is roughly equal to that of $m(g)$: by (7.5.5), $m - m(g) = O(\theta^{n-2-2q})$, which we can make as small as we like by choosing θ large enough. Now, for the second component we have $m\hat{c} \approx m(g)\hat{c}$, and as have assumed $m(g) \neq 0$, by varying $\hat{c}$, the center of mass of the rescaled Schwarzschild, we can cover the $o(1) = O(\gamma(\theta)/\theta)$ error from (7.5.9). To do this, we are varying m and $\hat{c}$ (equivalently c) together — the error terms depend on both parameters and on θ — and recall that we have assumed $|c| \leq \frac{\theta}{2}$, i.e., $|\hat{c}| \leq \frac{1}{2}$. The leading order term of this projection map is a local diffeomorphism for $m(g) \neq 0$, the error term is small, and as we noted earlier, the map is continuous with respect to θ, m and $c = \theta\hat{c}$ (Remark 7-92). Thus, applying the Brouwer fixed point theorem or a homological degree argument, we conclude that for some $m \approx m(g)$ and some $\hat{c}$ with $|\hat{c}| \leq \frac{1}{2}$, $R(\tilde{g}_\theta + h_\theta) = 0$. To illustrate, let $\Theta = \left\{ (m, \hat{c}) : |m - m(g)| \leq \frac{1}{2}|m(g)|, \ |\hat{c}| \leq \frac{1}{2} \right\}$. The function

$$F(m, \hat{c}) = \left(m(g) - \xi^0(m, \hat{c}), \ -\frac{\xi(m, \hat{c})}{m(g) - \xi^0(m, \hat{c})} \right)$$

defines a continuous map $F : \Theta \to \Theta$, since $\xi = o(1)$, uniformly for θ large and $m, \hat{c} \in \Theta$. The Brouwer fixed point theorem yields an $(m, \hat{c}) \in \Theta$ such that $F(m, \hat{c}) = (m, \hat{c})$, which translates into $\int_{A_1} x^k R(\tilde{g}_\theta + h_\theta)\, dx = 0$ for $k \in \{0, 1, \ldots, n\}$ as desired.

Upon rescaling the metric back out from A_1, the center of mass becomes $c = \theta\hat{c}$, which can be "large", but which is relatively smaller than the scale θ at which the gluing is happening; for instance in the case $n = 3$, $q = 1$, we have $c = O(\log \theta)$. This might be expected, since we did not assume g had a convergent center of mass. As in [66] (note there are some unfortunate differences between the normalization there and here), if the original metric g satisfies better asymptotics, we can work a bit harder to get the final center of mass c to be close to the original center $c(g)$. In [66], we assumed asymptotically Schwarzschild asymptotics, but the same would hold under Regge–Teitelboim asymptotics (7.2.7), cf. [69; 61].

Exercises

Exercise 7-93. Let (M, g) be complete, noncompact and connected. Suppose there is a compact set $\mathcal{C} \subset M$ such that $M \setminus \mathcal{C}$ is a disjoint union $\bigcup_\alpha \mathcal{E}_\alpha$ of *ends*, where each end $\mathcal{E}_\alpha$ is diffeomorphic to $\{|x| > 1\}$ in $\mathbb{R}^n$, and that there is a $C > 0$ with $|g_{ij}(x) - \delta_{ij}| \leq C|x|^{-q}$ $(q > 0, |x| > 1)$ on each end. Let $d : M \times M \to [0, \infty)$

be the distance function induced by g. By the Hopf–Rinow theorem, a set $S \subset M$ is d-bounded if and only if its closure $\bar{S}$ is compact.

a. Show that $S \subset M$ is d-bounded if and only if there is an $r_0 > 1$ such that $S \cap \mathcal{E} \subset \{p \in \mathcal{E} : |x(p)| \le r_0\}$ for each end $\mathcal{E}$.

b. Prove that even if we had *a priori* allowed infinitely many ends, the number k of ends of (M, g) must be finite and is well-defined.

c. Show that in the definition of an asymptotically flat manifold with ends $\mathcal{E}_1, \ldots, \mathcal{E}_k$, we can arrange for each $\overline{\mathcal{E}_j}$ as well as the compact set $C \subset M$ with $M \setminus C = \bigcup_{j=1}^k \mathcal{E}_j$ to be smooth manifolds-with-boundary, with the asymptotically flat coordinates on each $\mathcal{E}_j$ extending to a neighborhood of $\overline{\mathcal{E}_j}$, and with $\partial C = \bigcup_{j=1}^k \partial \mathcal{E}_j$ a disjoint union of (topological) spheres.

Before the next part, you might recall as an example the $\mathbb{RP}^n$-geon from Remark 2-39. We also recall that if M is nonorientable, there is a connected orientable double cover $\pi : \widehat{M} \to M$.

d. Suppose $\pi : \widehat{M} \to M$ is a connected double cover, with the covering metric $\hat{g}$. Show that $(\widehat{M}, \hat{g})$ is asymptotically flat, with $2k$ asymptotic ends. Indeed, let $C \subset M$ be as in part c., and let $\widehat{C} = \pi^{-1}(C) \subset \widehat{M}$. Use covering arguments (path lifting properties) to prove that the path components of $\widehat{M} \setminus \widehat{C}$ give the ends of $\widehat{M}$. It may be useful to prove the following: if $p \in M \setminus C$ is a point in an end of M, and if $\pi^{-1}(p) = \{\hat{p}_1, \hat{p}_2\} \subset \widehat{M} \setminus \widehat{C}$, then since any asymptotic end is simply connected, any path from $\hat{p}_1$ to $\hat{p}_2$ in $\widehat{M}$ must hit $\widehat{C}$.

Exercise 7-94. Refer to Remark 7-77 for the setting of the exercise. Suppose (M, g) is asymptotically flat and let $u = 1 + v$ be smooth with $0 < u < 1$ and $u \to 1$ at infinity in each asymptotic end, with v subharmonic and lying in a suitable weighted space, admitting an expansion $v(x) = A/|x|^{n-2} + O_1(|x|^{-(n-2)-\gamma})$ on each end, for some $\gamma > 0$. Recall that these conditions come from solving $\Delta_g u - \frac{n-2}{4(n-1)} R(g) u = 0$, where $R(g) \ge 0$ is nontrivial and has suitable decay. Let $\mathcal{E}$ be the manifold-with-boundary corresponding to $\{|x| \ge r_0\}$ in asymptotic coordinates on a chosen end. Suppose w solves the boundary value problem $\Delta_g w = 0$, $w\big|_{\partial \mathcal{E}} = \max_{\partial \mathcal{E}} v < 0$ and $w(x) \to 0$ as $|x| \to \infty$ in the end of $\mathcal{E}$, with $w(x) = B/|x|^{n-2} + O_1(|x|^{-(n-2)-\gamma'})$ for some $\gamma' > 0$. Argue that $A \le B$ and $B < 0$. (Hint: use Hopf's lemma; see [107, Lemma 3.4] or [86, Chapter 6].)

Exercise 7-95. a. Complete the proof sketched in Remark 7-78. To check the integration by parts argument, first give a reasonably direct argument for $p > n$; to extend to $p > \frac{n}{2}$, one can use a suitable exhaustion of the ends, choosing $r_i \nearrow \infty$ for which $r_i^{1-n} \int_{\{|x|=r_i\}} |\nabla w|^p \, d\sigma$ decays suitably.

b. Prove Corollary 7-68 in case $k = 2$ if the assumption on h is changed to $h \in W_{-\nu}^{0,r}$, with $n < p \le r$. Generalize this to $k \ge 2$.

c. Consider metrics g with $g_{ij} - \delta_{ij} \in W_{-\tau}^{2,p}$, $p > n$ and $\tau > \frac{n-2}{2}$, e.g. in case g is asymptotically flat of rate $q > \frac{n-2}{2}$ and order 2. The energy integral converges for such g with $R(g) \in L^1$ (Exercise 7-53). Prove an analogue of Proposition 7-76 holds for such metrics, carrying out the proof with $u - 1 = v \in W_{-\tau}^{2,p}$. Comment on the asymptotics of the resulting metric $\bar{g}$.

d. Prove that the ADM mass (energy) is nonnegative for the metrics g in part c. with nonnegative scalar curvature. (You may assume the metrics are smooth, so the condition is just specifying the decay; metrics with only $W^{2,p}$ regularity require extra care; see, e.g., [142, Theorem 3.43].)

e. Use Remark 7-77 together with parts c. and d. to argue that if the mass of an end vanishes, then so does the scalar curvature: the mechanism is to remove energy density (scalar curvature), but applying part c. can fail to preserve order 2. An alternative approach is to instead observe that you need not remove *all* the energy density: modify equation (7.3.1) by replacing g_θ with g, and ψ_θ with $1 - \psi_\theta$. What can you conclude?

Exercise 7-96. In this exercise you will develop some further properties of Euclidean harmonic functions on punctured domains in $\mathbb{R}^n$.

a. Show that if u is harmonic in a punctured ball around the origin, then the isolated singularity at $x = 0$ is in fact removable if $\lim_{x \to 0} |x|^{n-2} u(x) = 0$ in case $n > 2$, and in case $n = 2$, if $\lim_{x \to 0} \frac{u(x)}{\log |x|} = 0$.

b. Suppose $n > 2$ and Ω is a domain containing the origin. If u is harmonic on $\Omega \setminus \{0\}$, and $\liminf_{x \to 0} |x|^{n-2} u(x) > -\infty$, there exist v harmonic in Ω and $b \in \mathbb{R}$ such that $u(x) = b|x|^{2-n} + v(x)$ on $\Omega \setminus \{0\}$. Formulate an analogue in case $n = 2$.

c. What can you say about a positive harmonic function on $\mathbb{R}^n \setminus \{0, a\}$, $a \in \mathbb{R}^n \setminus \{0\}$?

Exercise 7-97. The goal of this exercise is to verify directly that the ADM energy-momentum vector of Schwarzschild spacetime transforms correctly under boosts as a Lorentz-invariant vector. We consider the four-dimensional Schwarzschild metric $\bar{g}_S$ given by

$$\bar{g}_S = -\frac{\left(1 - \frac{2m}{r}\right)^2}{\left(1 + \frac{2m}{r}\right)^2} dt^2 + \left(1 + \frac{m}{2r}\right)^4 (dx^2 + dy^2 + dz^2)$$

$$= -\left(1 - \frac{2m}{r}\right) dt^2 + \left(1 + \frac{2m}{r}\right)(dx^2 + dy^2 + dz^2) + O_\infty(r^{-2}),$$

with $r = \sqrt{x^2 + y^2 + z^2}$ (and $G = 1$, $c = 1$). If Σ is the $t = 0$ slice with induced metric g_S, then as we have seen, Σ is time-symmetric (totally geodesic) with two asymptotic ends, which have ADM energy m and vanishing linear momentum in the above coordinates. For $0 < \alpha < 1$, consider a boosted slice Σ_α given in an asymptotic end by $t = \alpha x$, with induced metric g. The corresponding Lorentz transformation is given by $\tau = (t - \alpha x)/\sqrt{1 - \alpha^2}$, $\xi = (-\alpha t + x)/\sqrt{1 - \alpha^2}$; the worldline of the corresponding (asymptotic) observer is $\xi = 0$, i.e., $x = \alpha t$, and Σ_α is given by $\tau = 0$.

a. Show that the future-pointing unit normal to Σ_α is given by

$$n = \left(1 + \tfrac{m}{2r}\right)^{-2} \frac{\psi(r)\frac{\partial}{\partial t} + \alpha\frac{\partial}{\partial x}}{\sqrt{\psi(r) - \alpha^2}} = \frac{\left(1 + \frac{2m}{r}\right)\frac{\partial}{\partial t} + \alpha\left(1 - \frac{2m}{r}\right)\frac{\partial}{\partial x}}{\sqrt{\left(1 + \frac{2m}{r}\right) - \alpha^2\left(1 - \frac{2m}{r}\right)}} + O(r^{-2}),$$

where $\psi(r) = \left(1 - \tfrac{m}{2r}\right)^{-2}\left(1 + \tfrac{m}{2r}\right)^6 = \left(1 + \tfrac{4m}{r}\right) + O(r^{-2})$.

b. Note that

$$\frac{\partial}{\partial \tau} = \frac{1}{\sqrt{1 - \alpha^2}}\left(\frac{\partial}{\partial t} + \alpha\frac{\partial}{\partial x}\right) \quad \text{and} \quad \frac{\partial}{\partial \xi} = \frac{1}{\sqrt{1 - \alpha^2}}\left(\frac{\partial}{\partial x} + \alpha\frac{\partial}{\partial t}\right).$$

Show that the metric $\bar{g}_S$ in the coordinates (τ, ξ, y, z) has the matrix representation

$$[\bar{g}_S] = \begin{bmatrix} \left(-1 + \frac{1+\alpha^2}{1-\alpha^2}\cdot\frac{2m}{r}\right) & \frac{\alpha}{1-\alpha^2}\cdot\frac{4m}{r} & 0 & 0 \\ \frac{\alpha}{1-\alpha^2}\cdot\frac{4m}{r} & \left(1 + \frac{1+\alpha^2}{1-\alpha^2}\cdot\frac{2m}{r}\right) & 0 & 0 \\ 0 & 0 & \left(1 + \frac{2m}{r}\right) & 0 \\ 0 & 0 & 0 & \left(1 + \frac{2m}{r}\right) \end{bmatrix} + O_\infty(r^{-2}).$$

This shows that these are asymptotically Minkowskian coordinates. We caution however that $\partial r/\partial \tau \neq 0$. Along Σ_α, we have $t = \alpha x$, and hence $\xi = x\sqrt{1 - \alpha^2}$, so that $r^2 = \alpha^2\xi^2/(1-\alpha^2) + \rho^2$, where $\rho^2 := \xi^2 + y^2 + z^2 = r^2 - \alpha^2 x^2$. Note that $(1 - \alpha^2) \leq \rho^2/r^2 \leq 1$. The coordinates (ξ, y, z) are thus seen to be asymptotically flat coordinates on Σ_α.

c. Using the observations in part b., compute the ADM energy using the asymptotically flat coordinates (ξ, y, z) along Σ_α, and show it is equal to $m/\sqrt{1 - \alpha^2}$.

d. Compute the second fundamental form

$$K_{ij} = \bar{g}_S\left(\nabla_{\frac{\partial}{\partial x^i}}\frac{\partial}{\partial x^j}, n\right)$$

of Σ_α in the asymptotically flat coordinates (ξ, y, z). It might be instructive to do this in two ways: (i) compute the relevant Christoffel symbols directly from the metric $\bar{g}_S$ in (t, x, y, z) coordinates above, and recall the above formula for

$\partial/\partial\xi$ when computing $\overline{\nabla}_{\frac{\partial}{\partial\xi}}\frac{\partial}{\partial x^j}$; and (ii) use the ADM equation (5.3.4) and the form of the metric above to compute K_{ij} from the lapse and shift form of the metric $\bar{g}_S$ in the (τ, ξ, y, z) coordinate; we again emphasize that $\partial r/\partial\tau \neq 0$. In either case you should find that the matrix representation of K in the (ξ, y, z) coordinates is

$$[K] = \begin{bmatrix} \dfrac{\alpha(\alpha^2-3)}{(1-\alpha^2)^2}\dfrac{m\xi}{r^3} & -\dfrac{\alpha}{1-\alpha^2}\dfrac{2my}{r^3} & -\dfrac{\alpha}{1-\alpha^2}\dfrac{2mz}{r^3} \\[2mm] -\dfrac{\alpha}{1-\alpha^2}\dfrac{2my}{r^3} & \dfrac{\alpha}{1-\alpha^2}\dfrac{m\xi}{r^3} & 0 \\[2mm] -\dfrac{\alpha}{1-\alpha^2}\dfrac{2mz}{r^3} & 0 & \dfrac{\alpha}{1-\alpha^2}\dfrac{m\xi}{r^3} \end{bmatrix} + O(r^{-3}),$$

so that the trace of K is $\operatorname{tr}_g K = -\dfrac{\alpha(1+\alpha^2)}{(1-\alpha^2)^2}\dfrac{m\xi}{r^3} + O(r^{-3})$.

e. Find the momentum tensor $\pi_{ij} = K_{ij} - (\operatorname{tr}_g K)g_{ij} = K_{ij} - (\operatorname{tr}_g K)\delta_{ij} + O(r^{-3})$ in the (ξ, y, z) coordinates, noting in particular that

$$\pi_{\xi\xi} = -\frac{2\alpha}{1-\alpha^2}\frac{m\xi}{r^3} + O(r^{-3}).$$

Now compute the ADM momentum, showing that $P_1 = -m\alpha/\sqrt{1-\alpha^2}$ and $P_2 = P_3 = 0$. Note that this is consistent with the Lorentz transformation of coordinates, and makes sense on physical grounds in terms of the relative motion of the two observers with worldlines $x = 0$ and $\xi = 0$ in the asymptotic region.

Exercise 7-98 (Lorentz invariance of ADM energy-momentum). Consider a Lorentzian manifold $(S, \bar{g})$. We will be interested in the case when $\bar{g}$ is asymptotically Minkowskian, so we will assume that S is $\mathbb{R}^{n+1}$ (or a suitable open subset), and let η be a background Minkowski metric, so that the coordinates x^μ $(x^0 = t, c = 1)$ on $\mathbb{R}^{n+1}$ are inertial coordinates for η:

$$\eta = \eta_{\mu\nu}dx^\mu dx^\nu = -(dx^0)^2 + (dx^1)^2 + \cdots + (dx^n)^2.$$

Given a symmetric tensor h, for any point in S, there is a neighborhood on which $\eta + \epsilon h$ is a Lorentzian metric for sufficiently small ϵ. We define

$$DG|_\eta(h) = \frac{d}{d\epsilon}\Big|_{\epsilon=0} G(\eta + \epsilon h),$$

where $G(\bar{g}) = \operatorname{Ric}(\bar{g}) - \frac{1}{2}R(\bar{g})\bar{g} =: \kappa T$ is the Einstein tensor. We likewise let $D\mathrm{Ric}_\eta(h)$ and $DR_\eta(h)$ be the respective linearizations of the Ricci and scalar curvatures.

a. Verify that in any inertial coordinate system (i.e., $\eta_{0\nu} = -\delta_{0\nu}$ and $\eta_{\mu j} = \delta_{\mu j}$ for $j \geq 1$), we have $(D\mathrm{Ric}_\eta(h))_{\mu\nu} = \frac{1}{2}\eta^{\lambda\alpha}(h_{\alpha\mu,\lambda\nu} - h_{\lambda\alpha,\mu\nu} + h_{\alpha\nu,\lambda\mu} - h_{\mu\nu,\lambda\alpha})$, and thus $DR_\eta(h) = \eta^{\mu\nu}\eta^{\lambda\alpha}(h_{\alpha\mu,\lambda\nu} - h_{\lambda\alpha,\mu\nu})$. Note then that $DG_\eta(h) = D\mathrm{Ric}_\eta(h) - \frac{1}{2}DR_\eta(h)\eta$.

Define τ by $\kappa\tau^{\mu\nu} = \eta^{\mu\alpha}\eta^{\nu\beta}(DG_\eta(h))_{\alpha\beta}$. Then $\kappa\tau$ is comprised of the terms in the Einstein tensor that are linear in h.

b. Show that $\mathrm{div}_\eta\,\tau = 0$, i.e., in any inertial coordinate system, $\tau^{\mu\nu}{}_{,\nu} = 0$. You can show this directly from the formula above, or use the fact that $\mathrm{div}_{\bar{g}}\,G(\bar{g}) = 0$.

c. Show that in any inertial coordinate system, for $i \in \{1, \ldots, n\}$,

$$\kappa\tau^{00} = \tfrac{1}{2}\sum_{i,j=1}^{n}(h_{ij,i} - h_{ii,j})_{,j}$$

$$\kappa\tau^{0i} = \sum_{j=1}^{n}\tfrac{1}{2}\Big(h_{0i,j} - h_{ij,0} - h_{0j,i} + \sum_{k=1}^{n}h_{kk,0}\delta_{ij}\Big)_{,j}$$

$$= \sum_{j=1}^{n}\tfrac{1}{2}\Big(h_{0i,j} - h_{ij,0} + \sum_{k=1}^{n}(h_{kk,0} - h_{0k,k})\delta_{ij}\Big)_{,j}.$$

Let $y^\mu = \Lambda^\mu{}_\nu x^\nu + a^\mu$ for some proper Lorentz matrix $\Lambda^\mu{}_\nu$, and some vector with components a^μ. Then the volume form is $\omega = dx^0 \wedge dx^1 \wedge \cdots \wedge dx^n = dy^0 \wedge dy^1 \wedge \cdots \wedge dy^n$. Consider spacelike hyperplanes M and $\widetilde{M}$ in $\mathbb{R}^{n+1}$, where M is given by $x^0 = 0$ and $\widetilde{M}$ by $y^0 = 0$. Let $C_R = \{x \in \mathbb{R}^{n+1} : \sum_{i=1}^{n}(x^i)^2 \leq R^2\}$. Let $M_R = M \cap C_R$, $\widetilde{M}_R = \widetilde{M} \cap C_R$; then C_R, M_R and $\widetilde{M}_R$ bound a solid spacetime region W_R, the boundary of which is the union of M_R, $\widetilde{M}_R$ and a timelike hypersurface $\Sigma_R \subset \partial C_R$. For each μ, let I_μ be the n-index increasing from 0 to n, but omitting μ, and let dx^{I_μ} be the corresponding n-form. For a vector field $X = X^\mu\,\partial/\partial x^\mu$, note that $i_X\omega = \sum_{\mu=0}^{n}(-1)^{\mu+1}X^\mu dx^{I_\mu}$.

d. For any λ, let $X = \tau(\,\cdot\,, dx^\lambda)$. Observe that along M, the pullback of $i_X\omega$ is $\tau(dx^0, dx^\lambda)\,dx^1 \wedge \cdots \wedge dx^n$, and along $\widetilde{M}$, the pullback of $i_X\omega$ is $\tau(dy^0, dx^\lambda)\,dy^1 \wedge \cdots \wedge dy^n$. What is $d(i_X\omega)$, and what does Stokes' theorem give you for $\int_{W_R} d(i_X\omega)$?

For the remainder of the problem, we consider an asymptotically Minkowskian metric $\bar{g}_{\mu\nu} = \eta_{\mu\nu} + h_{\mu\nu}$, with $h_{\mu\nu} = O_2(|x|^{-q})$ in η-inertial coordinates, $q > \frac{n-2}{2}$, where $|x|^2 = \sum_{i=1}^{n}(x^i)^2$. We further assume that $G(\bar{g}) = \kappa T = O(|x|^{-2q-2})$.

e. Observe that $\tau = O(|x|^{-2q-2})$, so that it is integrable over M and $\widetilde{M}$. Prove that (with dx and dy the induced Euclidean volume measures on M and $\widetilde{M}$, respectively)

$$\int_M \tau(dx^0, dx^\lambda)\,dx = \int_{\widetilde{M}} \tau(dy^0, dx^\lambda)\,dy.$$

Conclude that $\widehat{\mathbb{P}} := \left(\int_{\{x^0=0\}} \tau(dx^0, dx^\lambda)\, dx \right) \partial/\partial x^\lambda$ is Lorentz-invariant and that $\widehat{\mathbb{P}}^0 = \frac{1}{2\kappa} \lim_{r\to\infty} \int_{\{x^0=0,\ |x|=r\}} \sum_{i,j=1}^n (h_{ij,i} - h_{ii,j})v^j\, d\sigma = E$, the ADM energy.

f. Write the metric $\bar{g}$ in lapse-shift form

$$\bar{g} = -N^2 dt^2 + g_{ij}(dx^i + X^i dt) \otimes (dx^j + X^j dt),$$

so that $N - 1 = O_2(|x|^{-q})$ and $X^j = O_2(|x|^{-q})$. Show that along M,

$$\frac{1}{2}\left(h_{0i,j} - h_{ij,0} - h_{0j,i} + \sum_{k=1}^n h_{kk,0}\delta_{ij} \right)$$

$$= \pi_{ij} + (X^k{}_{,k}\delta_{ij} - X^\ell{}_{,i}\delta_{\ell j}) + O(|x|^{-2q-1})$$

$$= \frac{1}{2}\left(h_{0i,j} - h_{ij,0} + \sum_{k=1}^n (h_{kk,0} - h_{0k,k})\delta_{ij} \right) + \frac{1}{2}(X^k{}_{,k}\delta_{ij} - X^\ell{}_{,i}\delta_{\ell j}) + O(|x|^{-2q-1}),$$

where $\pi_{ij} = K_{ij} - (\mathrm{tr}_g K)g_{ij}$; cf. (5.3.4). Use these formulas to compute π_{ij} for a boosted slice Σ_α in Schwarzschild as in Exercise 7-97; in the notation of that problem, $\partial r/\partial \tau \neq 0$.

g. Prove that the ADM momentum flux integral converges, and is given by

$$P_i = \lim_{r\to\infty} \int_{\{x^0=0,\ |x|=r\}} \Theta_{ij}\, v^j\, d\sigma = \widehat{\mathbb{P}}^i,$$

where $\kappa\Theta_{ij}$ can be taken to be either $\frac{1}{2}\left(h_{0i,j} - h_{ij,0} - h_{0j,i} + \sum_{k=1}^n h_{kk,0}\delta_{ij} \right)$ or $\frac{1}{2}\left(h_{0i,j} - h_{ij,0} + \sum_{k=1}^n (h_{kk,0} - h_{0k,k})\delta_{ij} \right)$. (You might note that in $\mathbb{R}^3$, if $F = \langle 0, -X^3, X^2 \rangle$, then $\nabla \times F = \langle X^k{}_{,k} - X^1{}_{,1}, -X^2{}_{,1}, -X^3{}_{,1} \rangle$.)

h. Summarize the preceding to formulate the Lorentz invariance of the ADM energy-momentum vector for an asymptotically flat metric $\bar{g}$ with decay rates as above. Note that given the asymptotic spacetime end (which is assumed to contain a suitable family of spacelike surfaces obtained from boosting the asymptotically flat end given by $x^0 = 0$ in asymptotic coordinates), for a given pair of spacelike slices we want to compare, we can choose a suitable asymptotic region Ω containing the ends of the slices, and extend the metric $\bar{g}$ as a Lorentzian metric on $\mathbb{R}^{n+1}$; this will not affect the flux integrals defining the ADM energy-momentum, and allows us to apply the preceding arguments.

Exercise 7-99. We use the notation from Exercise 7-98. Assume in addition to the asymptotically flat assumptions in that exercise that there are asymptotically flat coordinates x^μ in which $\bar{g}$ and T are even to one order better than their asymptotic decay rates (and $\partial\bar{g}$ is odd to one order better):

$$\bar{g}_{\mu\nu}(x) - \bar{g}_{\mu\nu}(-x) = O(|x|^{-q-1}),$$
$$\bar{g}_{\mu\nu,\sigma}(x) + \bar{g}_{\mu\nu,\sigma}(-x) = O(|x|^{-q-2}),$$
$$T_{\mu\nu}(x) - T_{\mu\nu}(-x) = O(|x|^{-2q-3})$$

(cf. [61, Appendix E]). Compare to the Regge–Teitelboim conditions on initial data sets (7.2.7)–(7.2.8).

a. Show that $J^{\nu\lambda} := \lim_{r\to\infty} \int_{\{x^0=0,\,|x|\le r\}} (x^\nu \tau^{0\lambda} - x^\lambda \tau^{0\nu})\,dx$ converges. Show that if $y^\mu = x^\mu + a^\mu$ where $a^0 = 0$ (spatial translation, so $x^0 = y^0$),

$$\tilde{J}^{\nu\lambda} := \lim_{r\to\infty} \int_{\{y^0=0,\,|y|\le r\}} (y^\nu \tau^{0\lambda} - y^\lambda \tau^{0\nu})\,dy$$

$$= \lim_{r\to\infty} \int_{\{x^0=0,\,|x|\le r\}} (y^\nu \tau^{0\lambda} - y^\lambda \tau^{0\nu})\,dx \;=\; J^{\nu\lambda} + a^\nu \mathbb{P}^\lambda - a^\lambda \mathbb{P}^\nu.$$

b. Let $y^\mu = \Lambda^\mu_{\ \nu} x^\nu + a^\mu$, and let

$$M = \{x \in \mathbb{R}^{n+1} : x^0 = 0\}, \quad \tilde{M} = \{x \in \mathbb{R}^{n+1} : y^0 = \Lambda^0_{\ \nu} x^\nu + a^0 = 0\},$$

and define M_R and $\tilde{M}_R$ as in Exercise 7-98. Using the vector field $X^\mu \partial/\partial x^\mu$ given by $X^\mu = (x^\nu \tau(dx^\mu, dx^\lambda) - x^\lambda \tau(dx^\mu, dx^\nu))$, prove that

$$J^{\nu\lambda} = \lim_{r\to\infty} \int_{M_r} (x^\nu \tau(dx^0, dx^\lambda) - x^\lambda \tau(dx^0, dx^\nu))\,dx$$

$$= \lim_{r\to\infty} \int_{\tilde{M}_r} (x^\nu \tau(dy^0, dx^\lambda) - x^\lambda \tau(dy^0, dx^\nu))\,dy,$$

so that

$$J^{\nu\lambda} \frac{\partial}{\partial x^\nu} \otimes \frac{\partial}{\partial x^\lambda}$$

$$= \left(\lim_{r\to\infty} \int_{\tilde{M}_r} ((y^\alpha - a^\alpha)\tau(dy^0, dy^\beta) - (y^\beta - a^\beta)\tau(dy^0, dy^\alpha))\,dy \right) \frac{\partial}{\partial y^\alpha} \otimes \frac{\partial}{\partial y^\beta}$$

$$= (\tilde{J}^{\alpha\beta} - a^\alpha \tilde{\mathbb{P}}^\beta + a^\beta \tilde{\mathbb{P}}^\alpha) \frac{\partial}{\partial y^\alpha} \otimes \frac{\partial}{\partial y^\beta},$$

i.e., $\tilde{J}^{\alpha\beta} = \Lambda^\alpha_{\ \nu} J^{\nu\lambda} \Lambda^\beta_{\ \lambda} + a^\alpha \Lambda^\beta_{\ \lambda} \mathbb{P}^\lambda - a^\beta \Lambda^\alpha_{\ \nu} \mathbb{P}^\nu$.

c. For $k \in \{1, \ldots, n\}$, show that J^{k0} gives the center of mass integral from (7.2.5). Give a physical interpretation of the result of part b. in case the Lorentz transformation is the identity, and $a^i = 0$ for $i > 0$.

d. For $\ell, m \in \{1, \dots, n\}$, show that $J^{\ell m} = \lim_{r \to \infty} \int_{\{x^0 = 0, \, |x| = r\}} \Theta_j^{\ell m} \, v^j \, d\sigma$, where

$$\kappa \Theta_j^{\ell m} = \tfrac{1}{2}(x^\ell h_{0m,j} - x^\ell h_{mj,0} - x^m h_{0\ell,j} + x^m h_{\ell j,0} + h_{0\ell}\delta_{mj} - h_{0m}\delta_{\ell j})$$
$$= (x^\ell K_{mj} - x^m K_{\ell j}) + \mathcal{D}_j^{\ell m} + \mathcal{E}_j^{\ell m},$$

where in terms of the lapse N and shift X, $\mathcal{D}_j^{\ell m} = -\tfrac{1}{2}x^\ell X_{j,m} + \tfrac{1}{2}x^m X_{j,\ell} + \tfrac{1}{2}X_\ell \delta_{mj} - \tfrac{1}{2}X_m \delta_{\ell j}$, and $\mathcal{E}_j^{\ell m} = x * X * \Gamma + (N-1) * x * K$, where $*$ denotes a linear combination of products of components of the indicated factors. By the parity assumptions, $N - 1$ and X are each even up to respective terms of order $O(|x|^{-q-1})$, and the Christoffel symbols and second fundamental form are odd up to respective terms of order $O(|x|^{-q-2})$. Conclude that $\lim_{r \to \infty} \int_{\{x^0 = 0, \, |x| = r\}} \mathcal{E}_j^{\ell m} \, v^j \, d\sigma = 0$. Show that $\lim_{r \to \infty} \int_{\{x^0 = 0, \, |x| = r\}} \mathcal{D}_j^{\ell m} \, v^j \, d\sigma = 0$ as well, and conclude that in case $n = 3$, $J^{23} = \mathcal{J}_1$, $J^{31} = \mathcal{J}_2$ and $J^{12} = \mathcal{J}_3$ from (7.2.6).

Exercise 7-100. Let $n \geq 3$. Recall the Newtonian potential,

$$N_f(x) = (f * \Gamma)(x) = \int_{\mathbb{R}^n} \Gamma(x - y) f(y) \, dy.$$

For $R > 0$, let $\chi_R = \chi_{B_R(0)}$ be the characteristic function of the ball of radius R about the origin.

a. Show that for $p > \tfrac{n}{2}$, there are functions $f \in L^p(\mathbb{R}^n)$ for which the integral defining the Newtonian potential does not converge. What can you say about the borderline case $p = \tfrac{n}{2}$?

b. Observe that for $f \in L_{\mathrm{loc}}^\infty(\mathbb{R}^n)$, the convolution integral $(\chi_1 \Gamma) * f$ converges everywhere, whereas by Young's inequality, for $1 \leq p \leq \infty$ and $f \in L^p(\mathbb{R}^n)$, $(\chi_1 \Gamma) * f \in L^p(\mathbb{R}^n)$, so that in particular the convolution integral converges almost everywhere.

c. Prove that for $1 \leq p < \tfrac{n}{2}$ and $f \in L^p(\mathbb{R}^n)$, the convolution $((1 - \chi_1)\Gamma) * f$ converges everywhere to a bounded, measurable function.

d. Generalize Proposition 7-3: prove that for $1 \leq p < \tfrac{n}{2}$, $\Delta N_f = f$ distributionally.

Exercise 7-101. Suppose $f \in L_{\mathrm{loc}}^\infty(\mathbb{R}^n) \cap L^1(\mathbb{R}^n)$, and suppose we have established $\partial_j N_f(x) = \int_{\mathbb{R}^n} \partial_j \Gamma(x - y) f(y) \, dy$. Let $\Omega \subset \mathbb{R}^n$ be a bounded open set; the goal is to estimate

$$\partial_j N_f(x_1) - \partial_j N_f(x_2) = \int_{\mathbb{R}^n} (\partial_j \Gamma(x_1 - y) - \partial_j \Gamma(x_2 - y)) f(y) \, dy$$

for points x_1 and x_2 in Ω. For notational convenience, given a measurable set $E \subset \mathbb{R}^n$, let $I(E) = \int_E (\partial_j \Gamma(x_1 - y) - \partial_j \Gamma(x_2 - y)) f(y) \, dy$, and for $r > 0$ and

$a \in \mathbb{R}^n$, let $B_r(a) = \{x : |x - a| < r\}$. Let $\rho = 2|x_1 - x_2|$; for any $R \geq 2\,\mathrm{diam}(\Omega)$, we have $R > \rho$, and note that $\overline{\Omega} \subset B_R(x_1)$.

a. Argue that for any $r > 0$ and $a \in \mathbb{R}^n$, $\int_{B_r(a)} |y|^{-(n-1)}\, dy \leq \int_{B_r(0)} |y|^{-(n-1)}\, dy$. Conclude that there is a c depending only on n so that for $\rho > 0$, $|I(B_\rho(x_1))| \leq c|x_1 - x_2| \sup_{x \in B_\rho(x_1)} |f(x)|$.

b. Let $x^\lambda = (1 - \lambda)x_1 + \lambda x_2$. Show that for any y with $|y - x_1| \geq \rho$, and for all $\lambda \in [0, 1]$, $|y - x^\lambda| \geq \frac{1}{2}|y - x_1| \geq |x_1 - x_2|$.

c. Conclude that there are constants K and L, depending only on n, such that for all $x_1, x_2 \in \Omega$, $|y - x_1| > \rho$ and $\lambda \in [0, 1]$, we have $|x^\lambda - y| > 0$, and moreover $|x_1 - y|^n |\partial_{ij}^2 \Gamma|_{(x^\lambda - y)}| \leq K$, from which it follows for suitably chosen L that $|x_1 - y|^n |\partial_j \Gamma(x_1 - y) - \partial_j \Gamma(x_2 - y)| \leq L|x_1 - x_2|$.

d. Show that $|I(\mathbb{R}^n \setminus B_R(x_1))| \leq L R^{-n} |x_1 - x_2| \|f\|_{L^1(\mathbb{R}^n \setminus B_R(x_1))}$.

e. Finally, for $\rho > 0$, derive the following estimate for C depending only on n:
$$\left| I(B_R(x_1) \setminus B_\rho(x_1)) \right| \leq C \sup_{x \in B_R(x_1)} |f(x)| |x_1 - x_2| (\log R - \log \rho).$$
Conclude that $N_f \in C_{\mathrm{loc}}^{1,\alpha}(\mathbb{R}^n)$ for any $0 < \alpha < 1$.

Exercise 7-102. Suppose $1 \leq p < \infty$.

a. Suppose $\Delta u = 0$ on $\mathbb{R}^n$, and that $u \in L_{-\tau}^p(\mathbb{R}^n)$. Show that if $\tau \geq 0$, then u is identically zero. (Hint: Use the mean value property.)

b. If $\tau < 0$, then if u is a polynomial lying in $L_{-\tau}^p(\mathbb{R}^n)$, show that $\deg u < -\tau$.

Exercise 7-103. Let $K'(x, y) = |x|^{-a}|x - y|^{a+b-n}|y|^{-b}$, with $a + b > 0$. Let $p > 1$ and $p' = \frac{p}{p-1}$. Prove that if $T(u)(x) = \int_{\mathbb{R}^n} K'(x, y)u(y)\, dy$ defines a bounded linear operator T on $L^p(\mathbb{R}^n)$, then $a < \frac{n}{p}$ and $b < \frac{n}{p'}$. To do this, first observe that $f(x) = \int_{\{|y| \leq 1\}} K'(x, y)\, dy$ gives a function $f \in L^p(\mathbb{R}^n)$. Furthermore, argue that the map $u \in L^p(\mathbb{R}^n) \mapsto \int_{\{|x| \leq 1\}} \int_{\mathbb{R}^n} K'(x, y)u(y)\, dy\, dx$ defines a bounded linear functional on $L^p(\mathbb{R}^n)$, and conclude that $g(x) = \int_{\{|x| \leq 1\}} K'(x, y)\, dx$ defines a function $g \in L^{p'}(\mathbb{R}^n)$. Now examine the asymptotic rates of f and g. We remark that the same analysis holds for the integral kernel
$$\widetilde{K}(x, y) = (1 + |x|)^{-a}|x - y|^{a+b-n}(1 + |y|)^{-b}.$$

Exercise 7-104. Let m be a nonnegative integer and $1 < p < \infty$. Show that $\Delta : W_m^{2,p}(\mathbb{R}^n) \to L_{m-2}^p(\mathbb{R}^n)$ does not have closed range. (A duality argument then shows that $\Delta : W_{2,-n-m}^{2,p}(\mathbb{R}^n) \to L_{-n-m}^p(\mathbb{R}^n)$ does not have closed range cf. [155].)

(Hint: Let $\mathcal{K}_m$ be space of homogeneous harmonic polynomials on $\mathbb{R}^n$ of degree at most m, so that $\mathcal{K}_{-1} = \{0\}$. The kernel of Δ in $W_m^{2,p}(\mathbb{R}^n)$ is precisely $\mathcal{K}_{m-1}$.

Let $0 \le \varphi \le 1$ be smooth with compact support in $\{x : |x| \le 2\}$, and with $\varphi(x) = 1$ for all $|x| \le 1$, and let $u \in \mathcal{K}_m \setminus \mathcal{K}_{m-1}$, i.e., u is a nontrivial homogeneous harmonic polynomial of degree m. For $k \in \mathbb{Z}_+$, let $u_k(x) = (\log k)^{-1/p} \varphi(x/k) u(x)$. Show that $\inf_{k \in \mathbb{Z}_+} \inf_{h \in \mathcal{K}_{m-1}} \|u_k - h\|_{W_m^{2,p}} > 0$. To do this, first argue that there are positive constants c_1 and c_2 such that for all $k \in \mathbb{Z}_+$, $c_1 \le \|u_k\|_{L_m^p} \le c_2$. Show that if $\inf_{k \in \mathbb{Z}_+} \inf_{h \in \mathcal{K}_{m-1}} \|u_k - h\|_{W_m^{2,p}} = 0$, then for some subsequence $k_j \in \mathbb{Z}_+$, and for some $v \in \mathcal{K}_{m-1}$, $u_{k_j} \to v$ in $L_m^p(\mathbb{R}^n)$; then argue that $v = 0$, which gives a contradiction. Argue carefully that $\Delta u_k \to 0$ in $L_{m-2}^p(\mathbb{R}^n)$. From here, an elementary functional analysis argument shows that the range is not closed. After you have finished the proof, note where you used that u_k is harmonic, and check directly that if you tried to run an analogous proof in $L_\mu^p(\mathbb{R}^n)$ for $m - 1 < \mu < m$, you can renormalize u_k in a reasonable way to get c_1 and c_2 as above, but the analogous proof does not go through (the issue then must lie with the estimate of Δu_k).)

Exercise 7-105. In parts a. and b. you will prove Corollary 7-69.

a. Prove the claim for the weight range $0 < \tau < n - 2$, which follows along the lines of the proof of Corollary 7-68.

b. For the borderline case $\tau = n - 2$, you should first establish an estimate of the form $\|hu\|_{L^1} \le C \|h\|_{C_{-\nu}^0} \|u\|_{C_{-\beta}^0}$ for $\nu > 2$, for some $n - \nu < \beta < n - 2$. Conclude that $T_h(u) = hu$ gives a compact linear mapping $T_h : C_{2-n}^0 \to L^1$. Following [211], we let $D_{-n}^{k-2,\alpha} = C_{-n}^{k-2,\alpha} \cap L^1$, with $\|f\|_{D_{-n}^{k-2,\alpha}} = \|f\|_{D_{-n}^{k-2,\alpha}} + \|f\|_{L^1}$, and we let $E_{2-n}^{k,\alpha} = \{u \in C_{-n}^{k,\alpha} : \Delta_g u \in L^1\}$, with norm $\|u\|_{E_{2-n}^{k,\alpha}} = \|u\|_{C_{2-n}^{k,\alpha}} + \|\Delta_g u\|_{L^1}$. Use (7.2.19) to show $(\Delta_g - h) : E_{2-n}^{k,\alpha} \to D_{-n}^{k-2,\alpha}$ is an isomorphism.

c. We note that in [211], a slightly different norm for $E_{2-n}^{k,\alpha}$ is used. Consider a background metric $\mathring{g}$ as we have done before, where $\mathring{g}_{ij} = \delta_{ij}$ on the asymptotic charts for g on each end. Show that for $u \in C_{2-n}^2$, $\Delta_g u \in L^1$ if and only if $\Delta_{\mathring{g}} u \in L^1$, by estimating $\|\Delta_g u - \Delta_{\mathring{g}} u\|_{L^1}$. (This shows that the functions in the space $E_{2-n}^{k,\alpha}$ are the same for all metrics which satisfy the same asymptotically flat condition as g does in the asymptotic charts for g.) Conclude that one gets an equivalent norm on $E_{2-n}^{k,\alpha}$ by replacing $\|\Delta_g u\|_{L^1}$ in $\|u\|_{E_{2-n}^{k,\alpha}}$ with $\|\Delta_{\mathring{g}} u\|_{L^1}$.

Exercise 7-106 (expansion in harmonic asymptotics). Define the operator

$$(\mathcal{L}_g X)_{ij} = X_{i;j} + X_{j;i} - X^k_{;k} g_{ij}.$$

If γ is a metric on M^3 and $u > 0$, let $g_{ij} = u^4 \gamma_{ij}$ and let $\pi_{ij} = u^2 (\mathcal{L}_\gamma X)_{ij}$.

a. Compute the constraints map $\Phi(g, \pi) = (R(g) - |\pi|_g^2 + \frac{1}{2}(\mathrm{tr}_g \pi)^2, \mathrm{div}_g \pi)$, and in case $\gamma = g_{\mathbb{E}^3}$, show that the vacuum constraints $\Phi(g, \pi) = (0, 0)$ can be written, in a Cartesian coordinate system for the background $g_{\mathbb{E}^3}$, as follows (subscripts on operators at the flat metric are omitted):

$$8\Delta u = u\left(-|\mathcal{L}X|^2 + \tfrac{1}{2}(\operatorname{tr}(\mathcal{L}X))^2\right)$$

$$\Delta X^i + 4u^{-1}u_{,j}(\mathcal{L}X)^j_i - 2u^{-1}u_{,i}\operatorname{tr}(\mathcal{L}X) = 0.$$

b. Suppose the equations in part a. hold on an asymptotic end of an asymptotically flat manifold (M, g), where u and X have expansions

$$u(x) = 1 + A/|x| + O_2(|x|^{-2}), \quad X^i(x) = B^i/|x| + O_2(|x|^{-2}).$$

Show that

$$\pi_{ij} = -\frac{B^i x^j + B^j x^i}{|x|^3} + \sum_{k=1}^{3} \frac{B^k x^k}{|x|^3}\delta_{ij} + O(|x|^{-3})$$

and that $P^i = -\tfrac{1}{2}B^i$ is the ADM linear momentum.

c. Justify the expansions of u and X^i above, assuming $(u-1)$ and X^i are in $W^{2,p}_{-q}$ for some $p > n$ and $q > \frac{n-2}{2}$. To do this, use Proposition 7-72 to get initial expansions of u and X^i. To obtain the improved asymptotics of u, show you can subtract off a function w defined near infinity, with $w(x) = O(|x|^{-2})$ and $\Delta(u - w) = O(|x|^{-4-\gamma})$ for some $\gamma > 0$. Conclude from here using weighted Schauder estimates as in Proposition 7-63; cf. [157, Theorems 1 and 2]. See [78, Proposition 24] for a solution if you get stuck; as you you will see there, the case $n > 3$ follows more readily, whereas the $n = 3$ requires the argument outlined here. The claim for X^i follows analogously.

d. Show in manner similar to part c. that the odd parts of u and X admit expansions

$$u^{\text{odd}}(x) = \sum_{k=1}^{3} \frac{\beta^k x^k}{|x|^3} + O_2(|x|^{-3}),$$

$$(X^i)^{\text{odd}}(x) = \sum_{k=1}^{3} \frac{d^k_{(i)} x^k}{|x|^3} + O_2(|x|^{-3}).$$

e. We saw in (7-39a) and (7.2.5) how to relate β from part d. to the center of mass. Show the components of the ADM angular momentum can be expressed as linear combinations of components of $d_{(i)}$: $J^{23} = \mathcal{J}_1 = \tfrac{1}{2}(d^3_{(2)} - d^2_{(3)})$, $J^{31} = \mathcal{J}_2 = \tfrac{1}{2}(d^1_{(3)} - d^3_{(1)})$ and $J^{12} = \mathcal{J}_3 = \tfrac{1}{2}(d^2_{(1)} - d^1_{(2)})$. Use this along with an expansion of $X^i(y - a)$ to show that under translation of asymptotically flat coordinates $y = x + a$, the angular momentum changes by $\mathcal{J}_i \mapsto \mathcal{J}_i + (a \times P)_i$.

Exercise 7-107 (DEC and PMT). This example from [130] illustrates a difference between the dominant energy condition and the weak energy condition (recall Section 2.3.2); at the level of the constraints operator on initial data sets, we are comparing the condition $\rho \geq |J|_g$ versus $\rho \geq 0$.

Consider an initial data set $(\mathbb{R}^3, \mathring{g}, \pi)$, where $\mathring{g}$ is the Euclidean metric, and where the momentum tensor π is pure-trace, $\pi = \frac{p}{3}\mathring{g}$, where $\operatorname{tr}_{\mathring{g}}\pi = p$ is a chosen (smooth) function on $\mathbb{R}^3$. As usual we let $(2\rho, J) = \Phi(g, \pi)$ be the constraints operator.

a. Show that $\rho \geq 0$ for such an initial data set.

b. The ADM energy E of this initial data set vanishes. Give an example of p for which $\rho \in L^1(\mathbb{R}^3)$, and the ADM linear momentum P does not vanish; thus the conclusion of the positive mass theorem does not hold for such an initial data set.

c. Show that if the DEC holds, and if p is compactly supported, then p is identically zero. To do this, translate the DEC into a differential inequality in the radial direction. Derive the same conclusion if $\rho \in L^1(\mathbb{R}^3)$.

Exercise 7-108. This exercise refers to the proof of Proposition 7-75.

a. Finish the proof of Proposition 7-75 by establishing the expansion. Recall that $q + \tau + 2 > n$, and outside a compact set, we have

$$R((1+v)^{\frac{4}{n-2}} g) = R(g) + S, \quad \text{with } v \in W^{k,p}_{-\tau}, \ S \in W^{0,p'}_{-\delta'}.$$

Show that $\Delta_g v \in W^{0,s}_{-\beta}$ for some $s > \frac{n}{2}$ and $\beta > n$. (Note that $\frac{np}{n-(k-2)p} > \frac{n}{2}$ if $(k-2)p \leq n$ and $p > \frac{n}{k}$.)

b. In the proof of Proposition 7-75, we argue that a solution of $L^*_g f = 0$ must be harmonic, and thus in a suitable weighted space it must be trivial. More generally, suppose $f \in W^{0,r}_{\tau+2-n}$ ($r > 1, 0 < \tau < n-2$) solves a Hessian equation $\operatorname{Hess}_g f = f\xi$, where $\xi \in O_{k-2}(|x|^{-q-2})$, with $\tau < q$. Bootstrap the regularity and decay of f as in the proof of Proposition 7-70, but utilize the Hessian equation to show that for any θ, $f \in W^{k,r}_{-\theta}$. Thus if $kr > n$ (e.g., $k > n$), then f vanishes to all orders at infinity.

Exercise 7-109. Let $L = DR_{\mathring{g}_{\mathbb{E}^n}}$. Furnish another proof of Proposition 7-73, in the case $\tau \in (0, n-2) \setminus \{1, 2\}$, by splitting the function space of symmetric tensors as $W^{k,p}_{-\tau}(\mathbb{R}^n) = L^*(W^{k+2,p}_{-\tau+2}(\mathbb{R}^n)) \oplus \{h_0 \in W^{k,p}_{-\tau}(\mathbb{R}^n) : Lh_0 = 0\}$. Use the fact that $LL^* = (n-1)\Delta^2$. To show the image of L^* is closed, one might use a "scale-broken" estimate for the Laplacian (see [15, Theorem 1.10] and the comments after Proposition 7-61 above) to get an injectivity estimate for L^* on a space transverse to $W^{k+2,p}_{-\tau+2}(\mathbb{R}^n) \cap \ker L^* \subset \operatorname{span}\{1, x^1, \ldots, x^n\}$.

Exercise 7-110 (solving with localized support: a linear model problem). In this exercise, we present a model problem from [186], and recounted in [68] for understanding the mechanism of the proof of Theorem 7-84. Suppose f is smooth with compact support on a ball B in $\mathbb{R}^n$. Our goal is to solve $\operatorname{div} X = f$, so that X extends by zero in C^k across ∂B. A necessary condition is that $\int_B f \, dx = 0$, which you should interpret thus: f is orthogonal to the kernel of the adjoint of div. For a weight function ρ as considered on page 294, let $\mathcal{G}(u) = \int_B \left(\frac{1}{2} |\nabla u|^2 \rho - uf \right) dx$. Let $0 \le \zeta \le 1$ be a nontrivial smooth bump function of compact support in B.

a. Show that the restriction of $\mathcal{G}$ to the set $\left\{ u \in H_\rho^1(B) : \int_B u\zeta dx = 0 \right\}$ has a minimizer in this set. Is it unique? To do this, you should establish a suitable H_ρ^1-estimate, a weighted Poincaré inequality to play the role of the weighted injectivity estmate (7.5.3).

b. Using the minimizer u from part a., show that $X = -\rho \nabla u$ solves $\operatorname{div}(X) = f + \lambda \zeta$ for some $\lambda \in \mathbb{R}$. Argue that $\lambda = 0$. Write the resulting PDE as a second-order linear PDE in u.

c. Let $\rho = (d(\cdot, \partial B))^N$ near ∂B, for N to be chosen sufficiently large. Use interior Schauder estimates to get the required pointwise control on X. In particular, if $x \in B$ with $d(x, \partial B) = d$ is such that $B_{2d/3}(x)$ is outside the support of f, apply the interior Schauder estimate for $B_{d/3}(x) \subset B_{2d/3}(x)$.

Exercise 7-111. We thank Xin Zhou for posing the question addressed in this exercise.

a. Show that in the setting of Theorem 7-89, one can carry out the argument to achieve $\bar{g} = g$ for $|x| \ge 2\theta$, and $\bar{g}$ is Schwarzschild for $|x| \le \theta$ (and $x \ne c$).

b. Consider the metric g on $\mathbb{R}^n \setminus \{|x| \le 1\}$ given by

$$g(x) = \left(1 + \frac{x^k x^\ell}{|x|^{n+2}} \right)^{\frac{4}{n-2}} g_{\mathbb{E}^n},$$

for any $k \ne \ell$. Show that this contains an asymptotically flat end, of zero scalar curvature, with vanishing mass and center of mass.

c. The metric in part b. is exactly parity-symmetric in x, i.e., the map $x \mapsto -x$ is an isometry of g. Argue that the gluing construction in Theorem 7-89 can be achieved in this case. Observe that if the construction of part a. is carried out, the mass m of the attached Schwarzschild must be negative.

CHAPTER 8

On the center of mass and constant mean curvature surfaces of asymptotically flat initial data sets

8.1. Introduction

Many deep results in mathematical general relativity concern the interplay between globally conserved quantities and the geometric structure of initial data sets. Examples include the minimal surface approach by R. Schoen and S.-T. Yau [199; 203] and the spinor method by E. Witten [224] in the proof of the Riemannian positive mass theorem; the inverse mean curvature flow by G. Huisken and T. Ilmanen [125] and the conformal flow by H. Bray [26] in the proof of the Penrose inequality; and the constant mean curvature foliation by G. Huisken and S.-T. Yau [126] (cf. R. Ye [225]) in establishing a geometric notion of center of mass.

In a broad sense, this chapter is intended to introduce some aspects of the connections between globally conserved physical quantities, such as the center of mass and angular momentum, and the geometric structure of the manifold, using analysis of the scalar curvature, or more generally the full constraint equations derived from the spacetime Einstein equation. The chapter focuses on constant mean curvature foliations and the geometric center of mass of asymptotically flat initial data sets. This research program was initiated by Huisken and Yau in 1996 and has drawn great interest in recent years; see, for example, [31; 32; 33; 79; 80; 81; 117; 156; 166; 184; 225]. This chapter begins with a partial survey of the classical results of constant mean curvature surfaces and introduces the now standard concept of stability. We then discuss some recent progress on the constant mean curvature surfaces in asymptotically flat initial data sets and the geometric center of mass. In the last part, we adopt a more analytic approach to

The author, Lan-Hsuan Huang, was partially supported by NSF through grants DMS-1308837 and DMS-1452477. This chapter is based upon two mini-courses presented in the 2012 Summer School on Mathematical General Relativity at MSRI and the 2013 Summer School on Mathematical General Relativity in Cortona, Italy. The author is very grateful to the organizers Justin Corvino and Pengzi Miao for their warm hospitality, which made the summer schools memorable. Sincere appreciation goes to Justin Corvino for valuable comments.

study the center of mass and angular momentum from the Einstein constraint equations.

A *spacetime* is an $(n + 1)$-dimensional smooth manifold equipped with a Lorentzian metric g of signature $(-+\cdots+)$. The Einstein equation is the tensor equation (in appropriate units)

$$\mathrm{Ric}(g) - \tfrac{1}{2} R(g) g = T,$$

where the energy-momentum tensor T represents the energy-momentum density of matter. A spacetime is called *vacuum* if it satisfies the Einstein equation with $T = 0$. The prototype vacuum spacetime is Minkowski space $\mathbb{R}^{n+1}$ equipped with the Minkowski metric $g = -(dx^0)^2 + (dx^1)^2 + \cdots + (dx^n)^2$. For a general energy-momentum tensor T, we assume the *dominant energy condition*, which is known to hold for physically reasonable matter fields. When expressed in terms of local coordinates, the left hand side of the Einstein equation forms a second-order system in the metric components $g_{\alpha\beta}$. The seminal work of Choquet-Bruhat [49] proved that the left-hand side of the Einstein equation can be expressed as a nonlinear hyperbolic operator by using the so-called wave coordinates. Finding a spacetime that satisfies the Einstein equation can then be viewed as the evolution problem for a given initial data set. Thus, it is important to understand the physical and geometric structure of initial data sets.

An *initial data set* for the Einstein equation is a triple (M, g, k), where (M, g) is an n-dimensional Riemannian manifold and k is a symmetric $(0, 2)$ tensor on M. The Gauss–Codazzi equations for submanifolds, along with the Einstein equation, imply that if M is a submanifold in a spacetime with the induced metric g and the induced second fundamental form k, then (M, g, k) must satisfy the constraint equations

$$R(g) - |k|_g^2 + (\mathrm{tr}_g k)^2 = 2\mu \quad \text{and} \quad \mathrm{div}_g k - d(\mathrm{tr}_g k) = J,$$

where μ is the energy density and J is the momentum density. More specifically, let T be the energy-momentum tensor and let ν be the future-directed timelike normal to M. We define $\mu := T(\nu, \nu)$ and $J := T(\nu, \cdot)$. The dominant energy condition on the tensor T reduces to the inequality $\mu \geq |J|_g$ at each point of M. When $k \equiv 0$, (M, g) is called a *time-symmetric* (or *Riemannian*) initial data set. It is simple to see that in the time-symmetric case the system of constraint equations becomes a single equation $R(g) = 2\mu$. Thus the dominant energy condition coincides with the condition that the scalar curvature of g is nonnegative, which is a condition that naturally appears in Riemannian geometry. (However, for general k the dominant energy condition involves a system of equations and is more complicated.)

One family of commonly studied models of isolated gravitational systems is the set of asymptotically flat initial data sets. We say that an initial data set (M, g, k) is *asymptotically flat* (with one end) if there is a compact set $K \subset M$ and a coordinate diffeomorphism $x : M \setminus K \to \mathbb{R}^n \setminus B$ for some closed ball $B \subset \mathbb{R}^n$ such that, for $i, j = 1, 2, \ldots, n$,

$$g_{ij}(x) - \delta_{ij} = O(|x|^{-q}), \quad k_{ij}(x) = O(|x|^{-1-q}),$$

and such that

$$\mu(x) = O(|x|^{-n-q_0}), \quad J_i(x) = O(|x|^{-n-q_0}),$$

where $q > \frac{n-2}{2}$ and $q_0 > 0$. Here the expression $f(x) = O(|x|^{-q})$ stands for a function satisfying $|f(x)| \le C|x|^{-q}$ for a constant C depending only on g, k. When a subscript k appears in the expression $f = O_k(|x|^{-q})$, it indicates additional fall-off rates on the derivatives $|\partial^I f(x)| \le C|x|^{-q-|I|}$ for $|I| = 0, 1, \ldots, k$, but in this chapter we often omit the subscript k and avoid the discussion about the optimal assumption on regularity.

Note that by definition an asymptotically flat initial data set has trivial topology outside a compact set, but it was shown by J. Isenberg, R. Mazzeo, and D. Pollack [129] that there are no topological obstructions within the compact set.

It is known that asymptotically flat initial data sets possess globally conserved physical quantities. In 1962, R. Arnowitt, S. Deser, and C. W. Misner [10] proposed the definitions of the (total) *energy* E and the *linear momentum* P of an asymptotically flat initial data set (M, g, k) as follows, for $i = 1, 2, \ldots, n$:

$$E = \frac{1}{2(n-1)\omega_{n-1}} \lim_{r \to \infty} \int_{\{|x|=r\}} \sum_{j,k=1}^{n} \left(\frac{\partial g_{jk}}{\partial x^k} - \frac{\partial g_{kk}}{\partial x^j} \right) v_0^j \, d\mathcal{H}_0^{n-1},$$

$$P_i = \frac{1}{(n-1)\omega_{n-1}} \lim_{r \to \infty} \int_{\{|x|=r\}} \sum_{j=1}^{n} \pi_{ij} v_0^j \, d\mathcal{H}_0^{n-1}.$$

Here, the integrals are computed in the coordinate chart $M \setminus K \cong_x \mathbb{R}^n \setminus B$; $v_0^j = x^j/|x|$; $\pi_{ij} = k_{ij} - (\mathrm{tr}_g k)g_{ij}$; $\mathcal{H}_0^{n-1}$ is the $(n-1)$-dimensional Euclidean Hausdorff measure; and ω_{n-1} is the volume of the standard unit sphere in $\mathbb{R}^n$. The work of R. Bartnik [15] and P. T. Chruściel [57] showed that the scalar E and the vector $(P_1, \ldots, P_n)$ are geometric invariants. The celebrated positive mass conjecture asserts that $E \ge |P|$ [199; 202; 203; 78], so that in particular $E \ge 0$ and as well we can define the mass $m = \sqrt{E^2 - |P|^2}$. In the time-symmetric case, we may unambiguously use the ADM mass m to denote the energy E, since $|P| = 0$.

There are also the notions of center of mass and angular momentum for an asymptotically flat initial data set. T. Regge and C. Teitelboim [187] and R. Beig and N. Ó Murchadha [18] proposed the following definitions of the center of mass $\mathcal{C}_{\mathrm{BORT}}$ and the angular momentum $\mathcal{J}$ (if $E \neq 0$), for $k, \ell = 1, 2, \ldots, n$:

$$C_\ell = \frac{1}{2(n-1)E\omega_{n-1}} \lim_{r\to\infty} \int_{\{|x|=r\}} \left(x^\ell \sum_{i,j=1}^{n} \left(\frac{\partial g_{ij}}{\partial x^i} - \frac{\partial g_{ii}}{\partial x^j} \right) v_0^j - \sum_{i=1}^{n} (g_{i\ell} v_0^i - g_{ii} v_0^\ell) \right) d\mathcal{H}_0^{n-1}, \tag{8.1.1}$$

$$\mathcal{J}_{(k\ell)} = \frac{1}{(n-1)E\omega_{n-1}} \lim_{r\to\infty} \int_{\{|x|=r\}} \sum_{i,j=1}^{n} \pi_{ij} Y_{(k\ell)}^i v_0^j \, d\mathcal{H}_0^{n-1}, \tag{8.1.2}$$

where $Y_{(k\ell)} = x^k \, \partial/\partial x^\ell - x^\ell \, \partial/\partial x^k$ are the Euclidean rotational vector fields.[7]

To distinguish the above definitions from other notions of center of mass and angular momentum (e.g. [126; 48]), we refer to the integrals (8.1.1) and (8.1.2) as the BORT center of mass and the ADM angular momentum, respectively.

In contrast to the ADM energy-momentum, the integrals of $\mathcal{C}_{\mathrm{BORT}}$ and $\mathcal{J}$ are less well understood and may not even converge in general. In fact, explicit examples of asymptotically flat initial data sets such that the integrals diverge have been constructed [18; 43; 47; 46; 118]. Nevertheless, the author shows that if one assumes the following Regge–Teitelboim conditions, then (8.1.1) and (8.1.2) converge and transform correctly with respect to different coordinate charts [116].

An initial data set (M, g, k) is said to satisfy the *Regge–Teitelboim conditions* if it is asymptotically flat and, in the coordinate chart $M \setminus K \cong_x \mathbb{R}^n \setminus B$,

$$g_{ij}(x) - g_{ij}(-x) = O(|x|^{-1-q}), \quad k_{ij}(x) + k_{ij}(-x) = O(|x|^{-2-q}),$$

and

$$\mu(x) - \mu(-x) = O(|x|^{-n-q_0-1}), \quad J_i(x) - J_i(-x) = O(|x|^{-n-q_0-1}).$$

Example 8-1 (three-dimensional Schwarzschild manifolds). A fundamental example in general relativity is the Schwarzschild spacetime, which describes the exterior gravitational field of a static, spherically symmetric body. The totally geodesic time-slice outside the apparent horizon of the Schwarzschild spacetime of mass $m > 0$ can be expressed as a Riemannian manifold $M = (2m, \infty) \times \mathbb{S}^2$ endowed with the metric

$$(1 - 2ms^{-1})^{-1} ds^2 + s^2 g_{\mathbb{S}^2},$$

---

[7] In the literature (see [69], for example), the BORT center of mass and angular momentum are sometimes defined as $E\mathcal{C}_\ell$ and $E\mathcal{J}_{(k\ell)}$, respectively.

where $g_{\mathbb{S}^2}$ is the round metric on the unit sphere. One can readily check that M is the manifold interior of an asymptotically flat initial data set with a minimal boundary and one end, and it has zero scalar curvature. Mathematically one can extend M to a complete asymptotically flat initial data set of zero scalar curvature by "doubling" M across its minimal boundary. The complete two-ended asymptotically flat initial data set can be expressed as a conformally flat metric $(\mathbb{R}^3 \setminus \{a\}, g_{m,a})$, where $g_{m,a} = u^4 g_{\mathbb{E}}$ and

$$u(x) = 1 + \frac{m}{2|x - a|},$$

where $g_{\mathbb{E}}$ is the Euclidean metric. We generally suppress "a" from the notation and write $g_m = g_{m,a}$. One computes directly that m is the ADM energy and a is the BORT center of mass. The asymptotic expansion of g_m for $|x|$ large is

$$g_m = \left(1 + \frac{2m}{|x|} + \frac{2ma \cdot x}{|x|^3} + \frac{3m^2}{2|x|^2} + O(|x|^{-3})\right) g_{\mathbb{E}}.$$

It follows that m appears in the $|x|^{-1}$-term of the expansion and the BORT center of mass appears in the odd part of the $O(|x|^{-2})$-term. This demonstrates that appropriate assumptions need to be imposed on the leading-order terms of the data in order for the integrals (8.1.1) and (8.1.2) to converge. It explains the motivation behind the definition of the Regge–Teitelboim conditions. We also note that the BORT center of mass of a Schwarzschild manifold is not a point of the manifold. $\qquad\square$

This chapter is organized as follows. Section 8.2 is devoted to the classical Alexandrov theorem about embedded constant mean curvature surfaces in Euclidean space. In Section 8.3 we introduce variational formulas and stability of constant mean curvature surfaces, and then discuss the classical result of Barbosa and do Carmo about uniqueness of stable constant mean curvature surfaces in Euclidean space. In Section 8.4 we show existence of constant mean curvature surfaces in asymptotically flat initial data sets that are asymptotic to Schwarzschild. We also prove that the geometric center of mass, to be defined in (8.4.8), coincides with the BORT center of mass. Section 8.5 presents methods to analyze the spectrum of the stability operator and show that the constant mean curvature surfaces constructed in the previous section are stable and form a smooth foliation. In Section 8.6, we discuss density results for the Einstein constraint equations and an application to arbitrarily specifying the BORT center of mass and the ADM angular momentum.

8.2. Uniqueness of embedded CMC surfaces

A fundamental problem in differential geometry is to characterize the constant mean curvature hypersurfaces in a Riemannian manifold. A classical result due to Alexandrov asserts that the only embedded and closed constant mean curvature surfaces in Euclidean space are the round spheres. The original proof of Alexandrov is based on the arguments which came to be known as the method of moving planes. We instead present another proof due to S. Montiel and A. Ros [164, Section 6.4].

Let (M, g) be an orientable Riemannian manifold, and let $\Sigma^n \subset M^{n+1}$ be an immersed *two-sided* hypersurface, i.e., there exists a globally defined smooth unit normal vector field ν along Σ. We let $d\mu$ be the induced hypersurface measure. The *mean curvature* of Σ with respect to ν is defined by $H := \operatorname{div}_\Sigma \nu$. The mean curvature detects how the (extrinsic) normal vector varies along Σ. According to our convention, the mean curvature of a Euclidean n-sphere is n with respect to the *outward* unit normal vector. An immersed submanifold is said to be *closed* if it is compact and has no boundary, and is said to be *embedded* if in addition it has no self-intersection. We first review some basic integral formulas involving mean curvature for closed hypersurfaces.

A *conformal vector field* X is a vector field on M that satisfies

$$L_X g = 2fg,$$

for some function $f : M \to \mathbb{R}$, where L_X is the Lie derivative.

Example 8-2 (see [31]). Consider the n-dimensional Schwarzschild metric with ADM mass $m > 0$ of the form $g_m = (1 - 2ms^{2-n})^{-1}ds^2 + s^2 g_{\mathbb{S}^{n-1}}$ on $(s_0, \infty) \times \mathbb{S}^{n-1}$, where $s_0 = (2m)^{1/(n-2)}$. We change variables and set

$$r(s) = \int_{s_0}^{s} (1 - 2m\tau^{2-n})^{-\frac{1}{2}}\, d\tau.$$

Define $h(r) = s(r)$. We then rewrite the Schwarzschild metric in the form $g_m = dr^2 + h^2(r)g_{\mathbb{S}^{n-1}}$. Define the vector field $X = h(r)\, \partial/\partial r$. By direct computation,

$$\begin{aligned}
L_X g_m &= L_X(dr) \otimes dr + dr \otimes L_X(dr) + X(h^2)g_{\mathbb{S}^{n-1}} \\
&= 2h'(r)dr \otimes dr + 2(h(r))^2 h'(r)g_{\mathbb{S}^{n-1}} \\
&= 2h'(r)g_m.
\end{aligned}$$

Thus X is a conformal vector field that satisfies $L_X g_m = 2f g_m$, where

$$f(r) = h'(r) = \left(\frac{dr}{ds}\right)^{-1} = (1 - 2ms^{2-n})^{1/2}.$$

Also note that f satisfies the *static potential equation*

$$(\Delta_{g_m} f) g_m - \text{Hess}_{g_m} f + f \, \text{Ric}_{g_m} = 0. \qquad \square$$

In what follows, $d\mu$ will generally denote the induced surface measure on $\Sigma \subset (M, g)$.

Theorem 8-3 (generalized Minkowski integral formula [31, Proposition 2.3]). *Suppose that (M^{n+1}, g) has a conformal vector field X such that $L_X g = 2fg$ for some function f. Let Σ^n be a closed two-sided hypersurface in M, and let H be the mean curvature of Σ with respect to the unit normal vector v. Then*

$$\int_\Sigma (nf - Hg(X, v)) \, d\mu = 0.$$

Proof. Recall that the Lie derivative $L_X g$ in a local frame $\{e_1, \ldots, e_{n+1}\}$ has the expression

$$(L_X g)(e_i, e_j) = g(\nabla_{e_i} X, e_j) + g(e_i, \nabla_{e_j} X)$$

for $i, j = 1, 2, \ldots, n+1$. We decompose the conformal vector field along Σ into $X = X' + g(X, v)v$, where $X' \in T\Sigma$. Suppose further that $\{e_1, \ldots, e_n\}$ is a local orthonormal frame on Σ, and let ∇ be the covariant derivative of g. Then, at each point of Σ,

$$\text{div}_\Sigma X' + Hg(X, v) = \text{div}_\Sigma X' + g(X, v) \sum_{i=1}^n g(\nabla_{e_i} v, e_i) = \sum_{i=1}^n g(\nabla_{e_i} X, e_i)$$

$$= \frac{1}{2} \sum_{i=1}^n (L_X g)(e_i, e_i) = f \sum_{i=1}^n g(e_i, e_i) = nf.$$

Integrating on Σ and applying the divergence theorem yields the desired integral formula. $\qquad \square$

We finish this section with two theorems on closed embedded hypersurfaces Σ^n in $\mathbb{R}^{n+1}$. We recall that by the Jordan–Brouwer separation theorem, any such $\Sigma = \partial\Omega$, where Ω is a bounded open set; in particular, then, Σ is two-sided.

Theorem 8-4 (The Heintze–Karcher inequality; see [31, Theorem 3.5]). *Let Σ^n be a closed, embedded hypersurface in $\mathbb{R}^{n+1}$, with $\Sigma = \partial\Omega$, where Ω is a bounded region in $\mathbb{R}^{n+1}$ with volume $\text{Vol}\,\Omega$. Suppose that the mean curvature H is positive with respect to the outward unit normal. Then*

$$n \int_\Sigma \frac{1}{H} \, d\mu \ge (n+1) \text{Vol}\,\Omega,$$

with equality if and only if Σ is a round sphere.

Proof. Consider a deformation $F : \Sigma \times [0, \infty) \to \mathbb{R}^{n+1}$ given by

$$F(x, t) = x - t\nu(x),$$

where ν is the outward unit normal on Σ. Let $\Sigma_t := F(\Sigma, t)$, with surface measure $d\mu$ (suppressing the dependence on t). Let $d_\Sigma(p)$ denote the distance of $p \in \mathbb{R}^{n+1}$ to Σ. For t sufficiently small, $\Sigma_t = d_\Sigma^{-1}(t) \cap \overline{\Omega}$ is smooth, but Σ_t may begin to have self-intersection for some t. Hence, instead of working on Σ_t, we consider

$$\Sigma_t^* = \{F(x, t) : d_\Sigma(F(x, t + \delta)) = t + \delta \text{ for some } \delta > 0\}.$$

This is a smooth hypersurface contained in Σ_t. Since $\frac{\partial F}{\partial t} = -\nu$, the variational formulas (8.3.1) and (8.3.2) yield

$$\frac{\partial}{\partial t} d\mu = -H\, d\mu, \qquad \frac{\partial H}{\partial t} = |A|^2 \geq \frac{H^2}{n}, \qquad \frac{\partial H^{-1}}{\partial t} = -H^{-2}|A|^2 \leq -\frac{1}{n}.$$

Define $Q(t) := n \int_{\Sigma_t^*} H^{-1}\, d\mu$. Then

$$Q'(t) = n \int_{\Sigma_t^*} (-H^{-2}|A|^2 + H^{-1}(-H))\, d\mu$$

$$\leq n \int_{\Sigma_t^*} \left(-\tfrac{1}{n} - 1\right) d\mu = -(n+1) \int_{\Sigma_t^*} d\mu.$$

Thus, for $\tau \in (0, \infty)$, we have,

$$Q(0) - Q(\tau) = -\int_0^\tau Q'(t)\, dt \geq (n+1) \int_0^\tau \int_{\Sigma_t^*} d\mu\, dt = (n+1) \int_{\{x \in \overline{\Omega} : d_\Sigma(x) \leq \tau\}} dx.$$

Since $Q(\tau) \geq 0$, we obtain the desired inequality by letting $\tau \to \infty$. It is straightforward to verify that Σ is umbilic if the equality holds. $\qquad\square$

Theorem 8-5 (Alexandrov's theorem). *Let Σ^n be a closed, embedded, connected hypersurface in $\mathbb{R}^{n+1}$ with constant mean curvature. Then Σ is a round sphere.*

Remark 8-6. The theorem fails without the assumption that Σ is embedded. H. Wente [222] produced an immersed torus of constant mean curvature in $\mathbb{R}^3$. Immersed surfaces of higher genus have been constructed by N. Kapouleas [133].

Proof. The position vector field $X = (x^1, \ldots, x^{n+1})$ in $\mathbb{R}^{n+1}$ is a conformal vector field (with $f = 1$). By the Minkowski integral formula (Theorem 8-3) and the divergence theorem, with $\Sigma = \partial\Omega$,

$$\frac{1}{H} \int_\Sigma n\, d\mu = \int_\Sigma \langle X, \nu \rangle\, d\mu = \int_\Omega \operatorname{div}_{\mathbb{R}^{n+1}} X\, dx = (n+1)\operatorname{Vol}\Omega.$$

Thus, we obtain equality in the Heintze–Karcher inequality, which implies that Σ is a round sphere. $\qquad\square$

This theorem has been generalized to a large class of warped manifolds, including the Schwarzschild manifolds, by Brendle [31, Theorem 1.1].

8.3. Stable CMC surfaces

8.3.1. *Variational formulas.* Let Σ^n be a smooth closed two-sided hypersurface in (M^{n+1}, g). We are interested in how some geometric quantities on Σ, such as mean curvature, surface area, and enclosed volume, change under deformations of Σ. The relevant formulas are called the *variational formulas*. Consider a deformation of Σ along its normal direction $F : \Sigma \times (-\epsilon, \epsilon) \to M$ satisfying

$$\frac{\partial}{\partial t} F(x, t) = \eta(x, t) v(x, t),$$
$$F(x, 0) = x,$$

where $v(x, t)$ is a unit normal to $\Sigma_t := F(\Sigma, t)$. Define $F_t(x) = F(x, t)$. We further suppose that $F_t : \Sigma \to M$ is an immersion. By direct computation, the first variation formula says

$$\frac{d}{dt}\bigg|_{t=0} \mathcal{H}^n(\Sigma_t) = \int_\Sigma H \eta \, d\mu, \qquad (8.3.1)$$

where $H = \mathrm{div}_\Sigma v$. If one allows arbitrary deformations, then we can strictly decrease the volume of Σ by deforming Σ along $-Hv$, unless Σ is a minimal hypersurface ($H \equiv 0$). Thus, minimal hypersurfaces are the critical points of the area functional.

The second variation formula at a minimal hypersurface says

$$\frac{d^2}{dt^2}\bigg|_{t=0} \mathcal{H}^n(\Sigma_t) = \int_\Sigma \left(|\nabla^\Sigma \eta|^2 - (|A|^2 + \mathrm{Ric}(v, v))\eta^2 \right) d\mu, \qquad (8.3.2)$$

where A is the second fundamental form of Σ and Ric is the Ricci tensor of (M, g). Now define the stability operator

$$L_\Sigma := -\Delta_\Sigma - (|A|^2 + \mathrm{Ric}(v, v)).$$

A minimal hypersurface is said to be *stable* if $\int_\Sigma \eta L_\Sigma \eta \, d\mu \geq 0$ for all smooth functions η.

If we restrict our attention to a smaller class of deformations on Σ, hypersurfaces of constant mean curvature $H \neq 0$ can appear as the critical points of the functional $\mathcal{H}^n(\Sigma_t)$. Consider the $(n+1)$-dimensional *signed* volume $V(t)$

between Σ_t and Σ. The volume function satisfies the following variational formula.

Proposition 8-7. *Let Σ^n be a smooth closed two-sided hypersurface in (M^{n+1}, g). Let $F : \Sigma \times (-\epsilon, \epsilon) \to M$ satisfy*

$$\frac{\partial}{\partial t} F(x, t) = \eta(x, t) v(x, t),$$

$$F(x, 0) = x,$$

where $v(x, t)$ is a unit normal to $\Sigma_t := F(\Sigma, t)$. Define the volume function by

$$V(t) = \int_{\Sigma \times [0,t]} F^* d\mathrm{vol}_M.$$

Then

$$\frac{d}{dt}\bigg|_{t=0} V(t) = \int_\Sigma \eta \, d\mu.$$

Proof. Let $p \in \Sigma$. Let $\{v, e_1, \ldots, e_n\}$ be a local oriented orthonormal frame about $F(p, 0)$. Then $F^* d\mathrm{vol}_M = a(t, p) \, dt \wedge d\mu$, where

$$a(t, p) = F^* d\mathrm{vol}_M \left(\frac{\partial}{\partial t}, e_1, \ldots, e_n \right) = d\mathrm{vol}_M \left(\frac{\partial F}{\partial t}, dF_t(e_1), \ldots, dF_t(e_n) \right),$$

$$a(0, p) = g\left(\frac{\partial F}{\partial t}\bigg|_{(p,0)}, v(p, 0) \right) = \eta(p, 0).$$

It follows that

$$\frac{d}{dt}\bigg|_{t=0} V(t) = \int_\Sigma a(0, p) d\mu = \int_\Sigma \eta \, d\mu. \qquad \square$$

A variation such that $V(t) = V(0)$ for all $t \in (-\epsilon, \epsilon)$ is called *volume-preserving*. Proposition 8-7 shows that if a variation satisfies $\int_\Sigma \eta \, d\mu = 0$, then it preserves the volume between Σ_t and Σ "infinitesimally" at $t = 0$. Conversely, any smooth function $\eta(x)$ on Σ such that $\int_\Sigma \eta \, d\mu = 0$ gives rise to a volume-preserving variation [13; 14].

One can readily see that for volume-preserving variations, the hypersurfaces of constant mean curvature are the critical points of the area functional. The second variation formula for volume-preserving variations at hypersurfaces of constant mean curvature becomes

$$\frac{d^2}{dt^2}\bigg|_{t=0} \mathcal{H}^n(\Sigma_t) = \int_\Sigma \eta L_\Sigma \eta \, d\mu + H \frac{d^2}{dt^2}\bigg|_{t=0} V(t) = \int_\Sigma \eta L_\Sigma \eta \, d\mu.$$

Therefore, a hypersurface Σ of constant mean curvature is said to be *stable* if and only if

$$\int_\Sigma \eta L_\Sigma \eta \, d\mu \geq 0$$

for all $\eta \in C^\infty(\Sigma)$ such that $\int_\Sigma \eta \, d\mu = 0$. More specifically, define

$$\mu_0 := \inf \left\{ \int_\Sigma \eta L_\Sigma \eta \, d\mu : \eta \in C^\infty(\Sigma), \; \|\eta\|_{L^2(\Sigma)} = 1, \; \int_\Sigma \eta \, d\mu = 0 \right\}. \quad (8.3.3)$$

Then Σ is stable if $\mu_0 \geq 0$. From the discussion above, it follows that stable hypersurfaces are the local minimizers of the area functional $\mathcal{H}^n(\Sigma_t)$ among volume-preserving variations.

Example 8-8. The n-dimensional sphere S_r in $\mathbb{R}^{n+1}$ of radius $r > 0$ centered at the origin is a stable hypersurface of constant mean curvature n/r. The stability operator on S_r is

$$L_0 = -\Delta_0 - (|A|^2 + \mathrm{Ric}(\nu, \nu)) = -\Delta_0 - \frac{n}{r^2},$$

where Δ_0 is the Laplace operator on S_r. Because μ_0 is also an eigenvalue of $-\Delta_0$, by analyzing the eigenvalues of $-\Delta_0$, we obtain $\mu_0 = 0$ with the eigenspace spanned by the coordinate functions $\{x^1, \ldots, x^{n+1}\}$ restricted to S_r. Also note that L_0 is self-adjoint and its cokernel equals its kernel.

Example 8-9. Consider the Schwarzschild manifold $M = (\mathbb{R}^3 \setminus \{\mathcal{C}_{\mathrm{BORT}}\}, g_m)$, where $g_m = u^4 g_\mathbb{E}$ and

$$u(x) = 1 + \frac{m}{2|x - \mathcal{C}_{\mathrm{BORT}}|}.$$

For each $r > 0$, let $S_r = \{\mathbb{R}^3 : |x - \mathcal{C}_{\mathrm{BORT}}| = r\}$, a constant mean curvature sphere homologous to the minimal sphere.

We recall a transformation formula for conformal metrics. Let g_1, g_2 be metrics on an n-dimensional manifold, related by $g_2 = u^{4/(n-2)} g_1$. If ν_1 is a unit normal to a hypersurface with respect to g_1, then $\nu_2 = u^{-2/(n-2)} \nu_1$ is a unit normal with respect to g_2. The corresponding mean curvatures H_1 and H_2 of the hypersurface are related by

$$H_2 = u^{-\frac{2}{n-2}} \left(H_1 + \frac{2(n-1)}{n-2} u^{-1} \nabla_{\nu_1} u \right). \quad (8.3.4)$$

It is not hard to see that umbilicity is preserved under conformal transformation, so the sphere S_r is umbilic in M and has constant mean curvature $(2r - m)/(r^2 u^3)$. This implies that $S_{m/2}$ is a minimal surface and $S_{(2+\sqrt{3})m/2}$ has the largest mean curvature: the mean curvature of S_r is increasing in r for $m/2 < r \leq (2+\sqrt{3})m/2$, and decreasing in r if $r \geq (2+\sqrt{3})m/2$. The stability operator on S_r is given by

$$L_{S_r} = -\Delta_{S_r} - (|A|^2 + \mathrm{Ric}_{g_m}(\nu, \nu)) = -u^{-4}\Delta_0 + \frac{-4r^2 + 8rm - m^2}{2r^4 u^6}, \quad (8.3.5)$$

where Δ_0 is the Laplacian operator of the round sphere of radius r. The smallest eigenvalue of L_{S_r} is

$$\lambda_0 = \frac{-4r^2 + 8rm - m^2}{2r^4 u^6} = -\frac{2}{r^2} + \frac{10m}{r^3} + O(r^{-4}),$$

with the corresponding eigenspace spanned by constant functions. The next eigenvalues are

$$\lambda_1 = \lambda_2 = \lambda_3 = \frac{6m}{r^3 u^6} = \frac{6m}{r^3} + O(r^{-4}),$$

with corresponding eigenspace spanned by the coordinate functions $\{x^1, x^2, x^3\}$ restricted to S_r. Thus, if $m > 0$, S_r is a stable hypersurface of constant mean curvature (with respect to volume-preserving variations). The spheres $\{S_r\}$ form a smooth foliation of constant mean curvature spheres with common center C_{BORT}.

8.3.2. *Uniqueness of stable CMC surfaces.* A classical result of Barbosa and do Carmo [13] characterizes stable hypersurfaces in Euclidean space.

Theorem 8-10 (Barbosa–do Carmo [13]). *The only closed, stable, connected, two-sided hypersurfaces of constant mean curvature in Euclidean space are round spheres.*

Proof. Let Σ^n be a hypersurface in $\mathbb{R}^{n+1}$ of constant mean curvature H that satisfies the assumptions in the theorem. Consider the deformation $F : \Sigma \times (-\epsilon, \epsilon) \to \mathbb{R}^{n+1}$ given by

$$\frac{\partial F}{\partial t} = \eta v,$$

where $\eta = n - H\langle X, v\rangle$ and $X = (x^1, \ldots, x^{n+1})$ is the position vector of Σ. Then $\int_\Sigma \eta \, d\mu = 0$ by the Minkowski integral formula (with $f = 1$). Let $\{e_1, \ldots, e_n\}$ be a local orthonormal frame along Σ. Note that $\nabla_{e_i} X = e_i$, where ∇ is the ambient connection. For any point in Σ, we can choose the frame so that $\nabla^\Sigma_{e_i} e_j = 0$ at the point, at which we find

$$\Delta_\Sigma \langle X, v\rangle = \sum_{i=1}^{n} e_i e_i \langle X, v\rangle$$

$$= \sum_{i=1}^{n} e_i \left(\langle \nabla_{e_i} X, v\rangle + \langle X, \nabla_{e_i} v\rangle \right)$$

$$= \sum_{i=1}^{n} \left(\langle \nabla_{e_i} e_i, v\rangle + \langle e_i, \nabla_{e_i} v\rangle + \langle \nabla_{e_i} X, \nabla_{e_i} v\rangle + \langle X, \nabla_{e_i} \nabla_{e_i} v\rangle \right)$$

$$= \sum_{i=1}^{n} \left(\langle e_i, \nabla_{e_i} v\rangle + \langle X, \nabla_{e_i} \nabla_{e_i} v\rangle \right)$$

$$= H - |A|^2 \langle X, v\rangle, \tag{8.3.6}$$

where in the last step we have used the equality $\sum_{i=1}^{n}\langle e_k, \nabla_{e_i}\nabla_{e_i} v\rangle = 0$, which holds for each k because H is constant.

Let L_Σ be the stability operator on Σ. The computation above implies that with H constant and $\eta = n - H\langle X, v\rangle$,

$$L_\Sigma \eta = -\Delta_\Sigma \eta - |A|^2 \eta = H^2 - n|A|^2.$$

Since Σ is stable, and since H is constant and $\int_\Sigma \eta\, d\mu = 0$, we have

$$
\begin{aligned}
0 \le \int_\Sigma \eta L_\Sigma \eta\, d\mu &= \int_\Sigma (H^2 - n|A|^2)\eta\, d\mu \\
&= -n \int_\Sigma |A|^2 (n - H\langle X, v\rangle)\, d\mu \\
&= -n \int_\Sigma (n|A|^2 - H^2)\, d\mu,
\end{aligned}
$$

where in the last equality we used $\int_\Sigma (H - |A|^2 \langle X, v\rangle)\, d\mu = 0$, which is implied by (8.3.6). Because $n|A|^2 \ge H^2$ with equality if and only if Σ is umbilic, we conclude that Σ is a sphere. $\square$

Remark 8-11. The uniqueness result has been generalized to an ambient Riemannian manifold which is complete, simply connected with constant sectional curvature [14]. More precisely, the only stable closed hypersurfaces of constant mean curvature in a complete simply-connected Riemannian manifold with constant sectional curvature are the geodesic spheres.

8.4. Existence of CMC surfaces in asymptotically flat initial data sets

We have seen in Example 8-9 that the BORT center of mass of a Schwarzschild manifold of positive mass is the common geometric center of the (unique) foliation of the stable constant mean curvature surfaces. In 1996, motivated by the goal of finding a geometric description of the center of mass in general relativity, Huisken and Yau [126] initiated a program to study stable constant mean curvature surfaces in more general asymptotically flat initial data sets.

Throughout this section, we consider three-dimensional asymptotically flat initial data sets; in addition, we impose the Regge–Teitelboim conditions where C_{BORT} is used.

We recall that the three-dimensional Schwarzschild metric of mass m is denoted by $g_m = \left(1 + \frac{m}{2|x|}\right)^4 g_{\mathbb{E}}$. Here we are interested in the exterior region of the manifold, so the metric is valid for all $m \in \mathbb{R}$, and not only for $m > 0$. For most of the results presented here, we focus on an asymptotically flat manifold that is close to some Schwarzschild manifold in the following sense.

Definition 8-12. A three-dimensional asymptotically flat initial data set (M, g) is said to be C^k-asymptotic to Schwarzschild of mass m if there is a compact subset $K \subset M$ and a diffeomorphism $M \setminus K \cong_x \mathbb{R}^3 \setminus B$ for a closed ball $B \subset \mathbb{R}^3$, satisfying

$$\sum_{|I| \le k} |x|^{2+|I|} \left| \partial^I \left(g_{ij}(x) - (g_m)_{ij}(x) \right) \right| \le C$$

for $i, j = 1, 2, 3$ and some constant $C > 0$.

Remark 8-13. The assumptions on k for regularity C^k vary among different results that we discuss below, but we omit the precise assumptions on C^k in their statements.

Theorem 8-14 (Huisken–Yau [126]). *If (M, g) is asymptotic to Schwarzschild of mass $m > 0$, there exists a foliation of stable constant mean curvature surfaces in the exterior region of M.*

The proof by Huisken and Yau consists of two parts. For the existence part, they use the volume-preserving mean curvature flow to evolve a sufficiently round initial surface into a constant mean curvature surface. Next, using the estimates obtained from the flow, they analyze the eigenvalues of the stability operator and show that the constant mean curvature surfaces are stable and form a smooth foliation. We sketch the method of volume-preserving mean curvature flow in Section 8.4.1 and discuss the eigenvalue estimates in Section 8.5. A different approach by Ye [225] uses the inverse function theorem for the existence part, which we discuss in Section 8.4.2.

8.4.1. *Volume-preserving mean curvature flow.* The volume-preserving mean curvature flow is a normalized mean curvature flow. It was first introduced by Huisken in the Euclidean setting [124]. The flow is designed specifically to keep the enclosed volume the same and to decrease the surface area under the flow.

Let (M, g) be asymptotic to Schwarzschild. Denote by $S_r = \{x : |x| = r\}$ the coordinate sphere in M. For each r sufficiently large, we define the volume-preserving mean curvature flow $F_r : \mathbb{S}^2 \times [0, T) \to M$ as follows, for $t \ge 0$, $p \in \mathbb{S}^2$:

$$\frac{\partial}{\partial t} F_r(p, t) = (\overline{H} - H) \nu(p, t),$$

$$F_r(p, 0) = rp \in S_r, \tag{8.4.1}$$

where $\overline{H} = |\Sigma_t|^{-1} \int_{\Sigma_t} H \, d\mu$, $\Sigma_t = F_r(\mathbb{S}^2, t)$, and $|\Sigma_t|$ is the area of Σ_t. By Proposition 8-7, the flow keeps the signed volume between Σ_t and S_r the same.

Furthermore, the first variation formula implies

$$\frac{d}{dt}|\Sigma_t| = -\int_{\Sigma_t} (H - \bar{H})^2 \, d\mu.$$

Thus the area of Σ_t is strictly decreasing unless H is a constant. Therefore, if the flow exists for all time, Σ_t converges to a constant mean curvature surface.

Note that the volume-preserving mean curvature flow (8.4.1) is a quasilinear parabolic system, so it has a unique short-time solution for a smooth initial surface. However, the flow may develop singularities at a finite time. For surfaces in Euclidean space that are uniformly convex, the flow exists for all time and converges to a round sphere [124]. For surfaces in an initial data set that is asymptotic to Schwarzschild, we have the following result.

Theorem 8-15 (Huisken–Yau [126]). *Let (M, g) be asymptotic to Schwarzschild of mass $m > 0$. There exist positive constants r_0 and C, depending only on g, such that for all $r \geq r_0$, the volume-preserving mean curvature flow (8.4.1) has a unique smooth solution for all time. Furthermore, Σ_t converges exponentially fast to an embedded surface Σ of constant mean curvature H, and, for $x \in \Sigma$,*

$$\Big||x| - r\Big| \leq C \quad and \quad \left|H - \frac{2}{r} + \frac{4m}{r^2}\right| \leq Cr^{-2}.$$

The main ingredient of the proof is to show that the solution Σ_t stays in a class of sufficiently round surfaces, and hence it does not develop singularities along the flow.

For a surface Σ in (M, g), let g_Σ be the induced metric, and denote by $\mathring{A} := A - \frac{1}{2}Hg_\Sigma$ the traceless part of the second fundamental form A of Σ. Let r be large and B_0, B_1, B_2 be positive. Define the class $\mathfrak{B}_r(B_0, B_1, B_2)$ of smooth closed surfaces of genus zero in (M, g) by

$$\mathfrak{B}_r(B_0, B_1, B_2) =$$
$$\left\{\Sigma \subset M : |\mathring{A}| < B_1 r^{-3}, \ |\nabla\mathring{A}| \leq B_2 r^{-4}, \ and \ \Big||x| - r\Big| \leq B_0 \ for \ all \ x \in \Sigma\right\}.$$

By carefully choosing the constants B_0, B_1, B_2 (depending on the *a priori* estimates of $|\mathring{A}|$ and $|\nabla\mathring{A}|$ along the flow), one can find r_0 sufficiently large (depending on B_0, B_1, B_2 and g) such that for each $r \geq r_0$ the solution Σ_t to (8.4.1) remains in $\mathfrak{B}_r(B_1, B_2, B_3)$. This then implies long-time existence of the solutions. Details can be found in the original paper [126, Section 3].

We now explain the motivation behind the smallness assumptions on $|\mathring{A}|$ and $|\nabla\mathring{A}|$ in the definition of $\mathfrak{B}_r(B_0, B_1, B_2)$. Note that if $|\mathring{A}| = 0$, then all the principal curvatures at each point of the hypersurface are equal and hence the hypersurface is umbilic. It is known that the only closed umbilic hypersurfaces in Euclidean space are round spheres. (We applied this fact earlier in the proof of Theorem 8-10.) For surfaces that are *almost* umbilic, there are several quantitative estimates that measure how far the surfaces are from being round; see, for example, [72]. Below we provide a simple version of the quantitative estimates.

We define the *area radius* for a closed surface Σ in (M, g) by

$$r_\Sigma := \sqrt{\frac{|\Sigma|}{4\pi}}.$$

Proposition 8-16 [126, Proposition 2.1]. *There exists an absolute constant $C > 0$ such that the following holds. Let Σ be a closed surface in $\mathbb{R}^3$ of genus zero. Let B_1 and B_2 be positive numbers such that*

$$|\mathring{A}| \le B_1 r_\Sigma^{-3} \quad and \quad |\nabla \mathring{A}| \le B_2 r_\Sigma^{-4}.$$

If $r_\Sigma > C(\sqrt{B_1} + \sqrt{B_2})$, then the principal curvatures λ_1, λ_2 satisfy, for $i = 1, 2$,

$$\left| \lambda_i - \frac{\overline{H}}{2} \right| \le C(B_1 + B_2)r_\Sigma^{-3}.$$

Proof. In the proof, C is assumed to be an absolute constant and may change from line to line. As a consequence of the Codazzi equation [123, Lemma 2.2], we have

$$|\nabla A|^2 \ge \tfrac{3}{4}|\nabla H|^2.$$

Together with the assumption on $|\nabla \mathring{A}|$, this implies an upper bound on $|\nabla H|$:

$$|\nabla \mathring{A}|^2 = |\nabla A|^2 - \tfrac{1}{2}|\nabla H|^2 \ge \tfrac{1}{4}|\nabla H|^2.$$

Let $x_0 \in \Sigma$ be such that $H(x_0) = \overline{H}$. By the mean value theorem and the above bound on $|\nabla H|$, we have

$$|H(x) - \overline{H}| \le \sup_{x \in \Sigma} |\nabla H(x)| d \le 2B_2 r_\Sigma^{-4} d, \tag{8.4.2}$$

where d is the *intrinsic* diameter of Σ, which can be estimated in terms of the mean curvature [215, Theorem 1.1] as follows:

$$d \le C \int_\Sigma |H| \, d\mu \le C\left(|\overline{H}|r_\Sigma^2 + B_2 r_\Sigma^{-2} d\right),$$

where we applied (8.4.2) in the last inequality. Choosing $r_\Sigma \ge \sqrt{2CB_2}$ yields $d \le 2C|\overline{H}|r_\Sigma^2$. Using (8.4.2) again, we have for all $x \in \Sigma$

$$|H(x) - \overline{H}| \le CB_2 r_\Sigma^{-2}|\overline{H}|.$$

By the Gauss–Bonnet theorem and the assumption on $|\mathring{A}|$, we have

$$|\bar{H}| \leq |\Sigma|^{-\frac{1}{2}} \left(\int_\Sigma H^2 \, d\mu \right)^{\frac{1}{2}} \leq |\Sigma|^{-\frac{1}{2}} \left(\int_\Sigma 2|\mathring{A}|^2 \, d\mu + 4 \int_\Sigma K \, d\mu \right)^{\frac{1}{2}} \leq C r_\Sigma^{-1},$$

provided $r_\Sigma \geq \sqrt{B_1}$. Thus we obtain

$$|H(x) - \bar{H}| \leq C B_2 r_\Sigma^{-3}.$$

By the assumption that $|\mathring{A}| \leq B_1 r_\Sigma^{-3}$, we conclude, for $i = 1, 2$, that

$$\left| \lambda_i - \frac{\bar{H}}{2} \right| \leq C(B_1 + B_2) r_\Sigma^{-3}. \qquad \square$$

8.4.2. *The inverse function theorem.* An alternative method to construct a constant mean curvature surface is by graphically perturbing an initial surface whose mean curvature is almost constant.

Let (M, g) be asymptotic to Schwarzschild of mass m. Let

$$p_{ij}(x) := g_{ij}(x) - \left(1 + \frac{2m}{|x|} \right) \delta_{ij}.$$

For r sufficiently large, let $S_r(a)$ be a coordinate sphere defined by $S_r(a) = \{x \in M : |x - a| = r\}$. Set $\rho^i = \frac{x^i - a^i}{r}$. By direct computation [116, (5.1)], the mean curvature of $S_r(a)$ at x in $S_r(a)$ is

$$H_{S_r(a)} = \frac{2}{r} - \frac{4m}{r^2} + \frac{6m(x - a) \cdot a}{r^4} + \frac{9m^2}{r^3} + \frac{1}{2} p_{ij,k}(x) \rho^i \rho^j \rho^k + 2 \frac{p_{ij}(x)}{r} \rho^i \rho^j$$

$$- p_{ij,i}(x) \rho^j - \frac{p_{ii}(x)}{r} + \frac{1}{2} p_{ii,j}(x) \rho^j + O(r^{-4}(1 + |a|))$$

$$=: \frac{2}{r} - \frac{4m}{r^2} + \frac{6m(x - a) \cdot a}{r^4} + \frac{9m^2}{r^3} + G_r(x, a). \tag{8.4.3}$$

Throughout this section, we use the Einstein summation convention and sum over repeated indices, and a comma denotes a partial derivative. From (8.4.3), the mean curvature of the coordinate sphere is almost constant, up to terms of order $O(r^{-3})$. We show below that if $m \neq 0$, one can find a surface of constant mean curvature near $S_r(a)$ for a suitably chosen vector a (depending on r). Moreover, the center a will converge to the BORT center of mass as r tends to infinity.

Theorem 8-17 (Ye [225], Huang [116]). *Let (M, g) be asymptotic to Schwarzschild of mass $m \neq 0$. There exist positive constants r_0 and C, depending only on g, such that for each $r \geq r_0$, there exists a surface Σ_r of constant mean curvature $\frac{2}{r} - \frac{4m}{r^2}$, and Σ_r can be expressed as a normal graph over the coordinate sphere,*

$$\Sigma_r = \left\{ x + \phi(x) \nu_{S_r}(x) : x \in S_r(\mathcal{C}_{\text{BORT}}) \right\},$$

for some $\phi \in C^{2,\alpha}(S_r(\mathcal{C}_{\mathrm{BORT}}))$ satisfying

$$\sum_{|I|\leq 2} r^{|I|}|\partial^I \phi| + \sum_{|I|=2} r^{2+\alpha}[\partial^I \phi]_\alpha \leq C r^{-1},$$

where v_{S_r} is the outward unit normal vector on S_r with respect to g.

Sketch of proof. We will suppress the subscript r in Σ_r when the context is clear. Fix an asymptotically flat coordinate system in the exterior region of M. Let Σ be a graph over the coordinate sphere: for $\phi \in C^{2,\alpha}(S_r(a))$ suitably small,

$$\Sigma = \left\{ x + \phi v_{S_r} : x \in S_r(a) \right\}.$$

Fix r sufficiently large, which will be specified later. Denote by $\mathcal{H}_r(a, \phi)$: $\mathbb{R}^3 \times C^{2,\alpha}(S_r(a)) \to C^{0,\alpha}(S_r(a))$ the mean curvature operator that sends the function ϕ to the mean curvature of the normal graph Σ in (M, g). By Taylor expansion in the ϕ-component,

$$\begin{aligned}
\mathcal{H}_r(a, \phi) = {}& \mathcal{H}_r(a, 0) \\
& + d\mathcal{H}_r(a, 0)(\phi) + \int_0^1 \big(d\mathcal{H}_r(a, s\phi) - d\mathcal{H}_r(a, 0)\big)(\phi)\,ds, \quad (8.4.4)
\end{aligned}$$

where $d\mathcal{H}_r$ is the first Fréchet derivative with respect to the second component. Specifically, $d\mathcal{H}_r(a, 0)$ is the stability operator on $S_r(a)$ with respect to g:

$$d\mathcal{H}_r(a, 0) = -\Delta_{S_r(a)} - (|A|^2 + \mathrm{Ric}(v_{S_r}, v_{S_r})) =: L_{S_r(a)}.$$

The term $\mathcal{H}_r(a, 0)$ in (8.4.4) is the mean curvature of the coordinate sphere computed as in (8.4.3). Thus, solving $\mathcal{H}_r(a, \phi) = \frac{2}{r} - \frac{4m}{r^2}$ for some (a, ϕ) is equivalent to solving

$$\begin{aligned}
L_{S_r(a)}\phi = {}& -\frac{6m(x - a) \cdot a}{r^4} - \frac{9m^2}{r^3} \\
& - G_r(x, a) - \int_0^1 \big(d\mathcal{H}_r(a, s\phi) - d\mathcal{H}_r(a, 0)\big)(\phi)\,ds. \quad (8.4.5)
\end{aligned}$$

Since (M, g) is asymptotic to Schwarzschild, on the coordinate sphere $S_r(a)$ we have

$$|A|^2 = \frac{2}{r^2} + O(r^{-3}), \quad \mathrm{Ric}(v_{S_r}, v_{S_r}) = O(r^{-3}),$$

provided that $|a|$ is bounded by a constant independent of r. Hence we replace the stability operator $L_{S_r(a)}$ by the stability operator $L_0 := -\Delta_0 - \frac{2}{r^2}$ on the Euclidean round sphere, where Δ_0 is the Laplace operator on the Euclidean round sphere $S_r(a)$. We then rewrite (8.4.5) as a differential equation on the

Euclidean round sphere (compare (8.3.5)):

$$L_0\phi = -\frac{6m(x-a)\cdot a}{r^4} - \frac{9m^2}{r^3} - G_r(x,a) + O\big(r^{-1}|\partial^2\phi| + r^{-2}|\partial\phi| + r^{-3}|\phi|\big)$$

$$=: F_r(x,a,\phi,\partial\phi,\partial^2\phi). \tag{8.4.6}$$

A *necessary* condition for this equation to have a solution is that F_r be perpendicular to the cokernel of L_0, which is spanned by $\{x^1 - a^1, x^2 - a^2, x^3 - a^3\}$ restricted on $S_r(a)$ (see Example 8-8). By the following lemma, the parameter a can be chosen to accomplish this.

Lemma 8-18 (Huang [116, Lemma 5.1]). *Let (M, g) be asymptotic to Schwarzschild of mass m. There exists r_0 sufficiently large such that for each $r \geq r_0$ and, for each $i = 1, 2, 3$,*

$$\int_{S_r(a)} (x^i - a^i)G_r(x,a)\, d\mu_0 = -8\pi m\, \mathcal{C}^i_{\mathrm{BORT}} + O(r^{-1}),$$

where $G_r(x,a)$ is the remainder term in (8.4.3) and $d\mu_0$ is the area measure of the Euclidean round sphere.

Remark 8-19. Lemma 8-18 has been generalized to initial data sets with the Regge–Teitelboim conditions. We prove this in Lemma 8-22 below.

By Lemma 8-18 and direct computation, we obtain

$$\int_{S_r(a)} F_r(x,a,\phi,\partial\phi,\partial^2\phi)(x^i - a^i)\, d\mu_0 = -8\pi m(a^i - \mathcal{C}^i_{\mathrm{BORT}}) + O(r^{-1}\|\phi\|_{C^2}),$$

for $i = 1, 2, 3$. If $m \neq 0$, we choose $a = \mathcal{C}_{\mathrm{BORT}} + O(r^{-1}\|\phi\|_{C^2})$ such that the above integral vanishes. Thus, $F_r(x,a,\phi,\partial\phi,\partial^2\phi)$ belongs to the range of L_0.

Next we use the Schauder fixed point theorem (see [107, Chapter 11], for example) to find a solution to (8.4.6).

Theorem 8-20 (Schauder fixed point theorem). *Let $\mathcal{B}$ be a compact convex subset in a Banach space, and let $T : \mathcal{B} \to \mathcal{B}$ be a continuous map. Then T has a fixed point, that is, $Tx = x$ for some $x \in \mathcal{B}$.*

Define the convex subset $\mathcal{B} \subset C^2(S_r(a))$ by $\mathcal{B} := \{u \in C^2(S_r(a)) : \|u\|_{C^{2,\alpha}} \leq 1\}$. Note that $\mathcal{B}$ is compact by the Arzelà–Ascoli theorem. Given $w \in C^2(S_r(a))$, we have shown that there exists a vector a such that $F_r(x,a,w,\partial w,\partial^2 w)$ belongs to the range of L_0. Hence there exists a solution $v \subset C^{2,\alpha}(S_r(a))$ such that

$$L_0 v = F_r(x,a,w,\partial w,\partial^2 w). \tag{8.4.7}$$

Define the map $T : \mathcal{B} \to C^2(S_r(a))$ by $T(w) = v$, where v is the unique solution to (8.4.7) such that v is perpendicular to the kernel of L_0. One can verify that

T is continuous. By the Schauder estimates for solutions perpendicular to the kernel, we obtain

$$\|v\|_{C^{2,\alpha}(S_r(a))} \leq C \left\| F_r(x, a, w, \partial w, \partial^2 w) \right\|_{C^{0,\alpha}(S_r(a))} \leq Cr^{-1} \|w\|_{C^{2,\alpha}(S_r(a))},$$

where the constant C depends only on the metric g. Choose r such that $r \geq C$. It follows that T maps $\mathcal{B}$ into itself. Thus, by the Schauder fixed point theorem, T has a fixed point ϕ. Then ϕ solves the desired equation

$$\mathcal{H}_r(a, \phi) = \frac{2}{r} - \frac{4m}{r^2},$$

where $a = C_{\text{BORT}} + O(r^{-1}\|\phi\|_{C^2})$. $\square$

Let $\{\Sigma_r\}$ be the family of constant mean curvature surfaces constructed in the previous theorem, and let $\{x^1, x^2, x^3\}$ be the coordinate functions. The geometric center of mass of (M, g) proposed by Huisken–Yau is defined as follows, for $i = 1, 2, 3$:

$$C_{\text{Geom}}^i = \lim_{r \to \infty} \frac{\int_{\Sigma_r} x^i \, d\mu_0}{\int_{\Sigma_r} d\mu_0}, \tag{8.4.8}$$

where $d\mu_0$ is the Euclidean area measure.

Corollary 8-21 (Huang [116, Theorem 2]). *If (M, g) is asymptotic to Schwarzschild of mass $m \neq 0$, these two notions of the center of mass coincide:*

$$C_{\text{BORT}} = C_{\text{Geom}}.$$

This corollary has been generalized to the class of asymptotically flat initial data sets that satisfy the Regge–Teitelboim conditions [117].

8.4.3. *Another notion of center of mass.* In this section we show the following identity between the mean curvature of the coordinate spheres and the BORT center of mass.

Lemma 8-22 (Huang [117]). *Let (M, g, k) be an asymptotically flat initial data set satisfying the Regge–Teitelboim conditions. Given $a \in \mathbb{R}^3$, denote by $S_r(a) = \{x \in M : |x - a| = r\}$ the coordinate sphere centered at a of radius r. Then we have*

$$\int_{S_r(a)} (x^\alpha - a^\alpha)\left(H - \frac{2}{r}\right) d\mu_0 = 8\pi E(a^\alpha - C_{\text{BORT}}^\alpha) + O(r^{1-2q}) \tag{8.4.9}$$

for $\alpha = 1, 2, 3$, where H is the mean curvature of $S_r(a)$ with respect to g and $d\mu_0$ is the area measure of the Euclidean round sphere $S_r(a)$.

Proof. Write $h_{ij} = g_{ij} - \delta_{ij}$ and $\rho^i = (x^i - a^i)/r$. Employing the Einstein convention to sum over all repeated indices, we find by direct computation [117, Lemma 2.1] that, for $x \in S_r(a)$,

$$H(x) = \frac{2}{r} + \frac{1}{2} h_{ij,k}(x) \rho^i \rho^j \rho^k + 2 h_{ij}(x) \frac{\rho^i \rho^j}{r} - h_{ij,i}(x) \rho^j + \frac{1}{2} h_{ii,j}(x) \rho^j - \frac{h_{ii}(x)}{r} + E_0(x),$$

where $E_0(x) = O(r^{-1-2q})$ and $E_0(x) - E_0(-x) = O(r^{-2-2q})$. We will use the key identity

$$\int_{S_r(a)} (x^\alpha - a^\alpha) \tfrac{1}{2} h_{ij,k}(x) \rho^i \rho^j \rho^k \, d\mu_0$$
$$= \int_{S_r(a)} (x^\alpha - a^\alpha) \left(\frac{1}{2} h_{ij,i}(x) \rho^j - 2 h_{ij}(x) \frac{\rho^i \rho^j}{r} \right) d\mu_0$$
$$+ \int_{S_r(a)} \frac{1}{2} \left(h_{ii}(x) \rho^\alpha + h_{i\alpha}(x) \rho^i \right) d\mu_0. \quad (8.4.10)$$

Assuming (8.4.10), we obtain

$$\int_{S_r(a)} (x^\alpha - a^\alpha) \left(H(x) - \frac{2}{r} \right) d\mu_0$$
$$= -\frac{1}{2} \int_{S_r(a)} (x^\alpha - a^\alpha)(h_{ij,i} - h_{ii,j}) \rho^j \, d\mu_0 + \frac{1}{2} \int_{S_r(a)} (h_{i\alpha} \rho^i - h_{ii} \rho^\alpha) \, d\mu_0 + O(r^{1-2q})$$
$$= -\frac{1}{2} \int_{S_r(a)} \left(x^\alpha (g_{ij,i} - g_{ii,j}) \rho^j - (g_{i\alpha} \rho^i - g_{ii} \rho^\alpha) \right) d\mu_0$$
$$+ \frac{1}{2} a^\alpha \int_{S_r(a)} (g_{ij,i} - g_{ii,j}) \rho^j \, d\mu_0 + O(r^{1-2q})$$
$$= -\frac{1}{2} \int_{S_r(a)} \left(x^\alpha (g_{ij,i} - g_{ii,j}) \frac{x^j}{r} - \left(g_{i\alpha} \frac{x^i}{r} - g_{ii} \frac{x^\alpha}{r} \right) \right) d\mu_0$$
$$+ \frac{1}{2} a^\alpha \int_{S_r(a)} (g_{ij,i} - g_{ii,j}) \frac{x^j}{r} \, d\mu_0 + O(r^{1-2q}),$$

where we used the Regge–Teitelboim conditions in all the equalities. The desired identity follows from the definitions of the ADM energy and the BORT center of mass.

It remains to prove (8.4.10). Our original proof uses a density theorem (Theorem 8-34) which states that initial data sets with harmonic asymptotics are dense among initial data sets with the Regge–Teitelboim conditions in a suitable topology such that the ADM energy and the BORT center of mass vary continuously. It is then straightforward to verify that (8.4.10) holds for initial data sets with harmonic asymptotics. Eichmair and Metzger later gave the following proof [81]. For each α, define the vector field $X_{(\alpha)} = (x^\alpha - a^\alpha) h_{ij} \rho^i \partial_j$. By the first variation formula,

$$\int_{S_r(a)} \operatorname{div}_0 X_{(\alpha)}\, d\mu_0 = \int_{S_r(a)} H_0 \langle X_{(\alpha)}, \rho \rangle\, d\mu_0 = \int_{S_r(a)} 2(x^\alpha - a^\alpha) \frac{h_{ij}\rho^i \rho^j}{r}\, d\mu_0,$$

where div_0 is the divergence operator on the Euclidean round sphere $S_r(a)$. Then (8.4.10) follows from a direct computation:

$$\operatorname{div}_0 X_{(\alpha)} = (\delta_{ij} - \rho^i \rho^j)\partial_i X_{(\alpha)}^j$$
$$= h_{i\alpha}\rho^i + (x^\alpha - a^\alpha)\left(\frac{h_{ii}}{r} - 2\frac{h_{ij}}{r}\rho^i \rho^j + h_{ij,j}\rho^i - h_{ij,k}\rho^i \rho^j \rho^k \right). \qquad \square$$

Lemma 8-22 gives us a new notion of center of mass that involves the mean curvature: for $\alpha = 1, 2, 3$,

$$C_H^\alpha = \lim_{r \to \infty} -\frac{1}{8\pi E} \int_{S_r(0)} x^\alpha H\, d\mu_0,$$

where H is the mean curvature of the coordinate sphere $S_r(0)$ with respect to g, and $d\mu_0$ is the area measure of a Euclidean round sphere.

Corollary 8-23. *Let (M, g, k) be an asymptotically flat initial data set satisfying the Regge–Teitelboim conditions. Then these notions of center of mass coincide:*

$$C_H = C_{\mathrm{BORT}}.$$

8.5. Stability and foliations

After having obtained a family of constant mean curvature surfaces in Section 8.4, we now discuss their properties in this section. We continue to restrict our discussions to three-dimensional asymptotically flat initial data sets throughout this section, unless otherwise specified.

8.5.1. *Analyzing the stability operator.* We have shown in Section 8.4 the existence of a family of constant mean curvature surfaces in an initial data set (M, g) asymptotic to Schwarzschild. From the construction, each member of the family of constant mean curvature surfaces $\{\Sigma_r\}$ can be expressed as a normal graph over the corresponding coordinate sphere at a common center a:

$$\Sigma_r = \{x + \phi v_{S_r} : x \in S_r(a)\}, \tag{$\star$}$$

where $\phi \in C^{2,\alpha}(S_r(a))$ depends on r and

$$\sum_{|I| \le 2} r^{|I|}|\partial^I \phi| + \sum_{|I|=2} r^{2+\alpha}[\partial^I \phi]_\alpha \le Cr^{-1}.$$

In particular, each Σ_r satisfies

$$H = \frac{2}{r} - \frac{4m}{r^2}, \qquad\qquad |\Sigma_r| = 4\pi r^2 + O(r),$$

$$|A|^2 + \mathrm{Ric}(\nu, \nu) = \frac{2}{r^2} - \frac{10m}{r^3} + O(r^{-4}), \qquad K \geq \frac{1}{r^2} - \frac{2m}{r^3} - Cr^{-4}, \tag{$\star\star$}$$

where K is the Gaussian curvature of Σ_r and ν is the unit normal to Σ_r, and where $f = O(r^q)$ denotes a function satisfying $|f| \leq Cr^q$ on Σ_r, for all r, where $C > 0$ depends only on g.

We next show that for a family of surfaces satisfying the properties $(\star\star)$, there exists r_0 sufficiently large such that for each $r \geq r_0$, the constant mean curvature surface Σ_r is stable. We first recall a classical estimate on the first nonzero eigenvalue of the Laplace operator.

Lemma 8-24 (Lichnerowicz). *Let Σ be an n-dimensional closed Riemannian manifold. Let λ_{Lap} be the first nonzero eigenvalue of the Laplace operator $-\Delta$. If the Ricci curvature satisfies*

$$\mathrm{Ric}(\xi, \xi) \geq (n-1)\kappa |\xi|^2,$$

for some constant $\kappa > 0$ and all $\xi \in TM$, then $\lambda_{\mathrm{Lap}} \geq n\kappa$.

Remark 8-25. The equality $\lambda_{\mathrm{Lap}} = n\kappa$ holds in the preceding if and only if Σ is isometric to the n-sphere of constant sectional curvature κ [173].

Proof. Recall the Bochner–Lichnerowicz identity:

$$\tfrac{1}{2}\Delta|\nabla u|^2 = |\mathrm{Hess}\, u|^2 + \langle \nabla u, \nabla \Delta u\rangle + \mathrm{Ric}(\nabla u, \nabla u).$$

Let u be an eigenfunction corresponding to λ_{Lap} such that $-\Delta u = \lambda_{\mathrm{Lap}} u$. Integrating the identities over Σ and using $|\mathrm{Hess}\, u|^2 \geq (\Delta u)^2/n$, we obtain $\int_\Sigma |\nabla u|^2\, d\mu = \lambda_{\mathrm{Lap}} \int_\Sigma u^2\, d\mu$ and

$$0 = \frac{1}{2}\int_\Sigma \Delta|\nabla u|^2\, d\mu \geq \int_\Sigma \left(\frac{(\Delta u)^2}{n} + \langle \nabla u, \nabla \Delta u\rangle + \mathrm{Ric}(\nabla u, \nabla u)\right) d\mu$$

$$\geq \left(\frac{\lambda_{\mathrm{Lap}}^2}{n} + ((n-1)\kappa - \lambda_{\mathrm{Lap}})\lambda_{\mathrm{Lap}}\right)\int_\Sigma u^2\, d\mu.$$

Since $\lambda_{\mathrm{Lap}} > 0$, the desired inequality follows. $\square$

Theorem 8-26 (Huisken–Yau [126]). *Let (M, g) be asymptotic to Schwarzschild of mass m. Suppose $\{\Sigma_r\}$ is a family of surfaces satisfying the properties $(\star\star)$. Then there exists $C > 0$, depending only on g, such that for each Σ_r*

$$\mu_0 \geq \frac{6m}{r^3} - Cr^{-4},$$

where μ_0 is defined by (8.3.3). As a consequence, if $m > 0$, there exists r_0 sufficiently large such that for each $r \geq r_0$, we have $\mu_0 > 0$ and hence Σ_r is stable.

Proof. Applying Lemma 8-24 to a two-dimensional surface Σ yields $\lambda_{\text{Lap}} \geq 2\kappa$, where κ is the minimum of the Gauss curvature of Σ. On Σ_r we have, by $(\star\star)$,

$$\lambda_{\text{Lap}} \geq \frac{2}{r^2} - \frac{4m}{r^3} - Cr^{-4}.$$

The proposition follows from the definition of μ_0 and the properties $(\star\star)$. $\qquad\square$

8.5.2. Invertibility. To show that the stability operator is invertible, we analyze the eigenvalues of the operator.

Theorem 8-27 (cf. Huisken–Yau [126, Theorem 4.1]). *Let (M, g) be asymptotic to Schwarzschild of mass m. Suppose $\{\Sigma_r\}$ is a family of surfaces satisfying the properties $(\star\star)$. Let λ_0 and λ_1 be the lowest and next lowest eigenvalues of the stability operator L_{Σ_r} on Σ_r, respectively. Then there exists $C > 0$, depending only on g, such that*

$$\lambda_0 = -\frac{2}{r^2} + \frac{10m}{r^3} + O(r^{-4}) \quad \text{and} \quad \lambda_1 \geq \frac{6m}{r^3} - Cr^{-4}.$$

As a consequence, if $m > 0$, there exists r_0 sufficiently large such that for each $r \geq r_0$, the stability operator $L_{\Sigma_r} : C^{2,\alpha}(\Sigma_r) \to C^{0,\alpha}(\Sigma_r)$ is a linear isomorphism.

Proof. Let w be an eigenfunction for λ_0, so that

$$L_{\Sigma_r} w = -\Delta w - (|A|^2 + \text{Ric}(\nu, \nu))w = \lambda_0 w. \tag{8.5.1}$$

Multiplying by w and integrating over Σ_r, we have

$$\lambda_0 \int_{\Sigma_r} w^2 \, d\mu = \int_{\Sigma_r} \left(|\nabla w|^2 - (|A|^2 + \text{Ric}(\nu, \nu))w^2\right) d\mu$$
$$\geq \left(-\frac{2}{r^2} + \frac{10m}{r^3} - Cr^{-4}\right) \int_{\Sigma_r} w^2 \, d\mu,$$

where we used $|\nabla w| \geq 0$ and applied $(\star\star)$ to the term involving $|A|^2 + \text{Ric}(\nu, \nu)$. On the other hand, using a constant function in the Rayleigh quotient yields

$$\lambda_0 \leq -\frac{2}{r^2} + \frac{10m}{r^3} + Cr^{-4}.$$

Thus we have shown that

$$\lambda_0 = -\frac{2}{r^2} + \frac{10m}{r^3} + O(r^{-4}). \tag{8.5.2}$$

The function w is almost constant in the L^2 sense. Indeed, let $\overline{w} = |\Sigma_r|^{-1} \int_{\Sigma_r} w \, d\mu$ denote the mean value of w. Multiply (8.5.1) by $(w - \overline{w})$ and integrate to obtain

$$\int_{\Sigma_r} |\nabla(w - \overline{w})|^2 \, d\mu = \int_{\Sigma_r} (\lambda_0 + |A|^2 + \mathrm{Ric}(v, v))(w - \overline{w})^2 \, d\mu$$
$$+ \int_{\Sigma_r} (|A|^2 + \mathrm{Ric}(v, v))\overline{w}(w - \overline{w}) \, d\mu.$$

Using the estimates of λ_{Lap}, λ_0 and the properties $(\star\star)$, we see

$$\left(\frac{2}{r^2} - Cr^{-3}\right) \int_{\Sigma_r} |w - \overline{w}|^2 \, d\mu \leq Cr^{-4} \int_{\Sigma_r} \left(|w - \overline{w}|^2 + |\overline{w}||w - \overline{w}|\right) d\mu.$$

From the elementary inequality $|\overline{w}||w - \overline{w}| \leq \epsilon r^2 |w - \overline{w}|^2 + C(\epsilon)r^{-2}\overline{w}^2$, we obtain

$$\|w - \overline{w}\|_{L^2(\Sigma_r)} \leq Cr^{-2}|\overline{w}||\Sigma_r|^{\frac{1}{2}}. \tag{8.5.3}$$

In particular, if w is not constant, then $\overline{w} \neq 0$.

Let u be an eigenfunction with respect to λ_1. Then

$$L_{\Sigma_r}(u - \bar{u}) = \lambda_1(u - \bar{u}) + (\lambda_1 + |A|^2 + \mathrm{Ric}(v, v))\bar{u}.$$

We multiply the above identity by $(u - \bar{u})$ and integrate over Σ_r. Since $u - \bar{u}$ has zero mean value, we apply Theorem 8-26 and obtain

$$\left(\frac{6m}{r^3} - Cr^{-4}\right) \int_{\Sigma_r} |u - \bar{u}|^2 \, d\mu \leq \int_{\Sigma_r} (u - \bar{u})L_{\Sigma_r}(u - \bar{u}) \, d\mu$$
$$= \lambda_1 \int_{\Sigma_r} (u - \bar{u})^2 \, d\mu + \int_{\Sigma_r} (\lambda_1 + |A|^2 + \mathrm{Ric}(v, v))\bar{u}(u - \bar{u}) \, d\mu$$
$$\leq \lambda_1 \int_{\Sigma_r} (u - \bar{u})^2 \, d\mu + Cr^{-4} \int_{\Sigma_r} |\bar{u}||u - \bar{u}| \, d\mu$$
$$\leq \lambda_1 \int_{\Sigma_r} (u - \bar{u})^2 \, d\mu + Cr^{-4}|\bar{u}||\Sigma_r|^{1/2}\|u - \bar{u}\|_{L^2}.$$

To estimate the last term on the right, we note that

$$0 = \int_{\Sigma_r} uw \, d\mu = \int_{\Sigma_r} (u - \bar{u})(w - \overline{w}) \, d\mu + \int_{\Sigma_r} u\overline{w} \, d\mu.$$

By the Hölder inequality and the L^2 bound of $(w - \overline{w})$ in (8.5.3),

$$|\bar{u}||\Sigma_r| = \left|\int_{\Sigma_r} u \, d\mu\right| \leq |\overline{w}|^{-1}\|u - \bar{u}\|_{L^2}\|w - \overline{w}\|_{L^2} \leq Cr^{-2}|\Sigma_r|^{1/2}\|u - \bar{u}\|_{L^2}.$$

Putting these inequalities together, we have

$$\left(\frac{6m}{r^3} - Cr^{-4}\right) \|u - \bar{u}\|_{L^2}^2 \leq \lambda_1\|u - \bar{u}\|_{L^2}^2 + Cr^{-6}\|u - \bar{u}\|_{L^2}^2,$$

which shows that

$$\lambda_1 \geq \frac{6m}{r^3} - Cr^{-4}.$$

This implies that if $m > 0$, for r sufficiently large, $L_{\Sigma_r} : C^{2,\alpha}(\Sigma_r) \to C^{0,\alpha}(\Sigma_r)$ is injective. By the Fredholm alternative, L_{Σ_r} is surjective. Hence, it is a linear isomorphism. $\qquad\square$

8.5.3. _Foliations._ Let (M, g) be asymptotic to Schwarzschild of mass $m > 0$. Let Σ_r be a surface in the family $\{\Sigma_r\}$ that satisfy the properties $(\star)$ and $(\star\star)$. As before, we define the mean curvature operator $\mathcal{H} : C^{2,\alpha}(\Sigma_r) \to C^{0,\alpha}(\Sigma_r)$ to be the differential operator that maps ϕ to the mean curvature of the normal graph $\{x + \phi\nu : x \in \Sigma_r\}$. Theorem 8-27 says that the linearized operator $d\mathcal{H} = L_{\Sigma_r}$ is a linear isomorphism for r sufficiently large. We now recall:

Theorem 8-28 (inverse function theorem). _Let E and F be Banach spaces, and let U be an open subset of E. Suppose $f : U \subset E \to F$ is of class C^k, $k \geq 1$. Let $x_0 \in U$. Suppose that $Df(x_0)$ is a linear isomorphism. Then f is a C^k diffeomorphism of some neighborhood of x_0 onto some neighborhood of $f(x_0)$._

Fix r such that $d\mathcal{H}$ is a linear isomorphism on the surface Σ_r of constant mean curvature h_0. The inverse function theorem implies that there exists $\epsilon > 0$ such that for each constant $h \in (h_0 - \epsilon, h_0 + \epsilon)$, there is a _unique_ normal graph over Σ that has constant mean curvature h. This also implies that we can define a differentiable deformation $F : \Sigma_r \times (h_0 - \epsilon, h_0 + \epsilon) \to M$ by sending (Σ_r, h) to the unique normal graph over Σ_r that has constant mean curvature h. Let $H(h) = h$ denote the mean curvature of the normal graph $F(\Sigma_r, h)$. Since each surface has constant mean curvature, only the normal component $\partial F / \partial h$ contributes to the evolution of $H(h)$. Thus,

$$1 = \frac{d}{dh}\bigg|_{h=h_0} H(h) = L_{\Sigma_r}\phi, \tag{8.5.4}$$

where $\phi = g\big((\partial F/\partial h)\big|_{h=h_0}, \nu\big)$. In the following we show that ϕ has a sign, from which it follows that members of the family of constant mean curvature surfaces do not intersect, and in fact form a foliation.

We recall below a standard application of Moser iteration and include the proof since our setting is slightly different from [107, Theorem 8.17].

Proposition 8-29. _Let (Σ, g) be a two-dimensional closed Riemannian manifold. Let v be a C^2 solution to_

$$-\Delta_\Sigma v - Qv = f, \tag{8.5.5}$$

where Q and f are in $L^\infty(\Sigma)$ and $Q \geq 0$. Then, for any $p_0 \geq 2$,

$$\sup_\Sigma |v| \leq C_0\big(\|v\|_{L^{p_0}(\Sigma)} + k\big),$$

where $k = \|f\|_{L^\infty}$ and C_0 depends on p_0, Σ, g, $\|Q\|_{L^\infty}$.

Proof. Let $v^+ := \max_\Sigma\{v, 0\}$. Note $v^+ \in W^{1,2}(\Sigma)$ and

$$\nabla v^+ = \begin{cases} \nabla v & \text{if } v > 0, \\ 0 & \text{if } v \leq 0. \end{cases}$$

Let $w = v^+ + k$, where k is defined as in the proposition. For any real number $p \geq 1$, we multiply the differential equation (8.5.5) by w^p and integrate over Σ:

$$\begin{aligned}
\int_\Sigma |\nabla(w^{\frac{p+1}{2}})|^2 \, d\mu &= \frac{(p+1)^2}{4p} \int_\Sigma (Qvw^p + fw^p) \, d\mu \\
&\leq p \int_\Sigma (Q+1)w^{p+1} \, d\mu \\
&\leq p \max_\Sigma(Q+1) \int_\Sigma w^{p+1} \, d\mu, \tag{8.5.6}
\end{aligned}$$

where in the first inequality we used the assumption that $Q \geq 0$. This computation establishes an upper bound of the $W^{1,2}$-norm of $w^{(p+1)/2}$ by purely the L^2-norm of $w^{(p+1)/2}$. To begin the iteration procedure, we need to relate the higher-order L^q-norms to the $W^{1,2}$-norm. For manifolds of higher dimensions, the standard procedure is to apply the Gagliardo–Nirenberg–Sobolev inequality: for $1 \leq p < n$,

$$\|u\|_{L^{p^*}} \leq C_0 \|u\|_{W^{1,p}},$$

where $p^* = \frac{np}{n-p}$ and n is the dimension of the manifold. However, $n = 2$ in our case and $p = 2$ is the borderline case of the Gagliardo–Nirenberg–Sobolev inequality, so we use another inequality specifically for a two-dimensional manifold Σ [204, p. 193]: for any $1 \leq q < \infty$,

$$\|u\|_{L^q(\Sigma)} \leq C_0\sqrt{q}\,\|u\|_{W^{1,2}(\Sigma)}, \tag{8.5.7}$$

where C_0 depends on (Σ, g).

Apply (8.5.7) with $u = w^{(p+1)/2}$, and let $q = 2\kappa > 2$ be a fixed real number. Together with (8.5.6) and enlarging C_0 if necessary, we obtain, for any $1 \leq p < \infty$,

$$\|w\|_{L^{(p+1)\kappa}} \leq C_0^{\frac{1}{p+1}} (p+1)^{\frac{1}{p+1}} \|w\|_{L^{p+1}}, \tag{8.5.8}$$

where C_0 depends on κ, Σ, g, $\|Q\|_{L^\infty}$. Now we define a sequence of numbers

$$p_0 = p+1, \quad p_i = p_{(i-1)}\kappa = (p+1)\kappa^i, \quad i = 1, 2, \ldots.$$

The estimate (8.5.8) implies that

$$\|w\|_{L^{p_{(i+1)}}} \le C_0^{\sum_{j=0}^{i}\frac{1}{p_j}} \prod_{j=0}^{i} p_j^{\frac{1}{p_j}} \|w\|_{L^{p_0}}.$$

As i tends to infinity, $\|w\|_{L^{p_{(i+1)}}}$ converges to $\|w\|_{L^\infty}$ and the factor multiplying the product converges because $\kappa > 1$. This implies that, for κ fixed and for any $p_0 \ge 2$, and enlarging C_0 from one inequality to the next if necessary,

$$v \le \sup_{\Sigma} w \le C_0\|w\|_{L^{p_0}} \le C_0\left(\|v^+\|_{L^{p_0}} + k\right) \le C_0\left(\|v\|_{L^{p_0}} + k\right),$$

where C_0 depends on p_0, Σ, g, $\|Q\|_{L^\infty}$. Replacing v with $-v$ yields

$$-v \le C_0(\|v\|_{L^{p_0}} + k).$$

The desired estimate follows. $\qquad\qquad\square$

We now use a scaling argument to factor out the dependence of the constant C_0 in Proposition 8-29 from the family of surfaces $\{\Sigma_r\}$.

Proposition 8-30. *Let (M, g) be asymptotic to Schwarzschild of mass m. Suppose $\{\Sigma_r\}$ is a family of surfaces satisfying the property $(\star)$. For each r, let v be a C^2 solution to*

$$-\Delta_{\Sigma_r} v - Qv = f,$$

where Q and f are in $L^\infty(\Sigma_r)$ and $Q \ge 0$. Then, for any $p_0 \ge 2$,

$$\sup_{\Sigma_r} |v| \le C_0(r^{-\frac{2}{p_0}}\|v\|_{L^{p_0}(\Sigma_r)} + k),$$

where $k = \|f\|_{L^\infty}$ and C_0 depends on g, p_0, $\|Q\|_{L^\infty}$ (but is independent of r).

Proof. The property $(\star)$ gives a family of smooth diffeomorphisms $F_r : \mathbb{S}^2 \to \Sigma_r$ such that $|dF_r - r\,\mathrm{Id}| = O_2(1)$, where Id is the identification map from $T\mathbb{S}^2$ to $T\Sigma_r$. This implies the pullback metric satisfies

$$\|F_r^* g_{\Sigma_r} - g_{\mathbb{S}^2}\|_{C^2} \le C,$$

where C depends only on g. Considering the pullback of the differential equation for $v \circ F_r$ and applying Proposition 8-29 on the fixed geometry $(\mathbb{S}^2, g_{\mathbb{S}^2})$, we obtain

$$\sup_{\mathbb{S}^2} |v \circ F_r| \le C_0\left(\|v \circ F_r\|_{L^{p_0}(\mathbb{S}^2)} + k\right),$$

where C_0 depends on g, p_0, $k = \|f \circ F_r\|_{L^\infty(\mathbb{S}^2)}$, and Q. Using the area formula for the L^{p_0}-norm, we have the desired estimate. $\qquad\square$

Theorem 8-31. *Let (M, g) be asymptotic to Schwarzschild of mass $m > 0$. Suppose $\{\Sigma_r\}$ is a family of surfaces satisfying the properties $(\star)$ and $(\star\star)$. Let $u \in C^{2,\alpha}(\Sigma_r)$ satisfy*

$$L_{\Sigma_r} u := -\Delta_{\Sigma_r} u - \left(|A|^2 + \mathrm{Ric}(\nu, \nu)\right)u = c,$$

for some constant c. Then there exists r_0 large enough so that, for each $r \geq r_0$,

$$\sup_{\Sigma_r} |u - \bar{u}| \leq Cr^{-1}|\bar{u}|,$$

where C depends only on g. As a consequence, for r_0 sufficiently large, the solution u is either positive or negative for each $r \geq r_0$.

Proof. Note that $(u - \bar{u})$ satisfies the equation $L_{\Sigma_r}(u - \bar{u}) = c + (|A|^2 + \mathrm{Ric}(\nu, \nu))\bar{u}$. By Theorem 8-26 and the estimate on $|A|^2 + \mathrm{Ric}(\nu, \nu)$ from $(\star\star)$, we obtain

$$\left(\frac{6m}{r^3} - \frac{C}{r^4}\right)\int_{\Sigma_r} |u - \bar{u}|^2 \, d\mu \leq \int_{\Sigma_r} (u - \bar{u}) L_{\Sigma_r}(u - \bar{u}) \, d\mu$$

$$= \int_{\Sigma_r} (|A|^2 + \mathrm{Ric}(\nu, \nu))\bar{u}(u - \bar{u}) \, d\mu$$

$$\leq \frac{C}{r^4} \int_{\Sigma_r} |\bar{u}||u - \bar{u}| \, d\mu$$

$$\leq \frac{C}{r^4} \left(\int_{\Sigma_r} |u - \bar{u}|^2 \, d\mu\right)^{\frac{1}{2}} \left(\int_{\Sigma_r} |\bar{u}|^2 \, d\mu\right)^{\frac{1}{2}}.$$

This implies

$$\|u - \bar{u}\|_{L^2(\Sigma_r)} \leq Cm^{-1}r^{-1}|\bar{u}||\Sigma_r|^{\frac{1}{2}}.$$

By Proposition 8-30,

$$\sup_{\Sigma_r} |u - \bar{u}| \leq C\left(r^{-1}\|u - \bar{u}\|_{L^2(\Sigma_r)} + k\right),$$

where $k = \max_{\Sigma_r} |c + (|A|^2 + \mathrm{Ric}(\nu, \nu))\bar{u}|$ and the constant C depends on $\sup_{\Sigma_r}(|A|^2 + \mathrm{Ric}(\nu, \nu))$. To estimate c, we integrate $L_{\Sigma_r} u = c$ over Σ_r and use $(\star\star)$ to obtain

$$|\Sigma_r||c| \leq \left|\int_{\Sigma_r} (|A|^2 + \mathrm{Ric}(\nu, \nu))(u - \bar{u}) \, d\mu\right| + \left|\int_{\Sigma_r} (|A|^2 + \mathrm{Ric}(\nu, \nu))\bar{u} \, d\mu\right|$$

$$\leq C(r^{-4}\|u - \bar{u}\|_{L^2}|\Sigma_r|^{\frac{1}{2}} + |\bar{u}|).$$

By the above estimates and the properties $(\star\star)$, we have

$$\sup_{\Sigma_r} |u - \bar{u}| \leq Cr^{-1}|\bar{u}|. \qquad \square$$

Applying Theorem 8-31 to (8.5.4), we obtain the following result.

Corollary 8-32. *Let (M, g) be asymptotic to Schwarzschild of mass $m > 0$. Suppose $\{\Sigma_r\}$ is a family of surfaces satisfying the properties $(\star)$ and $(\star\star)$. Then there exists $r_0 > 0$ such that the family of surfaces $\{\Sigma_r\}$ for $r \geq r_0$ forms a foliation.*

8.6. Density theorems

8.6.1. *Weighted Sobolev spaces.* We introduce a topology on the space of asymptotically flat initial data sets using the following weighted norm. Let B be a ball in $\mathbb{R}^n$ centered at the origin. For $k \in \{0, 1, \dots\}$, $p \geq 0$, and $q \in \mathbb{R}$, we define the *weighted Sobolev space* $W^{k,p}_{-q}(\mathbb{R}^n \setminus B)$ to be the set of functions $f \in W^{k,p}_{\mathrm{loc}}(\mathbb{R}^n \setminus B)$ with

$$\|f\|_{W^{k,p}_{-q}(\mathbb{R}^n \setminus B)} := \left(\int_{\mathbb{R}^n \setminus B} \sum_{|I| \leq k} (|\partial^I f(x)| |x|^{|I|+q})^p |x|^{-n} \, dx \right)^{\frac{1}{p}} < \infty.$$

When $p = \infty$,

$$\|f\|_{W^{k,\infty}_{-q}(\mathbb{R}^n \setminus B)} := \sum_{|I| \leq k} \operatorname*{ess\,sup}_{\mathbb{R}^n \setminus B} |\partial^I f| |x|^{|I|+q}.$$

Suppose M is a smooth manifold such that there is a compact set $K \subset M$ and a diffeomorphism $M \setminus K \cong \mathbb{R}^n \setminus B$. Choose an atlas for M that consists of the diffeomorphism $M \setminus K \cong \mathbb{R}^n \setminus B$ and finitely many precompact charts on K. We define the $W^{k,p}_{-q}(M)$ norm on M by summing over the $W^{k,p}_{-q}$ norm on the noncompact chart and the $W^{k,p}$ norm on the precompact charts. The definition extends to the tensor bundles of M by considering the components with respect to these charts, and can also easily extend to an asymptotically flat manifold with a finite number of ends. We sometimes write $W^{k,p}_{-q}$ for $W^{k,p}_{-q}(M)$.

It is known that the ADM energy and linear momentum are continuous functions with respect to the appropriate weighted Sobolev topology.

Theorem 8-33. *Let $p > n \geq 3$, $q \in \left(\frac{n-2}{2}, n-2 \right)$, $q_0 > 0$. Let (g, k) and $(\bar{g}, \bar{k})$ be $C^2_{\mathrm{loc}} \times C^1_{\mathrm{loc}}$ asymptotically flat initial data sets such that*

$$(g - g_0, k), \ (\bar{g} - g_0, \bar{k}) \in W^{2,p}_{-q} \times W^{1,p}_{-1-q},$$

where g_0 is a smooth symmetric $(0, 2)$ tensor that coincides with $g_{\mathbb{E}}$ on $M \setminus K$, and such that

$$\mu, J, \bar{\mu}, \bar{J} \in W^{0,p}_{-n-q_0}.$$

Let $\epsilon > 0$. There exists $\delta > 0$ such that if

$$\|g - \bar{g}\|_{W^{2,p}_{-q}} \leq \delta \quad \text{and} \quad \|k - \bar{k}\|_{W^{1,p}_{-1-q}} \leq \delta,$$

then

$$|E - \bar{E}| < \epsilon \quad \text{and} \quad |P - \bar{P}| < \epsilon.$$

The proof of this fact goes back to [202, p. 50] for E only and to [69, p. 198] in the vacuum case. The proof of the general case can be found in [119, Proposition 2.4] and [78, Proposition 19].

On the other hand, the BORT center of mass and the ADM angular momentum may not be defined in general for asymptotically flat initial data sets since the integrals (8.1.1) and (8.1.2) may diverge [117; 47; 43; 46]. In fact, the BORT center of mass and the ADM angular momentum are *discontinuous* with respect to above topology (see Theorem 8-39 below). Nevertheless, if we consider a topology that incorporates the Regge–Teitelboim conditions, we have an analogous continuity result for the center of mass and angular momentum.

We let $f^{\mathrm{odd}}(x) = (f(x) - f(-x))/2$ and $f^{\mathrm{even}}(x) = (f(x) + f(-x))/2$ with respect to an asymptotically flat coordinate chart.

Theorem 8-34 (cf. Huang [119, Proposition 2.4], [116, Theorem 2.2]). *Let $p > n \geq 3$, $q \in \left(\frac{n-2}{2}, n - 2\right)$, $q_0 > 0$. Let (g, k), $(\bar{g}, \bar{k})$, (μ, J), $(\bar{\mu}, \bar{J})$ satisfy the assumptions in Theorem 8-33. Suppose they also satisfy*

$$(g_{ij}^{\mathrm{odd}}, k_{ij}^{\mathrm{even}}), (\bar{g}_{ij}^{\mathrm{odd}}, \bar{k}_{ij}^{\mathrm{even}}) \in W^{2,p}_{-1-q}(M \setminus K) \times W^{1,p}_{-2-q}(M \setminus K)$$

and

$$\mu^{\mathrm{odd}}, J_i^{\mathrm{odd}}, \bar{\mu}^{\mathrm{odd}}, \bar{J}_i^{\mathrm{odd}} \in W^{0,p}_{-n-q_0-1}(M \setminus K).$$

Let $\epsilon > 0$. There exists $\delta > 0$ such that if

$$\|g^{\mathrm{odd}} - \bar{g}^{\mathrm{odd}}\|_{W^{2,p}_{-1-q}(M \setminus K)} \leq \delta \quad \text{and} \quad \|k^{\mathrm{even}} - \bar{k}^{\mathrm{even}}\|_{W^{1,p}_{-2-q}(M \setminus K)} \leq \delta$$

then

$$|\mathcal{C}_{\mathrm{BORT}} - \bar{\mathcal{C}}_{\mathrm{BORT}}| < \epsilon \quad \text{and} \quad |\mathcal{J} - \bar{\mathcal{J}}| < \epsilon.$$

8.6.2. *Scalar curvature equation*[8]. We discuss a density result for the scalar curvature equation due to Schoen and Yau. The density argument is used in the proof of the Riemannian positive mass theorem and enables them to reduce the case of the general asymptotically flat metrics to the case that the metrics are scalar flat and conformally flat at infinity. In what follows, we consider an asymptotically flat manifold M of dimension $n \geq 3$.

---

[8]See p. 280–284 for more details on the results in this section, which we include in order to keep the chapter self-contained.

Theorem 8-35 (Schoen–Yau [202]). *Let (M, g) be an n-dimensional asymptotically flat initial data set with nonnegative scalar curvature and the ADM mass m. Given $\epsilon > 0$, there exists an asymptotically flat metric $\bar{g}$ with zero scalar curvature such that, outside a compact set of M, the metric has the form*

$$\bar{g}_{ij} = u^{\frac{4}{n-2}} \delta_{ij}$$

with $u = 1 + \frac{\bar{m}}{2} r^{2-n} + O(r^{1-n})$, where $\bar{m}$ is the ADM mass of $\bar{g}$ and

$$\bar{m} \leq m + \epsilon.$$

For the proof of the theorem, we establish the following lemma. The analysis can be carried out with $q_0 > 0$, with an error term that will in general have slightly weaker decay; see [78, Proposition 24]. This remark applies also to Proposition 8-37 below.

Lemma 8-36 (Schoen–Yau [199, Lemma 3.3]). *Let (M, g) be an n-dimensional asymptotically flat initial data set with the ADM mass m. Suppose the scalar curvature $R(g) \geq 0$ is positive somewhere, and $R(g) = O(|x|^{-n-q_0})$ for some $q_0 > 1$. Then there exist a constant $A < 0$ and a unique metric $\bar{g} = u^{4/(n-2)} g$ with zero scalar curvature such that*

$$u = 1 + \frac{A}{2} r^{2-n} + O(r^{1-n}).$$

Furthermore, the ADM mass $\bar{m}$ of $\bar{g}$ satisfies $\bar{m} = m + A < m$.

Proof. Let $\bar{g} = u^{4/(n-2)} g$. The scalar curvatures of g and $\bar{g}$ are related by

$$R(\bar{g}) = u^{-\frac{n+2}{n-2}} \left(R(g)u - \frac{4(n-1)}{n-2} \Delta_g u \right).$$

Denote the conformal Laplace operator by $L = \Delta_g - \frac{n-2}{4(n-1)} R(g)$. Let $p > n$, $q \in \left(\frac{n-2}{2}, n-2 \right)$. Then $L : W^{2,p}_{-q} \to W^{0,p}_{-2-q}$ is a Fredholm operator of index zero, by [15, Proposition 1.14]. To find a solution to the inhomogeneous equation $Lv = f$ for $f \in W^{0,p}_{-2-q}$, it suffices to prove that L has a trivial kernel by the Fredholm alternative. Let $v \in W^{2,p}_{-q}$ satisfy $Lv = 0$. Multiplying the equation $Lv = 0$ by v and applying the divergence theorem, we have

$$0 \leq \int_M |\nabla v|^2 \, d\sigma = -\frac{n-2}{4(n-1)} \int_M R(g)v^2 \, d\sigma + \lim_{r \to \infty} \int_{\{|x|=r\}} v \frac{\partial v}{\partial \nu} \, d\mu$$

$$= -\frac{n-2}{4(n-1)} \int_M R(g)v^2 \, d\sigma \leq 0,$$

where $d\sigma$ is the volume measure of (M, g) and we have used the fall-off rates of v and ∂v to compute the boundary term. We conclude that $v \equiv 0$. Therefore there is a unique solution to $Lv = \frac{n-2}{4(n-1)} R(g)$. Let $u = v + 1$. Then u satisfies

$Lu = 0$; moreover, $u > 0$ everywhere by the strong maximum principle. Thus $\bar{g} = u^{4/(n-2)}g$ is the desired metric. The asymptotic expansion of u follows from [15, Theorem 1.17]; compare [78, Proposition 24].

To show that $A < 0$, we integrate $Lu = 0$ over a large ball and apply the divergence theorem:

$$
\begin{aligned}
0 < \lim_{r \to \infty} \frac{n-2}{4(n-1)} \int_{\{|x| \le r\}} R(g)u \, d\sigma &= \lim_{r \to \infty} \int_{\{|x|=r\}} \frac{\partial u}{\partial \nu} \, d\mu \\
&= \lim_{r \to \infty} \int_{\{|x|=r\}} -\frac{(n-2)A}{2} r^{1-n} \, d\mu \\
&= -\frac{n-2}{2} \omega_{n-1} A.
\end{aligned}
$$

Proof of Theorem 8-35. By Lemma 8-36, we may assume that g has zero scalar curvature. For $\lambda \ge 1$ large, we define the cutoff metric

$$
\hat{g}_\lambda := \chi_\lambda g + (1 - \chi_\lambda) g_{\mathbb{E}},
$$

where $\chi_\lambda(x) = \chi(x/\lambda)$ and χ is a smooth cutoff function on $\mathbb{R}^n$ that is 1 on $\{|x| \le 1\}$ and 0 on $\{|x| \ge 2\}$. Note that the cutoff metric has zero scalar curvature everywhere except the interpolating region $\lambda \le |x| \le 2\lambda$ and $R(\hat{g}_\lambda) = O(\lambda^{-n})$ there. We would like to find a metric $\bar{g} = u^{4/(n-2)} \hat{g}_\lambda$ with zero scalar curvature in the conformal class of $\hat{g}_\lambda$. By the transformation formula of the scalar curvature, it suffices to find a positive function u that tends to 1 at infinity and satisfies

$$
\Delta_{\hat{g}_\lambda} u - \frac{n-2}{4(n-1)} R(\hat{g}_\lambda) u = 0.
$$

Note that $R(\hat{g}_\lambda)$ may not be nonnegative everywhere, so the proof of Lemma 8-36 cannot be applied. The approach of Schoen and Yau relies on a Sobolev inequality and requires $\| R(\hat{g}_\lambda)^- \|_{L^{\frac{n}{2}}(M)}$ sufficiently small, which is achieved by choosing λ sufficiently large. We refer the details to [199, Lemma 3.2].

The last statement that $\bar{m} \le m + \epsilon$ follows from the continuity of the ADM mass by Theorem 8-33. $\qquad\square$

8.6.3. *Einstein constraint equations.* Let (M, g, k) be an initial data set. Define the momentum tensor

$$
\pi = k - (\mathrm{tr}_g k) g.
$$

It can be convenient to express initial data in terms of π rather than k. We refer to (M, g, π) as an initial data set in this section and define the constraint map

$$
\Phi(g, \pi) = (2\mu, J) = \left(R(g) - |\pi|_g^2 + \tfrac{1}{n-1}(\mathrm{tr}_g \pi)^2, \, \mathrm{div}_g \pi \right).
$$

We say that (M, g, π) has *harmonic asymptotics* if there exist a smooth function u and a smooth vector field X such that $u \to 1$, $X \to 0$ at infinity and, outside a compact set of M, we have

$$g = u^{\frac{4}{n-2}} g_{\mathbb{E}} \quad \text{and} \quad \pi = u^{\frac{2}{n-2}} (\mathcal{L}_{g_{\mathbb{E}}} X),$$

where the operator $\mathcal{L}_g$ is defined by $\mathcal{L}_g X = L_X g - (\operatorname{div}_g X) g$ and $L_X g$ is the Lie derivative. Throughout this section, we denote by g_0 a smooth symmetric $(0, 2)$ tensor on M that coincides with $g_{\mathbb{E}}$ on $M \setminus K$.

The term "harmonic" follows from the following proposition that the leading-order terms of the function u and the vector field X are harmonic.

Proposition 8-37 (Corvino–Schoen [69]; see also [78, Proposition 24]). *Let $p > n$, $q \in \left(\frac{n-2}{2}, n-2\right)$, $q_0 > 1$. Suppose that (M^n, g, π) is an asymptotically flat initial data set that satisfies*

$$(g - g_0, \pi) \in W^{2,p}_{-q}(M) \times W^{1,p}_{-1-q}(M),$$

and

$$(\mu, J) \in W^{0,p}_{-n-q_0},$$

such that (g, π) has harmonic asymptotics:

$$g = u^{\frac{4}{n-2}} g_{\mathbb{E}}, \quad \pi = u^{\frac{2}{n-2}} \mathcal{L}_{g_{\mathbb{E}}} X, \tag{8.6.1}$$

outside a compact set, for some $(u - 1, X) \in W^{2,p}_{-q}$. Then (u, X) admits an expansion of the form

$$u(x) = 1 + a|x|^{2-n} + O(|x|^{1-n}),$$
$$X_i(x) = b_i|x|^{2-n} + O(|x|^{1-n}),$$

where $X = \sum_{i=1}^{n} X_i \frac{\partial}{\partial x^i}$.

A generalization of Theorem 8-35 to the full constraint equations states that initial data sets with harmonic asymptotics are dense among general asymptotically flat initial data sets:

Theorem 8-38 (Corvino–Schoen [69, Theorem 1]). *Let $p > n$, $q \in \left(\frac{n-2}{2}, n-2\right)$. Let (g, π) and $(\bar{g}, \bar{\pi})$ be vacuum asymptotically flat initial data sets satisfying*

$$(g - g_0, \pi) \in W^{2,p}_{-q} \times W^{1,p}_{-1-q}.$$

Let $\epsilon > 0$. There exists a vacuum asymptotically flat initial data set $(\bar{g}, \bar{\pi})$ with *harmonic asymptotics such that*

$$\|g - \bar{g}\|_{W^{2,p}_{-q}} < \epsilon, \quad \|\pi - \bar{\pi}\|_{W^{1,p}_{-1-q}} < \epsilon$$

and

$$|E - \bar{E}| < \epsilon, \quad |P - \bar{P}| < \epsilon.$$

Proof. For $\lambda \geq 1$ large, define the cutoff initial data

$$(\hat{g}_\lambda)_{ij} = \chi_\lambda g_{ij} + (1 - \chi_\lambda)\delta_{ij}, \quad \hat{\pi}_\lambda = \chi_\lambda \pi,$$

where $\chi_\lambda(x) = \chi(x/\lambda)$ and χ is a smooth cutoff function on $\mathbb{R}^n$ such that χ is 1 on $\{|x| \leq 1\}$ and 0 on $\{|x| \geq 2\}$. In the following, we suppress the subscript λ when the context is clear.

The Einstein constraint equations form an underdetermined system: the number of unknowns is greater than the number of the equations that determine them. One reason to introduce a function u and a vector field X in the expression of harmonic asymptotics is to obtain a determined-elliptic system. Let

$$\tilde{g} = u^{\frac{4}{n-2}}\hat{g}, \quad \tilde{\pi} = u^{\frac{2}{n-2}}(\hat{\pi} + \mathcal{L}_{\hat{g}}X).$$

We would like to find u tending to 1 and X tending to 0 at infinity such that $(\tilde{g}, \tilde{\pi})$ satisfies the vacuum constraints.

We define a map $T_{(\hat{g},\hat{\pi})} : (W^{2,p}_{-q} + 1) \times W^{2,p}_{-q} \to W^{0,p}_{-2-q}$ via the constraint operator $T_{(\hat{g},\hat{\pi})}(u, X) = \Phi(\tilde{g}, \tilde{\pi})$, and we define $T_{(g,\pi)} : (W^{2,p}_{-q} + 1) \times W^{2,p}_{-q} \to W^{0,p}_{-2-q}$ analogously; these are both smooth maps. The linearization of $T_{(\hat{g},\hat{\pi})}$ at $(1, 0)$ is

$$DT_{(\hat{g},\hat{\pi})}|_{(1,0)}(v, Z)$$

$$= \Big(\tfrac{-4}{n-2}\big((n-1)\Delta_{\hat{g}}v + \big(R_{\hat{g}} - |\hat{\pi}|^2_{\hat{g}} + \tfrac{1}{n-1}(\mathrm{tr}_{\hat{g}}\hat{\pi})^2\big)v\big) - 4Z_{k;\ell}\hat{\pi}^{k\ell} + \tfrac{2}{n-1}\mathrm{tr}_{\hat{g}}\hat{\pi}\,\mathrm{div}_{\hat{g}}Z,$$

$$\mathrm{div}_{\hat{g}}(\mathcal{L}_{\hat{g}}Z)_j + \tfrac{2(n-1)}{n-2}v_{,k}\hat{\pi}^k_j - \tfrac{2}{n-2}v_{,j}\mathrm{tr}_{\hat{g}}\hat{\pi} - \tfrac{2}{n-2}(\mathrm{div}_{\hat{g}}\hat{\pi})_j v\Big),$$

where indices are raised and covariant derivatives are taken with respect to $\hat{g}$. Because $q \in \big(\tfrac{n-2}{2}, n - 2\big)$ and $p > n$, $DT_{(\hat{g},\hat{\pi})}|_{(1,0)}$ and $DT_{(g,\pi)}|_{(1,0)}$ are Fredholm operators of index 0 for λ sufficiently large [15]. Instead of proving the linearization has a trivial kernel as in the proof of Theorem 8-35, which seems difficult for the system, we use the following argument.

Let K_1 be a complementing subspace for $\ker(DT_{(g,\pi)}|_{(1,0)})$ in $W^{2,p}_{-q} \times W^{1,p}_{-1-q}$. Since $DT_{(g,\pi)}|_{(1,0)}$ is Fredholm and the linearization $D\Phi|_{(g,\pi)} \cdot W^{2,p}_{-q} \times W^{1,p}_{-1-q} \to W^{0,p}_{-2-q}$ is surjective (see [69, Proposition 3.1]; cf. [120, Lemma 2.10]), we can find smooth compactly supported symmetric $(0, 2)$-tensors $\{(h_k, w_k)\}_{k=1}^N$ whose images $\{D\Phi|_{(g,\pi)}(h_k, w_k)\}_{k=1}^N$ form a basis for a complementing subspace of

$\operatorname{ran}(DT_{(g,\pi)}|_{(1,0)})$ in $W^{0,p}_{-2-q}$. Let $K_2 = \operatorname{span}\{(h_k, w_k)\}^N_{k=1}$. For $(u-1, X) \in K_1$ and $(h, w) \in K_2$, define the maps $\overline{T}_{(\hat{g},\hat{\pi})}$, $\overline{T}_{(g,\pi)}$ by

$$\overline{T}_{(\hat{g},\hat{\pi})}(u, X, h, w) = \Phi(u^{\frac{4}{n-2}}\hat{g} + h,\, u^{\frac{2}{n-2}}(\hat{\pi} + \mathcal{L}_{\hat{g}}X) + w),$$

$$\overline{T}_{(g,\pi)}(u, X, h, w) = \Phi(u^{\frac{4}{n-2}}g + h,\, u^{\frac{2}{n-2}}(\pi + \mathcal{L}_g X) + w).$$

By construction, $D\overline{T}_{(\hat{g},\hat{\pi})}|_{(1,0,0,0)}$ is an isomorphism for λ sufficiently large.

Using that $(\hat{g}, \hat{\pi})$ converges to (g, π) in $W^{2,p}_{-q} \times W^{1,p}_{-q-1}$ as $\lambda \to \infty$, it is easy to see that $D\overline{T}_{(\hat{g},\hat{\pi})}|_{(u,X,h,w)}$ converges to $D\overline{T}_{(g,\pi)}|_{(u,X,h,w)}$ as $\lambda \to \infty$, locally uniformly in (u, X, h, w) in the strong operator topology. By the inverse function theorem, for all $\lambda \geq 1$ sufficiently large, $\overline{T}_{(\hat{g},\hat{\pi})}$ restricts to a diffeomorphism defined on an open neighborhood of $(1, 0, 0, 0)$ (independent of $\lambda \geq 1$) and onto an open neighborhood containing a ball centered at $(0, 0)$ in $W^{0,p}_{-2-q}$. The preimage $\overline{T}^{-1}_{(\hat{g},\hat{\pi})}(0, 0)$ gives the desired solution.

Finally, it follows from Theorem 8-33 that

$$|E - \bar{E}| < \epsilon, \quad |P - \bar{P}| < \epsilon. \qquad \square$$

The vacuum assumption in Theorem 8-38 can be replaced by appropriate assumptions on (μ, J). In fact, using a more delicate perturbation argument, one can prove that if (g, π) satisfies the dominant energy condition, it is possible to obtain a strict dominant energy condition for the approximate data $(\bar{g}, \bar{\pi})$ with harmonic asymptotics [78, Theorem 18]. This fact is used in the proof of the spacetime positive mass theorem to reduce the general case of the theorem to the special case of initial data that has harmonic asymptotics with a strict dominant energy condition [78].

8.6.4. *Applications to the center of mass and angular momentum.* Generalizing the proof of Theorem 8-38, we show that one can arbitrarily specify the BORT center of mass and the ADM angular momentum.

Theorem 8-39 (Huang–Schoen–Wang [122, Theorem 3]). *Let (M, g, π) be a three-dimensional vacuum asymptotically flat initial data set satisfying the Regge–Teitelboim conditions and $E > |P|$. Given any constant vectors $\vec{\alpha}_0, \vec{\gamma}_0 \in \mathbb{R}^3$, and $\epsilon > 0$, there is a vacuum initial data set $(\bar{g}, \bar{\pi})$ satisfying the Regge–Teitelboim conditions such that*

$$\|g - \bar{g}\|_{W^{2,p}_{-q}} \leq \epsilon, \quad \|\pi - \bar{\pi}\|_{W^{1,p}_{-1-q}} \leq \epsilon$$

and

$$\bar{E} = E, \quad \bar{P} = P, \quad \bar{J} = J + \vec{\alpha}_0, \quad \bar{C}_{\mathrm{BORT}} = C_{\mathrm{BORT}} + \vec{\gamma}_0.$$

Analogous to the positive mass conjecture, there is a conjectured inequality between the ADM energy and angular momentum. It is known that $E \geq \sqrt{|\mathcal{J}|}$ for axially symmetric asymptotically flat black hole initial data sets ([58; 60; 63; 71; 206; 232]; see also [231]). However, Theorem 8-39 shows this inequality does not hold in general for asymptotically flat data sets without axial symmetry.

The proof of Theorem 8-39 goes along the same lines as that of Theorem 8-38, but different cutoff data sets are employed. Let σ, τ be symmetric $(0, 2)$-tensors on $\mathbb{R}^3$. Suppose further that σ, τ are compactly supported on $\{1 \leq |x| \leq 2\}$ satisfying the linearized constraint equations (at the trivial data)

$$\sum_{i,j} (\sigma_{ij,ij} - \sigma_{ii,jj}) = 0,$$

$$\sum_i \tau_{ij,i} = 0 \quad \text{for } j = 1, 2, 3.$$

Consider

$$\hat{g}_\lambda = g + \sigma_\lambda \qquad \text{and} \qquad \hat{\pi}_\lambda = \pi + \tau_\lambda,$$

where $\sigma_\lambda = \sigma(x/\lambda)$, $\tau_\lambda = \tau(x/\lambda)$. To specify the center of mass and angular momentum, the proof centers on constructing the tensors σ, τ with certain desired properties. For the angular momentum, we find σ, τ with components satisfying $\sigma_{ij}(x) = \sigma_{ij}(-x)$, $\tau_{ij}(x) = \tau_{ij}(-x)$ and so that, for a given $\vec{\alpha} = (\alpha_1, \alpha_2, \alpha_3) \in \mathbb{R}^3$,

$$\int_{\{1 \leq |x| \leq 2\}} \sum_{i,j,l} \left(\tfrac{1}{2} \tau_{ij,l} Y^l_{(k)} + \tau_{il} (Y^l_{(k)})_{,j} \right) \sigma^{ij}\, dx = \alpha_k$$

for each $k = 1, 2, 3$, where $Y_{(k)}$ is the rotation vector field $Y_{(k)} = \frac{\partial}{\partial x^k} \times \vec{x}$. To specify the center of mass, we construct a divergence-free and trace-free tensor σ such that for a given $\vec{\gamma} = (\gamma_1, \gamma_2, \gamma_3) \in \mathbb{R}^3$,

$$\int_{\{1 \leq |x| \leq 2\}} x^k \sum_{i,j,l} (\sigma_{ij,l})^2\, dx = \gamma_k,$$

for each $k = 1, 2, 3$. For the construction of those tensors, see Theorems 2.1 and 2.2 of [122].

CHAPTER 9

On the Riemannian Penrose inequality

9.1. Introduction

In 1973 Roger Penrose [180] proposed an inequality which states that, in an isolated gravitational system with nonnegative local energy density, the total mass of the system must be at least as much as that contributed by any black holes contained within. The Penrose inequality takes the form

$$m \geq \sqrt{\frac{A}{16\pi}}, \tag{9.1.1}$$

where m is the total mass of a spacelike slice of a spacetime and A is the total area of the black holes within the spacetime, as viewed from the spacelike slice. Penrose's original motivation for proposing this inequality is the fact that a counterexample of this inequality would produce a counterexample for cosmic censorship, one of the central open questions in relativity, related to the deterministic character of the theory [178; 180; 181].

The Penrose inequality in full generality remains an open problem. In this chapter we focus on proofs of an important special case of the statement, the so-called Riemannian Penrose inequality (RPI). This scenario arises when considering the inequality on time-symmetric spacetimes. (For a discussion of the full Penrose inequality for general spacetimes, see the survey paper by Marc Mars [150].) In this case, the Penrose inequality is formulated on an asymptotically flat, three-dimensional *Riemannian* manifold (M, g) with nonnegative scalar curvature that contains "black holes", i.e., closed minimal surfaces. The inequality retains the same form as (9.1.1), where m is the ADM mass of (M, g) and A is the area of the *outermost minimal surface* in M.

We consider three approaches to the Riemannian Penrose inequality. In 2001 Huisken and Ilmanen [125] proved it using Geroch's monotonicity formula for

This chapter arose from class notes for the minicourse "On the Penrose Inequality" taught by Fernando Schwartz in the 2012 MSRI Summer School on Mathematical Relativity and the 2013 Summer School in Cortona, Italy. Brian Allen attended the MSRI course and contributed to the writing, having carefully typed the contents of the course and improved the exposition, particularly in Section 9.4.

357

the *inverse mean curvature flow*. Their proof has an unavoidable limitation: the term A in (9.1.1) is the area of any single component of the outermost minimal surface. In the same year, Bray [26] proved the general statement of the RPI using a conformal flow and the positive mass theorem. (Bray's proof was later generalized by Bray and Lee to work up to dimension seven [28].) Finally, Lam [138] proved the RPI for graphical manifolds, in arbitrary dimensions, in 2010. Related results in different ambient spaces appear in [95].

In this chapter we present the main ideas of these three proofs of the RPI. We start with Lam's proof, which is the simplest; we move onto Huisken and Ilmanen's proof, which is the one that builds on Geroch's original argument; and we end with Bray's proof, which gives the most general result to date.

9.2. Preliminaries

The Riemannian Penrose inequality is an important special case of the general Penrose inequality. It arises when we consider Cauchy hypersurfaces M that are time-symmetric, i.e. those for which the second fundamental form of M inside a spacetime L is identically zero. If we also assume that the Lorentzian space-time L satisfies the dominant energy condition, a desirable physical hypothesis, then it can be shown using the Einstein constraint equations that M must have nonnegative scalar curvature. The Riemannian Penrose inequality deals with the effects of black holes on the total mass of an isolated system. A seemingly simple, and closely related, question is to determine the effect of nonnegative local mass density on the total mass of such an isolated system. It turns out that this question is quite subtle in itself, and for non-spin manifolds in arbitrary dimensions remained an open problem whose resolution was announced only relatively recently [205] (see also [147]). It is known as the *positive mass theorem*, the Riemannian version of which (in dimension three) is stated as follows.

Theorem 9-1 (positive mass theorem, PMT). *Let (M^3, g) be an asymptotically flat Riemannian manifold with nonnegative scalar curvature and ADM mass m. Then $m \geq 0$, with equality if and only if M is isometric to flat Euclidean space.*

The PMT was first proved by Schoen and Yau [199] using minimal surfaces. Their original proof works through dimension seven. The dimensional restriction comes from the fact that (stable) minimal surfaces can have singularities in high ambient dimensions, a well-known complication coming from geometric measure theory. A clever proof of the PMT for spin manifolds, which works in all dimensions, was found by Witten [224]. In this chapter we will provide a simple proof of the PMT by Lam [138], which covers the case when M is isometric to the graph of a function defined on Euclidean space.

The statement of the Penrose inequality in the time-symmetric case can be written in purely Riemannian terms. In the context of asymptotically flat Riemannian manifolds, the Penrose inequality deals with the outermost minimal surface, since this is the Riemannian counterpart of the event horizon of a black hole in this case. In analogy with the positive mass theorem, there is a precise characterization of the case of equality, as achieved by the Riemannian Schwarzschild manifold of mass m given by $\left(\mathbb{R}^3 \setminus B_{m/2}, \left(1 + \frac{m}{2r}\right)^4 \delta\right)$, where $r = |x|$, $B_{m/2} = \left\{x \in \mathbb{R}^3 : |x| < \frac{m}{2}\right\}$, and δ is the Euclidean metric. More precisely:

Theorem 9-2 (Riemannian Penrose inequality). *Let (M^3, g) be an asymptotically flat Riemannian manifold with nonnegative scalar curvature and ADM mass m. Let $\Sigma^2 \subset M$ be the outermost minimal hypersurface in M, and denote by $|\Sigma|$ its area. Then*

$$m \geq \sqrt{\frac{|\Sigma|}{16\pi}}.$$

Equality is achieved if and only if (M^3, g) is isometric (outside Σ) to the Riemannian Schwarzschild manifold of mass m.

Throughout this chapter we will discuss three different proofs of the Riemannian Penrose inequality that apply to three slightly different, precisely stated versions of the inequality above. Each one of the proofs gives us a different perspective on the problem, but only Bray's proof fully establishes Theorem 9-2.

In Section 9.3 we discuss Lam's proof of the PMT, as well as his argument for proving the RPI for a manifold that is a graph over $\mathbb{R}^n$. Both proofs work in arbitrary dimensions.

In Section 9.4 we discuss Huisken and Ilmanen's proof of the Riemannian Penrose inequality. Their proof develops a weak setting in which Geroch's monotonicity formula for the Hawking mass under inverse mean curvature flow can be applied. The techniques contain many interesting ideas of independent interest, some of which have had a broader impact in geometric analysis and general relativity (see e.g. [167; 213]).

Finally, in Section 9.5 we discuss Bray's proof of the Riemannian Penrose inequality. Bray's argument fully proves Theorem 9-2. It uses the PMT together with a novel conformal flow of the metric and a mass-capacity inequality.

Our goal in this chapter is to provide a gentle introduction to the main ideas, central arguments, and motivating calculations that go into the aforementioned proofs. In particular both Huisken and Ilmanen's proof and Bray's proof are quite involved and require deep knowledge and lengthy calculations from elliptic PDE and geometric measure theory, for instance. Thus, we shall not give a complete account of their arguments here. Rather, we hope that the interested reader will be intrigued enough by the sheer beauty of the results, ultimately visiting the original works in search of their full exposition.

9.3. Lam's proof of the RPI (and PMT) for graphs, in arbitrary dimensions

For this section we will adapt the definition of asymptotically flat manifolds to the setting of graphs over $\mathbb{R}^n$.

Definition 9-3. Let $\Omega \subset \mathbb{R}^n$, $n \geq 3$, be a bounded open set (possibly empty), and let $f : \mathbb{R}^n \setminus \Omega \to \mathbb{R}$ be a smooth function. Let $M = \{(x, f(x)) : x \in \mathbb{R}^n \setminus \Omega\}$ denote the graph of f in $\mathbb{R}^{n+1}$, and we denote partial derivatives with subscripts. We say that M is *asymptotically flat* if

$$f_i(x) = O(|x|^{-p/2}), \quad |x||f_{ij}(x)| + |x|^2|f_{ijk}(x)| = O(|x|^{-p/2}),$$

as $|x| \to \infty$, for some $p > \frac{n-2}{2}$.

Let us state now the positive mass theorem and the Riemannian Penrose inequality for the case of graphs.

Theorem 9-4 (PMT for graphs). *Let (M^n, g) be as in Definition 9-3, with $\Omega = \varnothing$. Then the ADM mass of M satisfies*

$$m = \frac{1}{2(n-1)\omega_{n-1}} \int_M \frac{R(g)}{\sqrt{1+|\nabla f|^2}} dV_g,$$

where $R(g)$ and dV_g are the scalar curvature and volume measure of (M, g), respectively, ω_{n-1} is the area of the $(n-1)$-dimensional unit sphere and $|\nabla f|^2 = f_1^2 + \cdots + f_n^2$.

Remark 9-5. If $R(g) \geq 0$ above, then $m \geq 0$. Also, assuming $R(g) \geq 0$, then $m = 0$ if and only if f is constant. In other words, $m = 0$ if and only if M is flat Euclidean space; see [121].

Lam's formula is quite general and does not need an assumption on the scalar curvature. This observation has been used for extending the PMT to different settings (see [160], for example). The RPI for graphs is the following inequality.

Theorem 9-6 (Riemannian Penrose inequality for graphs). *Let (M, g) be isometric to the smooth asymptotically flat graph of $f : \mathbb{R}^n \setminus \Omega \to \mathbb{R}$, where $\Omega \subset \mathbb{R}^n$ is a smooth bounded open set that is the union of its finitely many components, the closure of each of which is a convex smooth compact set. Assume that f is constant on each component of $\partial\Omega$ and $|\nabla f(x)| \to \infty$ as $x \to \Sigma := \partial\Omega$. Then if*

$|\Sigma|$ *is the area of* Σ, *and if* $R(g)$ *and* dV_g *are the scalar curvature and volume measure of* g, *the ADM mass of* M *satisfies*

$$m \geq \frac{1}{2}\left(\frac{|\Sigma|}{\omega_{n-1}}\right)^{\frac{n-2}{n-1}} + \frac{1}{2(n-1)\omega_{n-1}}\int_M \frac{R(g)}{\sqrt{1+|\nabla f|^2}}dV_g.$$

Hence, if $R(g) \geq 0$, *then* $m \geq \frac{1}{2}\left(|\Sigma|/\omega_{n-1}\right)^{\frac{n-2}{n-1}}$.

Remark 9-7. Lam's proof of the RPI does not deal with the case of equality. Nevertheless, Huang and Wu establish it in [121]. Theorem 9-6 can be generalized to the case of Σ mean-convex and outer-minimizing. This follows from estimates by Freire and Schwartz [95], as we will see at the end of this section.

We let the mean curvature H of a hypersurface Σ with respect to a smooth unit normal field ν be the divergence of ν taken along Σ (note the sign convention). A standard formula for the mean curvature of graphs [41] gives the mean curvature H of Σ with respect to the induced metric as

$$H = \frac{H_0}{\sqrt{1+|\nabla f|^2}},$$

where H_0 is the mean curvature of Σ inside Euclidean space. So the assumption that $|\nabla f| \to \infty$ on Σ guarantees that $H = 0$; i.e., Σ is a minimal surface.

Before we can prove the above theorems we will need some preliminary results, the proofs of which can be found in [41; 138]. In what follows, we will use the convention of summing over pairs of repeated indices.

Lemma 9-8. *The induced metric of a manifold* M^n *that is the graph of a function* $f : \mathbb{R}^n \to \mathbb{R}$ *is given by*

$$g_{ij} = \delta_{ij} + f_i f_j, \quad g^{ij} = \delta^{ij} - \frac{f_i f_j}{1+|\nabla f|^2}.$$

The Christoffel symbols of this metric are

$$\Gamma^k_{ij} = \frac{f_{ij}f_k}{1+|\nabla f|^2},$$

$$\Gamma^k_{ij,k} = \frac{f_{ijk}f_k}{1+|\nabla f|^2} + \frac{f_{ij}f_{kk}}{1+|\nabla f|^2} - \frac{2f_{ij}f_{kl}f_k f_l}{(1+|\nabla f|^2)^2}.$$

Combining the formula

$$R(g) = g^{ij}\left(\Gamma^k_{ij,k} - \Gamma^k_{ik,j} + \Gamma^l_{ij}\Gamma^k_{kl} - \Gamma^l_{ik}\Gamma^k_{jl}\right).$$

for the scalar curvature $R(g)$ in terms of the Christoffel symbols with Lemma 9-8 we obtain the following.

Lemma 9-9. *The scalar curvature $R(g)$ of a manifold M that is the graph of a function $f : \mathbb{R}^n \to \mathbb{R}$ is given by*

$$R(g) = \frac{1}{1+|\nabla f|^2}\left(f_{ii}f_{jj} - f_{ij}f_{ij} - \frac{2f_j f_k}{1+|\nabla f|^2}(f_{ii}f_{jk} - f_{ij}f_{ik}) \right).$$

Lemma 9-10. *The scalar curvature of a graph over $\mathbb{R}^n$ is given by*

$$R(g) = \mathrm{div}_0\left(\frac{1}{1+|\nabla f|^2}(f_{ii}f_j - f_{ij}f_i)\frac{\partial}{\partial x_j} \right),$$

where div_0 is the divergence with respect to the Euclidean metric.

The proof, a simple calculation using Lemma 9-9, is left as an exercise.

We will need one more lemma relating the volume form and mean curvature with respect to g to the volume form and mean curvature with respect to the Euclidean metric.

Lemma 9-11. *Let dV_g be the volume measure of (M, g), and let dV_0 be the Euclidean volume measure on $\mathbb{R}^n$. Using pullback under graphical coordinates, we have*

$$dV_g = \sqrt{1+|\nabla f|^2}\, dV_0.$$

Also, if Σ is a regular level set of a smooth function $f : \mathbb{R}^n \setminus \Omega \to \mathbb{R}$, its mean curvature H_0 with respect to the Euclidean metric is given by

$$H_0 = \frac{1}{|\nabla f|^2}(f_{ii}f_j - f_{ij}f_i)v_j,$$

where $v = v_j\dfrac{\partial}{\partial x_j}$ is a unit normal vector to Σ.

Proof. The first equation follows directly from the fact $dV_g = \sqrt{\det g}\, dV_0$. The second equality comes from the fact that, for a set defined as a level set of a function, a unit normal vector is given by $v = \pm\frac{\nabla f}{|\nabla f|}$, and the mean curvature is just $H_0 = \mathrm{div}_0(v)$. Hence we find

$$H_0 = \pm\frac{\partial}{\partial x_i}\frac{f_i}{|\nabla f|} = \pm\frac{1}{|\nabla f|^3}(f_{ii}f_jf_j - f_{ij}f_if_j) = \frac{1}{|\nabla f|^2}(f_{ii}f_j - f_{ij}f_i)v_j,$$

as desired. $\qquad\square$

Remark 9-12. That $|\nabla f(x)| \to \infty$ as $x \to \Sigma$ is a technicality that is not discussed in Lam's original paper. It can be dealt with by approximating Σ by (sufficiently close) parallel surfaces Σ_ϵ on its outside; see [121]. All integrals over Σ should actually be construed as limits of integrals over Σ_ϵ; we avoid further comment here for the sake of simplicity.

Proof of Theorems 9-4 and 9-6. We will prove both theorems at once where we note that for Theorem 9-4 we will use the assumption that M is a graph over all of $\mathbb{R}^n$, whereas for Theorem 9-6 we will use the assumption that M is a graph over $\mathbb{R}^n \setminus \Omega$, and $\Omega \neq \varnothing$. The mass of (M, g) is given by the following where $S_r = \{x \in \mathbb{R}^n : |x| = r\}$ and dS_r is the Euclidean surface measure associated to this set:

$$
\begin{aligned}
m &= \lim_{r \to \infty} \frac{1}{2(n-1)\omega_{n-1}} \int_{S_r} (g_{ij,i} - g_{ii,j}) v_j \, dS_r \\
&= \lim_{r \to \infty} \frac{1}{2(n-1)\omega_{n-1}} \int_{S_r} (f_{ii} f_j - f_{ij} f_i) v_j \, dS_r \\
&= \lim_{r \to \infty} \frac{1}{2(n-1)\omega_{n-1}} \int_{S_r} \left(\frac{f_{ii} f_j - f_{ij} f_i}{1 + |\nabla f|^2} \right) v_j \, dS_r,
\end{aligned}
$$

where the last equality follows from the fact that

$$
\frac{1}{1 + |\nabla f(x)|^2} = 1 + O(|x|^{-p)}) \quad \text{as } |x| \to \infty,
$$

by Definition 9-3. To prove Theorem 9-4 we use the divergence theorem. We notice that when M is a graph over all of $\mathbb{R}^n$ there is no boundary term. Applying the divergence theorem gives

$$
\begin{aligned}
m &= \frac{1}{2(n-1)\omega_{n-1}} \int_{\mathbb{R}^n} \mathrm{div}_0 \left(\frac{f_{ii} f_j - f_{ij} f_i}{1 + |\nabla f|^2} \frac{\partial}{\partial x_j} \right) dV_0 \\
&= \frac{1}{2(n-1)\omega_{n-1}} \int_{\mathbb{R}^n} \frac{R(g)}{\sqrt{1 + |\nabla f|^2}} \, dV_g,
\end{aligned}
$$

as desired. For the case where the asymptotic end of M is a graph over $\mathbb{R}^n \setminus \Omega$, with $\Omega \neq \varnothing$, we apply the divergence theorem and this time we get a boundary term. Since $|\nabla f(x)| \to \infty$ as $x \to \Sigma$, we must proceed carefully.

By virtue of Lemma 9-11 we see that

$$
\frac{f_{ii} f_j - f_{ij} f_i}{1 + |\nabla f|^2} v_j = \frac{|\nabla f|^2}{1 + |\nabla f|^2} H_0.
$$

The behavior of $|\nabla f(x)|^2$ as $x \to \Sigma$ is no cause for concern since $\dfrac{|\nabla f|^2}{1 + |\nabla f|^2} \to 1$. With v pointing into Ω along Σ, we find

$$
\begin{aligned}
m &= \frac{1}{2(n-1)\omega_{n-1}} \left(\int_{\mathbb{R}^n \setminus \Omega} \mathrm{div}_0 \left(\frac{f_{ii} f_j - f_{ij} f_i}{1 + |\nabla f|^2} \frac{\partial}{\partial x_j} \right) dV_0 + \int_{\Sigma} \frac{f_{ii} f_j - f_{ij} f_i}{1 + |\nabla f|^2} v_j \, d\mu_0 \right) \\
&= \frac{1}{2(n-1)\omega_{n-1}} \left(\int_{\mathbb{R}^n \setminus \Omega} \frac{R(g)}{\sqrt{1 + |\nabla f|^2}} \, dV_g + \int_{\Sigma} H_0 \, d\mu_0 \right),
\end{aligned}
$$

where $d\mu_0$ is the surface measure of Σ with respect to the Euclidean metric.

The proof now follows from the next result. $\square$

Theorem 9-13 (Alexandrov–Fenchel inequality [4; 5]). *Let Σ be a convex hypersurface in $\mathbb{R}^n$. If we let H denote its mean curvature (with respect to the outward normal), we have*

$$\frac{1}{2(n-1)\omega_{n-1}} \int_\Sigma H \, d\mu_0 \geq \frac{1}{2} \left(\frac{|\Sigma|}{\omega_{n-1}} \right)^{\frac{n-2}{n-1}}.$$

See [138] for a short proof. We will show in Theorem 9-26 a generalization of Theorem 9-13, due to Freire and Schwartz, which works for mean-convex, outer-minimizing domains Σ. It relates to Huisken and Ilmanen's proof of the RPI, in that it uses their work on inverse mean curvature flow in arbitrary dimensions [125]. Theorem 9-26 can be applied to Lam's result, readily generalizing Theorem 9-6.

9.4. Huisken and Ilmanen's proof of the RPI using IMCF

We now give an overview of Huisken and Ilmanen's proof of the Riemannian Penrose inequality using inverse mean curvature flow (IMCF).

Theorem 9-14 (Huisken–Ilmanen [125]). *Let (M^3, g) be a complete, asymptotically flat manifold with compact, minimal surface boundary ∂M, $R(g) \geq 0$ and ADM mass m. Then*

$$m \geq \sqrt{\frac{|\Sigma|}{16\pi}}$$

whenever Σ is a connected component of ∂M, and there are no other closed minimal surfaces in $M \setminus \partial M$. Equality is achieved if and only if (M^3, g) is isometric to the Riemannian Schwarzschild manifold.

Huisken and Ilmanen's proof uses an important geometric evolution equation called inverse mean curvature flow (IMCF), which is interesting in its own right. It was first introduced by Geroch [105], and later developed by Jang and Wald [131]. The precise definition is the following.

Let M^{n+1} and Σ^n be Riemannian manifolds. (Think of Σ^n as a submanifold of M^{n+1}.)

Definition 9-15. A *smooth solution to IMCF* is a one-parameter family of embeddings $F : \Sigma \times [0, T) \to M$ such that

$$\frac{\partial F}{\partial t} = \frac{\nu}{H}, \tag{9.4.1}$$

where ν is a smooth unit normal field to $\Sigma_t := F(\Sigma, t)$ and $H > 0$ is the mean curvature of Σ_t with respect to the choice of normal.

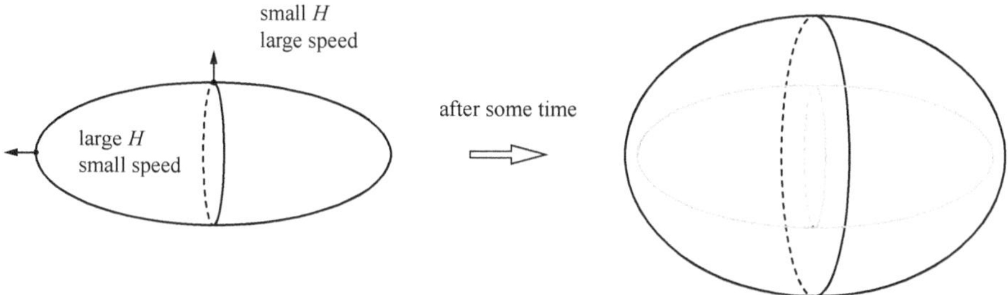

Figure 8. An ellipsoid becoming more round along IMCF.

For the purposes of this chapter we will always be dealing with $\Sigma \subset M$ a closed hypersurface, and in case Σ bounds a region we will choose ν to be the outward pointing normal vector. H will then be positive for (small) geodesic spheres in M under our convention.

IMCF defined in this way is a degenerate nonlinear parabolic PDE, since the denominator in the right side of (9.4.1) can vanish. Because of this, the IMCF should behave like the heat equation provided we can get control on the mean curvature of Σ_t. Heuristically, if we think of curvature as heat and the IMCF as the heat equation, then curvature (or heat) should become uniform over time along IMCF. This is to say, we can expect that under IMCF a suitably chosen initial condition Σ_0 should evolve into a surface of constant curvature as $t \to \infty$.

Example 9-16 (IMCF with a round sphere as initial condition). We start out by giving an example of how $\mathbb{S}^n(r) \subset \mathbb{R}^{n+1}$ (the round sphere of radius r) evolves under IMCF. Since the mean curvature of $\mathbb{S}^n(r)$ is constant and equal to $\frac{n}{r}$, it follows from (9.4.1) that a smooth solution to IMCF with initial condition $\mathbb{S}^n(r)$ is given by $\Sigma_t = \mathbb{S}^n(r(t))$, where $r(t)$ satisfies the ODE $r'(t) = \frac{r(t)}{n}$. (Check this!) This ODE, in turn, has solution $r(t) = e^{t/n}$. By uniqueness, the above solution is the only solution of the IMCF with initial condition $\mathbb{S}^n(r)$. In other words, the round sphere remains a round sphere under IMCF, and its radius expands exponentially fast in time.

A natural question that arises from this example is to determine which surfaces behave like spheres under IMCF. This is to say, for which initial conditions on Σ_0 do we find exponentially fast convergence to round spheres as $t \to \infty$? Since IMCF is a nonlinear and degenerate flow (see (9.4.2) below), we do not expect this behavior to hold for all initial hypersurfaces. For example, singularities such as $H \to 0$ or $|A|^2 \to \infty$ as $t \to T < \infty$ will cause the flow to no longer remain an embedding at finite values of t. Gerhardt [102] and Urbas [216] separately give certain sufficient conditions:

Theorem 9-17 (Gerhardt [102], Urbas [216]). *If $\Sigma_0 \subset \mathbb{R}^{n+1}$ is star-shaped with $H > 0$, the IMCF with initial condition Σ_0 has a smooth solution for all time, and the rescaled embeddings $\widetilde{F}(t) = e^{-t/n} F(t)$ converge exponentially fast to a smooth embedding $\widetilde{F}_\infty$ for which $\widetilde{F}_\infty(\Sigma) = \mathbb{S}^n(r_\infty)$, where $r_\infty = (|\Sigma_0|/\omega_n)^{1/n}$.*

Now we switch gears back to the Riemannian Penrose inequality, and define the Hawking mass of a hypersurface. This quantity is a crucial ingredient in Huisken and Ilmanen's proof of the RPI.

Definition 9-18. The Hawking mass of a hypersurface $\Sigma^2 \subset (M^3, g)$ (with surface measure $d\mu$) is

$$m_H(\Sigma) := \sqrt{\frac{|\Sigma|}{(16\pi)^3}} \left(16\pi - \int_\Sigma H^2 d\mu \right).$$

The outline of the proof of the RPI envisioned by Geroch [105] and further developed by Jang and Wald [131] is based on the key facts collected in the following proposition, from which Theorem 9-14 follows.

Proposition 9-19. *Suppose (M^3, g) is asymptotically flat with nonnegative scalar curvature and Σ_0 is a spherical hypersurface with smooth normal field ν. If the smooth IMCF exists for all time, then*

(i) *for $H = 0$ (that is, if Σ_0 is a minimal surface), we have $m_H(\Sigma_0) = \sqrt{|\Sigma_0|/(16\pi)}$;*

(ii) *$m_H(\Sigma_t)$ is monotone nondecreasing under IMCF; and*

(iii) *$\lim_{t \to \infty} m_H(\Sigma_t) = m_{ADM}(M)$.*

Hence

$$\sqrt{\frac{|\Sigma_0|}{16\pi}} = m_H(\Sigma_0) \le m_H(\Sigma_t) \le \lim_{t \to \infty} m_H(\Sigma_t) = m_{ADM}(M) = m.$$

Conclusion (i) follows by Definition 9-18, and (ii) follows from the Geroch monotonicity formula in Section 9.4.1 below. Underlying (iii) is a result (similar to Theorem 9-17 above) on weak convergence to a round sphere, in which Huisken and Ilmanen are able to relate the Hawking mass to a limit of integrals over coordinate spheres. From these three key facts the proof of the RPI follows, so long as the IMCF has a smooth solution for all time, which does not happen in general. Huisken and Ilmanen's contribution consists in the monumental task of developing a weak existence setting for which the above facts still remain true.

9.4.1. *Geroch's monotonicity formula.* We want to compute the evolution equation of the mean curvature of a hypersurface along the IMCF. For this, we use the Riccati equation together with the definition of IMCF and obtain (as in [125]):

$$\frac{\partial H}{\partial t} = -\Delta\left(\frac{1}{H}\right) - \frac{|A|^2}{H} - \frac{\overline{Rc}(v, v)}{H}, \qquad \frac{\partial}{\partial t} d\mu_t = d\mu_t,$$

where Δ is the Laplacian on Σ_t, A is the second fundamental form of Σ_t, $\overline{Rc}(\cdot, \cdot)$ is the Ricci curvature of (M, g), and $d\mu_t$ is the induced area measure on Σ_t. Notice that the second equation implies that $|\Sigma_t| = |\Sigma_0|e^t$. We also need the following formula, which is a consequence of the Gauss equation, where σ_Σ is the sectional curvature of $T_x\Sigma$ in Σ (i.e. the Gauss curvature of Σ), $\bar\sigma_\Sigma$ is the sectional curvature of $T_x\Sigma$ in M, $\bar R$ is the scalar curvature of M, and λ_1, λ_2 are the principal curvatures of Σ in M:

$$\sigma_\Sigma = \bar\sigma_\Sigma + \lambda_1\lambda_2 = \tfrac{1}{2}\bar R - \overline{Rc}(v, v) + \tfrac{1}{2}(H^2 - |A|^2).$$

This is a general formula that holds on Σ_t as well.

Putting all these together we get

$$\frac{\partial}{\partial t}\int_{\Sigma_t} H^2 d\mu_t = \int_{\Sigma_t}\left(2H\frac{\partial H}{\partial t}d\mu_t + H^2\frac{\partial}{\partial t}d\mu_t\right)$$

$$= \int_{\Sigma_t}\left(-2H\Delta\left(\frac{1}{H}\right) - 2|A|^2 - 2\overline{Rc}(v, v) + H^2\right)d\mu_t$$

$$= \int_{\Sigma_t}\left(-2\frac{|\nabla H|^2}{H^2} - |A|^2 - \bar R + 2\sigma_{\Sigma_t}\right)d\mu_t$$

$$= 4\pi\chi(\Sigma_t) + \int_{\Sigma_t}\left(-2\frac{|\nabla H|^2}{H^2} - \tfrac{1}{2}H^2 - \tfrac{1}{2}(\lambda_1 - \lambda_2)^2 - \bar R\right)d\mu_t$$

$$\le \frac{1}{2}\left(16\pi - \int_{\Sigma_t} H^2 d\mu_t\right),$$

where we integrated by parts and used the Gauss equation in the third equality, and then used Gauss–Bonnet and the fact that $|A|^2 = \tfrac{1}{2}H^2 + \tfrac{1}{2}(\lambda_1 - \lambda_2)^2$ in the fourth. The last inequality follows from the assumption $\bar R \ge 0$ and the fact that if (smooth) IMCF exists for all $t \subset [0, \infty)$, the topology of Σ_t remains the same for all t. Hence, if Σ_0 is a sphere (topologically), then Σ_t remains a topological sphere for all t; in particular it remains *connected*, and $\chi(\Sigma_t) = 2$ for all t; of course if Σ_t is a closed, connected, orientable surface, then $\chi(\Sigma_t) \le 2$ and the inequality would also persist.

With this, we deduce

$$\frac{\partial}{\partial t}\left(16\pi - \int_{\Sigma_t} H^2 d\mu_t\right) \geq -\frac{1}{2}\left(16\pi - \int_{\Sigma_t} H^2 d\mu_t\right).$$

Integrating this differential inequality we obtain that the quantity

$$e^{t/2}\left(16\pi - \int_{\Sigma_t} H^2 d\mu_t\right)$$

is nondecreasing in t. Using $|\Sigma_t|^{1/2} = |\Sigma_0|^{1/2}e^{t/2}$ in the preceding equation proves that the Hawking mass is nondecreasing along IMCF. This fact is known as *Geroch's monotonicity formula*.

Remark 9-20. If we could prove that a smooth solution of IMCF starting from a minimal surface exists for all time, we would have a proof of the RPI. This is the heuristic argument that originally motivated Geroch to pursue IMCF. Unfortunately, this cannot be expected, as there are counterexamples where smooth solutions of IMCF do not exist for all time. Indeed, if we start the flow with initial condition consisting of two spheres far apart (far away black holes), both surfaces will expand exponentially fast under IMCF, as we have seen. Thus, they eventually meet, and this produces a "jump" in the topology. More precisely, embeddedness of the flow no longer holds [125]. For another example, take for initial condition a thin torus — one for which the inner radius is much smaller than the outer radius. Then $H > 0$, so under IMCF the torus will expand, but it cannot expand forever. Eventually $H \to 0$ somewhere and $|A|^2 \to \infty$ [125].

9.4.2. *IMCF in the weak setting.* The weak setting developed by Huisken and Ilmanen in [125] produces a flow that exists for all time starting off with any reasonable initial condition. Their weak solution is unique and has "jumps", but nevertheless associated quantities such as the area of the flowing hypersurface remain continuous (except possibly at time zero). Other quantities, such as the total mean curvature, remain monotonic along the weak flow. Here we present the main ideas in the construction of weak solutions and mention some of its consequences, while omitting details and heavy calculations, such as those involving geometric measure theory.

The first step in defining weak solutions to IMCF is to rewrite the IMCF equation as a (degenerate) elliptic PDE. This is accomplished by defining a level set formulation for which Σ_t is the level set of a function. More precisely, we let $u : \Omega \subset M \to \mathbb{R}$, $\Omega \subset M$ open, and define

$$E_t := \{u < t\}, \quad \Sigma_t := \partial E_t, \quad E_t^+ := \mathrm{int}\{u \leq t\}, \quad \Sigma_t^+ := \partial E_t^+.$$

where ∂ and int represent the topological boundary and interior of a set.

If $x(t)$ is a path with $u(x(t)) = t$ (i.e., $x(t) \in \Sigma_t$), then as $v = \frac{\nabla u}{|\nabla u|}$, we see the normal speed of a moving level set is given by

$$\left| \left(\frac{dx}{dt} \right)^N \right| = \frac{1}{|\nabla u(x)|}.$$

If the level sets were to satisfy IMCF, so that $\frac{dx}{dt} = \frac{v}{H}$ with $x(t) = F(p, t)$, we would obtain

$$\mathrm{div}_g \left(\frac{\nabla u}{|\nabla u|} \right) = |\nabla u|, \qquad (9.4.2)$$

as the term on the left is the mean curvature of a level set.

Huisken and Ilmanen define weak solutions of IMCF by constructing a functional whose Euler–Lagrange equation is (almost) (9.4.2). Actually, they want to find a functional whose Euler–Lagrange equation "freezes" the right-hand side of (9.4.2), so to speak. The idea is the following. We note that the difference between Σ_t and Σ_t^+ from above arises when u is constant on a set with nonempty interior. Let us define the values $\hat{t}$ for which $\Sigma_{\hat{t}} \neq \Sigma_{\hat{t}}^+$, and call them *jump times* since Σ_t will not evolve smoothly into $\Sigma_{\hat{t}}^+$, rather $\Sigma_{\hat{t}}$ will jump (instantly) to $\Sigma_{\hat{t}}^+$ in the weak setting. With this in mind, we define weak solutions to the IMCF as follows. Fix a locally Lipschitz function u and a compact set $K \subset \Omega$, and consider the functional J_u^K defined as

$$J_u^K(v) = \int_K \left(|\nabla v| + v|\nabla u| \right) dv_g,$$

where v is locally Lipschitz, $\overline{\{v \neq u\}} \subset \Omega$ is compact and dv_g is the measure associated to (M, g). Then, it follows that the Euler–Lagrange equation for this functional is given by

$$\mathrm{div}_g \left(\frac{\nabla v}{|\nabla v|} \right) = |\nabla u|,$$

which is (9.4.2) with the right-hand side frozen (see [125]). From this, we see that if u minimizes J_u^K, i.e. if $J_u^K(u) \leq J_u^K(v)$ for all such v, then we have found a weak level-set solution to (9.4.2), as desired.

The problem has been reduced to finding a function u that minimizes $J_u^k(v)$. Huisken and Ilmanen show, using the co-area formula, that this is equivalent to having $E_t = \{u < t\}$ minimize the functional

$$J_u(F) = |\partial^* F \cap K| - \int_F |\nabla u| dv_g,$$

where F has locally finite perimeter. Here, $|\partial^* F \cap K|$ is the Hausdorff measure, and ∂^* stands for the reduced boundary; this quantity is defined for sets of locally finite perimeter, and coincides with the usual notion of boundary on smooth sets. (See [108] for more on these geometric measure theoretical notions.)

Huisken and Ilmanen are able to prove using standard regularity results from geometric measure theory that Σ_t is at least $C^{1,\alpha}$ [125] (ambient dimension less than eight). Thus, we will assume in what remains of this section that all sets we deal with are at least this regular. Now we will describe an intuitive, geometric way of understanding weak solutions of IMCF using the definitions from above.

Since equation (9.4.2) may be degenerate when $|\nabla u| = 0$, we need to regularize the PDE in order to prove existence of solutions. It turns out that this process also holds the key for showing that the monotonicity of the Hawking mass can be applied in the weak setting, and so it is important to understand how to accomplish this.

To regularize the PDE (9.4.2) we take $\epsilon > 0$, and we consider the equation

$$\operatorname{div}_g \frac{\nabla u_\epsilon}{\sqrt{|\nabla u_\epsilon|^2 + \epsilon^2}} = \sqrt{|\nabla u_\epsilon|^2 + \epsilon^2}. \tag{9.4.3}$$

If u_ϵ is a smooth solution of (9.4.3), and if we define $U_\epsilon : \Omega \times \mathbb{R} \to \mathbb{R}$ by $U_\epsilon(x, z) = u_\epsilon(x) - \epsilon z$, then U_ϵ is a smooth solution to (9.4.2), but in one dimension higher. In fact if we let $\widetilde{\Sigma}_t^\epsilon := \{U_\epsilon = t\}$, then one can show that

$$\widetilde{\Sigma}_t^\epsilon = \operatorname{graph}\left(\frac{u_\epsilon}{\epsilon} - \frac{t}{\epsilon}\right),$$

and hence $\widetilde{\Sigma}_t^\epsilon$ is a translating solution of IMCF inside $\Omega \times \mathbb{R}$ (see Figure 9).

Huisken and Ilmanen show that existence of solutions to (9.4.3) is guaranteed by the existence of a subsolution of (9.4.2) which can be taken to be an exponentially expanding coordinate sphere, e.g., $u(x) = C \log |x|$, for some $C > 0$.

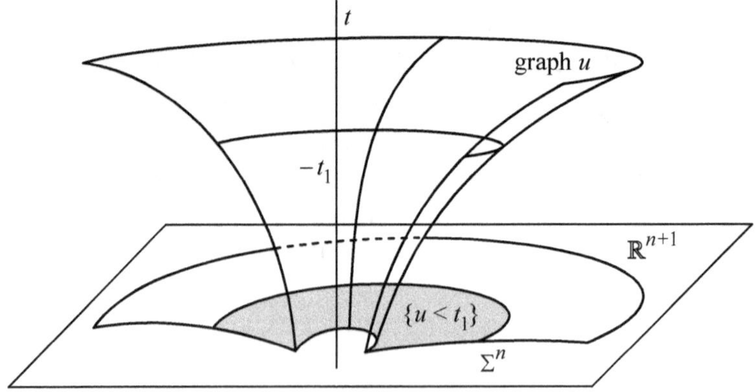

Figure 9. Level set graph $\widetilde{\Sigma}_t^\epsilon$ translating downwards [149].

(Since this is a strong subsolution it is defined in the usual way for (9.4.2).) The estimates in [125] also give that for a suitable subsequence $\epsilon_i \to 0^+$, we have that $U_{\epsilon_i}(x, z) \to u(x)$ as well as $\tilde{\Sigma}_t^{\epsilon_i} \to \Sigma_t \times \mathbb{R}$, locally in $C^{1,\alpha}$. This last fact is key for proving the monotonicity of the Hawking mass under weak solutions, as we see below.

Now we turn to developing an intuitive, geometric way of understanding weak solutions of IMCF. In order to do that, we define the following notion.

Definition 9-21. Let $\Omega \subset M$ be open. We say that $E \subset M$ is a *minimizing hull* if it minimizes area on the outside (in Ω). This is, E satisfies

$$|\partial^* E \cap K| \leq |\partial^* F \cap K|$$

for any F such that $E \subset F$ and $\overline{F \setminus E} \subset \Omega$ is compact, and any compact set $K \subset \Omega$ such that $F \setminus E \subset K$. Additionally, we say that E is a *strictly minimizing hull* if equality above implies that $F \cap \Omega = E \cap \Omega$ almost everywhere.

Minimizing hulls minimize area amongst all competing sets that contain them, and are thus called *outer minimizing sets*. Finding the minimizing hull containing some set is sometimes called the *shrink wrap problem*, since it amounts to finding the least area enclosure of a given region. By the first variation formula for the area, an outer-minimizing hull must have nonnegative (weak) mean curvature everywhere. (Otherwise, we could choose a compactly supported, outward-pointing deformation of the hull with lesser area.) In particular, an outer-minimizing, closed minimal surface is a minimizing hull. This suggests that we should be able to run the weak IMCF with such a surface as the initial condition, as it will immediately flow to a surface with positive (weak) mean curvature.

Proposition 9-22 (Huisken–Ilmanen [125, Proposition 1.4]). *Consider a weak solution to IMCF as above, and assume that E_t is precompact. Then:*

(i) *E_t is a minimizing hull in M for all $t > 0$.*

(ii) *E_t^+ is a strictly minimizing hull in M for all $t \geq 0$.*

(iii) *$|\Sigma_t| = |\Sigma_t^+|$ for all $t \geq 0$, provided that E_0 is a minimizing hull.*

Using this, we can portray a (heuristic) geometric characterization of weak solutions to the flow:

- E_t flows by the smooth IMCF as long as E_t is a strictly minimizing hull.

- E_t jumps to its strictly minimizing hull E_t^{+} when it is not a strictly minimizing hull.

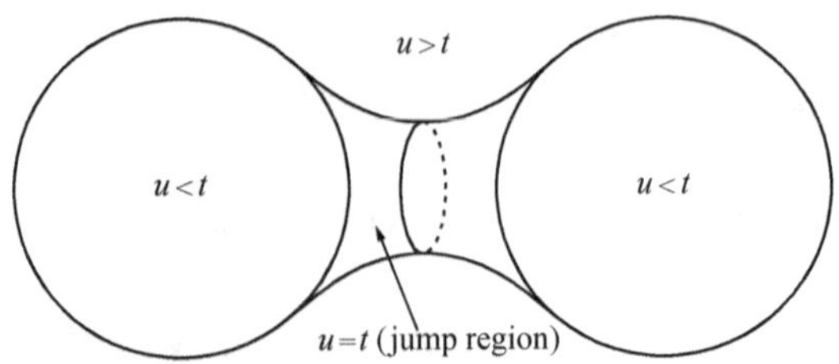

Figure 10. Time to jump [125].

In Figure 10 we see two spheres that are flowing under weak IMCF until
the moment when they can be enclosed by a "peanut" — two spherical caps
joined by a catenoidal bridge — of the same area. The weak solution will then
instantly jump across the region between these two surfaces, and resume from
the outermost surface.

(There is an alternative notion of weak solutions to IMCF using viscosity
solutions [53]. Such solutions will agree with smooth IMCF until a singularity
is formed; but the problem arises that, after a singular time, Σ_t may no longer
be a hypersurface anymore. Therefore, a viscosity solution of the IMCF is less
desirable, since it is not clear how to preserve — or even make sense of — the
monotonicity of the Hawking mass.)

From this intuitive geometric picture we can expect at jump times under
weak IMCF that (1) the area of the flowing hypersurface remains continuous
($|\partial E_t^+| = |\partial E_t|$), and (2) the mean curvature does not increase. Therefore, the
following should hold at jump times:

$$\int_{\partial E_t^+} H^2 \, d\mu_t \leq \int_{\partial E_t} H^2 \, d\mu_t.$$

Remark 9-23. To prove these statements one considers $\widetilde{\Sigma}_t^\epsilon$, the smooth translating solutions of IMCF in $M^3 \times \mathbb{R}$, for which the smooth monotonicity of the
Hawking mass calculation holds. With this, weak monotonicity of the Hawking
mass follows by taking a limit as $\epsilon \to 0^+$ *so long as* $\chi(\Sigma_t) \leq 2$. This, in turn, is
accomplished by showing that if Σ_0 is connected, the weak solution of IMCF
will remain connected for all times. See Sections 4 and 5 of [125] for details.

The proof of part (iii) of Proposition 9-19, in the weak setting, follows from a
weak convergence result for weak solutions of IMCF.

Theorem 9-24 (Huisken–Ilmanen [125, p. 417]). *Suppose (M^3, g) is asymptotically flat. Let $\Sigma_t \subset M$ be a weak solution to IMCF and let $r(t)$ be the area radius,
i.e., the quantity defined implicitly by $|\Sigma_t| = 4\pi r(t)^2$. Then $\dfrac{1}{r(t)} \Sigma_t$ converges in
C^1 to to the round unit sphere as $t \to \infty$.*

Note that since we are rescaling by the area radius, all rescaled solutions here converge to the unit sphere. This is in contrast to the result of Theorem 9-17, where we have rescaled to keep area constant to show that all star-shaped hypersurfaces converge to a sphere with the same area as Σ_0.

Using Theorem 9-24 it can be shown that $r(t)H_{\Sigma_t}$ converges in L^2 to the analogous quantity for the round sphere. This is helpful for showing that the Hawking mass of the flowing hypersurface (in the weak setting) converges to the ADM mass as $t \to \infty$, since the ADM mass is defined as the limit of integrals over spheres of increasing radius [15]. Generally speaking, the fact that $m_H(\Sigma_t)$ limits to $m_{ADM}(M)$ as $t \to \infty$ is expected, since these two quantities are equal in the case of the (Riemannian) Schwarzschild manifold. (Compare this to the case of asymptotically hyperbolic manifolds [167] where this asymptotic property is not true.) The proof of this statement is somewhat lengthy, though; we give only a brief overview here.

Let us denote various geometric quantities of Σ_t with respect to (M, g) by $H, A, v, d\mu_t, \nabla$, and the corresponding ones with respect to $(\mathbb{R}^3, \delta)$ by H_0, A_0, v_0, $d\mu_t^0, \nabla^0$. If we recall that $|p(x)| \to 0$ as $|x| \to \infty$, where $p_{ij} = g_{ij} - \delta_{ij}$ (because of our assumption of asymptotic flatness, which also gives us a specified rate of convergence $|p| \le C/|x|$), we obtain [125, p. 418]

$$H - H_0 = -h^{ik} p_{kl} h^{lj} A_{ij} + \tfrac{1}{2} H v^i v^j p_{ij} - h^{ij} \nabla_i p_{jl} v^l + \tfrac{1}{2} h^{ij} \nabla_l p_{ij} v^l$$
$$\pm C|p||\nabla p| \pm C|p|^2 |A|,$$
$$H_0^2 (d\mu_t - d\mu_t^0) = \left(\tfrac{1}{2} H^2 h^{ij} p_{ij} \pm |p|^2 |A|^2 \pm C|\nabla p|^2 \right) d\mu_t,$$

where h is the metric of Σ_t.

Now write

$$\int_{\Sigma_t} H^2 \, d\mu_t = \int_{\Sigma_t} H_0^2 d\mu_t^0 + H_0^2 (d\mu_t - d\mu_t^0) + \left(2H(H - H_0) - (H - H_0)^2 \right) d\mu_t.$$

The last term converges to zero by the L^2 convergence of H to H_0. Using that $\int_\Sigma H_0^2 d\mu^0 \ge 16\pi$ for $C^{1,\alpha}$ hypersurfaces in $\mathbb{R}^3$ [223], we obtain control of the first term in terms of the 16π factor that we encounter in the definition of the Hawking mass. After integrating by parts, the remaining terms cancel, leaving error terms as well as, remarkably, terms in the definition of the ADM mass. More specifically, using the above expressions the factor $h^{ij} \nabla_i p_{jl} v^l$ in $H - H_0$ recovers terms in the definition of the ADM mass, plus error terms that integrate to zero in the limit as $t \to \infty$, where we note that $\nabla_i^0 p_{jl} = \nabla_i^0 g_{jl}$ and $\nabla = \nabla^0 \pm C|p||\nabla p|$. This completes our overview of the proof of asymptotic convergence of the Hawking mass to the ADM mass in the weak setting. Using it, we obtain a proof of Theorem 9-14. The interested reader is directed to [125] for details.

Remark 9-25. One could try to run IMCF on a spacelike slice of spacetime with initial condition a codimension-two surface, in order to prove the full Penrose inequality. Huisken and Ilmanen point out in [125] that such a construction produces a flow that is a forward-backward system of PDE, and hence does not possess good existence theory.

9.4.3. *The generalized Alexandrov–Fenchel inequality.* Now we state and outline the proof of the promised generalization of Theorem 9-13.

Theorem 9-26 (Freire–Schwartz [95]). *Let $\Omega \subset \mathbb{R}^n$ be open, with smooth, mean convex, outer-minimizing boundary Σ with surface measure $d\mu$. Then*

$$\frac{1}{2(n-1)\omega_{n-1}} \int_\Sigma H \, d\mu \geq \frac{1}{2} \left(\frac{|\Sigma|}{\omega_{n-1}} \right)^{\frac{n-2}{n-1}}.$$

Equality is achieved if and only if Ω is a round ball.

Proof. The full proof of this theorem requires weak solutions to IMCF, so we will only prove this theorem for the case where IMCF starting from Σ is smooth for all time, e.g. when Σ is mean convex and star-shaped. For the rest of the proof in the weak case see [95].

Claim: *Let Σ be as above, and consider a (weak) solution of IMCF in $\mathbb{R}^n$ with initial condition $\Sigma_0 = \Sigma$. Then*

$$\int_{\Sigma_t} H \, d\mu_t \leq \left(\int_\Sigma H \, d\mu \right) e^{\frac{n-2}{n-1}t}.$$

To see this in the smooth case, since $(n-1)|A|^2 \geq H^2$, we have

$$H - \frac{|A|^2}{H} - \frac{n-2}{n-1}H = \left(H^2 - (n-1)|A|^2 \right) \frac{1}{(n-1)H} \leq 0.$$

Using this and the Riccati equation we obtain

$$\frac{d}{dt} \int_{\Sigma_t} H \, d\mu_t = \int_{\Sigma_t} \left(H - \frac{|A|^2}{H} \right) d\mu_t \leq \frac{n-2}{n-1} \int_{\Sigma_t} H \, d\mu_t.$$

Integrating gives the claim.

Now if we let $f(t) = |\Sigma_t|^{-\frac{n-2}{n-1}} \int_{\Sigma_t} H \, d\mu_t$, then using $\frac{d}{dt}|\Sigma_t| = |\Sigma_t|$, we find

$$f'(t) = \frac{d\,|\Sigma_t|^{-\frac{n-2}{n-1}}}{dt} \int_{\Sigma_t} H \, d\mu_t + |\Sigma_t|^{-\frac{n-2}{n-1}} \frac{d}{dt} \int_{\Sigma_t} H \, d\mu_t$$

$$\leq -\frac{n-2}{n-1} |\Sigma_t|^{-\frac{n-2}{n-1}} \int_{\Sigma_t} H \, d\mu_t + \frac{n-2}{n-1} |\Sigma_t|^{-\frac{n-2}{n-1}} \int_{\Sigma_t} H \, d\mu_t$$

$$= 0.$$

Then we see that

$$|\Sigma|^{-\frac{n-2}{n-1}} \int_{\Sigma} H \, d\mu = f(0) \geq \lim_{t \to \infty} f(t) = (n-1)\omega_{n-1}^{1/(n-1)},$$

where the last equality follows from the fact that IMCF converges to a round sphere. We are omitting the technical part of the argument, which is to show the convergence in the weak setting for the above quantities. This is not automatic since we are not integrating H^2 as in Huisken and Ilmanen's work, rather we are integrating H. See [95] for details. $\qquad\square$

The Freire–Schwartz theorem allows us to generalize Theorem 9-6 to the case where $\partial\Omega$ is a union of smooth, mean-convex, outer-minimizing sets. Related results in [95] include a generalized Pólya–Szegő inequality, as well as mass-capacity and volumetric Penrose inequalities for the conformally flat case. The proofs of these results work in arbitrary dimensions and are independent of the positive mass theorem. It is worth pointing out that the proof of the mass-capacity inequality is the only part of Bray's work [26] where the positive mass theorem is used; cf. Theorems 9-32 and 9-33 below.

9.5. Bray's proof of the RPI using PMT

We now turn our attention to an overview of Bray's proof of the Riemannian Penrose inequality. The precise statement is the following:

Theorem 9-27 (Bray [26]). *Let (M^3, g) be a complete, smooth, asymptotically flat Riemannian manifold with $R \geq 0$, with ADM mass m and with outer-minimizing minimal surface Σ with area $|\Sigma|$. Then*

$$m \geq \sqrt{\frac{|\Sigma|}{16\pi}},$$

and equality is achieved if and only if (M^3, g) is isometric (outside Σ) to the Riemannian Schwarzschild manifold of mass m.

The main advantage of Bray's proof is that it can handle *horizons* (outer-minimizing minimal surfaces) that are not connected. Another is that it does not use the Geroch monotonicity formula, which in turn relies on Gauss–Bonnet. Thus, it can be (and has been) generalized to higher dimensions [28]. On the downside, Bray's proof relies on the positive mass theorem and on geometric measure theory quite heavily, so proving the RPI in dimensions eight and above with this approach does not follow directly (and remains open, though recent developments [205] might prove promising in this direction).

Bray's proof of the RPI can be broken down into four main steps:

1. Define an appropriate conformal flow of metrics $\{g_t\}_{t\geq 0}$ on $M_t := M \setminus \Omega_t$, where Ω_t is the region inside the horizon of g_t.

2. Prove that the area of the horizon of g_t, given by $A(t)$, is constant.

3. Prove that the mass of M_t, given by $m(t)$, is nonincreasing.

4. Prove that g_t converges to the Schwarzschild metric as $t \to \infty$.

Using these steps, Bray concludes that

$$\sqrt{\frac{A(0)}{16\pi}} = \sqrt{\frac{A(\infty)}{16\pi}} = m(\infty) \leq m(0),$$

which proves the RPI. The second equality here comes from the fact that the RPI is an equality for the Schwarzschild metric.

We now go over some elements of the proofs of the steps outlined above. We begin with a definition (which can be extended to multiple asymptotic ends).

Definition 9-28. We say that (M, g) is *harmonically flat at infinity* if for some compact set $K \subset M$ the complement $M \setminus K$ has zero scalar curvature and is conformal to $(\mathbb{R}^3 \setminus B_1(0), \delta)$; i.e., $g = u_0^4 \delta$ on $M \setminus K$, with $u_0(x) \to a > 0$ as $|x| \to \infty$.

The justification for this name is that the conformal factor u_0 above is harmonic. This follows from the well-known formula for the transformation of the scalar curvature under conformal deformations:

$$R(u^4 g) = u^{-5} (R(g)u - 8\Delta_g u), \tag{9.5.1}$$

where Δ_g is the Laplacian in the metric g.

If we write $g = u_0^4 \delta$ outside a compact set, then the preceding formula gives that $\Delta_\delta u_0 = 0$. Using spherical harmonics, we obtain the expansion

$$u_0(x) = a + \frac{b}{|x|} + O\left(\frac{1}{|x|^2}\right),$$

where $a, b \in \mathbb{R}$ are constants, so that $u_0(x) \to a$ as $|x| \to \infty$. Using this, we readily compute that the mass of (M, g) is $2ab$.

The following lemma justifies our interest in harmonically flat manifolds.

Theorem 9-29 (Schoen–Yau [203]). *Let (M^3, g) be an asymptotically flat manifold with $R(g) \geq 0$. For any $\epsilon > 0$, there exists a metric g_0 on M that is harmonically flat at infinity, has $R(g_0) \geq 0$, and satisfies*

$$(1 - \epsilon)g(V, V) \leq g_0(V, V) \leq (1 + \epsilon)g(V, V)$$

for all $V \in T_p M$ and $p \in M$; moreover

$$|m_0 - m| \leq \epsilon,$$

where m_0 and m are the ADM masses of (M^3, g_0) and (M^3, g).

Using Theorem 9-29 above we see that the proof of the RPI can be reduced to the harmonically flat case [26]. Let us now define the conformal flow that Bray introduced to prove the RPI for these manifolds.

Steps 1 and 2 (see previous page). Let Σ_t be the outermost area-minimizing enclosure of $\Sigma = \Sigma_0$ in (M^3, g_t) and g_t be a one-parameter family of metrics defined by

$$g_t = u_t(x)^4 g_0 \text{ for } t \geq 0,$$

where $u_0(x) \equiv 1$, and $u_t(x)$ is given by

$$u_t(x) = 1 + \int_0^t v_s(x)ds.$$

Here, we choose $v_t(x)$ so that it satisfies

$$\begin{cases} \Delta_{g_0} v_t(x) = 0 & \text{outside } \Sigma_t, \\ v_t = 0 & \text{on } \Sigma_t, \\ \lim_{|x| \to \infty} v_t(x) = -e^{-t}. \end{cases}$$

Extending v_t by 0 inside Σ_t, we see v_t, and hence u_t, is superharmonic, and hence $u_t > 0$ on M. The transformation law (9.5.1) for the scalar curvature under conformal deformations yields $R(g_t) = u_t^{-5}(R(g_0)u_t - 8\Delta_{g_0} u_t) \geq 0$ outside Σ_t. Note also that g_t remains asymptotically flat, since $u_t(x) \to e^{-t}$ when $|x| \to \infty$.

The proof of existence and regularity of solutions of the conformal flow defined above is based on an approximating scheme, and uses deep notions from geometric measure theory. It is rather involved, and we omit it here. Along the way, Bray is able to prove that Σ_t is enclosed by (but disjoint from) Σ_{t^*} whenever $t < t^*$. From this and the definition of v_t he shows that the area of the horizon does not increase infinitesimally, concluding $|\Sigma_t| = |\Sigma_0|$ for all $t \in (0, \infty)$. This fact is vital to his proof of the RPI, but is technical in the weak setting.

Exercise 9-30. Prove that $d|\Sigma_t|/dt = 0$ in the smooth case; this calculation uses and motivates the definition of v_t.

Step 3. We next sketch the proof that $m(t)$ is nonincreasing. This involves a generalization of the classical notion of electrostatic capacity of bodies in Euclidean space (see [183]).

Definition 9-31. Let (M^3, g) be an asymptotically flat manifold with boundary Σ. The capacity of Σ is

$$C(\Sigma) = \inf_{\varphi \in \mathcal{M}_0^1} \left\{ \frac{1}{4\pi} \int_M |\nabla \varphi|^2 dV_g \right\},$$

where $\mathcal{M}_0^1$ denotes the space of smooth functions φ such that $\varphi|_\Sigma = 0$ and $\varphi \to 1$ as $|x| \to \infty$ and ∇ is the gradient operator of the metric g.

Bray uses a mass-capacity inequality to prove that $m(t)$ is nonincreasing, as we see below. The precise statement is the following.

Theorem 9-32 (mass-capacity inequality [26]). *Let (M^3, g) be an asymptotically flat manifold with nonnegative scalar curvature and boundary Σ, which is a minimal surface. Then*

$$m \geq C(\Sigma),$$

where m is the ADM mass of M. Equality is achieved if and only if (M^3, g) is isometric to the Riemannian Schwarzschild manifold.

The idea behind the proof of Theorem 9-32 is to use a trick due to Bunting and Masood-ul-Alam [36], which consists in reflecting the metric of M across the boundary of M in order to obtain a asymptotically flat metric with two ends. This reflection is nontrivial since, in general, the reflected metric is not smooth across the boundary. Bray overcomes this difficulty by solving an ODE system, as we see below. Once a suitable metric has been constructed in the reflected manifold, Bray applies the following theorem:

Theorem 9-33. *Let (M^3, g) be a Riemannian manifold with nonnegative scalar curvature, multiple asymptotically flat ends and no boundary. Let E be one of its ends. Then*

$$\tfrac{1}{2} m(E) \geq C(E) := \inf_{\varphi \in \mathcal{M}_0^1} \left\{ \frac{1}{4\pi} \int_M |\nabla \varphi|^2 dV_g \right\},$$

where $\mathcal{M}_0^1$ is the set of all smooth functions in M that approach 1 at infinity in the end E and approach 0 in all the other ends.

The proof depends on the following observation:

Claim. *The function $\phi \in \mathcal{M}_0^1$ that realizes the infimum of $C(E)$ has the special form $\phi(x) = 1 - C(E)/|x| + O(|x|^{-2})$.*

The Euler–Lagrange equation for the capacity is the Laplace equation, and indeed the infimum $C(E)$ is realized by the harmonic function with the appropriate end behavior (and analogously for $C(\Sigma)$). Since ϕ is harmonic and tends to 1 at infinity in the chosen end, we have

$$C(E) = \frac{1}{4\pi} \int_\Sigma \frac{d\phi}{dv} d\mu_g$$

after integrating by parts over the manifold. Here, Σ is the smooth compact boundary of a set containing all the ends of M except for the chosen end E, and v is the unit normal vector to Σ. Letting Σ be an arbitrarily large sphere in E, the claim follows.

Proof of Theorem 9-33. Let k be the number of ends of M other than E. Without loss of generality let us assume that (M, g) is harmonically flat at infinity, and consider the metric $\tilde{g}$ defined on M as $\tilde{g} = \phi(x)^4 \bar{g}$, where $\phi(x)$ is the function that realizes the infimum in the calculation of $C(E)$ [26]. Since ϕ goes to zero in all the ends except for E, we can use the removable singularity theorem to extend the metric $\tilde{g}$ to $\widetilde{M} = M \cup \{\infty_1\} \cup \cdots \cup \{\infty_k\}$. Thus $\widetilde{M}$ is the manifold obtained from compactifying all ends of M other than E. It follows that $\tilde{g}$ is a harmonically flat metric on $\widetilde{M}$ with one end, and nonnegative scalar curvature.

To keep track of the mass $\tilde{m}$ of $(\widetilde{M}, \tilde{g})$ in terms of $m = m(E)$, we use the expansion of u from above. More precisely, since g is harmonically flat, we know it has the form $g = u^4 \delta$, where $u(x) = 1 + \frac{1}{2} m/|x| + O(|x|^{-2})$. We can similarly write $\tilde{g} = \tilde{u}^4 \delta = \phi^4 u^4 \delta$, where $\tilde{u}(x) = 1 + \frac{1}{2} \tilde{m}/|x| + O(|x|^{-2})$. Using the spherical harmonic expansion for ϕ we find that

$$m - 2C(E) = \tilde{m} \geq 0,$$

where the last inequality follows from the positive mass theorem. Putting this together yields $m \geq 2C(E)$, as desired. See [26] for the case of equality. $\square$

Proof of the mass-capacity inequality, Theorem 9-32. Bray uses a reflection argument inspired by the work of Bunting and Masood-ul-Alam [36] in order to double M and reflect the metric smoothly across the boundary. The process is depicted in Figure 11.

The main idea is to double (M, g) using a specific construction. We glue Σ-cylindrical tubes of the form $(\Sigma \times [0, 2\delta], G)$ onto the two copies of the inner boundaries. We define coordinates (z, t) on the tube $\Sigma \times [0, 2\delta]$, so that points of the form $(z, 0)$ are glued to one copy of M, and points of the form $(z, 2\delta)$

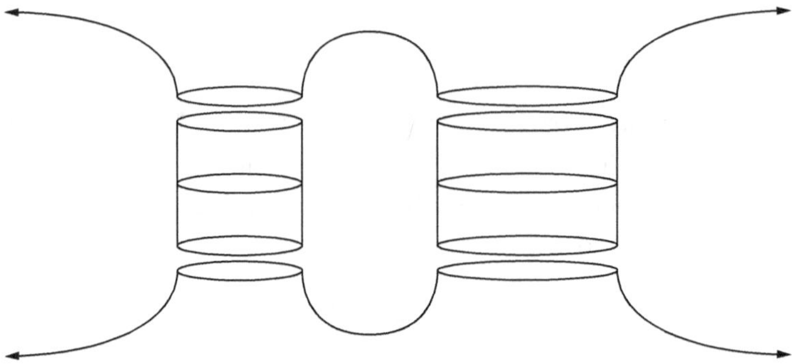

Figure 11. Bunting and Masood-ul-Alam's reflection trick [26].

land on the other copy. The difficulty of this process is to be able to define the metric G in such a way that the doubled manifold $(\widetilde{M}_\delta, \tilde{g}_\delta)$ not only has a smooth metric, but also its scalar curvature remains nonnegative. Furthermore, we want the mass of either end of $\widetilde{M}_\delta$ be arbitrarily close to m.

In order to do this, Bray solves an ODE, as we see below. We start by defining the following components of G:

$$G(\partial_t, \partial_t) = 1, \quad G(\partial_t, \partial_{z_1}) = 0, \quad G(\partial_t, \partial_{z_2}) = 0,$$

where $\{\partial_t, \partial_{z_1}, \partial_{z_2}\}$ are the coordinate vector fields. We still need to define the metric over all slices $\Sigma \times \{t\}$. To do this, consider the unique solution of the ODE system

$$\frac{d}{dt} G_{ij}(z, t) = 2G_{ik}(z, t) A^k_j(z, t),$$

with smooth initial conditions $G_{ij}(z, 0) = g_{ij}(z, 0)$, where $A_{ij}(z, 0)$ is the second fundamental form of Σ in M for i, j indexing $\{\partial_{z_1}, \partial_{z_2}\}$, and where $A^k_j(z, t) = -A^k_j(z, 2\delta - t)$ extends smoothly to $\widetilde{M}_\delta$. Then, using this definition of G, the second fundamental form of each slice is automatically $A_{ij}(z, t)$, and G is symmetric about $t = \delta$; i.e., $G_{ij}(z, t) = G_{ij}(z, 2\delta - t)$. Bray then conformally deforms the metric once more, and obtains a metric on $\widetilde{M}_\delta$ whose scalar curvature is nonnegative, so its mass $\tilde{m}_\delta$ converges to m as $\delta \to 0^+$.

Finally, let φ be the function that realizes $C(\Sigma)$, so that

$$\begin{cases} \lim\limits_{x \to \infty} \varphi(x) = 1, \\ \Delta_g \varphi = 0 \quad \text{in } M, \\ \varphi(x) = 0 \quad \text{on } \Sigma. \end{cases}$$

For an end E_δ of $\widetilde{M}_\delta$, let $\phi = \phi_\delta$ be the function which realizes $C(E_\delta)$. Then ϕ satisfies

$$\begin{cases} \lim_{x \to \infty} \phi(x) = 1, \\ \Delta_{\tilde{g}_\delta} \phi = 0 \quad \text{in } \widetilde{M}_\delta, \\ \lim_{x \to -\infty} \phi(x) = 0. \end{cases}$$

(Here we use $-\infty$ to denote the reflected end in $\widetilde{M}_\delta$.) Letting $\delta \to 0^+$ we find that, by symmetry, $\phi(x) = \frac{1}{2}$ on Σ. Hence by uniqueness of harmonic functions we find that $\phi(x) = \frac{1}{2}(\varphi(x) + 1)$ on M, and so $C(E) = \frac{1}{2}C(\Sigma)$, as required. The proof of the theorem follows from the fact that $m \geq 2C(E) = C(\Sigma)$ because of Theorem 9-33. The characterization of the case of equality can be found in [26]. $\qquad \square$

To finalize the proof of Step 3, we see that from the definition of u_t it follows that $du_t/dt = v_t$. The same argument for the claim on p. 379 yields

$$v_0(x) = -1 + \frac{C(\Sigma_0)}{|x|} + O\left(\frac{1}{|x|^2}\right).$$

We can write the initial metric in harmonically flat format as $g_0 = U_0^4 \delta$, where U_0 goes to 1 at infinity, and so that $U_0(x) = 1 + \frac{m(0)}{2|x|} + O\left(\frac{1}{|x|^2}\right)$. Using this we write g_t as both $g_t = U_t(x)^4 \delta$ and $g_t = u_t(x)^4 g_0$; this shows that $U_t(x) = u_t(x)U_0(x)$.

Now let $a(t)$ and $b(t)$ be the functions defined by $u_t(x) = a(t) + \frac{b(t)}{|x|} + O\left(\frac{1}{|x|^2}\right)$. Using the above calculations it follows that

$$U_t(x) = a(t) + \left(b(t) + \frac{m(0)}{2}a(t)\right)\frac{1}{|x|} + O\left(\frac{1}{|x|^2}\right),$$

so $m(t) = 2a(t)\left(b(t) + \frac{1}{2}m(0)a(t)\right)$. By the expansion of $u_t(x)$ from above, and the fact that $u_0(x) \equiv 1$, we obtain $a(0) = 1$, $b(0) = 0$, $a'(0) = -1$, and $b'(0) = C(\Sigma_0)$. From this, it follows that

$$m'(0) = 2C(\Sigma_0) - 2m(0) \leq 0,$$

as desired.

Step 4. The last step is to show that $(M, g(t))$ converges (in some sense) to the Riemannian Schwarzschild metric as $t \to \infty$. Bray is able to show that the rescaled horizon converges to a coordinate sphere of radius $\frac{m}{2}$ in $\mathbb{R}^3$ and that the rescaled solution of the flow $\tilde{u}_t$ converges to $1 + \frac{m}{2|x|}$ as $t \to \infty$. The interested reader is directed to Bray's paper for more details. $\qquad \square$

References

[1] R. Abraham, J. E. Marsden, and T. Ratiu, *Manifolds, tensor analysis, and applications*, 2nd ed., Applied Math. Sciences **75**, Springer, New York, 1988.

[2] R. A. Adams, *Sobolev spaces*, Pure and Applied Math. **65**, Academic Press, San Diego, 1975.

[3] S. Agmon, A. Douglis, and L. Nirenberg, "Estimates near the boundary for solutions of elliptic partial differential equations satisfying general boundary conditions, II", *Comm. Pure Appl. Math.* **17** (1964), 35–92.

[4] A. D. Alexandrov, "On the theory of mixed volumes, II: New inequalities between mixed volumes and their applications", *Mat. Sbornik (N.S.)* **2** (1937), 1205–1238. In Russian.

[5] A. D. Alexandrov, "On the theory of mixed volumes, III: Extension of two theorems of Minkowski on convex polyhedra to arbitrary convex surfaces", *Mat. Sbornik (N.S.)* **3** (1938), 27–46. In Russian.

[6] J. Anderson, J. Corvino, and F. Pasqualotto, "Multi-localized time-symmetric initial data for the Einstein vacuum equations", *J. Reine Angew. Math.* **808** (2024), 67–110.

[7] L. Andersson and V. Moncrief, "Elliptic-hyperbolic systems and the Einstein equations", *Ann. Henri Poincaré* **4**:1 (2003), 1–34.

[8] V. I. Arnold, *Mathematical methods of classical mechanics*, Graduate Texts in Math. **60**, Springer, New York, 1978.

[9] R. Arnowitt, S. Deser, and C. W. Misner, "Coordinate invariance and energy expressions in general relativity", *Phys. Rev. (2)* **122** (1961), 997–1006.

[10] R. Arnowitt, S. Deser, and C. W. Misner, "The dynamics of general relativity", pp. 227–265 in *Gravitation: an introduction to current research*, edited by L. Witten, Wiley, New York, 1962.

[11] A. Ashtekar, B. K. Berger, J. Isenberg, and M. MacCallum (editors), *General relativity and gravitation: a centennial perspective*, Cambridge Univ. Press, Cambridge, 2015.

[12] S. Axler, P. Bourdon, and W. Ramey, *Harmonic function theory*, Graduate Texts in Mathematics **137**, Springer, 1992.

[13] J. L. Barbosa and M. do Carmo, "Stability of hypersurfaces with constant mean curvature", *Math. Z.* **185**:3 (1984), 339–353.

[14] J. L. Barbosa, M. do Carmo, and J. Eschenburg, "Stability of hypersurfaces of constant mean curvature in Riemannian manifolds", *Math. Z.* **197**:1 (1988), 123–138.

[15] R. Bartnik, "The mass of an asymptotically flat manifold", *Comm. Pure Appl. Math.* **39**:5 (1986), 661–693.

[16] R. Bartnik, "Phase space for the Einstein equations", *Comm. Anal. Geom.* **13**:5 (2005), 845–885.

[17] R. Bartnik and J. Isenberg, "The constraint equations", pp. 1–38 in *The Einstein equations and the large scale behavior of gravitational fields*, edited by P. T. Chruściel and H. Friedrich, Birkhäuser, Basel, 2004.

[18] R. Beig and N. Ó Murchadha, "The Poincaré group as the symmetry group of canonical general relativity", *Ann. Physics* **174**:2 (1987), 463–498.

[19] R. Beig, P. T. Chruściel, and R. Schoen, "KIDs are non-generic", *Ann. Henri Poincaré* **6**:1 (2005), 155–194.

[20] M. Berger and D. Ebin, "Some decompositions of the space of symmetric tensors on a Riemannian manifold", *J. Differential Geometry* **3** (1969), 379–392.

[21] A. N. Bernal and M. Sánchez, "Globally hyperbolic spacetimes can be defined as 'causal' instead of 'strongly causal'", *Classical Quantum Gravity* **24**:3 (2007), 745–749.

[22] A. L. Besse, *Einstein manifolds*, Ergebnisse der Math. (3) **10**, Springer, Berlin, 1987.

[23] E. Bonning, P. Marronetti, D. Neilsen, and R. Matzner, "Physics and initial data for multiple black hole spacetimes", *Phys. Rev. D* (3) **68**:4 (2003), 044019, 17.

[24] J.-P. Bourguignon, D. G. Ebin, and J. E. Marsden, "Sur le noyau des opérateurs pseudo-différentiels à symbole surjectif et non injectif", *C. R. Acad. Sci. Paris Sér. A-B* **282**:16 (1976), Aii, A867–A870.

[25] H. L. Bray, *The Penrose inequality in general relativity and volume comparison theorems involving scalar curvature*, Ph.D. thesis, Stanford University, 1997, https://www.proquest.com/docview/304386501. arXiv 0902.3241

[26] H. L. Bray, "Proof of the Riemannian Penrose inequality using the positive mass theorem", *J. Differential Geom.* **59**:2 (2001), 177–267.

[27] H. L. Bray, "On dark matter, spiral galaxies, and the axioms of general relativity", pp. 1–64 in *Geometric analysis, mathematical relativity, and nonlinear partial differential equations*, Contemp. Math. **599**, Amer. Math. Soc., Providence, RI, 2013.

[28] H. L. Bray and D. A. Lee, "On the Riemannian Penrose inequality in dimensions less than eight", *Duke Math. J.* **148**:1 (2009), 81–106.

[29] H. Bray and F. Morgan, "An isoperimetric comparison theorem for Schwarzschild space and other manifolds", *Proc. Amer. Math. Soc.* **130**:5 (2002), 1467–1472.

[30] H. L. Bray and A. R. Parry, "Modeling wave dark matter in dwarf spheroidal galaxies", *J. Phy. Conf. Series* **615** (2015), art. id. 012001.

[31] S. Brendle, "Constant mean curvature surfaces in warped product manifolds", *Publ. Math. Inst. Hautes Études Sci.* **117** (2013), 247–269.

[32] S. Brendle and M. Eichmair, "Isoperimetric and Weingarten surfaces in the Schwarzschild manifold", *J. Differential Geom.* **94**:3 (2013), 387–407.

[33] S. Brendle and M. Eichmair, "Large outlying stable constant mean curvature spheres in initial data sets", *Invent. Math.* **197**:3 (2014), 663–682.

[34] S. Brendle and F. C. Marques, "Scalar curvature rigidity of geodesic balls in S^n", *J. Differential Geom.* **88**:3 (2011), 379–394.

[35] S. Brendle, F. C. Marques, and A. Neves, "Deformations of the hemisphere that increase scalar curvature", *Invent. Math.* **185**:1 (2011), 175–197.

[36] G. L. Bunting and A. K. M. Masood-ul-Alam, "Nonexistence of multiple black holes in asymptotically Euclidean static vacuum space-time", *Gen. Relativity Gravitation* **19**:2 (1987), 147–154.

[37] M. Cai and G. J. Galloway, "Rigidity of area minimizing tori in 3-manifolds of nonnegative scalar curvature", *Comm. Anal. Geom.* **8**:3 (2000), 565–573.

[38] J. J. Callahan, *The geometry of spacetime: an introduction to special and general relativity*, Springer, New York, 2000.

[39] M. Cantor, "Elliptic operators and the decomposition of tensor fields", *Bull. Amer. Math. Soc. (N.S.)* **5**:3 (1981), 235–262.

[40] M. Carfora and A. Marzuoli, *Einstein constraints and Ricci flow: a geometrical averaging of initial data sets*, Springer, Singapore, 2023.

[41] M. P. do Carmo, *Riemannian geometry*, Birkhäuser, Boston, 1992.

[42] S. Carroll, *Spacetime and geometry: An introduction to general relativity*, Addison Wesley, San Francisco, 2004.

[43] C. Cederbaum and C. Nerz, "Explicit Riemannian manifolds with unexpectedly behaving center of mass", *Ann. Henri Poincaré* **16**:7 (2015), 1609–1631.

[44] A. Chaljub-Simon, "Systèmes elliptiques linéaires dans des espaces de fonctions höldériennes à poids", *Rend. Circ. Mat. Palermo* (2) **30**:2 (1981), 300–310.

[45] A. Chaljub-Simon and Y. Choquet-Bruhat, "Problèmes elliptiques du second ordre sur une variété euclidienne à l'infini", *Ann. Fac. Sci. Toulouse Math.* (5) **1**:1 (1979), 9–25.

[46] P.-Y. Chan and L.-F. Tam, "A note on center of mass", *Comm. Anal. Geom.* **24**:3 (2016), 471–486.

[47] P.-N. Chen, L.-H. Huang, M.-T. Wang, and S.-T. Yau, "On the validity of the definition of angular momentum in general relativity", *Ann. Henri Poincaré* **17**:2 (2016), 253–270.

[48] P.-N. Chen, M.-T. Wang, and S.-T. Yau, "Conserved quantities in general relativity: from the quasi-local level to spatial infinity", *Comm. Math. Phys.* **338**:1 (2015), 31–80.

[49] Y. Choquet-Bruhat (as Fourès-Bruhat), "Théorème d'existence pour certains systèmes d'équations aux dérivées partielles non linéaires", *Acta Math.* **88** (1952), 141–225.

[50] Y. Choquet-Bruhat, "New elliptic system and global solutions for the constraints equations in general relativity", *Comm. Math. Phys.* **21** (1971), 211–218.

[51] Y. Choquet-Bruhat, *General relativity and the Einstein equations*, Oxford Univ. Press, 2009.

[52] Y. Choquet-Bruhat and J. W. York, Jr., "The Cauchy problem", pp. 99–172 in *General relativity and gravitation*, vol. 1, edited by A. Held, Plenum Press, New York, 1980.

[53] B. Chow and R. Gulliver, "Aleksandrov reflection and geometric evolution of hypersurfaces", *Comm. Anal. Geom.* **9**:2 (2001), 261–280.

[54] D. Christodoulou, "Global solutions of nonlinear hyperbolic equations for small initial data", *Comm. Pure Appl. Math.* **39**:2 (1986), 267–282.

[55] D. Christodoulou and S. Klainerman, *The global nonlinear stability of the Minkowski space*, Princeton Mathematical Series **41**, Princeton University Press, 1993.

[56] D. Christodoulou and N. Ó Murchadha, "The boost problem in general relativity", *Comm. Math. Phys.* **80**:2 (1981), 271–300.

[57] P. T. Chruściel, "Boundary conditions at spatial infinity from a Hamiltonian point of view", pp. 49–59 in *Topological properties and global structure of space-time* (Erice, 1985), NATO Adv. Sci. Inst. Ser. B Phys. **138**, Plenum, New York, 1986.

[58] P. T. Chruściel, "Mass and angular-momentum inequalities for axi-symmetric initial data sets. I. Positivity of mass", *Ann. Physics* **323**:10 (2008), 2566–2590.

[59] P. T. Chruściel, "Elements of causality theory", preprint, 2011. arXiv 1110.6706v1

[60] P. T. Chruściel and J. L. Costa, "Mass, angular-momentum and charge inequalities for axisymmetric initial data", *Classical Quantum Gravity* **26**:23 (2009), 235013, 7.

[61] P. T. Chruściel and E. Delay, *On mapping properties of the general relativistic constraints operator in weighted function spaces, with applications*, Mém. Soc. Math. Fr. (N.S.) **94**, 2003.

[62] P. T. Chruściel and J. D. E. Grant, "On Lorentzian causality with continuous metrics", *Classical Quantum Gravity* **29**:14 (2012), art. id. 145001.

[63] P. T. Chruściel, Y. Li, and G. Weinstein, "Mass and angular-momentum inequalities for axi-symmetric initial data sets. II. Angular momentum", *Ann. Physics* **323**:10 (2008), 2591–2613.

[64] P. T. Chruściel, G. J. Galloway, and D. Pollack, "Mathematical general relativity: a sampler", *Bull. Amer. Math. Soc. (N.S.)* **47**:4 (2010), 567–638.

[65] G. B. Cook, "Initial data for numerical relativity", *Living Rev. Relativ.* **3** (2000), art. id. 2000-5.

[66] J. Corvino, "Scalar curvature deformation and a gluing construction for the Einstein constraint equations", *Comm. Math. Phys.* **214**:1 (2000), 137–189.

[67] J. Corvino and L.-H. Huang, "Localized deformation for initial data sets with the dominant energy condition", *Calc. Var. Partial Differential Equations* **59**:1 (2020), Paper No. 42, 43.

[68] J. Corvino and D. Pollack, "Scalar curvature and the Einstein constraint equations", pp. 145–188 in *Surveys in geometric analysis and relativity*, Adv. Lect. Math. (ALM) **20**, International Press, Somerville, MA, 2011.

[69] J. Corvino and R. M. Schoen, "On the asymptotics for the vacuum Einstein constraint equations", *J. Differential Geom.* **73**:2 (2006), 185–217.

[70] J. Corvino, M. Eichmair, and P. Miao, "Deformation of scalar curvature and volume", *Math. Ann.* **357**:2 (2013), 551–584.

[71] S. Dain, "Proof of the angular momentum-mass inequality for axisymmetric black holes", *J. Differential Geom.* **79**:1 (2008), 33–67.

[72] C. De Lellis and S. Müller, "Optimal rigidity estimates for nearly umbilical surfaces", *J. Differential Geom.* **69**:1 (2005), 75–110.

[73] D. M. DeTurck, "Existence of metrics with prescribed Ricci curvature: local theory", *Invent. Math.* **65**:1 (1981/82), 179–207.

[74] D. M. DeTurck, "Deforming metrics in the direction of their Ricci tensors", *J. Differential Geom.* **18**:1 (1983), 157–162.

[75] B. S. DeWitt, "Quantum theory of gravity, I: The canonical theory", *Phys. Rev.* **160**:5 (1967), 1113–1148.

[76] A. Douglis and L. Nirenberg, "Interior estimates for elliptic systems of partial differential equations", *Comm. Pure Appl. Math.* **8** (1955), 503–538.

[77] M. Eichmair, "The Jang equation reduction of the spacetime positive energy theorem in dimensions less than eight", *Comm. Math. Phys.* **319**:3 (2013), 575–593.

[78] M. Eichmair, L.-H. Huang, D. A. Lee, and R. Schoen, "The spacetime positive mass theorem in dimensions less than eight", *J. Eur. Math. Soc.* **18**:1 (2016), 83–121.

[79] M. Eichmair and J. Metzger, "On large volume preserving stable CMC surfaces in initial data sets", *J. Differential Geom.* **91**:1 (2012), 81–102.

[80] M. Eichmair and J. Metzger, "Large isoperimetric surfaces in initial data sets", *J. Differential Geom.* **94**:1 (2013), 159–186.

[81] M. Eichmair and J. Metzger, "Unique isoperimetric foliations of asymptotically flat manifolds in all dimensions", *Invent. Math.* **194**:3 (2013), 591–630.

[82] A. Einstein, "Zur Elektrodynamik bewegter Körper", *Ann. der Physik* **17** (1905), 891–921. Translated as "On the electrodynamics of moving bodies", pp. 35–65 in *The principle of relativity: a collection of original memoirs on the special and general theory of relativity*, Methuen, London, 1923; reprinted Dover, New York, 1952. Essentially the same text is posted at http://www.fourmilab.ch/etexts/einstein/specrel/www/. A new translation appeared in pp. 123–160 of [212].

[83] A. Einstein, "Die Grundlage der allgemeinen Relativitätstheorie", *Ann. der Physik* **49**:7 (1916), 769–822. Translated as "The foundation of the general theory of relativity", pp. 109–164 in *The principle of relativity: a collection of original memoirs on the special and general theory of relativity*, Methuen, London, 1923; reprinted Dover, New York, 1952.

[84] A. Einstein, *Relativity: the special and the general theory*, Methuen, London, 1920. Reprinted by Crown Publishers, New York, 1961.

[85] A. Einstein, *The meaning of relativity*, Princeton Univ. Press, Princeton, 1922.

[86] L. C. Evans, *Partial differential equations*, Graduate Studies in Math. **19**, Amer. Math. Soc., Providence, RI, 1998.

[87] C. F. W. Everitt et al., "The Gravity Probe B test of general relativity", *Classical Quantum Gravity* **32**:22 (2015), art. id. 224001.

[88] A. E. Fischer and J. E. Marsden, "Linearization stability of the Einstein equations", *Bull. Amer. Math. Soc.* **79** (1973), 997–1003.

[89] A. E. Fischer and J. E. Marsden, "Deformations of the scalar curvature", *Duke Math. J.* **42**:3 (1975), 519–547.

[90] A. E. Fischer and J. E. Marsden, *The initial value problem and the dynamical formulation of general relativity*, edited by S. W. Hawking and W. Israel, Cambridge Univ. Press, Cambridge, 1979.

[91] A. E. Fischer and J. A. Wolf, "The structure of compact Ricci-flat Riemannian manifolds", *J. Differential Geometry* **10** (1975), 277–288.

[92] D. Fischer-Colbrie and R. Schoen, "The structure of complete stable minimal surfaces in 3-manifolds of nonnegative scalar curvature", *Comm. Pure Appl. Math.* **33**:2 (1980), 199–211.

[93] G. B. Folland, *Introduction to partial differential equations*, 2nd ed., Princeton University Press, 1995.

[94] T. Frankel, *Gravitational curvature: An introduction to Einstein's theory*, W. H. Freeman, San Francisco, 1979.

[95] A. Freire and F. Schwartz, "Mass-capacity inequalities for conformally flat manifolds with boundary", *Comm. PDE* **39** (2014), 98–119.

[96] A. P. French, *Special relativity*, W. W. Norton, New York, 1968.

[97] H. Friedrich, "On the existence of n-geodesically complete or future complete solutions of Einstein's field equations with smooth asymptotic structure", *Comm. Math. Phys.* **107**:4 (1986), 587–609.

[98] K. O. Friedrichs, "The identity of weak and strong extensions of differential operators", *Trans. Amer. Math. Soc.* **55** (1944), 132–151.

[99] G. J. Galloway, "Least area tori, black holes and topological censorship", pp. 113–123 in *Differential geometry and mathematical physics* (Vancouver, 1993), edited by J. K. Beem and K. L. Duggal, Contemp. Math. **170**, Amer. Math. Soc., Providence, RI, 1994.

[100] G. J. Galloway, "Stability and rigidity of extremal surfaces in Riemannian geometry and general relativity", pp. 221–239 in *Surveys in geometric analysis and relativity*, edited by H. L.

Bray and W. P. Minicozzi, Adv. Lect. Math. (ALM) **20**, International Press, Somerville, MA, 2011.

[101] G. J. Galloway, *Notes on Lorentzian causality: ESI-EMS-IAMP Summer School on Mathematical Relativity* (Vienna, 2014), 2014. http://hdl.handle.net/10385/2167.

[102] C. Gerhardt, "Flow of nonconvex hypersurfaces into spheres", *J. Diff. Geom.* **32** (1990), 299–314.

[103] R. Geroch, "What is a singularity in general relativity?", *Ann. Phys.* **48**:3 (1968), 526–540.

[104] R. Geroch, "Domain of dependence", *J. Mathematical Phys.* **11** (1970), 437–449.

[105] R. Geroch, "Energy extraction", *Ann. New York Acad. Sci.* **224** (1973), 108–117.

[106] G. W. Gibbons, "The time symmetric initial value problem for black holes", *Comm. Math. Phys.* **27** (1972), 87–102.

[107] D. Gilbarg and N. S. Trudinger, *Elliptic partial differential equations of second order*, 2nd ed., Grundlehren der Math. Wiss. **224**, Springer, Berlin, 1983.

[108] E. Giusti, *Minimal surfaces and functions of bounded variation*, Monographs in Math. **80**, Birkhäuser, Basel, 1984.

[109] A. Gray, *Tubes*, Addison-Wesley, Redwood City, CA, 1990.

[110] O. Grøn, "Space geometry in rotating frames: a historical appraisal", pp. 285–334 in *Relativity in rotating frames: relativistic physics in rotating reference frames*, edited by G. Rizzi and M. L. Ruggiero, Kluwer, Dordrecht, 2004.

[111] Q. Han, *A basic course in partial differential equations*, Graduate Studies in Mathematics **120**, American Mathematical Society, 2011.

[112] S. W. Hawking and G. F. R. Ellis, *The large scale structure of space-time*, Cambridge Monog. Math. Phys. **1**, Cambridge Univ. Press, London, 1973.

[113] S. W. Hawking and G. T. Horowitz, "The gravitational Hamiltonian, action, entropy and surface terms", *Classical Quantum Gravity* **13**:6 (1996), 1487–1498.

[114] L. Hörmander, "Pseudo-differential operators and non-elliptic boundary problems", *Ann. of Math.* (2) **83** (1966), 129–209.

[115] L. Hörmander, *The analysis of linear partial differential operators, I: Distribution theory and Fourier analysis*, Grundlehren der Math. Wiss. **256**, Springer, 1990.

[116] L.-H. Huang, "On the center of mass of isolated systems with general asymptotics", *Classical Quantum Gravity* **26**:1 (2009), 015012, 25.

[117] L.-H. Huang, "Foliations by stable spheres with constant mean curvature for isolated systems with general asymptotics", *Comm. Math. Phys.* **300**:2 (2010), 331–373.

[118] L.-H. Huang, "Solutions of special asymptotics to the Einstein constraint equations", *Classical Quantum Gravity* **27**:24 (2010), 245002, 10.

[119] L.-H. Huang, "On the center of mass in general relativity", pp. 575–591 in *Fifth International Congress of Chinese Mathematicians*, AMS/IP Stud. Adv. Math. **51**, pt. 1, Amer. Math. Soc., 2012.

[120] L.-H. Huang and D. A. Lee, "Equality in the spacetime positive mass theorem", *Comm. Math. Phys.* **376**:3 (2020), 2379–2407.

[121] L.-H. Huang and D. Wu, "The equality case of the Penrose inequality for asymptotically flat graphs", *Trans. Amer. Math. Soc.* **367**:1 (2015), 31–47.

[122] L.-H. Huang, R. Schoen, and M.-T. Wang, "Specifying angular momentum and center of mass for vacuum initial data sets", *Comm. Math. Phys.* **306**:3 (2011), 785–803.

[123] G. Huisken, "Flow by mean curvature of convex surfaces into spheres", *J. Differential Geom.* **20**:1 (1984), 237–266.

[124] G. Huisken, "The volume preserving mean curvature flow", *J. Reine Angew. Math.* **382** (1987), 35–48.

[125] G. Huisken and T. Ilmanen, "The inverse mean curvature flow and the Riemannian Penrose inequality", *J. Differential Geom.* **59**:3 (2001), 353–437.

[126] G. Huisken and S.-T. Yau, "Definition of center of mass for isolated physical systems and unique foliations by stable spheres with constant mean curvature", *Invent. Math.* **124**:1-3 (1996), 281–311.

[127] J. Isenberg, "Constant mean curvature solutions of the Einstein constraint equations on closed manifolds", *Classical Quantum Gravity* **12**:9 (1995), 2249–2274.

[128] J. Isenberg and V. Moncrief, "A set of nonconstant mean curvature solutions of the Einstein constraint equations on closed manifolds", *Classical Quantum Gravity* **13**:7 (1996), 1819–1847.

[129] J. Isenberg, R. Mazzeo, and D. Pollack, "On the topology of vacuum spacetimes", *Ann. Henri Poincaré* **4**:2 (2003), 369–383.

[130] P. S. Jang, "On the positive energy conjecture", *J. Mathematical Phys.* **17**:1 (1976), 141–145.

[131] P. S. Jang and R. M. Wald, "The positive energy conjecture and the cosmic censor hypothesis", *J. Math. Phys.* **18** (1977), 41–44.

[132] T. Kaluza, "Zur Relativitätstheorie", *Physikal. Z.* **11** (1910), 977–978. Translated in https://en.wikisource.org/wiki/Translation:On_the_Theory_of_Relativity_(Kaluza).

[133] N. Kapouleas, "Constant mean curvature surfaces constructed by fusing Wente tori", *Invent. Math.* **119**:3 (1995), 443–518.

[134] J. L. Kazdan and F. W. Warner, "A direct approach to the determination of Gaussian and scalar curvature functions", *Invent. Math.* **28** (1975), 227–230.

[135] J. L. Kazdan and F. W. Warner, "Existence and conformal deformation of metrics with prescribed Gaussian and scalar curvatures", *Ann. of Math.* (2) **101** (1975), 317–331.

[136] J. L. Kazdan and F. W. Warner, "Prescribing curvatures", pp. 309–319 in *Differential geometry* (Stanford, 1973), vol. 2, Proc. Sympos. Pure Math. **27**, 1975.

[137] D. Kleppner and R. J. Kolenkow, *An introduction to mechanics*, McGraw Hill, New York, 1973.

[138] M.-K. G. Lam, "The graph cases of the Riemannian positive mass and Penrose inequalities in all dimensions", preprint, 2010. arXiv 1010.4256v1

[139] H. D. Lawson, Jr. and M. L. Michelsohn, *Spin geometry*, Princeton Math. Series **38**, Princeton Univ. Press, 1989.

[140] J. M. Lee, *Riemannian manifolds: an introduction to curvature*, Graduate Texts in Math. **176**, Springer, New York, 1997.

[141] J. M. Lee, *Introduction to smooth manifolds*, Graduate Texts in Math. **218**, Springer, New York, 2003.

[142] D. A. Lee, *Geometric relativity*, Graduate Studies in Mathematics **201**, Amer. Math. Soc., Providence, RI, 2019.

[143] J. M. Lee and T. H. Parker, "The Yamabe problem", *Bull. Amer. Math. Soc.* (*N.S.*) **17**:1 (1987), 37–91.

[144] G. Leoni, *A first course in Sobolev spaces*, Graduate Studies in Math. **105**, Amer. Math. Soc., Providence, RI, 2009.

[145] A. Lichnerowicz, "L'intégration des équations de la gravitation relativiste et le problème des *n* corps", *J. Math. Pures Appl.* (9) **23** (1944), 37–63.

[146] J. Lohkamp, "Scalar curvature and hammocks", *Math. Ann.* **313**:3 (1999), 385–407.

[147] J. Lohkamp, "The higher dimensional positive mass theorem, II", preprint, 2017. arXiv 1612.07505v2

[148] D. Lovelock and H. Rund, *Tensor, differential forms, and variational principles*, Wiley, New York, 1975.

[149] T. Marquardt, *The inverse mean curvature flow for hypersurfaces with boundary*, Ph.D thesis, Freie Universität Berlin, 2012.

[150] M. Mars, "Present status of the Penrose inequality", *Proc. Amer. Math. Soc.* **132**:1 (2004), 217–222.

[151] D. Maxwell, "Rough solutions of the Einstein constraint equations on compact manifolds", *J. Hyperbolic Differ. Equ.* **2**:2 (2005), 521–546.

[152] D. Maxwell, "Solutions of the Einstein constraint equations with apparent horizon boundaries", *Comm. Math. Phys.* **253**:3 (2005), 561–583.

[153] D. Maxwell, "The conformal method and the conformal thin-sandwich method are the same", *Classical Quantum Gravity* **31**:14 (2014), 145006, 34.

[154] D. Maxwell, "Initial data in general relativity described by expansion, conformal deformation and drift", *Comm. Anal. Geom.* **29**:1 (2021), 207–281.

[155] R. C. McOwen, "The behavior of the Laplacian on weighted Sobolev spaces", *Comm. Pure Appl. Math.* **32**:6 (1979), 783–795.

[156] J. Metzger, "Foliations of asymptotically flat 3-manifolds by 2-surfaces of prescribed mean curvature", *J. Differential Geom.* **77**:2 (2007), 201–236.

[157] N. Meyers, "An expansion about infinity for solutions of linear elliptic equations", *J. Math. Mech.* **12** (1963), 247–264.

[158] P. Miao, "Quasi-local mass via isometric embeddings: a review from a geometric perspective", *Classical Quantum Gravity* **32**:23 (2015), art. id. 233001.

[159] P. Miao and L.-F. Tam, "Evaluation of the ADM mass and center of mass via the Ricci tensor", *Proc. Amer. Math. Soc.* **144**:2 (2016), 753–761.

[160] H. Mirandola and F. Vitório, "The positive mass theorem and Penrose inequality for graphical manifolds", *Comm. Anal. Geom.* **23**:2 (2015), 273–292.

[161] C. W. Misner, K. S. Thorne, and J. A. Wheeler, *Gravitation*, Freeman, San Francisco, 1973.

[162] C. Møller, *The theory of relativity*, Clarendon Press, Oxford, 1952.

[163] V. Moncrief, "Spacetime symmetries and linearization stability of the Einstein equations, I", *J. Mathematical Phys.* **16** (1975), 493–498.

[164] S. Montiel and A. Ros, *Curves and surfaces*, vol. 69, Second ed., Graduate Studies in Mathematics, American Mathematical Society, Providence, RI; Real Sociedad Matemática Española, Madrid, 2009. Translated from the 1998 Spanish original by Montiel and edited by Donald Babbitt.

[165] J. R. Munkres, *Elements of algebraic topology* (Menlo Park, CA), Addison-Wesley, 1984.

[166] C. Nerz, "Foliations by stable spheres with constant mean curvature for isolated systems without asymptotic symmetry", *Calc. Var. Partial Differential Equations* **54**:2 (2015), 1911–1946.

[167] A. Neves, "Insufficient convergence of inverse mean curvature flow on asymptotically hyberbolic manifolds", *J. Differential Geom.* **84**:1 (2010), 191–229.

[168] L. Nirenberg and H. F. Walker, "The null spaces of elliptic partial differential operators in $\mathbf{R}^n$", *J. Math. Anal. Appl.* **42** (1973), 271–301. Collection of articles dedicated to Salomon Bochner.

[169] J. Norton, "What was Einstein's principle of equivalence?", *Stud. Hist. Philos. Sci.* **16**:3 (1985), 203–246.

[170] J. D. Norton, "General covariance and the foundations of general relativity: eight decades of dispute", *Rep. Progr. Phys.* **56**:7 (1993), 791–858.

[171] J. D. Norton, "Mach's principle before Einstein", pp. 9–57 in *Mach's principle: from Newton's bucket to quantum gravity*, edited by J. Barbour and H. Pfister, Einstein Studies **6**, Birkhäuser, 1995.

[172] N. Ó Murchadha and J. W. York, Jr., "Initial-value problem of general relativity, I: General formulation and physical interpretation", *Phys. Rev. D* (3) **10** (1974), 428–436.

[173] M. Obata, "Certain conditions for a Riemannian manifold to be isometric with a sphere", *J. Math. Soc. Japan* **14** (1962), 333–340.

[174] B. O'Neill, *Semi-Riemannian geometry, with applications to relativity*, Pure and Applied Math. **103**, Academic Press, New York, 1983.

[175] T. Parker and C. H. Taubes, "On Witten's proof of the positive energy theorem", *Comm. Math. Phys.* **84**:2 (1982), 223–238.

[176] R. Penrose, "Asymptotic properties of fields and space-times", *Phys. Rev. Lett.* **10** (1963), 66–68.

[177] R. Penrose, "Gravitational collapse and space-time singularities", *Phys. Rev. Lett.* **14** (1965), 57–59.

[178] R. Penrose, "Gravitational collapse: the role of general relativity", *Nuovo Cimento* **1**:(numero speciale) (1969), 252–276.

[179] R. Penrose, *Techniques of differential topology in relativity*, CBMS Regional Conf. Series Appl. Math. **7**, Society for Industrial and Applied Mathematics, Philadelphia, 1972.

[180] R. Penrose, "Naked singularities", *Ann. New York Acad. Sci.* **224** (1973), 125–134.

[181] R. Penrose, "Some unresolved problems in classical general relativity", pp. 631–668 in *Seminar on Differential Geometry*, edited by S.-T. Yau, Annals of Math. Stud. **102**, Princeton University Press, Princeton, NJ, 1982.

[182] P. Petersen, *Riemannian geometry*, 2nd ed., Graduate Texts in Math. **171**, Springer, 2006.

[183] G. Pólya and G. Szegö, *Isoperimetric inequalities in mathematical physics*, Annals of Mathematics Studies **27**, Princeton University Press, Princeton, NJ, 1951.

[184] J. Qing and G. Tian, "On the uniqueness of the foliation of spheres of constant mean curvature in asymptotically flat 3-manifolds", *J. Amer. Math. Soc.* **20**:4 (2007), 1091–1110 (electronic).

[185] J. Qing and W. Yuan, "On scalar curvature rigidity of vacuum static spaces", *Math. Ann.* **365**:3-4 (2016), 1257–1277.

[186] W. Qiu, "Interior regularity of solutions to the isotropically constrained Plateau problem", *Comm. Anal. Geom.* **11**:5 (2003), 945–986.

[187] T. Regge and C. Teitelboim, "Role of surface integrals in the Hamiltonian formulation of general relativity", *Ann. Physics* **88** (1974), 286–318.

[188] R. Resnick, *Introduction to special relativity*, Wiley, 1968.

[189] W. Rindler, *Introduction to special relativity*, 2nd ed., Oxford Univ. Press, New York, 1991.

[190] H. Ringström, *The Cauchy problem in general relativity*, European Mathematical Society, Zürich, 2009. Errata at https://people.kth.se/~hansr/errata.html.

[191] H. Ringström, *On the topology and future stability of the universe*, Oxford Univ. Press, 2013.

[192] H. P. Robertson, "Postulate versus observation in the special theory of relativity", *Rev. Modern Phys.* **21** (1949), 378–382.

[193] H. L. Royden, *Real analysis*, 3rd ed., Macmillan, New York, 1988.

[194] W. Rudin, *Real and complex analysis*, 3rd ed., McGraw-Hill, New York, 1987.

[195] W. Rudin, *Functional analysis*, 2nd ed., McGraw-Hill, New York, 1991.

[196] R. Schoen, "Conformal deformation of a Riemannian metric to constant scalar curvature", *J. Differential Geom.* **20**:2 (1984), 479–495.

[197] R. M. Schoen, "Variational theory for the total scalar curvature functional for Riemannian metrics and related topics", pp. 120–154 in *Topics in calculus of variations (Montecatini Terme, 1987)*, Lecture Notes in Math. **1365**, Springer, 1989.

[198] R. Schoen and S. T. Yau, "Existence of incompressible minimal surfaces and the topology of three-dimensional manifolds with nonnegative scalar curvature", *Ann. of Math.* (2) **110**:1 (1979), 127–142.

[199] R. Schoen and S. T. Yau, "On the proof of the positive mass conjecture in general relativity", *Comm. Math. Phys.* **65**:1 (1979), 45–76.

[200] R. Schoen and S. T. Yau, "On the structure of manifolds with positive scalar curvature", *Manuscripta Math.* **28**:1-3 (1979), 159–183.

[201] R. M. Schoen and S. T. Yau, "Complete manifolds with nonnegative scalar curvature and the positive action conjecture in general relativity", *Proc. Nat. Acad. Sci. U.S.A.* **76**:3 (1979), 1024–1025.

[202] R. Schoen and S. T. Yau, "The energy and the linear momentum of space-times in general relativity", *Comm. Math. Phys.* **79**:1 (1981), 47–51.

[203] R. Schoen and S. T. Yau, "Proof of the positive mass theorem, II", *Comm. Math. Phys.* **79**:2 (1981), 231–260.

[204] R. Schoen and S.-T. Yau, *Lectures on differential geometry*, Conference Proceedings and Lecture Notes in Geometry and Topology, I, International Press, Cambridge, MA, 1994.

[205] R. Schoen and S.-T. Yau, "Positive scalar curvature and minimal hypersurface singularities", pp. 441–480 in *Surveys in differential geometry, 2019: Differential geometry, Calabi–Yau theory, and general relativity*, vol. 2, Surv. Differ. Geom. **24**, International Press, Boston, 2022. arXiv 1704.05490

[206] R. Schoen and X. Zhou, "Convexity of reduced energy and mass angular momentum inequalities", *Ann. Henri Poincaré* **14**:7 (2013), 1747–1773.

[207] B. F. Schutz, *A first course in general relativity*, Cambridge Univ. Press, Cambridge, 1990.

[208] J. M. M. Senovilla and D. Garfinkle, "The 1965 Penrose singularity theorem", *Classical Quantum Gravity* **32**:12 (2015), 124008, 45.

[209] L. Simon, *Theorems on regularity and singularity of energy minimizing maps*, Birkhäuser, Basel, 1996.

[210] L. Simon, "Schauder estimates by scaling", *Calc. Var. Partial Differential Equations* **5**:5 (1997), 391–407.

[211] B. Smith and G. Weinstein, "Quasiconvex foliations and asymptotically flat metrics of non-negative scalar curvature", *Comm. Anal. Geom.* **12**:3 (2004), 511–551.

[212] J. Stachel, *Einstein's miraculous year: five papers that changed the face of physics*, Princeton Univ. Press, 1998.

[213] J. D. Streets, "Quasi-local mass functionals and generalized inverse mean curvature flow", *Comm. Anal. Geom.* **16**:3 (2008), 495–537.

[214] M. E. Taylor, *Partial differential equations, III: Nonlinear equations*, Applied Math. Sciences **117**, Springer, New York, 1997.

[215] P. Topping, "Relating diameter and mean curvature for submanifolds of Euclidean space", *Comment. Math. Helv.* **83**:3 (2008), 539–546.

[216] J. Urbas, "On the expansion of starshaped hypersurfaces by symmetric functions of their principal curvatures", *Math. Z.* **205** (1990), 355–372.

[217] J. W. Vick, *Homology theory: an introduction to algebraic topology*, Pure and Applied Math. **53**, Academic Press, New York, 1973.

[218] R. M. Wald, *General relativity*, University of Chicago Press, 1984.

[219] F. W. Warner, *Foundations of differentiable manifolds and Lie groups*, Graduate Texts in Mathematics **94**, Springer-Verlag, New York-Berlin, 1983. Corrected reprint of the 1971 edition.

[220] R. H. Wasserman, *Tensors and manifolds, with applications to mechanics and relativity*, Oxford Univ. Press, New York, 1992.

[221] S. Weinberg, *Gravitation and cosmology*, Wiley, New York, 1972.

[222] H. C. Wente, "Counterexample to a conjecture of H. Hopf", *Pacific J. Math.* **121**.1 (1986), 193–243.

[223] T. Willmore, *Total curvature in Riemannian geometry*, Wiley, 1982.

[224] E. Witten, "A new proof of the positive energy theorem", *Comm. Math. Phys.* **80**:3 (1981), 381–402.

[225] R. Ye, "Foliation by constant mean curvature spheres on asymptotically flat manifolds", pp. 369–383 in *Geometric analysis and the calculus of variations*, International Press, Cambridge, MA, 1996.

[226] J. W. York, Jr., "Gravitational degrees of freedom and the initial-value problem", *Phys. Rev. Lett.* **26** (1971), 1656–1658.

[227] J. W. York, Jr., "Role of conformal three-geometry in the dynamics of gravitation", *Phys. Rev. Lett.* **28**:16 (1972), 1082–1085.

[228] J. W. York, Jr., "Conformally invariant orthogonal decomposition of symmetric tensors on Riemannian manifolds and the initial-value problem of general relativity", *J. Mathematical Phys.* **14** (1973), 456–464.

[229] J. W. York, Jr., "Boundary terms in the action principles of general relativity", *Found. Phys.* **16**:3 (1986), 249–257.

[230] W. Yuan, "Brown-York mass and compactly supported conformal deformations of scalar curvature", *J. Geom. Anal.* **27**:1 (2017), 797–816.

[231] X. Zhang, "Angular momentum and positive mass theorem", *Comm. Math. Phys.* **206**:1 (1999), 137–155.

[232] X. Zhou, "Mass angular momentum inequality for axisymmetric vacuum data with small trace", *Comm. Anal. Geom.* **22**:3 (2014), 519–571.

Index